SMOOTHER
PEBBLES

SMOOTHER PEBBLES

Essays in the Sociology of Science

JONATHAN R. COLE

Columbia University Press
New York

Columbia University Press
Publishers Since 1893
New York   Chichester, West Sussex
cup.columbia.edu

Copyright © 2024 Columbia University Press
All rights reserved

Library of Congress Cataloging-in-Publication Data
Names: Cole, Jonathan R., author, editor. | Cole, Stephen, 1941– contributor.
Title: Smoother pebbles : essays in the sociology of science / Jonathan R. Cole.
Description: New York : Columbia University Press, [2024] |
   Includes bibliographical references and index.
Identifiers: LCCN 2023028504 (print) | LCCN 2023028505 (ebook) |
   ISBN 9780231212601 (hardback) | ISBN 9780231212618 (trade paperback) |
   ISBN 9780231559378 (ebook)
Subjects: LCSH: Science—Social aspects—United States. | Academic freedom—
   Social aspects—United States. | Education, Higher—Social aspects—
   United States. | Women in science—Social aspects—United States.
Classification: LCC Q175.55 .C65 2024 (print) | LCC Q175.55 (ebook) |
   DDC 306.4/50973—dc23/eng/20231012
LC record available at https://lccn.loc.gov/2023028504
LC ebook record available at https://lccn.loc.gov/2023028505

Printed and bound by CPI Group (UK) Ltd, Croydon, CR0 4YY

Cover design: Julia Kushnirsky
Cover image: Shutterstock

In memory of Robert K. Merton and Stephen Cole, and for my colleague, collaborator, and friend, Harriet A. Zuckerman

CONTENTS

Acknowledgments xi

Introduction 1

PART I. VALUES AND REWARDS IN SCIENCE

1. Scientific Output and Recognition: A Study in the Operation of the Reward System in Science (1967) 27
   STEPHEN COLE AND JONATHAN R. COLE

2. Visibility and Structural Bases of Awareness in Science (1968) 48
   STEPHEN COLE AND JONATHAN R. COLE

3. Patterns of Intellectual Influence in Scientific Research (1970) 74
   JONATHAN R. COLE

4. Measuring the Quality of Sociological Research: Problems in the Use of the *Science Citation Index* (1971) 101
   JONATHAN R. COLE AND STEPHEN COLE

5. The Emergence of a Scientific Specialty: The Self-Exemplifying Case of the Sociology of Science (1975) 115
   JONATHAN R. COLE AND HARRIET ZUCKERMAN

6. The Reputations of American Medical Schools (1977) 153
   JONATHAN R. COLE AND JAMES LIPTON

7. Age and Scientific Performance (1979) 178
   STEPHEN COLE

8. Balancing Acts: Dilemmas of Choice Facing Research Universities (1993) 199
   JONATHAN R. COLE

viii   Contents

9. Robert K. Merton, 1910–2003 (2004)   229
JONATHAN R. COLE

10. Why Elite-College Admissions Need an Overhaul (2016)   234
JONATHAN R. COLE

11. The Pillaging of America's State Universities (2016)   240
JONATHAN R. COLE

PART II. FREEDOM AND UNFREEDOM: THE CASE OF WOMEN IN SCIENCE

12. Women in American Science (1975)   247
JONATHAN R. COLE AND HARRIET ZUCKERMAN

13. Women in Science (1981)   271
JONATHAN R. COLE

14. The Productivity Puzzle: Persistence and Change in Patterns of Publication of Men and Women Scientists (1984)   286
JONATHAN R. COLE AND HARRIET ZUCKERMAN

15. Marriage, Motherhood, and Research Performance in Science (1987)   329
JONATHAN R. COLE AND HARRIET ZUCKERMAN

16. A Theory of Limited Differences: Explaining the Productivity Puzzle in Science (1991)   344
JONATHAN R. COLE AND BURTON SINGER

17. Freedom Gained and Freedom Lost in American Science (2022)   380
JONATHAN R. COLE AND DARIA FRANKLIN

PART III. CONSENSUS IN SCIENCE: JUDGMENT AND CHOICES

18. Peer Review and the Support of Science (1972)   415
JONATHAN R. COLE, STEPHEN COLE, AND LEONARD RUBIN

19. The "Ortega" Hypothesis (1972)   432
JONATHAN R. COLE AND STEPHEN COLE

20. Chance and Consensus in Peer Review (1981)   450
STEPHEN COLE, JONATHAN R. COLE, AND GARY A. SIMON

21. NSF Peer Review Continued (1982)   464
STEPHEN COLE, JONATHAN R. COLE, AND GARY A. SIMON

Contents   ix

22. Experts' Consensus and Decision-Making at
the National Science Foundation (1985)   472
JONATHAN R. COLE AND STEPHEN COLE

23. Testing the Ortega Hypothesis: Milestone or Millstone? (1987)   506
STEPHEN COLE AND JONATHAN R. COLE

24. Dietary Cholesterol and Heart Disease: The Construction of
a Medical "Fact" (1988)   516
JONATHAN R. COLE

25. Two Cultures Revisited (1996)   545
JONATHAN R. COLE

26. Intellectual Diversity in the United States: To What End? (2006)   555
JONATHAN R. COLE

PART IV. ACADEMIC FREEDOM AND FREE INQUIRY: THE ENABLING VALUE

27. The Patriot Act on Campus: Defending the University
after 9/11 (2003)   563
JONATHAN R. COLE

28. Academic Freedom under Fire (2005)   574
JONATHAN R. COLE

29. The New McCarthyism (2005)   589
JONATHAN R. COLE

30. Defending Academic Freedom and Free Inquiry (2009)   591
JONATHAN R. COLE

31. The Chilling Effect of Fear at America's Colleges (2016)   616
JONATHAN R. COLE

32. The Triumph of America's Research University (2016)   621
JONATHAN R. COLE

33. Academic Freedom as an Indicator of a Liberal Democracy (2017)   625
JONATHAN R. COLE

34. Academic Freedom under Fire (2021)   634
JONATHAN R. COLE

Index   637

ACKNOWLEDGMENTS

Over the past half-century, an enormous number of friends and colleagues have helped me to develop ideas about the sociology of science. Many of these colleagues became friends and were also trusted advisers during my time as Provost and Dean of Faculties at Columbia. Here are a few of those people who had a special influence on my thinking and writing.

I start, of course, with the three people to whom this collection is dedicated. We tend to forget our past and those who influenced their disciplines enormously but who for reasons of obliteration through incorporation, or because of our limited focus on the present, tend to forget to teach our students about the great minds of the past. My mentor and my intellectual model in my salad days was Robert K. Merton. He had an enormous influence on my thinking and as harsh a critic as he could be about the construction of an argument or a sentence. He was, for me, the ideal of the original mind with great depth and breadth who also cared enormously for those students who he felt deserved attention. Bob, as I would come to call him, could be an unapologetic critic—that intimidated some of his students—but if you took his criticism and suggestions to heart, there were pearls to be found in them. He was a great editor. I do recall, for example, misusing the word "stipend." Bob handed back my manuscript that contained this error along with a four-page, single-spaced, history and usage of the term. He was right, of course, about my error and I don't think I ever again misused stipend. He loved language and concepts and he hoped that you too would appreciate them more after learning from him. This was evident in his book *On the Shoulders of Giants—A Shandean Postscript* (1985) and his book, with Elinor Barber, *The Travels and Adventures of Serendipity* (2006). Bob was more than a reference point when addressing theoretical problems in science or elsewhere in society. He was a master teacher (many Columbia faculty members would sit in on his lectures) and as his assistant for several years I saw a mind at work in his lectures on analysis of social structures, or in his course on the history of sociological theory. So much of Mertonian theory was woven into my early work and collaborative work. There were also many brilliant ideas that never saw the light of day because he was not yet satisfied with their coherence or presentation. As I advanced in my own career, Bob became not only an admired colleague but a close friend. We would take in Knicks games together or just enjoy each other's company. When I had issues that concerned me while I was Provost, I would often consult Bob. Those discussions always included his important insights, questions, and useful suggestions.

With Bob and Steve, Harriet A. Zuckerman was my closest collaborator and developer of the sociology of science. While she was early on working on her classic book on Nobel laureates, I was working with Steve on aspects of the stratification system of science—using quantitative methods of the time to test theoretical ideas. And then Harriet and I, after I had finished my book *Fair Science: Women in the Scientific Community*, took on a big qualitative project interviewing in a semi-systematic way well over a hundred female and male scientists. We oversampled very distinguished women, many of whom had come of age in the first quarter or half of the twentieth century. We not only collected a plethora of background information on the professional careers of these scientists, but we placed them in a stratified set of categories. The idea was to interview these scientists in person, record the interviews, analyze the results using primitive forms of qualitative analytic tools, and publish papers on the comparisons of male and female scientists. We spent thousands of hours together: creating a workable interview guide that would allow us to make comparisons, collecting quantitative data on these scientists, and recording their published productivity and citations to their work over the first fifteen or twenty years of their post-doctoral careers. The men and women were matched by the year of their PhD—roughly then by age. Data were gathered on their family histories—marriage and motherhood, as well as other activities that consumed their time. We explored how they worked with detailed discussions about how the men and women "did" science. We traveled all over the country to gather these data; most of the time we both participated in the interviews; occasionally we conducted the interviews separately. These interviews still exist, but many of those interviewed have died. There is rich material in these interviews that have never been fully mined, as Harriet moved to the Mellon Foundation and I to academic administration. Finally, we spent countless hours together preparing material for publication. I could not have had a better colleague. Her influence on my essays is evident in our collaborative works present in this volume. We both benefited from generous support from the National Science Foundation.

Last, but hardly least, was the tremendous influence my brother Stephen had on my thinking and work. We were very different intellectual types. Steve, who died in 2018, was brilliant, contrarian, and tended to see the world in stark contrasts of black and white—with few grays in between. He had great insight, but his strong views and sometimes off-putting personality rubbed against some of the powers that existed. Yet, he was an exceptionally gifted and rigorous teacher and a wonderful mentor to scores of graduate students. I'm convinced that Steve, who had a distinguished career in sociology, lost out on many opportunities and forms of recognition because of his fierceness and deep commitment to his own beliefs.

There are few brother combinations in academic life—at least few that I know of. But we made an excellent pair, in part because I tended to think more in terms of shades of gray and was more hesitant at drawing extreme conclusions based on what I believed was insufficient evidence. Not that Steve wasn't rigorous, imaginative, and provocative in his intellectual character. He was also fun to work with. And I had the advantage of knowing him my entire life. I knew how far he would go. Meeting in the middle, working on testing theoretical ideas with empirical data, fit our personality differences well. And that combination led to many co-authored papers in the early decades of the development of our specialty. While Steve and I worked seemingly endlessly on aspects of the sociology of science, Steve had other strings to his bow. He wrote successful texts on sociology; he wrote about the unionization of teachers; he wrote critiques of features of our discipline, and he authored some of the best papers produced early on in our specialty. I included, as you no doubt have gathered, one of those solo authored essays, which addresses the age-old question of whether scientists, particularly mathematicians, do their most important work at an early age, only to experience a sharp age decrement in creativity thereafter.

Many others influenced the content of these essays, although the results and errors are attributable to my limitations. I start with my former undergraduate and graduate students who had an interest in this growing discipline and who asked critical and perceptive questions as well as producing their own essays on science. Let me mention a few individuals who influenced my thinking and the content of my work. These included Bernard Barber, Elinor Barber, Joseph Ben David, Akeel Bilgrami, John Bruer, Michael Crow, Yehuda Elkana, John Elster, Mary Frank Fox, Daria Franklin, Eugene Garfield, Thomas Gieryn, Charles Gillespie, Warren Hagstrom, Gerald Holton, Herbert H. Hyman, Paul F. Lazarsfeld, Joshua Lederberg, Walter Metzger, Derek J. de Solla Price, Margaret Rossiter, Richard Shweder, Burton Singer, Stephen Stigler, and Norman Storer.

xiv Acknowledgments

I also want to thank the scientists that Harriet Zuckerman and I interviewed. People like Salome Waelsch, who we interviewed three times, went above and beyond what we could have asked for. Others, like Andrea Dupree, acted similarly. They gave freely of their time and their ideas, which proved essential for our work. In fact, we published in truncated form a number of these interviews. Without their cooperation and openness, the project on women in science, sponsored by the NSF, could never have gotten off the ground. A number of scientists that I interviewed after the formal end to our initial project contributed greatly to my increasing understanding of the position of women in science. Cori Bargmann was always helpful in giving her thoughts about science and the place of women in the scientific community. Special thanks go to Bonnie Bassler who has taught me much about laboratory life and how the creative process operates in her laboratory—one where her graduate students and postdocs represent part of her family. I shall always respect bacteria and viruses as being very clever because of what Bonnie has taught me.

Special thanks also go to members of Columbia University Press. The Press has been enriched by its current brilliant leadership of Jennifer Crewe. I particularly thank Eric Schwartz, who has shepherded this project through the publication process and has offered useful suggestions from the initiation of the project to its publication.

Finally, this set of articles and essays could never have happened without the support of my family. My wife, Joanna, has been there to discuss ideas from the beginning. Her willingness to lend her amazing critical eye on the topics of these essay has always, it seems to me, improved them. I've been very fortunate to have two fabulous children, Daniel and his husband, Nick, and Susanna and her wife, Adrienne, as well as the two best granddaughters, Lydia and Charlotte—all whom live in New York, so I can see them enough to learn from each of them and to whom I can express my love in person.

SMOOTHER
PEBBLES

INTRODUCTION

"I do not know what I may appear to the world; but to myself I seem to have been only like a boy playing on the seashore, and diverting myself in now and then finding a smoother pebble or a prettier shell than ordinary, whilst the great ocean of truth lay all undiscovered before me."

Attributed to Sir Isaac Newton

From the great scientific revolution in seventeenth-century England until the third decade of the twentieth century, most scientists and others believed in the great man theory of scientific discovery.[1] Science was produced by individual geniuses, like Newton, who had a "sacred spark." Keen observers of scientific growth did not think of science as a product of a community of individuals guided by the values and norms of an institution. The social aspects of science were largely neglected. Henry Oldenburg's creation in 1665 of the *Philosophical Transactions* as a scholarly journal publishing the work of members of the Royal Society, one of the oldest learned societies in the world, was perhaps the first organized movement that recognized that scientific advances came from creations by a structured community.

Members of the Columbia School, or program, wanted to alter the idea that science and the growth of knowledge was largely the product of individual genius. We wanted to demonstrate how the institution of science, particularly academic science, was the product of a variety of social structural arrangements, beliefs, and values. While the program at Columbia emphasized the structure of scientific institutions, other schools, particularly in France and Great Britain, as well as the United States (such as Bruno Latour and Steven Woolgar's *Laboratory Life*, published by Princeton University Press in 1986), emerged in the 1980s and tried to examine the content of knowledge itself or the ideological influences on the growth of scientific knowledge. One consequence of these different points of view was that the emerging specialty became fractured rather than a forum for a productive dialogue. This damaged attempts to broaden and make more cohesive the sociology of science community.

The putative father of the social study of science was Robert K. Merton, one of Columbia's own and one of the giants of twentieth-century sociology. We can perhaps date the origin of important work in the sociology of science with the 1938 *Osiris* publication of Merton's Harvard dissertation, *Science, Technology and Society in Seventeenth Century England*. This remains, more than eighty years later, a landmark work. Merton, of course, had many strings to his bow. During the first several decades of his career, for example, he wrote pathfinding theoretical works on deviance in American society. During these years, he fathered and developed, among others, concepts like "the unintended consequence of purposive social action," "the focused interview," and "the self-fulfilling prophecy." These ideas have become part of our lexicon although most people are unaware that they originated with Merton. These are self-exemplifying cases of what Merton called "obliteration through incorporation."

Merton maintained an abiding interest in science. After achieving renown for his work in already accepted fields of sociology, Merton looked for a propitious moment to develop a program that would craft ideas about science into a school of thought and into a legitimate sociological specialty—one recognized by both sociologists and members of the scientific community. He did that in his 1957 presidential address to the American Sociological Association: "Priorities in Scientific Discoveries: A Chapter in the Sociology of Science" (*American Sociological Review* 22 [1957]: 635–59). The group began to take shape with Merton persuading three of his graduate students to focus their dissertations or early work on empirical studies related to his theoretical ideas. The group developed the specialty through focused, empirical work on a particular set of theoretical ideas—illustrations of what Merton called "theories of the middle range." With the help of an initial large, five-year grant from the National Science Foundation (NSF), the group's published research began (initially as a part of a graduate seminar) to have an impact in sociology. The questions addressed in these early papers and essays were recognized by colleagues inside and outside the field as worthy of extended sociological analysis. A key element in the institutionalization was the recognition among scientists (who had been resistant to the idea) that studying science as a social system had merit. Many of the essays that appear in this volume were published in either professional or specialized journals or in works by invitation—such as *Festschriften*. Bringing these essays together, I hope, will enable you to have a sense of the early history of the specialty through the content that was produced by members of the group. The coherence of the themes addressed is also one of my objectives—as well as to suggest the need to extend these early studies.

From roughly 1965 to 1990, Columbia University's Department of Sociology was arguably the leading center, both nationally and internationally, in creating and developing the sociology of science. Before there was anything identifiable as "the Columbia School," Merton wrote, as I've suggested, some influential essays on the subject, and his work was complemented by the essays and books produced by Bernard Barber, who had studied at Harvard with Talcott Parson and Merton. Barber had an exceptional nose for a good problem—he was a "truffle dog," who put himself and others onto interesting problems in the sociology of science. He was also highly supportive of his student's efforts. After gaining some traction within the profession, our work was often referred to as part of the Merton School. One cannot overestimate the influence of Merton's own extraordinary scholarship, his meticulous and clear writing, and his choice of strategic problems for the acceptance of his work and the group's as well. Sociology was often an object of ridicule in those early days—accused of using excessive jargon to describe commonsensical ideas. But Merton was very different in his scholarship, his writing, and his criticism of his students and others. By using his extensive network of scientists and other scholars, he helped crack open the door to a greater acceptance of sociology as a discipline—and science as one of its subdisciplines. Merton, Harriet Zuckerman, Stephen Cole, and I produced a foundation for the specialty and thus paved the way for later students studying social aspects of science. This collection of my papers and essays represents some of the work produced by that school over the past fifty years. Many of these essays are coauthored with the group's members—to which this volume is dedicated.

Our short-term goal was to explore the norms, values, and structure of science. The core values of the scientific and university system and the norms attached to them consisted of deeply held beliefs as well as principles that determined how individuals acted. They were organizing principles designed to support the institution in meeting its goals and mission. They influenced the types of social structures that developed to carry out the activities of scientists.

The essays in this volume show clearly that our attention was not on the production of individual pieces of knowledge but on the social systems that made that production possible. We tested ideas about science with both quantitative and qualitative data. Merton's theories formed the basis for a variety of empirical studies that were designed to examine and extend his theoretical ideas. Early interest focused on Merton's four norms governing scientific behavior.[2]

Many early articles examined the reward system of science. Others outlined and investigated the system of social stratification in science. Did it operate within the definition of universalism? Was it a fair system? Did it reward sheer quantity of publications over their quality? Did you have to publish or perish? Were truth claims judged independently from a person's religion, gender, nationality, race, or other ascribed characteristics? What were the differences in the careers of scientific elites compared with rank-and-file scientists? What was the relationship between age and scientific and mathematical performance? Was the widespread belief accurate that a physicist or mathematician was "over the hill" by age forty? How do you measure the quality of scientific research? Could citation counts prove to be a better measure of *quality* than counting the sheer number of a scientist's publications? How did the system and culture of science treat women and minorities—could we identify cases of unfreedom and the violation of science's meritocratic ideals? What were some of the ways that science failed to live up to its normative injunctions?

Some of these later quantitative and qualitative studies focused sharply on the careers of male and female scientists, many of whom were giants in twentieth-century science. The qualitative work was based on over 140 extended qualitative interviews conducted by Harriet Zuckerman and me. These data gave nuance to some of the quantitative work on women in science that I published in *Fair Science: Women in the Scientific Community.*

Our studies of "fairness" extended also to an analysis of the peer review system—the method used by leading journals to determine what papers would be published. Peer review is fundamental to the organization of science: not only for publication of papers but also for applications for grant funds from the major sources of science funding and for the system of academic promotions. Stephen Cole and I did extensive studies for the National Academy of Sciences that were designed to test the hypothesis that the national peer review system was essentially an "old boy's network." Members of Congress were periodically claiming that this male in-group prevented funds at the NSF from going to lesser universities in their state. Senator Proxmire of Wisconsin (whose major university in Madison received substantial research funding from NSF), used to publicly call out one grant that he tagged the "Golden Fleece of the Month Award"—he judged it to be an example of a rip-off of public funds. Other members of Congress would, for example, question grants in chemistry in which the title included the term "radical ions." Consequently, working scientists were told to omit the word "radical" from their grant applications.

Steve and I were asked to examine in depth the NSF peer review system in terms of its structure, and in a second study, its quality and fairness—much of that work continues to be highly referenced by scientists and students of the social studies of science. It was the most extensive examination of the peer review system to that date. Part of the study involved an experiment that found, unexpectedly, that a substantial amount of luck or chance was involved in whose grants were funded. This sent nervous chills down the spines of some members of the Academy Committee overseeing the study. They not only thought that the findings must be wrong but also that it would harm the funding of science by Congress. So adamant were some members of COSPUP (the Committee on Science and Public Policy) that the Committee insisted that we redo the experiment from scratch to see if the findings held up. We redid the entire experiment. A year later, when the results were in, we had almost identical results. The Committee yielded, but some members did not want us to publish the results. We refused to yield to that request and pointed out that by written agreement we had the right to publish our results. At the end of the day, the NSF survived and thrives; some of our findings led to policy changes within the NSF. And we published our results in *Science* magazine and elsewhere.

In the early days we also focused on methodological problems. We explored scientific influence through the analysis of citation networks. This was a precursor to today's far more sophisticated use of network analysis. And, of course, we helped develop the use of citations as a measure of the impact of a piece of scientific work.

These were just some of the questions and problems explored by the Columbia School. An important and influential collection of some of Merton's papers on science was published in 1973. This collection of papers and essays will add to a coherent understanding of the Columbia School and the problems we addressed over the last half-century. I discuss below what I might do differently if I were writing some of the papers and essays in this volume today, given the growth of knowledge over the past half-century. A by-product of this discussion is the hope that I can present a clearer picture of how science is done—far from the way most people understand the growth of scientific discovery.

The concepts of fairness, justice, and freedom have been the focus of much of my work since I began publishing in 1967—some fifty-four years ago. Most of these essays and papers attend to one or another of these themes. One of the central norms in the scientific community is universalism. It enjoins members of the community to reward and recognize people based solely on the assessed quality of their work. Personal qualities such as a scientist's political views, race, nationality, gender, or other ascribed characteristics should not affect decisions in hiring, promotion, tenure, gaining resources for research, and various forms of honorific recognition. Universalism is distinguishable from meritocracy, which is an increasingly ambiguous concept and is problematic in the way it is currently defined. The group spent a good deal of effort seeking evidence that would test whether universalism was approximated within the scientific community.

Related to work on universalism was the research we did on how early recognition produced later recognition. In fact, a good deal of effort was focused on elites within science. That effort was led by Harriet Zuckerman, beginning

6 Introduction

with her seminal work on Nobel Prize winners. We empirically explored the idea of cumulative advantage and disadvantage in science and the presence of what Merton called "The Matthew Effect," which was a feature of the accumulation of advantage and disadvantage.

There was a significant hiatus in my publications on science while I was Vice President of Arts and Sciences (1987–1989) followed by Provost and Dean of Faculties at Columbia from 1989 to 2003. After stepping down as Provost, I turned to work on the modern American research universities. Those books and essays were informed significantly by my earlier work on science. For example, the value of free expression and free inquiry—of academic freedom— lies at the very heart of both science and institutions of higher learning.

Retrospective

When I look back at some of the essays collected here, I often think, "Did I write that?" I find the questions and analyses continue to interest me. At other times I wonder why I did not include ideas that were not explored. Given my current knowledge I might have addressed some additional issues or spent more time developing studies of other important features of science.

Of course, scientific knowledge has grown exponentially since I wrote these essays. In some physics specialties, for example, knowledge has been doubling every year. The growth has been less dramatic in the social sciences, but it still has expanded at a formidable pace. I'm also not the same person I was then. My work with my colleagues was produced at the interstices between the end of Little Science and the full flowering of Big Science. Some of the School's later students, such as Thomas Gieryn, and many others in the history, philosophy, and sociology of science made formidable contributions to the growth of the discipline and its expansion.

Historical context here is valuable. At Columbia in the 1960s, sociology students were always asked: What question do you want to answer? What ideas do you want to examine? Why is this an interesting problem? Don't start with data; start with an interesting question about social life. This is what we tried to do. As the Nobel laureate Stanley B. Prusiner opined in his autobiography, "Every scientist must choose a problem; it is probably the single most important decision of his or her professional life."[3] Paul F. Lazarsfeld (PFL, which is the way we referred to him), who was renowned for his innovations in sociological methods, was, in fact, often particularly interested in theories that could be tested with empirical evidence. Correlatively, Merton (RKM) was always searching for ways for his students to test ideas (often his own).

When our group of sociologists of science was formed and received funding from the NSF, we worked in a decrepit, mildewy brownstone just off Broadway and 115th Street called the Bureau of Applied Social Research. Formed by PFL, "the Bureau," as it was affectionately called, was created primarily to house his projects. Other than the small laboratories occupied by psychologists, there was virtually no space on the campus to conduct most other empirical projects (economists tended to work through the National Bureau of Economic Research—the NBER had space in New York at the time), so the Bureau was the home for empirical research carried out by other faculty and for his and Merton's most promising graduate students. We met there almost every day to discuss ideas or at Tom's Restaurant on Broadway for lunch (made famous by *Seinfeld*). Professional conversations took place at Tom's most days over lunch. I almost always ordered an egg salad on rye, side of french fries, and a Coke for fifty-five cents; Harriet was more abstemious, often ordering her favorite cookie and that was it. Merton was never at Tom's and very rarely at the Bureau, but in many ways he held the Bureau together, especially when PFL was on leave in Europe or at some other place in the United States. When we decided on a research strategy, we began to collect data to test ideas, and we had some tools to do so at the Bureau.

Several examples will suffice to provide an idea of what these tools were like in the late 1960s and 1970s compared with those you would undoubtedly find familiar today. When we needed to calculate correlation coefficients, we did so with a hand calculator. A simple correlation matrix could take forever to complete. That was painstaking work to say the least. When we wanted to amass and analyze larger data sets (perhaps several hundred cases) we coded the data on sheets of paper. Then we transferred those data to an IBM 80 by 10 or 12 rows and column punch cards that recorded the data on each variable that we were studying. A hole was punched in each column depending on what the response was to a particular variable. At the beginning, we sorted these cards into piles by putting a special pick into the card for the relevant variable and then shaking the cards to see which ones fell out. That was one response on that variable, and it was placed in a pile of similar cards. We did so for the other responses on that same variable. Finally, we would count the cards in each category and jot the number down to create a table of responses. That told us, for example, how many men or women responded to our questions in a certain way. Then a minor revolution occurred that produced "counter-sorters," which could be used electronically to divide the cards into different piles and give us a count for each category of a variable. The Bureau hired a person to run tables for the researchers and if he was not available, we would often do it ourselves. We thought technology had made a giant leap forward.

Of course, the real revolution began for us in the 1970s when IBM introduced digital computers at university campuses. Columbia was selected as a test site for the development of these engines of innovation. Big machines, which occupied very large "machine rooms," were positioned on campus and one could submit "jobs" to the computer center. Data were placed on computer tapes, which were used to spin out jobs. These social science jobs often took a day to complete. The output, or results, could be represented in hundreds of pages of "printouts." If you were not careful in properly entering the coding instructions for the computer to provide a set of correlations or correlation matrices or to run some tabular or regression analyses, you could find yourself with hundreds of pages of zeros as output. Some of us were permitted, after the computer center staff went home, to run or "spin" our own tapes very late at night or early in the morning. We were delighted that such a revolutionary machine had been created that permitted us to analyze vastly more data than we could have imagined just a few years earlier. IBM continued to expand the power of these giant computers so that social scientists (as well as other scientists) could rapidly calculate empirical relationships. New startup companies were formed, like SPSS or SAS, to provide software that allowed us to use these upgraded computers for analytic purposes. We could not imagine then that the day was not far off when a desktop or laptop computer would use silicon chips that would vastly outperform those giant machines. Steve Jobs and Steve Wozniak, working out of Job's garage in Palo Alto, would create and sell their first Macintosh (Apple 1) desktop computer in the mid-1970s; IBM sold its first in 1981. A new age for social science analysis had dawned. The quality of the technology expanded rapidly; but that did not affect the quality of the questions, the data, or the theory that were critical to producing significant new knowledge about science as an institution.

We would be wise not to underestimate the role that technological innovations play in advancing knowledge. Merton discussed this in his dissertation; the Columbia group, however, did not spend as much time as it might have on the interactions between technology, science, and the growth of knowledge. More importantly, for our purposes, was the way these breakthrough technologies influenced the social organization of science and its reward system. But extremely important other questions could have been posed about how technological innovation affected scientific thinking. Some technological advances become disruptive of existing paradigmatic structures. There are, of course, classic examples, such as Galileo's discovery of the telescope or, hundreds of years later, the technological implications of the Hubble Space Telescope and now the promise of the James Webb Space Telescope for studying the origins of the universe and astrophysical phenomenon. Back on earth, innovations such as the cloud chamber, the development of cyclotrons initially by E. O. Lawrence in 1929 and 1930, the discovery of recombinant DNA in the early 1970s, or now CRISPR technology, allow science to develop not only beyond its former constraints, but the technology also has a profound effect on where young scientists choose to study and what types of science are apt to be rewarded at a specific time. When Thomas Kuhn discussed scientific revolutions, he paid little attention to the role that technological advances had in creating anomalies in existing scientific paradigms. But consider the role cyclotrons or rocket and computer technology had in disrupting science. Part of this absence of recognition of technology is the general conflation in the public mind between science and technology. After all, our ability to send men to the moon or rockets into deep space is more a result of technological advances than pure scientific discoveries.

Another example is the Wilson cloud chamber, which was produced in the early twentieth century. It was an instrument that allowed physicists to detect particles passing through the chamber. It also was the most heavily cited work in physics for some years. In medical science, the development of computer tomography (the CT scan) and MRI technology opened a new era for medical diagnostics, as has biomedical engineering. Imaging techniques have enabled cognitive neuroscientists to explore brain function that was not possible previously—and yet that imaging remains in a neonatal state of development. Historians of science and technology, individuals like Peter Galison, have provided us with superb discussions of how instruments and technologies have propelled science forward. Of course, a small group trying to bring life to a new specialty could not study all these questions. But this was not part of the portfolio of problems examined in the early days by the Columbia School.

Perhaps equally important has been the role that technology has played in the exchange of ideas and in collaborations. A hundred years ago, scientists shared the same laboratories or worked at the same universities. Consider, for example, Thomas Hunt Morgan's "fly group" at Columbia that worked on early genetics or the maize group at Cornell. Technology has altered the social arrangements of science and consequently the patterns of collaboration, data sharing, and the speed of innovative work. We studied almost exclusively American university–based science rather than the national and international collaborations that were beginning to take shape. While we were at the forefront of studying scientific networks, with work by Daniel Sullivan and others, that is, networks of communications and of social and intellectual distance between schools of scientific thought, our work on co-citation matrices was primitive and perhaps premature compared with the way network analysis is now carried out by social scientists. One trait of a good deal of our work, which has been largely lost, was the search for and analysis of individual "outliers" or anomalies in our datasets that we took seriously rather than as "mistakes."

The Columbia school was not isolated and did value enormously the new technologies and various forms of collaborations with others working on the history and philosophy of science. In fact, there was a moment during the early years when we had the idea of forming a collaboration that would bring together Thomas Kuhn, at Berkeley then Princeton, and Charles Gillispie, who was also at Princeton with Merton and his students. Serious discussions took place about organizing some structure at Columbia that would represent the history, philosophy, and sociology of science with the three prominent leaders. That idea was squashed by a hostile reaction at Columbia among some very prominent scientists who were not able to see the value in the work of the three leaders of the would-be group. This effort failed, but on a much more limited scale and scope, the Columbia group produced a master questionnaire for Gillispie, who was working at the time as the editor-in chief of the *Dictionary of Scientific Biography* (sixteen volumes published between 1970 and 1980). The questionnaire was to be presented to the biographers with the anticipation that all biographies of the scientists would include some common elements that could be used for analysis.

We were also influenced by many outside Columbia. Derek de Solla Price, Joshua Lederberg, Warren Hagstrom, Joseph Ben-David, Diane Crane, John Ziman, Lowell Hargens, Arnold Thackray, Paul Allison, and Scott Long, among many others, affected the development of both methods and the testing of the "middle-range" theories that we were working on. This was particularly true with Lederberg, Price, and Eugene Garfield, when the Columbia group was conducting early methodological work on the use of citation counts as a better indicator of quality of research than sheer publication counts. Most people familiar with the citation index, and those who do not know of its history, tend to forget or don't know that Garfield's creation of the *Science Citation Index* (SCI) was initially, and still is, a bibliographic tool that allowed scientists working in an area of research to navigate the extensive literature and identify others (who cited them) as potential members of the same "invisible college." That term was first used to describe the British Royal Society, a group of scholars and academics dating back to 1660 who met regularly to exchange research and ideas. In the 1960s, the historian of science Derek Price and the sociologist Diana Crane revived the term to describe the informal communities of scholars and professionals who communicate and share research and ideas.[4]

Within a few years of its initial publication, the SCI became known for citation counts. In the early days of our work with citations, there were no online compilations as there are now. There were only printed copies in books that listed the name of the cited author (actually the last name and first initial) and in a very small font the citation to that author as well as its source. If we didn't count the citations by hand ourselves, we hired research assistants to count them. It was a most hazardous task. We were literally quite worried about the mental health of our young students who would spend hours counting citations, omitting self-citations, and registering the numbers on sheets of paper. We were convinced that they could suffer from mental fatigue or worse.

The invention of the citation count turned out to be a mixed blessing. While we demonstrated that citations were a better predictor of quality than paper counts, we also opened Pandora's box. People started to compare individual academics in terms of their citation counts and they reified small differences. I sat in and evaluated over 750 tenure cases while Provost at Columbia. I can't tell you how many chairs of departments included in the tenure dossier the citation count of the candidate compared with other individuals—reifying these differences—holding that a person with eighty citations, for example, has produced work of higher quality than an individual who had seventy-five citations. Citation counts were intended to compare groups, not individuals, and they were not intended as precise counts of individual differences. Fortunately for us, the shape of the distribution of citations to scientists is highly skewed. Most receive few if any; a few receive the lion's share of citations. Consequently, in most of our aggregated counts, small differences meant nothing—including very different patterns of citation in different fields.

One important aspect of science that we almost totally ignored was industrial science. We did treat the Bell Labs, for example, as a virtual university, but by and large we did not examine the stratification system or the social structure of industrial companies dedicated to scientific and technological advance. This was a conscious decision. We found it far easier to study universities since their social arrangements were more similar than in industry. They allowed for easier comparisons. It was far easier to collect data on university scientists. Even today, too little work is conducted on non-university-based research despite its importance. Companies such as Google, Facebook, Apple, or Microsoft play critical roles, both in the collection of data and its analysis. The amount of data these private companies produce, hold, and analyze is enormous and has created a double-edged sword. The data allow social scientists to analyze important social problems at a scale unheard of previously. But simultaneously it creates serious issues related to the way the data are used for commercial purposes and for influencing individual choices. There is much to be concerned about in the tyranny of the algorithm. Infringements of individual privacy rights, the misuses of information, and ethics of creating algorithms tailored to an individual's "likes" and "dislikes" are a few examples of the risks of collecting this amount of proprietary data.[5] Nonetheless, much can be learned about significant social questions by analyzing these massive data sets. Social control of these media companies was not even a question that we dealt with since, of course, they did not come into existence until we were far along with our studies of science.

I think we gave short shrift to concepts and ideas drawn from other disciplines. Historians of science, like Gerald Holton, had a significant impact on our work, as did Kuhn and other philosophers of science. Kuhn's work was particularly influential in its emphasis on science as a community holding in common theories and methods—paradigms—that had potential for elaboration and that were dependent on social consensus. His work was, of course, open to many interpretations, but its implication was surely that scientific growth depended on social variables and that paradigms also limited what was acceptable as science within the paradigmatic framework. The Columbia group was very interested in multidisciplinary collaborations, but we rarely developed them. Our primary goal, beyond producing intrinsically interesting studies, was to convince our audience of the value of the sociology of science. As I suggested, many scientists of great renown were at first highly skeptical of its value—especially if discussed by those not working in the so-called "hard sciences." To demonstrate that sociology had something useful and undiscovered to say about science, we neglected, like most other social scientists of the time, to take advantage of the added value of a multidisciplinary perspective. Sociology was not alone in this failure. Economists began to study particularly sociological problems but did so for the most part using their own theoretical models and methods. Gary Becker studied sociological problems, but he developed concepts, such as human capital, using tools rooted in economics. His work, along with Jacob Mincer's and later James Heckman's, attacked sociological problems and was brilliant—but left some of the most interesting questions unattended to in the residuals of their equations. Economics began to overlap more with both sociology and psychology with the rise of behavioral economics and the acknowledgement of some valuable sociological concepts, such as Merton's concept of the self-fulfilling prophecy. The work of the psychologists Amos Tversky and Daniel Kahneman on decision-making under conditions of uncertainty questioned some of the core assumptions of economics and has been incorporated increasingly, through the work of Richard Thaler and others, into the fabric of economics. Much talk and action related to the interaction of people whose work really required collaborations with those in other sciences came later around the turn of the twenty-first century. Physicists and mathematicians required each other, as did biologists and computer scientists or biologists and engineers, among many combinations. The social sciences had grown to the point where work in one discipline really required cross-disciplinary collaboration. Today this is rather standard practice.

Our work on the scientific community and the mobility of scientists would have benefited, for example, from a greater use of the concept of culture and its differential influence on the careers of men and women. Much of my early work on female scientists and the essays that I coauthored with Harriet Zuckerman compared the careers of men and women largely after the race had already been run for most women. Had we continued our studies of women in science, after collecting massive amounts of qualitative interview data in the late 1970s and early 1980s, I think this lacuna would have been filled. The dynamics of cultural influences on social and self-selection of women into science, their travels in science, and their responses to success and failure would have deepened our understanding of career histories and scientific choices. Today I would place more emphasis on the preprofessional experiences and responses to those experiences for both male and female scientists. I think we would have had a more nuanced view of both universalism and particularism—at the institutional and individual level—had we done so. Much of this has been achieved through relatively recent autobiographical works by female scientists—as well as new works by feminist sociologists.

Further work on concepts like accumulation of advantage and disadvantage, and empirical work to test the theory of limited differences, which Burton Singer and I developed (see chapter 16), would have enhanced our understanding of the dynamics of social successes and failures on time-dependent careers of male and female scientists. Further development of "the Matthew Effect" would have told us more about the reward system of science. That was halted when I began my administrative work at Columbia in 1987 and Zuckerman took a high position at the Mellon Foundation. In short, we made substantial progress in understanding careers in science, but far more can still be done to understand with greater precision the mechanisms that influence career decisions and outcomes.

We spent little energy looking at the linkages between the two cultures: science and the humanities. I wrote only one essay on this topic. It has taken on greater importance over the years. If scientists can use sociologists to help understand the dynamic systems that produce novel discoveries, it should have been more apparent to us that some of the most important discoveries in science, such as stem cell research and CRISPR technology, would require us to answer historical, ethical, and moral questions that are often the foci of attention of humanists. The relationship between science and the humanities remains understudied.

Looking back, the one intellectual problem that I would have liked to have addressed more was the relationship between so-called positivist thinking about science and social constructivist and interest theory perspectives about scientific work and its relations to discovery and nature. Kuhn and others opened the door to thinking hard about drawing lines about where sociological variables entered the discovery process and where it yielded to nature—or a "there out there."

In retrospect, I believe the Columbia group did not sufficiently engage in a dialogue with the group of social constructivists and interest theorists, who were highly critical of what they believed was a positivist and empiricist orientation at Columbia—and some of its followers. Some of the lack of engagement can be attributed to our involvement in our own work with little time to pursue a dialogue about the nature of nature—and whether there really was a "there out there." But part of the group was conflict averse and refused to be drawn into what we felt was an unproductive dialogue with the early works of, for example, Bruno Latour, Karen Knorr-Cetina, Harry Collins, Steven Shapin,

David Bloor, Andrew Pickering, Barry Barnes, and even some of the Columbia School's own graduates.

Each different school of thought had its following. Here is not the place to detail our intellectual disagreements, but I believe that in the end, Stephen Cole and I believed that, of course, there were important social variables that affect choices of experiments, the technologies used, and evidence garnered, as well as how results were presented. That had already been well documented in works, for example, by Gerald Holton on Einstein and R. A. Millikan. The system and outcomes were highly influenced by sociological variables (totally consistent with long-standing views in the sociology of knowledge); there was no end to the search for "truth" in nature, but at some point there is an objective part of nature that theories and experimental work must acknowledge, conform, and explain. Effective vaccines, for example, are not just socially constructed. Neither are more abstract theories. But the social system of science is influenced by consensus and the lack thereof, the influence of scientific authority, and individual biases and presuppositions held by scientists. And at the end of the day, as Noam Chomsky has suggested, there just may remain mysteries that humans cannot comprehend—mysteries that only after long periods of time may move from mysteries to problems.

Following Kuhn's inversion of theory and research—that theory precedes research in the process of discovery—the "interest theorists" such as Barry Barnes asserted, in a crisp summary by the extraordinary historian of science Peter Galison, "Knowledge is to be accounted for in terms of social interests just as surely as 'conventional forms of artistic expression.'"[6] As Galison also notes, "In its strong form the interest-theory account denigrates the role of nature and supposes that scientists' presuppositions—bolstered by their interests—condition the admissible phenomena in such a way as to render a particular theory and its associated experiments closed and self-referential. In this view, a theoretical outlook, with the experiments that its advocates determine to be relevant, will be entirely divorced from the combination of theory and experiment that succeeds it." Some in the "interest group," such as Bloor, struck a similar note in claiming little space for nature in the scientific process of discovery. The idea that scientific knowledge was essentially little more than a form of social forces set the construction—the product of what scientists could reach consensus on—had a strong appeal to many social scientists of science. It had its critics, of course, but we shied away from an extended dialogue with the British and European scholars who produced this view of the growth of knowledge.

Pickering, a substantial figure in the social studies of science, went still further. As he put it, "Scientific communities tend to reject data that conflict with group commitments and, obversely, to adjust their experimental techniques and methods to 'tune in' on phenomena consistent with those commitments."[7]

In one of Stephen Cole's last books, *Making Science: Between Nature and Society*, he did directly address these groups of scholars. While accepting the idea that sociological variables play a significant role in the process of discovery, Cole claimed that the social constructivists failed to make direct causal connections between these sociological factors and the specific outcome of theoretical or experimental scientific work. There he discusses these various angles of vision and tries to strike a balance between them through an analysis of consensus in science, which we both worked on, and his own concept of the "core" and "periphery" of knowledge.[8]

Cole described it this way:

> The core consists of a small set of theories, analytic techniques, and facts which represent the given at any particular point in time. If we were to look at the content of courses taught advanced undergraduate students or first-year graduate students in a field such as physics, we would acquire a good idea of the content of physics' core of knowledge. The core is the starting point, the knowledge which people take as a given from which new knowledge will be produced. Generally the core of a research area contains a relatively small number of theories, on which there is substantial consensus. . . .
>
> The other component of knowledge, the research frontier, consists of all the work currently being produced by all active researchers in a given discipline. The research frontier is where all new knowledge is produced. Most observers of science agree that the crucial variable differentiating core and frontier knowledge is the presence or absence of consensus.[9]

Science has a great tolerance for ambiguity at the research frontier. The core of knowledge is linked to the research frontier through the evaluation process. But both Steve and I claimed in several essays contained in this volume that there was an objective reality that would not yield to individual beliefs, biases, or presuppositions. This conversation with those who disagreed with us would have been fruitful for the subdiscipline of the social studies of science. I sometimes wonder if their prior positions have evolved.

Major Themes in This Collection

Universalism and Fairness in Science

Ever since I was a young boy, I had a deep concern and interest in the idea of freedom and liberty, in fairness and justice. Undoubtedly, this interest, other than my obsession with playing organized baseball and basketball, was influenced by personal events. Both of my parents, like so many young socially

16   Introduction

conscious people who came of age in the 1930s and 1940s, were intrigued by Marxism and by social experiments built on the idea of greater freedom for workers. My father, who came from a poor, immigrant family, was an avid reader of left-wing literature and journalism (although he would religiously read Walter Lippman every week); less so my mother, who was far more of a literary person. My mother, who typically had the lion's share of family-related obligations, nonetheless had time to enter the PhD program at Columbia in English and study with Lionel Trilling until she had to quit to help keep our family afloat. My father was trained as a lawyer but gave it up immediately after earning his degree for what he really loved, acting. In the days of radio and early television, he was moderately successful. His deep bass voice made him perfect for the villains in the scripts. Unfortunately, villains didn't last long in these shows, so he was always bumped off after a few episodes—and was again looking for work. But he had continuing interest in organizing actors and others in the industry. Consequently, there was much dinner table talk about injustice, lack of freedom, and unfairness in "the business." Because of his union activities, I believe, my father was blacklisted during the McCarthy period. Unlike writers who could work under aliases, actors could not hide. FBI agents, in their Burberry raincoats, would come to our door periodically to question my father—to show him pictures of people they wanted him to identify. He never gave in to them and finally got rid of the agents by hiring a renowned civil rights lawyer, Arthur Kinoy, who had helped defend Ethel and Julius Rosenberg in their espionage trial.

I took these strong values to Columbia College in 1960. They were reinforced by most of the historians and humanists who were my teachers there. I was an organized dilettante as a student. I wanted to study with the great minds of Columbia, regardless of field. So, I took (or sat in on) many courses: the modern novel with Lionel Trilling; Impressionism with Meyer Schapiro; classical and modern drama, offered by Eric Bentley and Robert Brustein; aspects of American history with Richard Hofstadter, Walter Metzger, and James Shenton; physics with Polykarp Kush; and sociology with Daniel Bell, Merton, Lazarsfeld, and many other distinguished teachers. The values of these professors, all white males, reinforced my beliefs. When I met Merton and decided that I wanted to do graduate work with him, I carried with me baggage of concern with the ideas of fairness and justice. When I joined the sociology of science group of four (I was the last to do so, —influenced by my brother Stephen— I immediately wanted to study the norms of science, particularly the norm of universalism). I also wanted to study issues of fairness and justice in science.

Consequently, a substantial number of my early papers and essays focused directly or indirectly on five core themes: the core values in the scientific community and, through an examination of them, an assessment of whether the scientific community was approximating its ideals. For example, early work examined the reward system of science and addressed the fundamental question of whether scientists were being rewarded for the merit of their work or for other extraneous factors. We looked to answer questions such as these: Is it true that if scientists don't publish in great quantity, they perish? Is it true that the sheer number of papers is more important than their assessed quality? How was honorific recognition, like election to the National Academy of Science, the American Philosophical Society, or the American Academy of Arts and Sciences, and prestigious other forms of recognition, such as the Nobel Prize, distributed in the scientific community? How do scientists gain visibility within their specialty or discipline? Was the reward system dominated by a relatively small group of elites? Is there a process of accumulation of advantage that is observable in science? How does recognition at one point in time influence the probability of earning even greater recognition in the future? Were scientists at the most prestigious universities disproportionately rewarded compared to comparably productive scientists at lesser institutions? We focused perhaps more than we should have on elites. Zuckerman's work on Nobel Prize winners remains the definitive source for the effects of the prize on different recipients.

Consensus in Science: Judgment and Choices

I've suggested that consensus and the lack of it is central to our understanding of both the reward system of science as well as what is accepted at any point in time as a fact. Steve and I, along with a set of other collaborators, spent a great deal of time exploring the evaluation system of science and how that system affected grant applications for federal resources as well as recognition within the academic reward system. We argued that levels of consensus differed by discipline and therefore what were taken as fact also differed, especially between the physical, biological, and social sciences. Some of these findings, as I noted before, were controversial and met with resistance from scientists. Most of our findings have been replicated and some have proved to be the point of departure by members of a variety of scientific disciplines, particularly astronomy. The essays that continue to have an impact are contained in this section of the collection.

Unfreedom in Science: The Limitation of Options—
Freedom Gained and Freedom Lost

As noted, a substantial amount of research and publishing by the group examined forms of unfreedom in science—particularly the truncated opportunities and their consequences for women aspiring to become scientists. Unfreedom as I'm using it here refers to a limitation of options that are available to scientists, which limits their freedom in non-coercive ways. These unfreedoms are often taken for granted. These limitations are cultural values embedded in society, as well as in the social organization of university science. This is not just a matter of coercion but cultural, normative expectations that have been formed in the general society about appropriate behavior. They are part of the *zeitgeist*. These norms and values become internalized by members of the university and scientific communities. They exist, and may differ, depending on the power and authority arrangements at a university or another form of organization. For example, the cultural norms may be held by powerful administrative authorities—like trustees and presidents of universities. They may or may not be at variance with the beliefs of professors. The dynamics of how these cultural norms change is something that we did not study, but if we did, we could add a great deal to our understanding of the history of science.

I've included essays that relate to freedom and unfreedom in science. The focus was primarily on women in science and the various opportunities they had and even more those that were unavailable to them for the first half of the twentieth century—as well as on limited options they continue to have in the scientific community. I've produced independent works on the subject, but a good amount of the effort was collaborative with Harriet Zuckerman. The final essay in this section is an essay produced by Daria Franklin and me about women in the first half of the twentieth century. It has some elements that I did not address in early essays on women in science. It suggests new pathways to explore when studying limits on the options of women in science—in the past and present. There is virtually no focus on minority scientists since throughout the time that we were studying disparities in treatment between men and other social groups there were few minorities in academic science. As noted, our work would have benefited by taking a step back and looking at both the early childhood expectation of minority group members and the cultural impact on their choices.

Academic Freedom and Free Inquiry

A substantial number of the essays collected here focus on aspects of academic freedom and free inquiry at universities. Science cannot thrive without both. Academic freedom is the enabling value that allows for the activation of the other core values in science. This essentially means that the compact between science and the government must be one of mutual trust with each partner to the compact adhering to its freedom associated with its role. The bargain that was worked out in the late 1940s and 1950s was simple in principle but very difficult in execution. For the first time in American history, the federal and state governments would provide public taxpayer monies to support science and the growth of knowledge. And they would do so at arm's length. The government would allow scientists to judge what projects were worthy of support, the scientists would carry out those projects principally through universities, and the government would not interfere with the scientists by overly constraining what they should or could do. In turn, university scientists would produce new discoveries—and ones that could have a substantial positive effect not only on military uses but also on the productivity of the national and local economies. They would transfer knowledge to students and prepare them for the higher-skilled jobs of the future. Finally, universities would attempt to create more knowledgeable and active citizens upholding the core values of the nation. Those, in short, were the critical elements in the contract. To paraphrase Emile Durkheim, the nonessential elements of a contract are its most important part. Trust was key. In this science pact there was an essential tension in the bargain.[10] Those who held the purse strings too often attempted to intrude into the workings of the university's scientific community and to fetter some controversial areas of knowledge growth—like stem cell research or social science research that was viewed as too radical. The government also used its financial power to pressure universities to act in ways that were contrary to university values—like acknowledging the rights of gay people at the university. Finally, the government over several decades pulled back from its agreement to cover the full costs of university research and placed caps on the indirect costs of scientific projects. This required universities to cross-subsidize fundamental research.

Government efforts to interfere with university research have taken many forms over the past hundred years. The government has tried to dictate what scientific problems should not be studied; has accused scientists, like J. Robert Oppenheimer or Linus Pauling, of espionage or of "un-American activities"; and has tried to dictate curricular changes. Scientists and scholars have been required to sign loyalty oaths and have been charged with violations of the USA PATRIOT Act (2001) or the Public Health Security and Bioterrorism Preparedness and Response Act (2002). There have also been great defenders of academic freedom and freedom of inquiry. The University of Chicago has led the way in this defense. From William Rainey Harper, its first president, to Robert Hutchins, Edward Levi, Hanna Holborn Gray, and Robert Zimmer, its most recent, these individuals recognized that we cannot have truly distinguished universities without free inquiry in science and scholarship. I came late to an understanding of the centrality of academic freedom and free inquiry to the achievements of American science and scholarship. My undergraduate thesis adviser, Walter Metzger, was at the time writing about the history of academic freedom. I was always curious as to why such a talented historian should devote his energy to this topic, when he could be looking at the origins of the Civil

War, of Reconstruction, or of American progressivism. Only years later did I see the extraordinary value that his work with Richard Hofstadter on academic freedom had. I turned some of my effort to locating free inquiry and academic freedom at the center of scientific and university values. Several of my essays on academic freedom appear in this volume.

The American research university has withstood these attacks to become the greatest producer of scientific discovery perhaps in history. The structure of the way science is done in the United States, especially the attachment of funding to the universities—and with relative freedom of inquiry—has produced the most dynamic engine of innovation that we have ever seen from organized science. Much of that has to do with the structure of scientific research—and is detailed in some of these essays and in some of the books that I've published about American universities and their successes.[11] I've included in this volume a set of very short essays that I produced for *Atlantic.com* and a few other journals that deal with the pipeline for scientists in the academic community.

I. Values and Rewards in Science

Cole, Stephen, and Jonathan R. Cole. "Scientific Output and Recognition: A Study in the Operation of the Reward System in Science." *American Sociological Review* 32, no. 3 (1967): 391–403.

Cole, Stephen, and Jonathan R. Cole. "Visibility and Structural Bases of Awareness in Science." *American Sociological Review* 33, no. 3 (1968): 397–413.

Cole, Jonathan R. "Patterns of Intellectual Influence in Scientific Research." *Sociology of Education* 43, no. 4 (1970): 377–403.

Cole, Jonathan R., and Stephen Cole. "Measuring the Quality of Sociological Research: Problems in the Use of the Science Citation Index." *The American Sociologist* 5, no. 1 (1971): 23–29.

Cole, Jonathan R., and Harriet Zuckerman. "The Emergence of a Scientific Specialty: The Self-Exemplifying Case of the Sociology of Science." In *The Idea of Social Structure: Papers in Honor of Robert K. Merton*, ed. Lewis Coser, 139–74. New York: Harcourt, 1975.

Cole, Jonathan R., and James Lipton. "The Reputations of American Medical Schools." *Social Forces* 56, no. 31 (1977): 662–684. Table A-4, which reproduces the reputational ranks, is reprinted in Logan Wilson's *American Academics: Then and Now* (New York: Oxford University Press, 1979), 270–71.

Cole, Stephen. "Age and Scientific Performance." *American Journal of Sociology* 84, no. 4 (1979): 958–77.

Cole, Jonathan R. "Balancing Acts: Dilemmas of Choice Facing Research Universities." *Daedalus* 122, no. 4 (1993): 1–36.

——. "Robert K. Merton, 1910–2003." *Scientometrics* 60, no. 1 (2004): 37–40.

—. "Why Elite-College Admissions Need an Overhaul." *The Atlantic*, February 14, 2016.

—. "The Pillaging of America's State Universities." *The Atlantic*, April 10, 2016.

II. Freedom and Unfreedom: The Case of Women in Science

Cole, Jonathan R., and Harriet Zuckerman. "Women in American Science." *Minerva* 13, no. 1 (1975): 82–102.

Cole, Jonathan R. "Women in Science." *American Scientist* 69, no. 4 (1981): 385–91. Reprinted as "Women in Science: Past, Present, and Future." Chapter 16 in *Scientific Excellence: Origins and Assessment*, ed. Douglas N. Jackson and J. Philippe Rushton. Newbury Park, CA: Sage, 1986.

Cole, Jonathan R., and Harriet Zuckerman. "The Productivity Puzzle: Persistence and Change in Patterns of Publication of Men and Women Scientists." In *Advances in Motivation and Achievement: A Research Journal*, ed. Marjorie W. Steinkamp and Martin L. Maehr, 217–258. Greenwich, CT: JAI, 1984.

Cole, Jonathan R., and Harriet Zuckerman. "Marriage, Motherhood, and Research Performance in Science." *Scientific American* 255, no. 2 (1987): 119–25.

Cole, Jonathan R., and Burton Singer. "A Theory of Limited Differences: Explaining the Productivity Puzzle in Science." Chapter 13 in *The Outer Circle: Women in the Scientific Community*, ed. Harriet Zuckerman, Jonathan R. Cole, and John T. Bruer, 277–310. New York: Norton, 1991.

Cole, Jonathan R., and Daria Franklin, forthcoming. "Freedom Gained and Freedom Lost in American Science" (2023).

III. Consensus in Science: Judgment and Choices

Cole, Jonathan R., and Stephen Cole. "The Ortega Hypothesis." *Science* 178, no. 4059 (1972): 368–375.

Cole, Stephen, and Jonathan R. Cole. "Testing the Ortega Hypothesis: Milestone or Millstone?" *Scientometrics* 12, nos. 5–6 (1972): 345–53.

Cole, Jonathan R., Stephen Cole, and Leonard Rubin. "Peer Review and the Support of Science." *Scientific American* (1977): 34–41.

Cole, Stephen, Jonathan R. Cole, and Gary A. Simon. "Chance and Consensus in Peer Review." *Science* 214, no. 4523 (1981): 881–86.

Cole, Stephen, Jonathan R. Cole, and Gary A. Simon. "NSF Peer Review Continued." *Science* 22, no. 215 (1982): 346, 348.

Cole, Jonathan R., and Stephen Cole. "Experts' Consensus and Decision-Making at the National Science Foundation." Chapter 3 in *Selectivity in Information*

Systems: Survival of the Fittest, ed. Kenneth S. Warren, 27–63. New York: Praeger, 1985.

Cole, Jonathan R. "Dietary Cholesterol and Heart Disease: The Construction of a Medical 'Fact.'" Chapter 20 in *Surveying Social Life Papers in Honor of Herbert H. Hyman*, ed. Hubert J. O'Gorman, 437–466. Middletown, CT: Wesleyan University Press, 1988.

Porter, Roy. "The Two Cultures Revisited." *The Cambridge Review* 115, no. 2 (1994): 1–17. *The Bridge*, National Academy of Engineering, Volume 26, Number 3–4, Fall/Winter, pp. 16–21. Reprinted in Albert H. Teich, Stephen D. Nelson, Celia McEnaney (editors), *AAAS Science and Technology Policy Yearbook*, Committee on Science, Engineering, and Public Policy. Washington, DC: American Association for the Advancement of Science, 1997. 89–100.

Cole, Jonathan R. "Intellectual Diversity in the U.S.: To What End?" *The Journal of Higher Education: Academic Matters* (2006): 13–16.

IV. Academic Freedom and Free Inquiry: The Enabling Value

Cole, Jonathan R. "The Patriot Act on Campus: Defending the University after 9/11." *Boston Review*, June 1, 2003.

——. "Academic Freedom under Fire." *Daedalus* 134, no. 2 (2005): 5–17.

——. "The New McCarthyism." *The Chronicle Review*, September 9, 2005.

——. "Defending Academic Freedom and Free Inquiry." *Social Research* 76, no. 3 (2009): 811–44.

——. "The Chilling Effect of Fear at America's Colleges." *The Atlantic*, June 9, 2016.

——. "The Triumph of America's Research University." *The Atlantic*, September 20, 2016.

——. "Academic Freedom as an Indicator of a Liberal Democracy." *Globalizations* 14, no. 6 (2017): 862–68.

——. "Academic Freedom under Fire." *Science* 374, no. 6573 (2021).

Notes

1. Throughout this volume, minor editorial changes have been made in the text of the essays in order to improve clarity. However, many of these papers were published before the academic community changed norms about gendered language as well, and changes to such gendered language are made less frequently.

2. The four norms included universalism, organized skepticism, disinterestedness, and communism. For a full definition of these four norms, see Robert K. Merton, "Science and Technology in a Democratic Order." The essay was later reproduced in *The*

Sociology of Science: Theoretical and Empirical Investigations (Chicago: University of Chicago Press, 1973).

3. Stanley B. Prusiner, *Madness and Memory: The Discovery of Prions—A New Biological Principle of Disease* (New Haven, CT: Yale University Press, 2014), 23.

4. In 1972, Diana Crane published *Invisible Colleges: Diffusion of Knowledge in Scientific Communities* (Chicago: University of Chicago Press, 1972), which explores the way social structures and scholarly communication influence the development of ideas.

5. Shoshanna Zuboff, *The Age of Surveillance Capitalism: The Fight For A Human Future At The New Frontier of Power* (New York, Public Affairs, 2019).

6. Peter Galison, *How Experiments End* (Chicago: University of Chicago Press, 1987), 10.

7. Andrew Pickering, "The Hunting of the Quark," *Isis* 72, no. 2 (1981): 236.

8. Stephen Cole, *Making Science: Between Nature and* Society (Cambridge, MA: Harvard University Press, 61, 10–30.

9. Cole, *Making Science*, 15–16.

10. See Vannevar Bush, *Science, the Endless Frontier* (Princeton, NJ: Princeton University Press, 1945).

11. Jonathan R. Cole, *The Great American University: Its Rise to Preeminence, Its Indispensable National Role, and Why It Must Be Protected* (New York: Public Affairs, 2009); Jonathan R. Cole, *Toward a More Perfect University* (New York: Public Affairs, 2016).

PART I

VALUES AND REWARDS IN SCIENCE

CHAPTER 1

SCIENTIFIC OUTPUT AND RECOGNITION

A Study in the Operation of the Reward System in Science (1967)

STEPHEN COLE AND JONATHAN R. COLE

The relationship between the quantity and quality of scientific output of 120 university physicists was studied. Although these two variables are highly correlated, some physicists produce many papers of little significance and others produce a few papers of great significance. The responses of the community of physicists to these distinct patterns of research publication were investigated. Quality of output is more significant than quantity in eliciting recognition through the receipt of awards, appointment to prestigious academic departments, and being widely known to one's colleagues. The reward system operates to encourage creative scientists to be highly productive, to divert the energies of less creative physicists into other channels, and to produce a higher correlation between quantity and quality of output in the top departments than in the weaker departments.

Sociologists of science have for some time investigated the relationship between the sheer quantity of scientists' published research and its quality. A high or moderate correlation between the two has been found in several fields of science.[1]

American Sociological Review 32, no. 3 (June 1967): 377–390. Revision of a paper read at the annual meeting of the American Sociological Association in August 1966. This study was supported by grant number NSF-GS-960 from the National Science Foundation to the program in the Sociology of Science, Columbia University, Robert K. Merton, director. We want to thank Professor Merton for his helpful suggestions and Professor David Caplovitz for his criticism of an earlier draft of this paper. This may be identified as publication number A-460 of the Bureau of Applied Social Research, Columbia University.

But since the correlation is not perfect, this means, of course, that some scientists produce only a small number of papers which are judged to have contributed a great deal to their discipline while others have produced a long list of publications which have contributed relatively little.

This raises the question of how the community of scientists responds to these distinct patterns of research publication. Is it the case, as the "publish or perish" doctrine implies, that scientists publishing a long string of trivial papers will be rewarded while those who produce only a few papers, though of high quality, will be deprived of recognition? Or does the reward system operate somewhat more effectively to reward excellence where it is found?

This paper presents data, drawn from a larger study of university physicists, which illustrate a way of assessing the extent to which physicists receive recognition for the quantity of their published work and its quality. A methodological by-product of the paper may move us toward a better measure of the qualitative aspects of scientific output.

One set of data in this study consists of the scientific output of 120 university physicists, chosen from a sampling frame in which the population of university physicists was stratified along four dimensions: age, prestige rank of their university department, productivity, and number of honorific awards. The sample, selected for its suitability for several research problems, is not representative: it heavily overrepresents eminent scientists.[2] We have collected the following information for the 120 physicists: the number of papers they have published; the number of citations to their work; the nationally assessed rank of their departments; and the number of their awards.[3] A second set of data is designed to find out the relative prestige of awards received by members of the sample and the extent to which the 120 physicists are known within the national community of physicists. Questionnaires were sent to the 2,036 physicists who work in departments that have granted two or more PhDs in each of the last five years, and 1,281 usable ones were returned.[4]

Since we want first to investigate the relationship between the quantity and quality of research published by the 120 physicists, we must have reliable measures of both. As a measure of the quantity of scientific output, we take the total number of scientific papers by each physicist as listed in *Science Abstracts*.[5] The problem of assessing the "quality" of scientific publications is more difficult and has long been a major impediment to progress in the sociology of science. Most of us have typically paid homage to the idea that quantity of output is not the equivalent of quality and have then gone ahead to use publication counts anyway.[6] There seemed to be no practicable way to measure the quality of large numbers of papers or the total work of large numbers of scientists.[7] The invention of the *Science Citation Index* (SCI) a few years ago provides a new tool which yields a reliable and valid measure of the significance of individual scientists' contributions in certain fields of science.[8] Starting in 1961, the SCI has listed all citations to scientific papers appearing in an increasingly large number of journals. Thus, it is possible to count the number of citations made in 1961 and certain later years to any paper or group of papers by physicists. The number of citations is taken to represent the relative scientific significance or "quality" of papers in each field.

There is some supporting evidence for this assumption and procedure. In what is perhaps the most thorough study of measures of scientific output, Kenneth E. Clark asked a panel of experts in psychology to list the psychologists who had made the most significant contributions to their field.[9] He then investigated the correlation between the number of choices received by psychologists and other indices of eminence. The measure most highly correlated with number of choices was the number of journal citations to the man's work ($r = .67$). Clark concludes that the citation count is the best available indicator of the "worth" of research work by psychologists.

Consider another kind of validating evidence for this measure. Recipients of the Nobel Prize in the aggregate can be regarded as having contributed greatly to the advance of their fields in the physical and biological sciences, even though the great scarcity of Nobel Prizes means that there are probably other like-sized aggregates of eminent scientists who may have contributed as much.[10] Nevertheless, the laureates as a group can be safely assumed to have made outstanding contributions. The average number of citations in the SCI to the lifework of Nobel laureates (who won the prize in physics between 1955 and 1965) was fifty-eight, compared to an average of 5.5 citations to the work of other scientists cited in 1961.[11] Only 1.08 percent of the quarter of a million scientists who appear in the 1961 SCI received fifty-eight or more citations.[12] This evidence offers further support for the use of number of citations as an indicator of the scientific significance of published work.

Before analyzing the relationship between the number of published papers and the number of citations, we should consider two problems in the use of citations as a measure of quality. It is possible that the total number of citations to a man's work is not a completely independent indicator of quality since scientists who publish a large number of papers, each of which receives only a few citations, might accumulate as many citations as those who have published only a few papers which are heavily cited.[13] We therefore decided to take the number of citations to the three most heavily cited contributions by each physicist as an indicator of the impact of his best work. Since a contribution in physics does not typically take the form of a single paper but is usually presented in a series of papers, we have used citations to the year's output rather than the single paper as our unit of measure.[14]

Another problem in the use of citations as an indicator of quality is the extreme contemporaneity of science.[15] Papers in physics now have a half-life of no more than five years; that is, at least half the citations in any year are to work published in the five preceding years. We must take this into account in comparing the work of physicists who made their most important contributions at different times. To facilitate this comparison, we have developed a technique of weighting citations. Since *Physical Review* is the leading journal in physics, we have used the time distribution of citations in this journal as the basis of the weighting technique.[16] A study by M. M. Kessler found that 70 percent of all citations appearing in the journal for 1957 are to work published within the five previous years; 17 percent to work published 6 to 10 years before; 9 percent, 11 to 20 years before; and only 4 percent to work published more than 20 years before.[17] From this distribution, we can see the need for weighting citations if we are to compare research that is currently having its greatest impact with that which had its greatest impact at various times in the past. Thus, from Kessler we judge that research in physics published within the past 5 years is about 17 times more likely to be cited today than research of more than 20 years ago (i.e., 70/4). Following this model, we have assigned the following weights:

Citations to work 20 or more years before: 17
Citations to work 11–19 years before: 8
Citations to work 6–10 years before: 4
Citations to work 0–5 years before: 1

Throughout this paper, the average number of weighted citations to a physicist's research in his three most heavily cited years will be used as the measure of quality of his research.[18]

On the basis of these measures, table 1.1 presents the relationship between the "quantity" and "quality" of the research papers produced by the 120 physicists. This table serves two purposes: first, it indicates, as expected from earlier investigations, a correlation between the quantity and quality of the research published by these physicists ($r = .72$). Second, and of continuing interest to us

TABLE 1.1 **Percentage Distribution of 120 Physicists by Quantity and Quality of Published Research**

Quantity	High	Low
	Type I	Type II
High	33	12
	Type III	Type IV
Low	18	37

Note: Quantity refers to the number of published papers. "High" quantity is 30 or more; "low" quantity is less than 30. Quality refers to the average number of (weighted) citations to the three most cited years of the physicists' output. "High" quality is more than 60; "low" quality is less than 60. Sum of the percentages is 100.

throughout the rest of this paper, it generates four types of physicists, roughly described in terms of the production of scientific papers and their quality (as assessed by citations). Type I, comprising 33 percent of this selected sample, is the *prolific physicist*, in the dual sense of producing an abundance of papers which also tend to be fruitful (i.e., often used by others in the field). At the other extreme, Type IV, comprising 37 percent of the sample, is the relatively *silent physicist*, who produces comparatively few papers and, judging from the paucity of citations to them, they do not matter much to the field of physics. The other two types remind us that the sheer quantity of published papers is not always correlated with their quality. Type II is the undiscriminating *mass producer* in physics: the 12 percent in this sample publish a relatively large number of papers of little consequence. As a type, these physicists seem geared to getting many papers into print without much regard for their scientific significance. And finally, Type III might be described as the *perfectionist:* physicists who publish comparatively little but what they do publish has a considerable impact on the field. This type may include physicists who elect not to publish work which, in their own (possibly mistaken) judgment, does not measure up to their high standards. As a result, they are not the prolific researchers of Type I.

It is obvious that these four types are only crudely approximated by the particular data used here. A much larger sample would be needed to identify the extreme types by more precise criteria, e.g., the silent physicists of Type IV being those who had published no more than, say, two or three papers all told and the prolific ones of Type I, say, a hundred papers or more. The same can be said about the arbitrary cutoff points on the number of citations. Furthermore, a larger sample would enable us to identify intermediate and transitional types. All this is evident and is the basis for our saying that we are dealing here with only rough approximations to the four types. But, as we shall soon see, even these approximations permit us to get on with the main purpose of this paper: to analyze the recognition accorded the four types of physicists in the reward system of science.

The Reward System of Science

As Robert K. Merton pointed out some time ago, the institution of science has developed a reward system designed to give "recognition and esteem to those (scientists) who have best fulfilled their roles, to those who have made genuinely original contributions to the common stock of knowledge."[19] The graded forms of such recognition in science are many and, among these, we want to examine three kinds, as these are distributed among the several types of research physicists. In this way, we move toward an examination of how the reward system actually works.

32 VALUES AND REWARDS IN SCIENCE

The first form of recognition is the granting of honorific awards and memberships in honorific societies. Members of the twenty top-ranked physics departments listed more than 150 different awards after their names in *American Men of Science* (1960 edition). A total of ninety-eight of these awards were ranked in prestige from a high of 5 to a low of 1 by a sample of nearly 1,300 physicists.[20] Two other options were included in the questionnaire: physicists could report that they had heard of the award but did not have enough information to rank its prestige or that they had never heard of the award. Prestige scores were computed by taking the mean of ranks assigned by the sample of physicists; the visibility of each award was indicated by the percentage of physicists who knew enough about it to rank it at all.

Most of the ninety-eight awards are not highly visible to the national community of physicists. Only twenty-two were ranked by as many as half and only forty-two by as many as one-fifth of them. Evidently, a large number of the awards in which recipients take enough pride to list after their names are local honors which, though not a part of the national reward system, may nevertheless confer prestige in local environments. For this investigation, we adopted the convention that awards unknown to more than 80 percent of the physicists would be regarded as thoroughly parochial and were excluded from further analysis. Several aspects of the distribution of these honorific awards are evident. Of the forty-two awards meeting the criterion of national visibility— familiar to at least 20 percent of the physicists—two stand out above all the rest. They are, of course, the Nobel Prize (with a prestige score of 4.98 of a possible 5.0) and membership in the National Academy of Sciences (4.22). What is more, physicists at these highest levels of eminence monopolize all the other most prestigious awards (awards with scores of 4.01 or better). In our sample there was not a single recipient of the Fermi Award, Royal Astronomical Society Gold Medal, Albert Einstein Medal, or Fritz London Award, and not a single member of the Académie Française or Royal Society who was not either a Nobel laureate or a member of the National Academy. However we interpret it, the fact is that the awards indicating top-most eminence in physics are closely confined to a comparatively small group of physicists.

As table 1.2 shows, the prestige of the highest award received by physicists is correlated with the total number of their awards ($r = .70$). By way of anticipation, we note here that the total number of honorific awards correlates as highly as the prestige of highest award with almost every other indicator of recognition of scientific accomplishment.[21] In due course, we shall examine the distribution of total number of honorific awards among the four types of research physicists.

Prestigious awards and membership in honorific societies are, almost by definition, in short supply and thus are inadequate in providing recognition of lesser degrees of scientific accomplishment. Even among the 632 physicists employed by the top twenty departments of physics in 1960 (at the rank of

Scientific Output and Recognition

TABLE 1.2 Percentage Distribution of Physicists by Prestige of Highest Award and Total Number of Awards

Prestige of Highest Award	Number of Awards					
	4 or more	3	2	1	Total	N
Nobel laureates	45	36	19	—	100	(11)
Member of National Academy of Sciences	27	23	36	14	100	(22)
Awards with scores of 3.00–3.99	—	12	55	33	100	(33)
Awards with scores of less than 3.00	—	—	17	83	100	(24)

Note: Of this sample of physicists, 30 had received no awards.

assistant professor or higher), only one-third listed any awards (in *American Men of Science*). A second form of recognition for scientific work, one which is more widely distributed, is the granting of positions at top-ranked departments. Physicists who would never dream of winning a Nobel Prize or being elected to the National Academy may nevertheless aspire to a tenure position in a major department. By drawing upon Allen Cartter's recent study which ranks eighty-six departments of physics, we can determine the extent to which the work of our four types of physicists has been rewarded by appointment to departments of varying rank.[22]

The third kind of recognition we consider here is the most widespread and, in the view of Alan Waterman, operates as a greater incentive for scientists than more formal recognition, like awards and prizes. This is "the kind and degree of attention [one's] research receives from the scientific community."[23] Citations provide one feedback mechanism enabling scientists to gauge the extent to which their research is being used by others in the field. In this paper, we introduce an additional and perhaps more exacting measure. The sample of nearly 1,300 physicists was asked to describe the extent of their firsthand familiarity with the work of each of the 120 physicists in our more restricted sample, and if they had not read his work, to indicate whether or not they had heard of him at all. We have adopted the percentage of the community of physicists who reported knowing any part of a physicist's work at firsthand as a measure of the "scope of his scientific reputation."[24] This provides us with our third measure of the extent of recognition of the four types of research physicists.[25]

Types of Scientific Role Performance and the Reward System

Having mapped certain components in the reward system of science, we are ready to examine the extent and kinds of recognition given the research of our sample of physicists. Table 1.3 indicates that the quantity and the quality

TABLE 1.3 Coefficients of Correlation between Quantity and Quality of Research and Three Measures of Recognition

| Quantity and Quality of Research | Measures of Recognition | | | |
| | Awards | | | Percent of Community of Physicists Familiar with Individuals' Research |
	Prestige of Highest Award	Number of Awards	Rank of Department	
1. Quantity	.35	.46	.24	.49
2. Number of papers per year	.28	.32	.19	.43
3. Quality	.41	.67	.33	.64

Note: For definitions of quantity and quality, see note to table 1.1.

of research are correlated with all three kinds of recognition. But there is a consistently higher correlation between quality of research and the three types of recognition than between quantity of research and recognition. Substantively, this is the first specific indication that the reward system in physics operates to accord greater recognition to the quality of physicists' research than to its sheer bulk. Procedurally, table 1.3 suggests that although the mere count of published papers as a measure of the recognized significance of scientists' research will not introduce gross error, the count of citations provides a better measure.

In table 1.4, we turn to the extent of recognition through formal awards accorded the four types of research physicists. We should begin by noting that the *absolute size* of the percentages in this table is in part an artifact of the sampling frame by which the 120 physicists were chosen and of the arbitrary cutoff points for classifying their research. Some physicists classified here as relatively "low" on both quantity and quality have probably produced more and better papers than the average physicist. The significant finding is not in the absolute percentages indicating that most of this particular sample of physicists received some recognition through awards. Rather, it is in the pattern of *differences* between the percentages. And in this light the story is reasonably clear. It is the quality of research rather than its sheer amount that is most often recognized through honorific awards.[26] Although they have published fewer papers than the prolific Type I physicists, the Type III ("perfectionist") physicists are just as apt to be accorded recognition, and both these quality producers are far more likely to receive awards than the Type II mass producers who publish large numbers of papers indiscriminately.[27]

The conclusion drawn on the basis of the data in table 1.4 is supported by the results of a multiple regression analysis.[28] As we have noted, the zero-order

TABLE 1.4 Percentage of Physicists Having at Least One Award by Quantity and Quality of Published Research

	Quality	
Quantity	High	Low
	Type I	Type II
High	90 (40)	64 (14)
	Type III	Type IV
Low	91 (22)	57 (44)

Note: For definitions of quantity and quality, see note to table 1.1. Base of percentage is shown in parentheses.

correlation coefficient between number of awards and production of papers is .46. Thus, the sheer number of papers explains 19 percent of the variance in number of awards. When quality of work, as measured by citations, is introduced into the equation, the percentage of variance explained increases to 45 percent. If we reverse the order of introducing the independent variables, 44 percent of the variance is explained by quality alone. When we then introduce quantity, the amount of variance explained does not increase. Thus, once we know the quality of physicists' research, we need not know how much research they have produced to predict their "eminence." However, if we know only the quantity of their research, we will be greatly helped in predicting their eminence by knowing the quality of this work.[29] (Incidentally, it should also be noted that the regression analysis provides more reliable findings than those based on cross-tabulations since they are based upon ungrouped continuous data while the categories in the cross-tabulations are dependent upon arbitrary cutting points.)

The second kind of recognition for scientific work is appointment to a major academic department. Believers in the prevalence of the "publish or perish" policy hold that the mere number of publications determines appointments and that even in the top-ranking universities, the academic man who has published only a few papers, albeit significant ones, will typically be passed over in favor of the mass producer of trivia. Our data, presented in table 1.5, do not support this belief: Type III physicists (publishing relatively few papers which are widely cited in the field) are the most likely to be in the "distinguished" (top ten) departments, even more so than the prolific Type I physicists.[30] But the mass producer Type II physicists fare no better than those who produce fewer relatively undistinguished papers. The top departments, at least, prefer to choose their physicists on the basis of quality of research rather than on mere quantity.

36 VALUES AND REWARDS IN SCIENCE

TABLE 1.5 Percentage of Physicists in Top Ten Departments of Physics by Quantity and Quality of Published Research

Quantity	Quality	
	High	Low
	Type I	Type II
High	58 (40)	29 (14)
	Type III	Type IV
Low	77 (22)	27 (44)

Note: For definitions of quantity and quality, see note to table 1.1. Base of percentage is shown in parentheses.

Data bearing on our third form of recognition—familiarity with one's research among fellow scientists—were obtained by asking subsets of the nearly 1,300 physicists to indicate the extent of their firsthand acquaintance with the work of our prime sample of 120 physicists. The extent of such direct knowledge we call the "scope of reputation." As table 1.6 shows, the perfectionist Type III physicists are more likely to be known to their colleagues than the mass producer Type II physicists. Of all three forms of recognition, scope of reputation is the most influenced by quantity. Type I physicists are still more likely to be known than Type III, and Type II physicists are much more likely to be known than Type IV. This is exactly what we would expect. The other two forms of recognition—receiving awards and holding positions at top departments—are tied more closely to a *positive* evaluation of the scientist's work than is scope of reputation. Although scope of reputation is highly correlated with quality ($r = .63$), it is easier for the mass producer to make himself known through persistent exposure than it is for him to have his work positively evaluated.

TABLE 1.6 Percentage of Physicists Having at Least Fifty Percent of Fellow Physicists Familiar with Their Work, by Quantity and Quality of Published Research

Quantity	Quality	
	High	Low
	Type I	Type II
High	68 (40)	29 (14)
	Type III	Type IV
Low	55 (22)	5 (44)

Note: For definitions of quantity and quality, see note to table 1.1. Base of percentage is shown in parentheses.

But the high producers, although somewhat more likely to have at least part of their work read, are no more likely to have their reputation spread beyond the circle of those who have actually read it. We can divide physicists who have not read any of the work of a particular physicist into two groups: those who have heard of him and those who have not. The number of those who have heard of a physicist taken as a percentage of all those who are not directly familiar with his work is described as his "reputational visibility." Thus, if one hundred physicists have never read a particular physicist but fifty of them have at least heard of him, his reputational visibility is 50 percent. As we see in table 1.7, quality exceeds quantity in making for reputational visibility. The reputation of physicists who have produced research of consequence travels far within their national community regardless of the number of publications they have produced. Table 1.7 also suggests that the mass producer (Type II) is likely to be visible only to physicists who have had to look at some of his work because of the subject matter. Finally, the comparatively "silent physicists" of Type IV are practically unknown to the national community of physicists.

From the data presented in tables 1.4 to 1.7, it appears that the reward system in physics operates to give all three kinds of recognition primarily to *significant* research, whether this is found in the work of high or low producers. Mere quantity of published research seldom makes for equivalent recognition. To this extent, the reward system of physics approximates the often-expressed norm that excellence of research is what truly matters.

The data also testify that, by and large, the high producers *tend* to publish the more consequential research. There are at least two basic factors creating the high correlation between quantity and quality of work. The first is that engaging in a lot of research is in one sense a "necessary" condition for the production of high-quality work. As scientists of the first rank remind us, producing significant science is a risky business, full of uncertainties.[31] There is seldom a guarantee that a program of research will produce important results, and do

TABLE 1.7 **Percentage of Physicists Having Reputational Visibility (Scores of 50 Percent or More) by Quantity and Quality of Published Research**

	Quality	
Quantity	High	Low
	Type I	Type II
High	42 (40)	21 (14)
	Type III	Type IV
Low	41 (22)	7 (44)

Note: For definitions of quantity and quality, see note to table 1.1; for "reputational visibility," see text. Base of percentage is shown in parentheses.

so in short order. A physicist will try an idea out and sometimes it will work, but more often it will not. It is rare for scientists to have such a keen eye for crucial problems that they can limit their energies solely to an investigation of these problems. Even the average top scientists must make many experiments before they get an exciting result. We believe that unless a physicist makes a large number of attempts (i.e., has high productivity) the probability of making a significant discovery will be low.

A second reason why quality and quantity of work are so highly correlated is that the reward system operates in such a way as to encourage the creative scientists to be productive and to divert the energies of the less creative scientists into other channels. In the last part of this paper, we shall analyze the processes through which the reward system works to create and then reinforce the correlation between quantity and quality of work.

Reinforcement of Research Activity by the Reward System

Most physicists have been trained in the major departments. Fifty-six percent of the physicists among the nearly 1,300 we queried received their doctorates from the top-ranked fifteen departments; 44 percent were from the top ten. We assume that, as students, they internalized the norm prevailing in these departments of doing research. Soon after they receive their degrees (and sometimes before), these young physicists begin to publish their research, whether solo or as part of a research team. Their papers must first pass through the evaluation system. The first screening of this work is by the referees associated with the journals; the prime journal in the field, *Physical Review*, for example, has an especially elaborate system of refereeing papers. Standards are high and even the manuscripts of eminent scientists are sometimes rejected. Once the paper passes through this screening and is published, it is then informally evaluated by the national and international community of physicists. Sometimes it is largely ignored, with few citations to it, or it may be identified as a significant contribution and put to use in other published research. If the reward system, in the form of recognition by citation, does affect research productivity, we assume that the greater such collegial recognition of this early research by physicists, the greater the probability that they will continue to be productive. We hypothesize that few scientists will continue to engage in research if they are not rewarded for it.[32]

To test this hypothesis, we have traced the sequence of publication patterns among the sample of 120 physicists. We divided them into two broad categories: the early producers who published three or more papers within five years after their PhD and the others who published fewer than three. We then examined the collective responses to these early publications as measured

Scientific Output and Recognition 39

TABLE 1.8 Percentage of Physicists Who Continue to Be Productive, by Early Recognition and Early Productivity

Early Recognition (Number of weighted citations received in first five years after PhD)	Early Productivity			
	Wrote 3 or more papers in five years after PhD		Wrote less than 3 papers in five years after PhD	
0–25	30	(27)	15	(27)
26–100	48	(38)	33	(3)
101 or more	76	(25)	—	—

Note: Base of percentage is in parentheses. Since the percentages are based upon relatively small numbers, we computed chi square for the first column. $\chi^2=11.35$, d.f. 2, P>.01. Continued productivity was defined as producing 1.5 or more papers per year after the immediate postdoctoral years.

by the number of citations they received within the same five-year span.[33] Finally, we compared the later productivity of the physicists receiving differing amounts of recognition in these early years. The results are presented in table 1.8. Three-quarters of these physicists began their professional careers by publishing at least three papers soon after their doctorates. There are few "late bloomers": only five of the thirty physicists who started off slowly ever became highly productive (averaging 1.5 or more papers a year).[34]

Consider, next, the sequence of publication patterns among physicists who started their research careers by being productive. The more citations their early work received, the more often they continued being productive. Only 30 percent of those who received up to twenty-five weighted citations went on to continued high productivity, in contrast to the 76 percent with more than one hundred citations. These findings suggest that when a scientist's work is used by his colleagues he is encouraged to continue doing research and that when a scientist's work is ignored, his productivity will tail off.[35] Of course, it may be that the early starters who were also doing quality work, as indicated by the frequent use of their research, would have continued to produce even if they had not received recognition.[36]

These findings also suggest that the Type II researchers—those who publish many papers which are largely ignored—probably get their rewards from other parts of the system. For, as we have suggested before and now see more definitely, the criteria utilized in that system are not evenly distributed through every sector of academia. Table 1.9 indicates that the correlation between the quantity and quality of research output is higher in the better physics departments than in the weaker ones. Among the high producers in the top ten departments, 85 percent had written high-quality papers compared to 62 percent and 45 percent in the weaker departments.

40 VALUES AND REWARDS IN SCIENCE

TABLE 1.9 Percentage of Physicists Producing High Quality Research, by Quantity and Prestige of Department

Prestige of Department	Quantity	
	High	Low
High	85 (27)	59 (29)
Intermediate	62 (16)	14 (21)
Low	45 (11)	13 (16)

Note: Base of percentage is shown in parentheses. "High quality" research is defined in note to table 1.1.

The results of table 1.9 can be interpreted in at least two different ways. The reported distribution may result from a process of social selection whereby the outstanding departments recruit more able researchers with a better sense for the significant research problem. On the average, then, their research has a greater impact on the field. The weaker departments, further removed from the springs of scientific advance, tend to recruit less able investigators who gradually lose contact with the rapidly advancing frontiers of physics and produce work of less significance. Some of the physicists in these departments continue to publish, and it is in these departments, which find it difficult to recruit faculty who do research at all, that the sheer number of publications is more apt to be an important criterion for promotion. Thus, the reward system of the weaker departments more often makes for that displacement of goals which is expressed in the policy of "publish or perish."

The findings from table 1.9 might also be interpreted to result from an imperfect communication network in science. This interpretation starts from the assumption that the flow of scientific information moves principally in a one-directional path from the major to the minor centers of scientific inquiry. Physicists in the weaker departments are apt to know more of the work of men in the stronger departments while men in the stronger departments less often monitor the work of physicists in the weaker departments. Thus, work of "equal quality" produced in departments of differing rank will be differentially recognized and cited in the field. An extension of this view of the communication network maintains that the leading journals of physics are "controlled" by the same group who control the top-rated departments. The journals more readily publish papers by members of the in-group and their students who tend to cite the work of others in that group. This results in differing citation rates for research of comparable significance published by physicists located in departments of differing prestige.

To study these tentative interpretations, we follow two lines of investigation. We first examine differences in the visibility of scientific work produced by scientists of differing location in the social structure of science. For example, are physicists at the higher-ranked departments (e.g., Berkeley or Harvard) as familiar with "high-quality" research in their field carried out at less prestigious departments (e.g., Georgetown or the University of Kentucky) as are physicists in the lower-ranked departments with comparable work in the higher-ranked ones? Second, does the reward system uniformly operate to reward excellence of research by physicists irrespective of the rank of their departments?[37]

On this latter point, some preliminary evidence suggests rank-of-department differentials in the working of the reward system. When we take honorific awards as the dependent variable and introduce quality of research (weighted citations) into the regression equation, we account for 44 percent of the variance (as we have previously noted). When we introduce rank of department into the equation, we increase the percent of variance explained to 53. This indicates that high-quality research in the higher-ranked departments is more often recognized in the form of awards than comparable work in the lower-ranked departments. (The partial correlation between rank of department and number of awards when quality is held constant is .41.) That this results at least in part from the wider reputation of the physicists in the top departments is suggested by the fact that when scope of reputation is introduced into the regression equation prior to rank of department, the latter variable explains only an additional four percent of the variance in number of awards. The implications of these preliminary findings need to be pursued.

Summary

The quantity and quality of research by physicists tend to be related (generating the types we have described as the "prolific" and the "silent" physicists). When there is an inconsistency between quantity and quality of work—as in the cases of the "mass producers" and the "perfectionists"—quality proves to be a more significant correlate of the amount of recognition accorded research physicists. This is the case for three forms of recognition: honorific awards, appointments to top-ranking departments, and having one's research known in the national community of physicists. To a degree, the reward system operates to reinforce the patterns of work by the more creative scientists and tends to confirm the observed relation between quantity and quality of research. Some preliminary evidence suggests, however, that the reward system does not operate uniformly among all academic departments of physics. There are indications that the sheer quantity of publications is more likely to be used as a criterion of promotion in the less prestigious departments, and that quality research is more often rewarded when it is produced by physicists in high-ranking departments.

Notes

1. See, among others, Kenneth E. Clark, *America's Psychologists: A Survey of a Growing Profession* (Washington, DC: American Psychological Association, 1957), chap. 3; Wayne Dennis, "The Bibliographies of Eminent Scientists," *Scientific Monthly* 79 (September 1954): 180–83; Bernard M. Meltzer, "The Productivity of Social Scientists," *American Journal of Sociology* 55 (July 1949): 25–29; and Peggy Thomasson and Julian C. Stanley, "Exploratory Study of Productivity and 'Creativity' of Prominent Psychometricians," unpublished paper, University of Wisconsin, 1966.

2. The distribution of the 120 physicists among the various specialties closely parallels the distribution of the population of university physicists. The only exception is found in solid state physics. Twenty-one percent of American university physicists compared to 10 percent of our sample list solid state physics as their specialty.

3. The number of papers published was based on *Science Abstracts*; the number of citations on the *Science Citation Index*; the rank of departments on the study by Allan M. Cartter, *An Assessment of Quality in Graduate Education* (Washington, DC: American Council on Education, 1966), which is briefly described below; and the number of awards on listings in the tenth edition of *American Men of Science*. We are indebted to Dr. Cartter for making information available to us before publication of his book.

4. Of the total questionnaires sent to 2,036 physicists, 1,333 or 65.5 percent were returned. Twelve of these were incomplete or otherwise not usable; forty were returned after the cutoff date for coding. This left 1,281 usable questionnaires. Physicists in highly ranked departments responded to the questionnaire in the same proportions as those in lower-ranked departments; those with tenure rank to the same extent as those without tenure. So far as we can tell, the 1,281 physicists returning questionnaires are representative of physicists in American universities.

5. Two aspects of this decision should be noted. First, only papers and not books are included in research output, since physicists almost invariably publish their original research in papers, unlike the humanities and the social sciences. Second, we shall be using the total scientific output (the cumulative number of papers published by each physicist) rather than productivity rates (average number of papers per year). We find that both measures exhibit the same patterns of relation to other variables examined in this paper. (See note 25.)

6. See, among many others, Derek Price, *Little Science, Big Science* (New York: Columbia University Press, 1963), 40; Myron B. Coler, ed., *Essays on Creativity in the Sciences* (New York: New York University Press, 1963), xvi; Logan Wilson, *The Academic Man: A Study in the Sociology of a Profession* (New York: Octagon, 1964), 110; and Diana Crane, "Scientists at Major and Minor Universities: A Study of Productivity and Recognition," *American Sociological Review* 30 (October 1965): 699–714.

7. Researchers also have had difficulty in estimating the significance of even a small number of papers. Although a panel of judges is often used, problems of standardization of evaluation criteria and the individual biases of the evaluators have frequently been encountered. Cf. David Nasatir and David Elesh, "The Measurement of Quality in Educational Research: The Development and Validation of an Instrument," Bureau of Applied Social Research, Columbia University, 1965.

8. The *Science Citation Index* is compiled under the direction of Eugene Garfield. In 1961 it listed all citations made in 613 journals; 1962 and 1963 have not yet been indexed. In 1964, 700 journals were covered, and in 1965, 1,147 journals.

9. See Clark, *America's Psychologists*, for another study using number of citations as a measure of quality of scientific work; see also Alan E. Bayer and John Folger, "Some Correlates of a Citation Measure of Productivity in Science," *Sociology of Education* 39 (Fall 1966): 381–90.

10. See Harriet A. Zuckerman, "Nobel Laureates in Science: Patterns of Productivity, Collaboration and Authorship," *American Sociological Review* 32, no. 3 (June 1967): 391–403.

11. We thought it possible that winning the prize might make a physicist more visible and lead to a greater number of post-prize citations than the quality of work warranted. We therefore divided the laureates into two groups: those who won the prize five or fewer years before 1961 and those who won the prize after. The pre-1961 laureates were cited an average of forty-two times in the 1961 SCI; the post-1961 prize winners were cited an average of sixty-two times. Since the prospective laureates were more often cited than the actual laureates, we conclude that the larger number of citations primarily reflects the high quality of the work rather than the visibility gained by winning the prize. These statistics are based upon the work of twenty-four of the twenty-eight living laureates who won the prize in physics. The four living laureates who won the prize more than five years before 1961 were excluded so as not to introduce an age bias. Included in this computation are the non-American laureates; when they are excluded, the average number of citations to the work of American laureates is sixty-eight.

12. Irving H. Sher and Eugene Garfield, "New Tools for Improving and Evaluating the Effectiveness of Research," presented at the Second Conference on Research Program Effectiveness, Washington, DC, July 27–29, 1965. We would like to thank Dr. Sher for making available some unpublished data. By way of emphasizing the difference between the number of citations to the work of laureates and to the work of the average scientist, we should also point out that many scientists do not appear in the SCI and that the modal number of citations to the work of men who do appear is one.

13. Although this is possible, our data suggest that it is not probable. The correlation between total number of papers and total number of citations is .60. The correlation between total number of papers and a measure of quality which is not influenced by quantity (see note 15) is .72. This is exactly the opposite of what we would find if the total number of citations were merely a function of the total number of papers. We conclude that the total number of citations could serve as an adequate indicator of quality, but we have used a refined measure because it seems substantively more suitable.

14. The year is also an arbitrary unit since physicists do not, of course, arrange their related papers to fit the calendar year. For a more exacting procedure, it would be necessary to identify the series of papers representing an integrated contribution, a requirement extremely difficult to meet in dealing with large numbers of working scientists. Without such detailed information, it would seem preferable to use a period of time as a unit rather than single papers. See also the recent study of scientific productivity by Crane, "Scientists at Major and Minor Universities," which treats a series of four papers on the same topic as a "major" publication and single papers as "minor" publications.

15. There are other problems in the use of the SCI as a measure of quality: (1) Work of the highest significance often becomes common knowledge very quickly and is referred to in papers without being cited; (2) Citations may be critical rather than positive; (3) The various scientific fields differ in size. If we wish to compare the work of scientists in different fields, we must take into consideration the number of people actively working in these fields; and (4) The significance of scientific work is not always recognized by contemporaries (e.g., Mendel). New ideas are sometimes ignored or resisted. When citations are used to measure quality, scientific work that is currently "resisted"

44 VALUES AND REWARDS IN SCIENCE

or inadequately judged will be misclassified. For a discussion of resistance to scientific discovery, see Bernard Barber, "Resistance by Scientists to Scientific Discovery," *Science* 134, no. 3479 (1961): 596–602.For a further discussion of limitations of citations as a measure of quality and ways of dealing with these, see Leonard Ornstein et al., "Research, Scholarly Publication and Teaching: The Development of Some Objective Measures of a Scientist's Impact on Science," Mount Sinai School of Medicine, forthcoming.

16. The importance of *Physical Review* is illustrated by the high percentage of citations to articles appearing in this journal. Forty-seven percent of citations in *Physical Review* are to other articles in this journal; 28 percent of citations in *Physical Review Letters* are to articles in *Physical Review;* and 34 percent of citations in *Proceedings of Physical Society* (London) are to articles appearing in *Physical Review.* M. M. Kessler, "The MIT Technical Information Project," *Physics Today* 18 (March 1965): 28–36.

17. See M. M. Kessler, "Technical Information Flow Patterns," paper presented at the IRE-AIEE-ACM conference (Cambridge: Massachusetts Institute of Technology, 1961), 253. Kessler found similar patterns for other major journals of physics. In the Russian *Journal of Experimental and Theoretical Physics*, 63 percent of the work cited in 1957 had been published within the preceding five years; 20 percent, 6–10 years before; 10 percent, 11–20 years before; and 7 percent more than 20 years before. Citations in *Physica* were slightly less contemporaneous; the corresponding figures were 55 percent, 19 percent, 8 percent, and 18 percent. Similar patterns have been found in two leading journals of biology. Paul Weiss, "Knowledge: A Growth Process," *Science* 131 (June 1960): 1716–19.

18. There is evidence that this procedure indexes the utilization or "significance" of the research by the 120 physicists. For one thing, the average number of weighted citations to the work of the Nobel laureates is 335. For another, every table in this paper was run using other citation measures with similar substantive results. There is a .96 correlation between the total of weighted citations and the average number of weighted citations in the three highest years and a .80 correlation between totals of weighted and of unweighted citations. These high correlation coefficients once again illustrate the interchangeability of indices. See Paul F. Lazarsfeld, "Evidence and Inference in Social Research," *Daedalus* 87 (Fall 1958): 91–130. Since the number of a physicist's weighted citations are correlated very highly with the number of his unweighted citations, our substantive conclusions would probably not be changed if we did not use the weighting technique. Although weighting is methodologically dispensable, we use it because substantively it seems to yield a more accurate indication of the impact of older works when these are compared with more recent publications. In deciding whether the weighting technique is valid we must consider whether the principle of weighting is logically sound and then the accuracy of the particular weights used. First, let us consider the logic of weighting. Jeffrey Reitz at Columbia University has shown in a study of nineteenth century British scientists that the number of citations at one time period enables us to predict the number of citations at an earlier time. Reitz has shown, for example, that those scientists, who published papers in the 1830s and who received any citations in 1870, were four times more likely to have received citations in the ten years after their work was published.

Derek Price, in *Little Science, Big Science*, has suggested that "although half the literature cited will in general be less than a decade old, it is clear that, roughly speaking, any paper once it is published will have a constant chance of being used at all subsequent dates" (81). This means that papers published in 1930 which received an average of ten citations in 1961 received ten citations on average in 1931. The weighting technique is in no way at odds with the model suggested by Price. The weighting technique is not meant to predict the *number* of citations received by papers in the past but to control for the increasing total number of citations. Due to the exponential growth in science, top papers today are receiving many more citations than top papers did in the past. A paper which receives ten citations today is not among the most heavily cited. But a paper which received ten citations in the nineteenth century would have been one with relatively great impact. Therefore, in comparing work written in different periods we must standardize for the total number of citations being made.

This raises the question of how this standardization should be accomplished or what weights to assign old papers. We have described one such technique in the text. Another would be to compute the total number of citations being made in a field at different times and multiply old citations by the factor which would be required to bring them up to the total of current citations. More research is needed to determine what the most suitable weights would be. However, we must point out that weighting is only an approximation since it is based on averages—with individual papers having citation patterns diverging from the average. Also, when we group years together we are being inaccurate since the weights should be continuous. Finally, if the logic of weighting is accepted, it probably makes very little difference what weights are actually assigned.

19. Robert K. Merton, "Priorities in Scientific Discovery: A Chapter in the Sociology of Science," *American Sociological Review* 22 (December 1957): 635–59, esp. 639–47. The significance of recognition in a social institution of science has been further examined by F. Reif, "The Competitive World of the Pure Scientist," *Science* 134 (December 1961): 1957–62; Barney G. Glaser, *Organizational Scientists: Their Professional Careers* (Indianapolis, IN: Bobbs-Merrill, 1964), esp. chap. 1; Crane, "Scientists at Major and Minor Universities"; Warren O. Hagstrom, *The Scientific Community* (New York: Basic Books, 1965), esp. chap. 2; Norman W. Storer, *The Social System of Science* (New York: Holt, Rinehart & Winston, 1966).

20. The list of ninety-eight awards was assembled from *American Men of Science* and *Physics Today*. Many of the awards listed in AMS were of such limited local significance that they could not be identified and were omitted from the study. All the awards listed in AMS which could be identified were included. Added to this list were awards which appeared frequently in *Physics Today*. Thus, our list represents a large sample of all awards granted to physicists.

Since it is not feasible to ask each physicist to rank ninety-eight awards, we used five different forms of the questionnaire. Ten awards were included on all five forms. Since the difference between the scores of these awards which appeared on all five forms was statistically insignificant, we conclude that the score received by each award is representative of its prestige among academic physicists. As an example of the closeness of the ratings on the five forms, membership in the National Academy of Science received ratings of 4.28, 4.32, 4.03, 4.24, and 4.27 on the five forms. This was the wording of the question:

> The following thirty awards represent a sample of several kinds of awards. For those which are known to you, we ask that you indicate your judgment of its prestige by circling one of the five rankings. You may not have heard of many of these awards, as most are not widely known. If you have heard of an award, but do not know enough about it to evaluate its prestige, please circle No. 6. If you have never heard of the award, circle No. 7. The circling of either 6 or 7 provides useful information as a ranking since it will indicate which awards are least known among physicists.

21. See table 1.3 and note 25.

22. Cartter, *An Assessment of Quality in Graduate Education.*
23. Alan T. Waterman, *Science* 151 (January 1966): 61–64.
24. The names of twenty-four different physicists appeared on each of the five forms of our questionnaire. Thus, the scope of the reputation of each physicist was determined by a random sample of approximately 250.
25. These are the correlations among the three types of recognition:

	Prestige of Highest Award	Number of Awards	Scope of Reputation
Number of awards	.70	—	—
Scope of reputation	.63	.63	—
Rank of department	.56	.50	.57

The "deviant" cases in which one form of recognition is not accompanied by another should be studied for further clues to the working of the reward system.

26. The results were the same when the average number of papers published per year, rather than total number of papers published, were used.
27. We were interested in seeing what results would be obtained by weighting the stratified sample so that it would be representative of the *top twenty departments.* There are not enough cases in the sample of 120 which can be weighted to make it representative of the entire population of university physicists. We therefore converted the sample back to a simulated random sample of the top twenty departments of physics by weighting each man according to the percentage of the population he represented in the stratified sample. Thus, a man drawn from a group representing 20 percent of the population was weighted twice as heavily as a man drawn from a group representing only 10 percent. As can be seen in the following table, this weighting procedure gives a pattern of results similar to those of table 1.4. This suggests that the finding—quality is a far more frequent correlate of eminence than quantity—may be generalized at least to the population of physicists in the twenty most prestigious departments. The absolute percentages here are also closer to the distribution that we suppose would be found if we used more precise criteria of quantity and quality.

Type of Physicist	Percent with One or More Awards
Type I	80 (129)
Type II	53 (63)
Type III	91(91)
Type IV	29 (401)

28. We want to thank John Shelton Reed Jr., for his aid in using the Coregn regression program.
29. The same kind of regression analysis applied to several different measures of citations and also to measures of productivity (average papers per year) yields similar results in every case.
30. A possible explanation of this finding is that those physicists who make a few important contributions along with many run-of-the-mill contributions may dilute their reputations. It may also be that in the top departments, papers are more often circulated among colleagues with the result that they are sometimes not published at all.

31. H. A. Zuckerman, "Nobel Laureates in the United States," unpublished PhD dissertation, Columbia University, 1965, for the reports of laureates on the "pedestrian research" that often precedes scientific results of significance; see also Michael Polanyi, *Personal Knowledge* (London: Routledge & Kegan Paul), 1958.

32. Norman W. Storer suggests the importance of recognition in reinforcing motivation for scientific research. "Recognition is frequently interpreted by [the scientist] . . . not only as confirming the validity and significance of his work but more generally as an affirmation of his own personal worth, and it thereby gains meaning as an intrinsically satisfying reward. Further, . . . the act of creativity does not seem to be complete without some feedback from others, and in science, this feedback takes the form of recognition. Recognition is thus the appropriate response to creativity and is of *significant importance in the desire to engage in research*" (italics added). See his "Institutional Norms and Personal Motives in Science," presented at the annual meetings of the Eastern Sociological Society, April 1963.

33. Citations are being used in a slightly different sense here. From the standpoint of the system of science, citations indicate the impact of a piece of research; from the standpoint of the individual scientist, citations to his work provide a type of recognition.

34. See Robert K. Merton, "The Recognition of Excellence," in *The Recognition of Excellence*, ed. Adam Yarmolinsky (Glencoe, IL: Free Press, 1960), 297–328.

35. Zuckerman, "Nobel Laureates in the United States," chap. X, provides qualitative evidence for the reinforcing character of recognition in the early years of research.

36. Table 1.8 shows how the correlation between quantity and quality is reinforced by the least institutionalized part of the reward system—the use of one's work by colleagues. We believe that the more institutionalized parts of the reward system work in the same way. Thus, we would hypothesize that those scientists who have received some award in the early part of their careers are more likely to continue being productive than their colleagues who have not received awards. Also, those young scientists whose doctoral and immediate postdoctoral work is recognized by appointment as an assistant professor at a top university are more likely to continue producing than their colleagues who become assistant professors at lesser departments.

37. Crane, "Scientists at Major and Minor Universities."

CHAPTER 2

VISIBILITY AND STRUCTURAL BASES OF AWARENESS IN SCIENCE

(1968)

STEPHEN COLE AND JONATHAN R. COLE

This paper contains an analysis of several aspects of the communication process in science. Using data obtained from printed sources and questionnaires mailed to university physicists, the conditions making for high visibility of a scientist's work are studied. Four strong determinants of visibility were found: the quality of work, as measured by citations; the honorific awards received for work in physics; the prestige of the physics department to which the scientist belongs; and specialty. Quantity of output, age, and name-ordering patterns on collaborative papers have no independent effect on visibility. Just as some physicists may be easily seen (i.e., have high visibility), others are in positions where they may easily see. This latter characteristic is called "awareness." The data indicate that awareness is high in all sectors of the population studied. Variables such as age, rank of department, and quality of work made for only minor differences in awareness. We conclude that the communication system in physics operates efficiently.

A dvance in science depends, at least in part, upon efficient communication of ideas. In an ideally efficient communication system, each scientist would know all the relevant work of the other investigators in his field. Of course, this ideal is never approached in reality.

American Sociological Review 33, no. 3 (June 1968): 397–413.

The many instances of multiple discoveries attest to gaps in the communication system of science. In histories of science, we often read of scientists who jubilantly announce "new" discoveries only to find that a colleague already had made the same discovery. The more efficient the scientific communication system, the less there is unnecessary duplication of research and the greater the commonly held cultural base on which advance is dependent.[1]

Communication is necessary for scientific advance and is also the basis for the operation of the reward system. Thus, it is crucial in motivating individual scientists. One of the greatest rewards that a scientist can receive is the knowledge that his work has been read and used by his colleagues—that it has made a difference (cf. Waterman, 1966: 61–64). The individual scientist therefore becomes deeply concerned with the visibility of his work.[2]

The importance of communication for scientific advance and for the operation of the reward system in science suggests the need for a detailed study of the communication process in science. For communication to be successful, ideas must be written up, circulated, and then effectively utilized. To study effective utilization is difficult, and we shall not attempt to do so here.[3] We take a more elementary problem. Here we study only the extent to which the work of a sample of 120 physicists is familiar to the community of university physicists. Specifically, we are interested in those properties, both individual and contextual, which influence the visibility of the work of scientists.

If the investigation had been limited to the determinants of visibility, we would have been studying only one aspect of the communication process. Just as there are conditions influencing the visibility of a man's work, there may also be conditions influencing the capability of his audience to see his work. Just as some men can be easily seen (i.e., have high visibility), others are in positions where they can easily see. This latter characteristic has been called "awareness." The difference between visibility and awareness is the difference between being passive and active. Visibility characterizes the men being looked at; awareness, the men who are looking. The first part of this paper deals with visibility and the second part with awareness.[4]

Visibility

A sample of 120 physicists was chosen from a sampling frame in which the population of university physicists was stratified along four dimensions: age, prestige-rank of their department, productivity, and number of honorific awards.[5] We wanted to find out how well known each of these 120 physicists was. Since it was impractical to ask physicists how familiar they were with the work of the entire list of 120, five forms of the questionnaire, each form containing twenty-four names, were used.[6] One form of the questionnaire was sent to all physicists who worked in a university department that granted one or more PhDs per year from July 1952 through June 1962. A total of 2,036 questionnaires was sent out, and 1,308 usable ones were returned.[7] We also collected the following objective data from published sources for the 120 physicists: the number of papers they have published; the number of citations to their work; the nationally assessed rank of their departments; and the number of awards for scientific work each one received.[8]

The extent to which each of the 120 was known to his colleagues is called his visibility score. A scientist's visibility score is simply the percentage of respondents who said they were familiar with his work.[9] Thus, each scientist's visibility score was determined by a random sample of about 250 of his colleagues. This measure of visibility is dependent upon the subjective reports of the 1,308 physicists who returned our questionnaire. However, since we are more interested in relative differences in visibility than in the absolute scores, we believe that the exaggeration of some physicists' knowledge about the work of these men will not influence our conclusions.

Visibility of scientific research may be seen as a continuum ranging from one extreme occupied by the physicist who has produced no papers, and whose work is virtually invisible, to the other limiting case of the Wigners or Weisskopfs whose work is known to almost everyone in physics. Between these extremes lie the large majority of scientists who have published some research papers. The distribution of visibility scores is presented in table 2.1.[10] We now

TABLE 2.1 Visibility Scores of the 120 Physicists

Visibility Scores	No. of Men	Percent
90–100	13	(11)
80–89	8	(7)
70–79	7	(6)
60–69	8	(7)
50–59	9	(7)
40–49	11	(9)
30–39	17	(14)
20–29	12	(10)
10–19	20	(17)
0–9	15	(12)
	—	—
	120	100%

Mean=43.4.
Standard Deviation=29.3.
Coefficient of Variation=68%.

turn to a central question of this paper: Why is the work of some physicists more visible than that of others?

We shall consider the probability that the work of different types of scientists will become known to the scientific community. Consider, for the moment, only four such types. Take the scientist who has consistently produced high-quality work but is a member of a low-prestige department, far from the springs of scientific advance. To what extent does his academic location hinder the recognition of his work? Or, correlatively, take the scientist located at an extremely prestigious institution whose work has not reached the standard set by his colleagues at that department. Does a halo shine over him by virtue of his location, giving his work greater renown than its quality might merit? Further, take the case of the scholar who assumes he must publish or perish and so rushes into print at every available opportunity in search of recognition and, in passing, perhaps a good tenure position. Does this "mass producer" become a highly visible scientist? And finally, what of the Mossbauers, who produce only a few papers, all of which have been highly evaluated by the community of physicists?

Table 2.2 presents a matrix of zero-order correlation coefficients among visibility and several potential determinants of visibility. These several aspects of a physicist's scientific output and social location are highly related to visibility. The more awards a physicist has received and the greater his prestige, the more widely he is known.[11] The more papers he has published and the more citations his work has received, the more widely he is known.[12] And finally, the higher the rank of the department at which he works, the higher his visibility. In short, those men who have produced the best research and who work at the best departments have higher visibility scores. We now want to see whether all these variables have independent effects on visibility.

Quantity and Quality of Output. One might expect that the sheer number of papers published by a scientist would influence his visibility. And, in fact, quantity of published articles is correlated with visibility. However, when we control for the quality of a physicist's work, the quantity of that work has almost

TABLE 2.2 Intercorrelation Matrix

Variable	x_1	x_2	x_3	x_4	x_5	x_6
x_1 Visibility	..	.64	.64	.64	.57	.49
x_2 Prestige of Highest Award	..	..	.70	.42	.54	.34
x_3 Number of Awards	..	..	..	.68	.50	.45
x_4 Quality of Output	..	..	..	..	.33	.72
x_5 Rank of Department	..	..	..	..	..	.24
x_6 Quantity of Output	..	..	..	..	..	..

no independent effect on his visibility. An original correlation of $r = .49$ is reduced to a partial correlation of .06, when the quality of the scientist's work is held constant. On the other hand, the partial correlation between visibility and quality with quantity held constant is .47. Thus, we may conclude that if a physicist has produced high-quality papers, it makes little difference whether his bibliography is relatively long or short. This indicates that the publication of scientific papers does not at all assure that their author will gain visibility. Papers not judged significant by fellow physicists, who therefore do not make use of them in their work, are functionally almost invisible.[13]

Honorific Awards. The publication of high-quality work increases the visibility of a scientist. We now want to examine the extent to which formal recognition of a man's work through the presentation of honorific awards increases his visibility. We have seen that there is a high correlation between a man's visibility and the prestige of the highest award he has received. The average visibility score of the eleven Nobel laureates in our sample was 85 percent. There is a considerable gap between this group and even such highly esteemed members of the physics community as the members of the National Academy, whose average visibility score was 72. A still sharper drop is seen when we look at the recipients of awards of lower prestige; their average score is 38. To complete the picture, those scientists without any honorific awards had an average score of 17.

Awards are given for high-quality work. Will a scientist who has produced high-quality work, but who has not been formally recognized through the receipt of an honorific prize, be just as visible as his colleague who has been formally recognized? The correlation analysis indicates that receiving formal recognition does add to one's visibility. Prestige of highest award has a partial correlation of .53 with visibility, even after we have controlled for the quality of work.

Rank of Department. So far the picture is reasonably clear. Physicists who do the best work and who receive formal recognition for it are most visible to their colleagues. Both quality of work and having honorific awards are variables characterizing the individual scientist. We now want to see how the physicist's visibility is influenced by the context in which he works. In a highly efficient communication system, work of high quality would be visible irrespective of where it was produced.[14]

The correlation between visibility and the rank of department of the physicist is .57. This alone, however, does not indicate that the stratification of departments impedes the smooth flow of information through the system.[15] The high correlation between visibility and rank of department may be due to a concentration of physicists who have produced high-quality work at high-prestige departments. To test this hypothesis, we must look at the relationship between rank of department and visibility, with quality of research output held constant. The data suggest that there is an independent and strong effect of both quality of work and the prestige-rank of department on the visibility of physicists.[16]

When quality of work is held constant, the partial correlation between rank of department and visibility is .50. Likewise, when rank of department is held constant, the partial correlation between quality of work and visibility is .58.

There are many possible explanations of the fact that being at a prestigious department enhances the visibility of scientific work. There is the interest of scientists on the periphery in keeping up with the developments at the center. Work being done at the major universities is therefore given more attention than its scientific significance might dictate. Most physicists, regardless of where they teach, received their PhD degrees from one of the major departments.[17] Even alumni who have moved to less prestigious departments are more likely to maintain an interest in what is being done at their alma mater. An additional explanation might be that the physicists doing high-quality work at top departments are more likely to work in highly visible specialties.[18] The data allow us to reject this interpretation. The relationship between rank of department and visibility remains, even when specialty is controlled. Finally, another explanation lies in possible inaccuracies in our indicator of quality. It is possible that the men at lower-quality departments are publishing their papers and being cited in the less significant journals. There is some evidence against the validity of this explanation. When we use prestige of highest award as our measure of quality, we still find that rank of department has an independent effect on visibility, indicating that the finding is not an artifact of inaccuracy in measurement.

The findings of the above analysis are summarized in table 2.3, in which we show the cumulative effect of quality, prestige of highest award, and rank of department on visibility. These three variables explain 61 percent of the variance on visibility. No other variable in the study increased the percentage of variance explained beyond this point. But perhaps this statistical procedure is masking some relevant substantive results.

Specialty. In science, as in most other institutions, certain activities have greater prestige than others. In physics, some specialties are more prestigious than others, and the subject matter of a man's work might influence his visibility. We compared

TABLE 2.3 Stepwise Multiple Correlation Coefficients of Quality, Rank of Department, and Prestige of Highest Award with Visibility

Independent Variables	Dependent Variable	r/R	Total Variance Explained (%)
Quality of Output	Visibility	.639	39.9
Quality of Output and Rank of Department	Visibility	.739	54.7
Quality and Prestige of Highest Award	Visibility	.751	56.4
Quality and Rank of Department and Prestige of Highest Award	Visibility	.782	61.1

the visibility of men working in the four largest specialties: atomic and molecular, elementary particles, nuclear, and solid state. Solid state physicists had the lowest average visibility score, while physicists working in elementary particles had the highest average score.[19] This relationship was maintained even when quality of work was taken into consideration. (See table 2.4.) Men working in particle and nuclear physics still have the highest visibility.[20] Atomic and molecular physicists have somewhat less visibility, and solid state physicists have the least visibility. Physicists who have produced "low quality" work in elementary particles are just as visible as solid state physicists who have produced "high quality" work. These data indicate that reputation for work done in so-called "hot fields" can more easily permeate the boundaries of specialty than work in less prestigious fields.

Age. The correlation between age and visibility is .03. We thought that this negligible correlation might be the result of a curvilinear relationship between the two variables. This turns out to be the case. Physicists under forty have a mean visibility score of 36. Visibility seems to increase with age and reaches a peak of 54 in the early sixties. However, those physicists who were sixty-five years or older had an average score of 37. These data might indicate that there tends to be a substantial time lag in the growth of scientific reputation. A young man

TABLE 2.4* Specialty, Quality, and Visibility

| | Mean Visibility Score | |
| | Quality | |
Specialty	High (60+ Citations)	Low (0–59 Citations)
Elementary Particles	64	40
Nuclear	61	32
Atomic and Molecular	50	16
Solid State	39	14

* The number of physicists in each group and the total number of possible identifications of men in each group are presented below. The mean visibility score for any group can be seen as the percentage of positive identifications.

| | Quality | | | |
| | High | | Low | |
	No. of Men	Total No. of Identifications	No. of Men	Total No. of Identifications
Elementary Particles	10	2,658	11	2,861
Nuclear	12	3,163	26	4,357
Atomic and Molecular	13	3,418	1	272
Solid State	6	1,584	6	1,547

may produce significant work, but his work might not become highly visible for several years. Table 2.5 shows that this is not the case. Among those scientists who have produced high-quality work, age has no effect on visibility.[21] Younger men who have published high-quality work are just as visible as their older colleagues who have produced high-quality work. This suggests that the curvilinear relationship between age and visibility results from the correlation between age and quality in our sample ($r = .14$).[22] The fact that young men who have contributed to their field are just as visible as their older colleagues who have made contributions of comparable quality is another indication of the highly effective communication system in physics. Table 2.5 also indicates that even the scientist nearing or actually in retirement retains his visibility if he has produced high-quality work. On the other hand, the scientist who has not produced such significant work loses visibility rapidly as he approaches retirement. These scientists, who had some visibility due to the sheer volume of their activity in the field, fade into anonymity once they are no longer active, and younger scientists no longer experience them as significant parts of their environment.

TABLE 2.5* Age, Quality, and Visibility

| | Mean Visibility Score | |
| | Quality | |
Age	High (60+ Citations)	Low (0–59 Citations)
65 or older	60	13
60–64	62	35
50–59	60	29
40–49	59	31
Under 40	58	24

* The number of physicists in each group and the total number of possible identifications of men in each group are presented below.

| | Quality | | | |
| | High | | Low | |
	No. of Men	Total No. of Identifications	No. of Men	Total No. of Identifications
65 or older	11	2,893	11	2,864
60–64	10	2,662	6	1,543
50–59	21	5,444	13	3,387
40–49	12	3,190	12	3,126
Under 40	8	2,073	16	4,210

Since our data measure visibility at only one point in time, we could not directly study how different career patterns influence visibility. However, we were able to divide those scientists who have produced high-quality work into those whose work is still significant and those whose work is not.[23] The first group had an average visibility score of 64 and the latter an average score of 46. This suggests that just as a man may become visible at an early age if he has made a substantial discovery, his visibility declines if his old work loses significance and his current work does not meet the standards set by his earlier performance.

Collaboration. Another variable possibly influencing visibility is the position of an author's name in a series of collaborating authors. Today, collaborative research is the modal type in physics.[24] It is possible that a physicist who published few non-collaborative papers or whose name rarely appeared first among collaborators might be less visible to his colleagues. Recent research by Zuckerman (1967a) has shown that Nobel laureates are apt to claim that name ordering of authors is an insignificant matter. However, they often add that it is a source of stress and conflict among coworkers. This problem is often obviated in physics through the alphabetical arrangement of names.[25] However, some scientists feel that being first or last in a series of names will make their work more visible. Here, we should like to focus on only one question related to name ordering and recognition: In the end, does a Cain's visibility to his fellow physicists suffer at the hands of an Abel? This question is far more complex than it at first appears. For example, consideration should be given the varying eminence of the collaborating authors and to various alternative forms of name ordering (Merton, 1968: 56–63). Nevertheless, tentative evidence leads us to believe that name ordering is of more concern to the authors of papers than to the people who read them and has little effect upon the ultimate visibility of a scientist. Table 2.6 shows that regardless of how often a scientist's name appears alone or first among a group of collaborators, his visibility will be high only if the quality of his research is high.[26] Let us stress that these data deal only with the visibility of a scientist's entire body of work and not with any specific piece. Although name ordering may have negligible effect on a scientist's ultimate visibility, it may be important in assigning credit for a particular piece of work.[27]

Briefly, canvassing ground already covered, we have identified four strong determinants of visibility: the quality of work as measured by citations; the honorific awards received for work in physics; the prestige of the physics department to which the scientist belongs; and specialty. Quantity of output, age, and name-ordering patterns appear to have little independent effect on visibility.

This next question must be asked in the study of visibility: What are the consequences for scientists of differing degrees of visibility? Much research is

Visibility and Structural Bases of Awareness in Science (1968)

TABLE 2.6* Name-Ordering, Quality, and Visibility

Percent of Papers Solo or on Which Author's Name Appeared First	Mean Visibility Score	
	Quality	
	High (60+ Citations)	Low (0–59 Citations)
Under 40	57	23
40–69	61	29
70 or more	60	24

* The number of physicists in each group and the total number of possible identifications of men in each group are presented below.

Percent of Papers Solo or on Which Author's Name Appeared First	Quality			
	High		Low	
	No. of Men	Total No. of Identifications	No. of Men	Total No. of Identifications
Under 40	16	4,186	17	4,417
40–69	22	5,736	28	7,317
70 or more	24	6,331	13	3,405

needed in which visibility is used as an independent variable. We would want to know whether or not, and in what ways, visibility influences the scientist's chance of getting research funds or of having his further papers accepted for publication. Holding quality constant, are scientists of low visibility more likely to have their papers rejected by journals?[28] We would want to know how visibility influences career patterns and academic mobility. High visibility may aid upward mobility and, at the same time, hinder productivity. The highly visible scientist probably has a much larger role and correspondingly greater demands on his time than does his less visible colleague. Indeed, it has been found that those men at the pinnacle of visibility, Nobel laureates, experience productivity declines during the period immediately after receiving the prize (Zuckerman, 1967b: 400).[29] Finally, we would want to know how visibility affects the work of scientists. Do the interaction patterns of the highly visible differ from those of the less visible? At scientific meetings, do "visibles" talk predominantly with other "visibles," or do they make themselves available for discussion with less visible scientists? All these questions may have different answers for different fields. For example, the extent to which egalitarian norms prevail in a discipline may influence the interaction patterns of the highly visible. These questions, and the many others that could be raised on the consequences of visibility, deserve our attention.

Awareness of Scientific Research

So far, we have been considering only one-half of our problem: the characteristics of the physicists that make them visible. We now turn our attention to the characteristics of the scientific audience, those to whom communications are directed. Not all scientists have an equally broad knowledge of work being done in their field.[30] This introduces the question of whether these differences in knowledge are due to idiosyncratic factors or to socially structured differences in awareness. "Awareness" is here defined as the extent to which social characteristics of the individual, or the context in which he works, affect the extent to which he knows of the work of other scientists. Thus, a scientist with high awareness is one who has distinctive personal attributes or is located in a position that enables him to have great knowledge of the work that is being done in his field.

In a highly efficient communication system, there would be few structural bases of awareness. All scientists would have equal access to "knowledge" regardless of their individual or contextual social characteristics.[31] In such a system, physicists at top-ranking departments would have no more knowledge of what was going on in their field than those at the lesser departments. Men working at Eastern universities would be as thoroughly familiar with work being done at Western universities as with work done at universities closer to home. Physicists doing high-quality work would have no greater awareness than those doing less significant work. However, in a less efficient communication system, all these variables and others might influence the amount of physicists' knowledge of their field. Let us now turn to the data.

Awareness was measured by the same question that yielded visibility scores. Here, however, we are concerned with how many of the twenty-four men on the questionnaire each physicist was familiar with. A physicist's awareness score is the number of men whose work he knew about.[32] The distribution of awareness scores is presented in table 2.7. It is clear that awareness varies less

TABLE 2.7 Awareness Scores

Number of Men Known	Number of Observers
0–5	122 (–9%)
6–8	268 (21%)
9–12	531 (41%)
13–15	265 (20%)
16–24	122 (–9%)
Total	1308 (100%)

Mean=10.5.
Standard Deviation=3.73.
Coefficient of Variation=35%.

than visibility. In fact, the coefficient of variation for visibility is .68, compared with .35 for awareness. The majority of physicists knew the work of about half the men on their questionnaire. The relatively small amount of variance on awareness was the first sign that we would find the communication system of physics to be fairly efficient and that we were unlikely to discover great differences in socially structured awareness of scientific research.

Our hypothesis based on the marginal distribution presented in table 2.7 was confirmed by a multiple regression analysis. All of the independent variables could explain only 19 percent of the variance on awareness.[33] The correlation coefficients between awareness and several of the variables we thought might influence it are presented in table 2.8.[34] These correlation coefficients indicate that there are definite structured differences in awareness but that these differences are not great. Let us look at the data more closely.

Age and professorial rank are most highly correlated with awareness. These two variables are, of course, related to each other ($r = .53$). Only 7 percent of the physicists who were fifty or older were not full professors. Age has little independent effect on awareness, yielding a partial correlation of .09 when professorial rank is held constant. Professorial rank and awareness, however, have a partial correlation of .20 when age is held constant. The data indicate that it is not so much the accumulation of years of experience that makes for high awareness but, rather, it is the reaching of "professional maturity"—the achievement of a high academic rank. Full professors are more likely to supervise large numbers of doctoral dissertations, review applications for grants, and be on the editorial boards of journals.[35] These activities probably contribute to an awareness of scientific research.

Perhaps most significant is the relatively small differences that both rank and age make in explaining variations in awareness. These two variables together have a multiple correlation coefficient of .29 with awareness. They explain only 8 percent of the variance on awareness of scientific research. Thus, even the two variables most highly correlated with awareness make for little differentiation,

TABLE 2.8 Coefficients of Correlation between Awareness and Its Potential Producers

Independent Variable	Correlation with Awareness
Professorial Rank	.28
Age	.22
Prestige of Highest Award	.19
Number of Awards	.18
Rank of Current Department	.17
Total Number of Citations	.15

TABLE 2.9 Rank of Department, Quality of Work, and Awareness

| | Mean Awareness Score | | | | | |
| | Quality | | | | | |
Rank of Department	High (40 + Citations)	No. of Men	Medium (39–10 Citations)	No. of Men	Low (0–9 Citations)	No. of Men
Distinguished	11.8	(135)	10.6	(88)	11.0	(56)
Strong	11.0	(82)	10.8	(107)	10.3	(75)
Good, Adequate+	11.0	(95)	10.5	(162)	10.4	(163)
Fair, Poor	11.2	(45)	9.8	(95)	9.0	(205)

Note: The mean awareness score for any group is the average number of men identified. The highest possible score would be 24.

indicating once again the highly efficient operation of the communication system in physics. Measures of eminence of physicists and the quality of their work yielded the same small differences on the extent of their awareness.

The variable we were most interested in was rank of department. We thought that physicists who taught at top-ranked departments would more likely be tuned in on the communication system of physics and thus have greater awareness. Our data show a difference in the expected direction, but it's a small one ($r = .17$). Professors at the top departments had slightly greater awareness of research in physics than their colleagues at lower-ranking departments. Is this weak but nonetheless positive correlation a result of location in different departments, or can it be explained by the differing personal characteristics of scientists at departments of varying prestige? The correlation data indicate that rank of department has an independent effect on awareness. There is virtually no reduction in the correlation between these two variables when the number of awards held by a scientist is controlled (partial $r = .15$). However, this statistical procedure covers up a specification of the relationship that tabular data reveal. When we computed the mean number of men identified by physicists at the different departments and then controlled for the number of awards, we found no difference in awareness of men at different ranked departments among those who had more than one award.[36] Thus, the really eminent members of the community of physicists seem to be tied in to the communication system regardless of where they are located. Similarly, our correlation data indicate that knowing the quality of work done by men in different departments does not significantly reduce the correlation between department rank and awareness (partial $r = .14$). However, when the data are presented in tabular form (see table 2.9), we learn that when we control for the quality of the work done by men in the different departments, we reduce the influence of rank of department on awareness for those physicists who produced work of high or medium quality.[37] Those physicists who have done high-quality work seem to have relatively high awareness even if they work in undistinguished departments. An initial difference of mean awareness score of 1.8 is reduced to .6 among high-quality physicists. We would guess that productive men at low-ranking departments have cosmopolitan reference groups and thus find it easier to keep up with what is going on in their field than their colleagues whose work has been less significant.

So far we have reported data only on physicists' awareness of the research of other physicists. We wanted to see if similar results would be obtained for another aspect of the social structure of science. The questionnaire asked the physicists to rank a series of thirty awards and prizes. Two categories of answers were included, which indicated that the responding physicists did not have enough information to rank the award or had never heard of it. A physicist's awareness of awards score was the number of awards he ranked, and it is an indicator of his awareness of this aspect of the reward system.[38] It seemed plausible that those physicists who were eminent and had received prestigious awards would have greater knowledge of this aspect of the reward system than those who had not received awards. (They would at least know of their own awards when these were on the list of thirty.) The data are presented in table 2.10. As expected, men who have received awards have greater awareness of the reward system than those who have not. But here, as with awareness of colleagues, the difference is not great. It would seem that there are relatively few awards that physicists deem significant and that knowledge of these awards is fairly widespread. Of course, the handful of eminent men who have received four or more awards do have greater awareness of awards. This points out that awareness perhaps may be profitably studied qualitatively. We have been examining the influence on awareness of statuses occupied by a large number of scientists. The data in table 2.10 suggest that high awareness is likely to be an attribute of lofty statuses that are occupied by very few scientists. Here we see that the very eminent have greater awareness of this aspect of the reward system. Other statuses occupied by a limited number of men and likely to have

TABLE 2.10 Number of Awards Held and Awareness of Awards

Number of Awards Held	Mean Number of Awards Ranked	No. of Men
0	10.8	958
1	12.1	198
2	13.1	86
3	14.3	34
4 or more	15.5	32

high awareness are editors of key journals, officials of the National Science Foundation, and leaders of scientific societies. We might subsume all these statuses under the category of "gate-keepers," those who control access to rewards and other significant parts of the institution of science.

The data on awareness indicate that communication of various types of information from one sector of the community of physicists to another is apparently highly efficient. One further avenue of investigation remains. Although there are only minor differences in awareness, it is possible that these differences will increase for specific characteristics of the objects being observed. Until now, we have been considering what might be called "gross awareness"—awareness of the system as a whole. In the last part of our analysis, we shall vary characteristics of the observed scientist and the observers simultaneously.

Let us begin with an analysis of geographical location. In a perfect communication system, ideas would flow from research centers in one part of the country to research centers in all other parts.[39] People in the East will be just as familiar with work done in the West as will people who work in the West. As can be seen in table 2.11, this is approximated in the case of physics. Region makes for no systematic differences in awareness. (This may be seen by looking across the rows of table 2.11.) We thought it possible, however, that if we further divided the physicists being observed into those who had produced

TABLE 2.11 Geographical Location of the Observed and Observers

	Mean Visibility Score and Standardized Mean Awareness Score							
	Region of the Observers							
Region of 120 Observed	East	Total No. of Identifica-tions	South	Total No. of Identifica-tions	Midwest	Total No. of Identifica-tions	Far West	Total No. of Identifica-tions
East	50	(5,174)	44	(2,278)	48	(4,136)	47	(1,637)
South	18	(928)	19	(378)	22	(699)	17	(306)
Midwest	31	(3,703)	31	(1,587)	38	(2,874)	36	(1,222)
Far west	58	(2,294)	55	(945)	62	(1,759)	65	(699)
Total	43	(12,099)	40	(5,188)	46	(9,468)	44	(3,864)

Note: The mean visibility score in tables 2.11–2.14 is exactly the same as the mean visibility score described above (see table 2.4). However, in these tables we had to use a slightly different measure of awareness from that used in table 2.9. Because of the five forms of the questionnaire, not all physicists ranked the same number of men in each category. For example, the physicists receiving Form I ranked nine physicists from Eastern universities and physicists receiving Form II ranked fifteen physicists from Eastern universities. For this reason mean awareness scores would not be comparable. The scores were standardized by taking the total number of possible identifications in each group and dividing this into the actual number of men identified. This standardized awareness score is in fact the same number as the mean visibility score, emphasizing that visibility and awareness are analytically distinct aspects of an empirically unitary process.

Visibility and Structural Bases of Awareness in Science (1968)

TABLE 2.12 Quality and Region of the Observed and Region of the Observers

		Mean Visibility Score and Standardized Mean Awareness Score							
		Region of the Observers							
Quality of Observed	Region of Observed	East	Total No. of Identifications	South	Total No. of Identifications	Midwest	Total No. of Identifications	Far West	Total No. of Identifications
High	East	67	(2,928)	59	(1,269)	63	(2,316)	66	(913)
	South	50	(103)	67	(43)	65	(84)	64	(33)
	Midwest	45	(1,785)	42	(776)	50	(1,439)	55	(580)
	Far west	65	(1,479)	64	(597)	72	(1,118)	72	(446)
Low	East	29	(2,246)	26	(1,009)	27	(1,820)	22	(724)
	South	19	(825)	12	(335)	17	(615)	11	(273)
	Midwest	17	(1,918)	20	(811)	27	(1,435)	18	(642)
	Far west	46	(815)	40	(348)	46	(641)	53	(253)

high-quality work and those who had not, we might find some differences. It might be that high-quality work would be clearly visible in all parts of the country regardless of where it was produced, but low-quality work might be more visible to the men who worked in the region where it was produced. After all, they might have greater opportunities to meet one another at regional professional meetings and to read their work in regional journals. The data presented in table 2.12 enable us to reject this hypothesis. High-quality work is known throughout the country, regardless of where it is produced. Low-quality work is less visible but, again, is no more likely to be visible in the section of the country where it was produced than in other sections of the country.[40]

This type of analysis, in which we simultaneously vary characteristics of the observed and observers, enables us to ask two questions: (1) What types of physicists are more visible to some segments of the community than others; and (2) What types of physicists have greater awareness of some men than others? To answer these questions thoroughly we would have to present a large number of complex tables. Instead, we present one table that summarizes our findings. Dividing the 120 observed physicists by their visibility scores makes it possible to summarize the findings for all those variables that are highly associated with visibility: quality, honorific awards, rank of departments, and specialty. In the summary table, we characterize the 1,308 observers by the variable we are most interested in: rank of department. We hypothesized that the highly visible physicists would probably be visible to others in all departments and that the physicists with low visibility would be uniformly unseen. But what

64 VALUES AND REWARDS IN SCIENCE

TABLE 2.13 Visibility of Observed and Rank of Department of Observers

| | Mean Visibility Score and Standardized Mean Awareness Score | | | | | |
| | Visibility of 120 Physicists | | | | | |
Rank of Department of 1308 Observers	Low	Total No. of Identifications	Medium	Total No. of Identifications	High	Total No. of Identifications
Distinguished	16	(2,567)	47	(2,032)	86	(2,007)
Strong	15	(2,486)	44	(1,944)	84	(1,906)
Good, Adequate Plus	15	(3,969)	44	(3,081)	80	(3,078)
Fair, Poor	13	(3,185)	37	(2,584)	77	(2,463)

of those physicists with medium visibility? Here we might find that location at top-ranked departments increased awareness. The data are presented in table 2.13. Again, we must reject the hypothesis. Physicists are almost uniformly visible to all sectors of the community. Rank of department still fails to make for large differences in awareness, even when the objects being observed were divided by their visibility scores. *No matter how we classified the 120 observed physicists and the 1,308 observers, we found the same results: a man's visibility was equal in all segments of the community and there were only slight differences in awareness of scientific research.* The data lead to the conclusion that information of this kind is communicated efficiently through the social structure of physics.

There was only one variable on which we might expect these conclusions to be reversed, and that is specialty. For it is plausible that even in an efficient communication system, physicists working in one specialty will be more familiar with work in that specialty than they are with work in others. Table 2.14 presents data for the four largest specialties: atomic and molecular, elementary particles, nuclear, and solid state. As we would expect, the work of physicists in each of these specialties is more visible to their colleagues in the same specialty than to colleagues outside it.[41] This is true even when we control for the quality of work. (This may be seen by looking across the rows of table 2.14.) The differences, however, are considerably smaller than expected. For example, solid state physicists who have produced high-quality work are only slightly better known to other solid state physicists than to atomic and molecular physicists. This indicates that the communication system of physics extends across the boundaries of specialties. High-quality work done in one specialty is known to almost all physicists, regardless of their specialty.

Now let us turn to the significance of these data for our analysis of awareness of a scientist's research. Here the pattern of findings is not clear. Only in three of the eight comparisons do the data go in the expected direction, i.e., with physicists having greater awareness of work done in their own specialty.

TABLE 2.14 Specialty of Observed and Specialty of Observers

Quality of Observed	Specialty of Observed	Mean Visibility Score and Standardized Mean Awareness Score							
		Specialty of Observers							
		Atomic and Molecular	Total No. of Identifications	Elementary Particles	Total No. of Identifications	Nuclear	Total No. of Identifications	Solid State	Total No. of Identifications
High (60+ Citations)	Atomic and Molecular	63	(274)	54	(807)	50	(626)	51	(700)
	Elementary Particles	54	(218)	81	(594)	68	(574)	54	(528)
	Nuclear	61	(265)	70	(759)	57	(570)	48	(597)
	Solid State	42	(138)	35	(390)	36	(289)	49	(289)
Low (0–59 Citations)	Atomic and Molecular	25	(20)	11	(64)	15	(46)	21	(58)
	Elementary Particles	36	(246)	59	(684)	41	(542)	26	(554)
	Nuclear	24	(371)	41	(1,040)	38	(837)	23	(911)
	Solid State	12	(133)	11	(408)	15	(291)	23	(261)

66 VALUES AND REWARDS IN SCIENCE

(This may be seen by looking down the columns of table 2.14.) But even here the differences are not great, and we must conclude that communication of this type of information in physics easily crosses the boundaries of specialty—that physicists in any one specialty have relatively high awareness of high-quality work in all specialties.

We have concluded that awareness is not greatly influenced by physicists' individual or contextual characteristics. It is high in all sectors of the community of physicists, and this is taken as a sign of the relatively effective communication system in physics on the matters examined here.[42] Since one of the main instruments of communication in science is scientific journals, physicists in our sample were asked to list the three journals that they read most frequently. The most widely read journal is *Physical Review*, with a full 77 percent of our sample reporting that they frequently read it. The second most widely read journal is *Physical Review Letters*; 59 percent reported that they frequently read it. No other journal was mentioned by more than 25 percent of the sample.

We thought that those physicists who read the most "important" journals might have greater awareness.[43] It turns out that those physicists who listed the two most widely read journals had a mean awareness score of 10.8, whereas those physicists who did not list either of them had a mean score of 8.8. Although "journal reading" only explains a small amount of the variance on awareness, it is more powerful than other variables we have discussed. Those who work in poor departments but read the two top journals have greater awareness than men who work at the distinguished departments but do not read either. Men who have done low-quality work who read the two top journals have greater awareness than men who have done high-quality work but do not read them. In fact, the combined effect of journal reading and age gives us the greatest differentiation on awareness. Those who have been active in physics for a long time and who still keep up with new developments by reading the

TABLE 2.15 Journal Reading, Age, and Awareness

	Mean Awareness Score					
Age	Reads 2 Top Journals	No. of Men	Reads 1 of 2 Top Journals	No. of Men	Reads Neither of 2 Top Journals	No. of Men
60 or over	13.2	(18)	11.6	(34)	10.1	(37)
59–40	12.4	(239)	11.4	(223)	9.3	(103)
Under 40	9.8	(414)	9.6	(189)	6.9	(51)

Note: For measurement of the dependent variable, see note to table 2.9.

key journals have the highest awareness. Young men who do not read the key journals have the lowest awareness.[44] See table 2.15.

Conclusion

It is now clear that characteristics which make for high visibility do not make for high awareness. Quality, possession of honorific awards, rank of department, and specialty combine to explain a high proportion of the variance on visibility. The same variables, however, explain only a small amount of the variance on awareness. This is the case for physics. It remains to be seen whether these conclusions hold for other scientific disciplines and even the humanities. This research has been a case study of a single science. It seems probable that different results might be obtained in disciplines that are not as highly institutionalized as physics. Compared to some other fields, physics has been fractionalized less often and less severely. In physics, there may be greater consensus on what is or is not significant than in many other fields. We would expect that in less institutionalized fields we would find more differences in the awareness of scientists located in various sectors of the social structure of their field. We must also point out that we have been dealing solely with physicists in graduate departments of physics. This population is itself an elite group. It is likely that had we sent our questionnaire to physicists teaching at undergraduate colleges and to those working in industry, we would have found greater differences in knowledge.

Notes

1. Not all duplication of research is unnecessary; see the discussion of the functions of duplication of research in Merton, 1968: 56–63.
2. In most cases, the widespread knowledge of a man's research indicates positive evaluation, but sometimes notoriety rather than admiration accompanies high visibility. Nearly all scientists know the work of the Russian biologist Trofim Lysenko, who has been discredited, even in the former Soviet Union.
3. In an effort to identify possible variations in the utilization of ideas by physicists variously located in the social structure of science, we are currently studying citation patterns.
4. Visibility and awareness have been shown to be concepts relevant for the analysis of many types of social activity and not, of course, only for the study of communication systems in science. For a discussion of other types of human behavior for which these concepts are relevant, see Merton, 1957, Chapter IX. In this chapter, Merton uses the terms "visibility" and "observability." We substitute the term "awareness" for "observability." In other work in progress at Columbia, the terms are being used interchangeably.

68 VALUES AND REWARDS IN SCIENCE

5. The sample, selected for its suitability for several research problems, is not representative; eminent scientists are disproportionately represented.

6. There were only small differences in response rate to the five forms of the questionnaire. Having five forms of the questionnaire was equivalent to having the same study replicated five times. Although there were differences in magnitude of findings from form to form, in all cases the findings went in the same direction.

7. Of the total questionnaires sent to 2,036 physicists, 1,333 or 65.5 percent were returned. Twenty-five of these were incomplete or otherwise not usable. Physicists in leading departments responded to the questionnaire in the same proportions as those in lower-ranked departments; those with tenure rank, to the same extent as those without tenure. So far as we can tell, the 1,308 physicists returning questionnaires are representative of physicists in American universities producing doctorates in physics.

8. The number of papers was obtained from *Scientific Abstracts;* the number of citations from the *Science Citation Index;* the rank of departments from the study by Cartter, 1966; and the number of awards from listings in the tenth and eleventh editions of *American Men of Science.*

9. The exact question was "These twenty-five physicists are drawn from various universities, institutes and fields of investigation. You may not know the work of some of these men, but please indicate, for each case, the extent of your familiarity with their work by circling the appropriate number: (1) familiar with most of his work, (2) familiar with some of his work, (3) familiar with a small part of his work, (4) unfamiliar with his work but have heard of him, or (5) have never heard of him." We did not differentiate between the first three answers since we thought that scientists might use different frames of reference in estimating how much of a man's work they were familiar with. To check the adequacy of this decision, we computed mean visibility scores in which we did differentiate between all responses. The measure of visibility yielded correlation coefficients with the independent variables almost exactly like those presented in table 2.2.

   Perhaps the most serious difficulty concerning our measure of visibility is that of determining the extent to which the physicists exaggerated their knowledge. As a crude gauge of the validity of response, a total of five fictional names was included in the list of physicists, one on each form of the questionnaire. The number of physicists reporting that they were familiar with the work of one of the fictitious men was fifty-one, or 4 percent of the sample. Furthermore, of these fifty-one, all but three were names which resemble closely the names of actual physicists.

10. Since our sample heavily overrepresents eminent scientists, the average visibility score of university physicists generally can be assumed to be considerably lower than the mean score of forty-three for our sample.

11. The method of assessing the prestige of awards is reported in Cole and Cole, 1967: 383.

12. Here we use the total number of papers listed under the author's name in *Science Abstracts.* Only papers and not books are included in research output since physicists almost invariably publish their original research in papers, unlike the practice often followed in the humanities and the social sciences. We use total scientific output (the cumulative number of papers published by each physicist) rather than productivity rates (average number of papers per year). We find that both measures exhibit the same patterns of relation to other variables examined in this paper.

   The use of citations to measure "quality" of work is extensively discussed in Cole and Cole, 1967. The measure used in this paper is the average number of citations (weighted by age, with older citations counting more heavily than citations to current work) to work in the scientist's three "best" years. This index is highly correlated (r=.8O) with a straight count of citations. We use the index because substantively it seems to yield a more accurate indication of the impact of older works when these are compared with more recent publications.

13. Similar results were found in Cole and Cole, 1967. There we found that quality was a far more significant variable than quantity in influencing the receipt of honorific awards, attaining a position in a prestigious department, and in being widely known. Whether the same results would be found in other sciences or in the humanities remains to be studied.

14. This employs the familiar technique of contextual analysis in which the relationship between the two variables characterizing individuals is examined in different social contexts. See Paul F. Lazarsfeld and Wagner Thielens Jr., 1958, Chapter X.

15. It should be emphasized that we are dealing not with the communication of ideas but with a less specific type of information. One limitation of this study is that we cannot specify the precise character of information about the work of physicists.

16. Similar results were found by Crane, 1965: 710 in her study of biologists, political scientists, and psychologists. Crane shows that rank of department and productivity have an independent influence on recognition. She sees this relationship as a result of the increased visibility of scientists at major universities.

17. Fifty-six percent of the physicists among the 1,308 who returned our questionnaire received their doctorates from the top-ranked fifteen departments and 44 percent were from the top ten.

18. See pp. 10–11 for a discussion of the visibility of physicists in different specialties.

19. This result could possibly be due to an underrepresentation of solid state physicists among our respondents. That this is unlikely is suggested by table 2.14, in which we show the visibility of men working in different specialties to colleagues in their specialty and those in other specialties.

20. Mean visibility score was computed by counting the number of positive identifications of scientists in the group and dividing by the total number of possible identifications. Specialty, of course, could not be included in the correlation analysis due to the difficulty in ordering this variable.

21. We thought that this finding might be due to the dichotomy on quality. Perhaps young physicists of the first rank become immediately visible and the good (but not brilliant) young physicist gains visibility with age. However, when we split our high-quality group into those with more than 100 citations and those with 60 to 99 citations, the results did not change. Quality, not age, determines visibility. A more striking demonstration of this is obtained by using prestige of highest award as an indicator of quality. The few young physicists who had received top honors had a considerably higher average visibility score than their older peers. The data are presented below:

Mean Visibility Score

Age	Prestige of Highest Award		
	High	Medium	No Awards
65 or older	72	24	12
60–64	74	46	37
50–59	77	32	14
40–49	77	46	21
Under 40	94	36	39

22. This correlation is totally a result of the sampling criteria used to choose the 120 physicists. Since we were interested in including many scientists who possessed honorific awards, we included many elder eminent physicists. Among the 1,308 physicists who returned our questionnaire, there was a slight negative correlation between age and quality.

23. We took all physicists who registered high on our quality index (see note 12) and then divided them into those who had received a total of eighty or more citations in the years 1961, 1964, and 1965, and those with less than eighty. The weighted citation index is a relatively accurate indicator of the quality of the man's best work, regardless of when it was done. The number of current citations is a relatively good indicator of the current significance of a man's work. There was a high correlation between these two indicators. Of the 62 physicists who registered "high" on our citation index, 49 also had more than 80 current citations.

24. About two-thirds of papers published in the major journals in physics in this decade have two or more authors. See Zuckerman, 1965: 81.

25. About 60 percent of multi-authored papers listed in the *Abstracts* since 1920 have alphabetized name orders. See Zuckerman, 1967a.

26. We are aware of the difficulty involved in including single authored papers with collaborative papers on which the author's name appears first. However, it is not possible to separate the amount of visibility due to each of these types of papers and therefore is necessary to consider them jointly.

27. The distinction between the visibility of a physicist's entire body of work and the visibility of any particular piece of work is the same as the distinction we have been making between the communication of specific ideas and the communication of more general knowledge of a man's work. Throughout this paper, we deal only with the latter type of communication.

28. R. K. Merton and H. A. Zuckerman are currently investigating the reviewing practices of a major journal in physics.

29. Decrease in productivity among the highly visible may result from the loss of privacy and a corresponding hesitancy to publish. If an unknown scientist publishes a paper with an error, it may go unnoticed and unremarked. However, if a scientist of high visibility were to publish a paper with an error, it would be brought to his attention by a host of readers. The Nobel laureate Eugene Wigner, being interviewed long after his work had become well known, talked to this problem: "Before the war I was an obscure professor who taught classes and wrote articles; I never thought a reporter would come from New York to talk to me. But now? I no longer can talk nonsense and have it go unnoticed; it could have serious consequences. For better or worse, people pay attention to what I say. It is a very heavy responsibility" (press release, Department of Public Information, Princeton University).

30. We emphasize that we are dealing merely with scope, not depth, of knowledge about a physicist's work. Two physicists say they are very familiar with the work of another physicist. The first may have a detailed understanding and appreciation of his work; the second has only superficial knowledge.

31. Such a communication system is, of course, a contrary-to-fact construct akin to the physical concept of frictionless motion.

32. For a discussion of the validity of this measure of awareness, see note 43.

33. The fact that we were unable to explain a large amount of variance on awareness could mean that we didn't employ the relevant sociological variables or that the small amount of variance on this variable is the result of idiosyncratic factors.

34. The data on awareness could be presented in three different ways. The highest possible awareness score was 24. This score would be given to the man who said he was familiar with the work of all the physicists on his questionnaire. We could have taken some arbitrary cutoff point, such as twelve, and called all men who knew twelve or more physicists high "observers." Using this method we would then compute the percentage of men with "high awareness" in each group under consideration. When this method is used, we obtain percentage differences in our two variable tables which closely parallel the size of the correlation coefficients presented in table 2.8. A second method of presenting the data, and the one used here, is to take the mean number of men known by all the men in any group under consideration. In our opinion, these means summarize better the actual extent of differences in awareness. We might also have taken the percent of men known by each physicist. Had we used percentages such as this, we would have created what we believe to be seemingly larger differences than those actually obtained. For example, a difference in means of three would have appeared as a percentage difference of twelve.

35. It would, of course, be useful to have these data. If we knew which physicists actually did supervise a large number of doctoral dissertations, reviewed applications for grants, and were on the editorial boards of journals, we could then empirically test the validity of the interpretation given here.

36. The correlation analysis did not reveal this because there was, of course, a high correlation between number of honorific awards and rank of department. Only twenty-one men in the lowest-ranking departments had received more than one award. Since statistically rare situations are often the most theoretically interesting, we believe it is most fruitful to examine data in both statistical and tabular form.

37. The quality of work of the 1,308 physicists was measured by a straight citation count. The amount of work involved in computing index scores for this large sample was prohibitive. By using a straight citation count, we are excluding citations of collaborative papers and papers on which the respondent's name did not appear first. In individual cases, this may make for inaccurate measures of the impact of a physicist's work. When we consider the group as a whole, however, omission of these citations has no discernible effect. We are able to show this for our 120 physicists for which we have a full range of citation data. The correlation between straight citation count and total citations (including citations to collaborative work on which the physicist was not the first author) is .96. We arranged the 120 physicists in two ranked lists, one according to straight citation count and the other according to total citations; the Spearman rank-order correlation coefficient is .85.

38. The mean number of awards ranked by our sample was 11.4; the distribution had a standard deviation of 4.88 and a coefficient of variation of .43. For a discussion of the validity of this measure of awareness see note 43.

39. We must re-emphasize here that we are not really dealing with ideas but rather with the communication of a diffuse kind of knowledge—familiarity with the work of others in the field.

40. Physicists from far Western universities who have produced "low-quality" work are more visible than their counterparts from other regions. We suspect that this is due to the fact that some of the physicists in our sample from the far West have relatively few citations to their work but have gained visibility through administration of important laboratories.

41. An exception is found in nuclear physics, where the work of nuclear physicists is more visible to physicists working in elementary particles than to those in nuclear physics.

VALUES AND REWARDS IN SCIENCE

42. The small differences found on awareness could conceivably be the result of differential tendencies to exaggerate knowledge or systematic differentials in the criteria the physicists used in determining whether or not they were familiar with the work of a particular man. For instance, if physicists who had produced low-quality work used less stringent criteria in identifying others than their colleagues who had produced high-quality work, this might account for the small differences in awareness between these two groups. We have already pointed out that the total amount of exaggeration of knowledge seems to be negligible (note 9), and now we present data on the differential probability of exaggeration. There were no systematic differences in exaggeration. Four percent of physicists who had published high-quality work, medium-quality work, and low-quality work identified a fictitious name. Four percent of physicists from top ranked departments, and 4 percent of physicists from poor departments, said they were familiar with the work of a fictitious man. Since exaggeration was limited in general, and since we found no correlates of exaggeration among the variables in our data, it is probable that exaggeration was the result of randomly distributed error.

Our questionnaire also contained a rough gauge of the extent to which there was exaggeration of knowledge of awards. We included one fictitious award on each form of the questionnaire. (One was called the "Richard Saunders Gold Medal-Canadian Roentgenray Society.") Only 6 percent of the physicists, or seventy-nine, ranked one of the fictitious awards. Of these, forty-nine were rankings of the "Pupin Gold Medal," a fictitious award named after an eminent physicist. Only nine identified both a fictitious man and a fictitious award. There were no systematic differences in the extent to which fictitious awards were ranked. For example, 6 percent of men in all quality categories ranked a fictitious award.

To check the validity of our results further, we ran some tables excluding those men who had identified fictitious entries. As may be seen in the table below, the results were not significantly affected by the inclusion of those who exaggerated their knowledge. When the exaggerators are excluded, ones who have two or more awards ranked an average of three more awards than those who had no awards. When the exaggerators are included, the same results were obtained.

Number of Awards Held	Mean Number of Awards Ranked (Whole Sample)	Mean Number of Awards Ranked (Exaggerators Excluded)
0	10.8	10.5
1	12.1	11.6
2 or more	13.9	13.3

It is true that the exaggerators had higher awareness scores than the non-exaggerators. For example, the seventy-nine exaggerators on awards ranked an average of seventeen awards; the non-exaggerators ranked an average of eleven. However, since we were able to find no correlates of exaggeration among the variables in our data, we conclude that exaggeration probably had no significant influence on the conclusions we have reached.

43. We do not mean to imply that journals which are not widely read are not significant journals. However, there is other evidence that the two most widely read journals in physics also are the ones where the most significant research is generally published. See M. M. Kessler, 1965: 28–36.

44. An analysis of variance was performed on tables 2.9 and 2.10 in order to estimate the significance of the difference between the various group means obtained. Both of these tables showed insignificant F ratios, lending support to our interpretation that there are only minor differences in awareness among university physicists. An analysis of variance performed on table 2.15 yielded an F ratio of 2.02, significant at the .05 level.

References

Cartter, Allan M. 1966. *An Assessment of Quality in Graduate Education*. Washington, DC: American Council of Education.

Cole, Stephen, and Jonathan R. Cole. 1967. "Scientific Output and Recognition: A Study in the Operation of the Reward System in Science." *American Sociological Review* 32 (June).

Crane, Diana. 1965. "Scientists at Major and Minor Universities: A Study of Productivity and Recognition." *American Sociological Review* 30 (October): 710.

Kessler, M. M. 1965. "The M.I.T. Technical Information Project." *Physics Today* 18 (March): 28–36.

Lazarsfeld, Paul F., and Wagner Thielens Jr. 1958. *The Academic Mind*. Glencoe, IL: Free Press.

Merton, Robert K. 1957. *Social Theory and Social Structure*. Glencoe, IL: Free Press.

——. 1968. "The Matthew Effect in Science." *Science* 159 (January): 56–63.

Waterman, A. T. 1966. *Science* 151 (January): 61–64.

Zuckerman, Harriet A. 1965. "Nobel Laureates in the United States." Unpublished PhD dissertation, Department of Sociology, Columbia University.

——. 1967a. "Patterns of Name-Ordering among Authors of Scientific Papers." Read at Annual Meeting of the American Sociological Association in 1967.

——. 1967b. "Nobel Laureates in Science: Patterns of Productivity, Collaboration, and Authorship." *American Sociological Review* 32 (June): 400.

CHAPTER 3

PATTERNS OF INTELLECTUAL INFLUENCE IN SCIENTIFIC RESEARCH

(1970)

JONATHAN R. COLE

A widespread conception of the development of science holds that the great discoveries are a result of the cumulative work of a vast number of scientists. Those historians and philosophers of science who express this point of view see the scientist who produces pedestrian research as an integral part of the developmental process. The great men of science stand atop a pyramid of less distinguished and, to a large extent, invisible scientists. An alternative hypothesis holds that relatively few scientists are responsible for advance in science and that, in the broader historical perspective, most of the eminent scientists, even of the calibre of Nobel laureates and National Academy members of today, are the "pedestrians" of history.

This paper attempts to put these conflicting ideas to empirical test for the field of physics. Three independent sets of data are analyzed: one is drawn from a stratified random sample of American academic physicists, a second from a subjective evaluation of significant contributions to recent physics, a third from a set of papers cited in Physical Review. All three sets of data indicate that there is a sharp stratification in the use of work published by various types of scientists. The data support the hypothesis that

---

Sociology of Education 43, no. 4 (Autumn 1970): 377–403. Revision of a paper read at the annual meeting of the American Sociological Association, in August 1968. This research was supported by a grant from the National Science Foundation (GS 2736) to the program in the Sociology of Science, Columbia University. Stephen Cole aided in the design and analysis; Robert K. Merton helped improve an earlier draft of this paper. This paper may be identified as reprint no. A 616 of the Bureau of Applied Social Research.

the physicists who produce important discoveries depend almost wholly on the research produced by a relatively small number of scientists. The implications of these findings for the social structure of science are discussed and areas for necessary future research are suggested.

Until recently, historians and philosophers of science have attributed much of the growth of science to the work of the average scientist who, it is suggested, has paved the way with his "small" discoveries for the men of genius—the great discoverers. This hypothesis is asserted in many sources, but perhaps no more clearly than in the words of José Ortega y Gasset (1960):

> For it is necessary to insist upon this extraordinary but undeniable fact: experimental science has progressed thanks in great part to the work of men astoundingly mediocre, and even less than mediocre. That is to say, modern science, the root and symbol of our actual civilization, finds a place for the intellectually commonplace man and allows him to work therein with success. In this way the majority of scientists help the general advance of science while shut up in the narrow cell of their laboratory, like the bee in the cell of its hive, or the turnspit of its wheel.[1]

The imagery implied by this conception of scientific development is clear. Average scientists, working on relatively unambitious projects, make minor contributions; but without these lesser discoveries by a mass of scientists, the breakthroughs of the truly inspired scientist would not be possible. Thus, the great men stand atop a pyramid of lesser men and their view is clearer to a large extent because of the foundation laid by those of less distinction. There are, of course, a number of assumptions in this position. Consider two: first, it is assumed that the ideas of the average scientist are both visible and used by the outstanding scientist; second, it is assumed that the minor work is necessary for the production of major contributions. In short, it is proposed that the work of the average scientist is indispensable if science is to advance. Little empirical evidence exists to substantiate these widely held beliefs.

Why has this notion been so long lived without substantial questioning of its authenticity? What functions has its survival served for individual members of the scientific community? These broad queries, for the most part, remain beyond the scope of this paper. It is sufficient to say that an image of a social system in which every man's work counts toward reaching a set of common goals can have significant consequences for the stability of that system and for the motivation of its members. In a certain sense—at least in the eyes of many observers—an acceptance of this imagery necessitates our looking upon science as a more ideal or utopian social system than other parts of the society—a system where common goals are paramount and where widespread communality exists.

Our aim is to examine data bearing upon the validity of this view of scientific progress. In order to make an empirical test of this conception manageable, we confine ourselves to one of its several aspects and to only one field of science. We shall examine the work of various types of physicists and analyze what work these men built upon in making their discoveries.

We do not intend to argue that great discoveries in science by an Einstein or a Lee and Yang are not preceded by numerous "smaller" discoveries, or, that great discoveries do not in turn stimulate a multitude of lesser ones. (Cf. Sorokin and Merton, 1935; Sorokin, 1962; and Ben-David, 1960, for detailed and informative discussions of fluctuations in rates of discovery in the history of science. Also see Sarton [1931] for a qualitative treatment of the same idea.) We will suggest that even the men who make these "smaller" discoveries come principally from the top strata of the scientific community. In the proper perspective of the history of science, "normal science" as Kuhn refers to it, is not done by the average scientist but by the elite scientists (cf. Kuhn [1962] for a discussion and outlines of a theory of scientific development). Indeed, in the longer perspective, men of the stature of many of today's outstanding scientists, such as Nobel laureates and National Academy members, are often history's pedestrians (cf. Whitehead's discussion of "the fallacy of misplaced concreteness" in 1925). The question that we consider is not how many discoveries are being made, all told, in a specific period of time, but how many scientists are contributing to the movement of science, and how many are not?

With recent advances in the sociology of science, the idea of science as a highly stratified system, with skewed distributions of productivity and of rewards for outstanding performance, is no longer novel. For example, Price (1963) has suggested that the square root of the population of scientists produces 50 percent of scientific discoveries—a figure which in itself indicates a highly stratified system.[2] What Price does not deal with is the extent to which this distinct minority of scientists who produce 50 percent of the research publications are dependent on the overwhelming majority of research scientists and the 50 percent of the total research they produce.[3]

In an earlier paper we have shown that the visibility of scientific research is itself highly stratified (cf. Cole and Cole, 1968). High-quality work by esteemed physicists located at the most prestigious academic departments is visible throughout the system; less significant research produced at low-prestige departments remains largely unknown. In this paper we deal with patterns of actual use of work by other scientists. If the bulk of the scientific community produces work that is rarely used, that is, infrequently cited in the work of outstanding scientists, then this indicates that their work does not materially advance the development of science. This is our basic question: What are the intellectual sources of influence on the production of scientific research of varying quality? If Ortega is correct, then the vast majority of physicists will be cited in what the field itself generally regards as major discoveries.

This paper presents data, drawn from a larger study of university physicists, which illustrates the citation practices of varying types of academic physicists. One set of data consists of the citations made by eighty-four university physicists in their paper most heavily cited in the 1965 *Science Citation Index* (SCI). We consider this to be the physicist's outstanding piece of work as gauged in 1965.[4] These physicists were chosen from a sampling frame in which the population of university physicists was stratified along four dimensions: age, prestige rank of their university department, productivity, and number of honorific awards.[5] This sample overrepresents eminent physicists.[6] Information collected for these physicists included the number of papers they have published; the number of citations to their work; the nationally assessed rank of their departments; and the number and prestige of their various awards. A second set of data consists of information on a one-third random sample of the scientists who were cited in the best paper of each of the eighty-four physicists. Social and individual characteristics similar to those collected for the citers were gathered for this sample of 385 cited authors.[7]

A basic assumption in this paper is that the research that a scientist cites in producing his own papers represents a roughly valid indicator of influence on his work. Of course, not all citations represent direct and specific influence of this kind. We all know about cases of a scholar paying an intellectual debt to a mentor through a ceremonial citation. More difficult to ascertain is the extent to which a cited work represents a significant, even necessary, antecedent to the scientist's discovery as opposed to a tangentially relevant piece of work, in which the author merely is demonstrating his "knowledge of the literature." However, a reasonable case can be made that citations generally represent an authentic indicator of influence (cf. Sher and Garfield, 1965; Clark, 1957; Bayer and Folger, 1966; and Cole and Cole, 1967, for detailed discussion of what citations do measure). If we put aside attempts to measure various types of influence and examine only use, then citations become a meaningful indicator. The norms of science require scientists to cite the work that they have found useful in pursuing their own research, and for the most part they abide by these norms, which have strong sanctions. Moreover, the audience of the work generally takes citations as such an indicator. We only have to think of the number of times we have taken a quick glance at the acknowledgments and indexes of books and papers with the intent of noting the influence on a piece of work to realize that at the very least citations do indicate intellectual connections.

78 VALUES AND REWARDS IN SCIENCE

Furthermore, although some citations that do appear in print may not, in fact, indicate influence, this does not imply the converse, that work not cited at all is nevertheless influential.

Since citations also are used in this paper as a measure of the quality of a scientist's work, let us note a number of citation patterns of various types of scientists (cf. Cole and Cole, 1971). A number of brief examples can illustrate the frequency with which the truly outstanding scientists are cited, and correlatively, the paucity of citations to the average scientist. The average number of citations in the SCI (1961) to the life work of Nobel laureates (who won the prize between 1955 and 1965) was fifty-eight, compared to an average of 5.5 citations to the life work of all other scientists. Only 1.08 percent of the quarter of a million scientists who appear in the 1961 SCI received fifty-eight or more citations. On the average, members of the National Academy of Sciences had over forty citations to their life's work in the 1961 index. Of course, there are other scientists, not as heavily rewarded, whose work has received attention equal to that of the laureate and Academy members.

Our definition of "high-quality" work (life's work) in this paper is exacting: to have been cited a total of sixty times or more in the 1965 SCI.[8] This measure pegs the standard for "high-quality" work very high, about equivalent to the citation rate to the Nobel laureates. Even the work here classified as "medium quality" has received more than twice the number of citations received by the average scientist listed in the index, since in this category the number of citations range from fifteen to fifty-nine.

Consider another indication of the frequency of citations to physicists: citations to the relatively highly productive and eminent group of physicists found in university departments. Among a representative random sample of 1,308 academic physicists,[9] 8 percent had over sixty citations in the 1965 SCI; 85 percent had under forty, with 67 percent of these having fewer than fifteen.[10] In short, very few physicists were cited heavily by their colleagues. The heavily cited scientists are in this operational sense part of the "scientific elite."

To give these data greater comparative relevance, consider the number of men whose work potentially could be used by an academic physicist. Using only the members of the American Physical Society in our estimate of the number of working physicists, and neglecting the many industrial and foreign physicists who are not members, some twenty-five thousand scientists can be classified as members of the community of American physics. Of course, many of these scientists have never published a scientific paper; most of those who have published at all have published only a few papers.[11] Scientists who do not publish any scientific work are not making direct substantive contributions to the advance of science, though they may be contributing indirectly through training or administration. Nevertheless, the published work of thousands of physicists is available for the use of their colleagues and could be cited by them.

To be sure, when this aggregate is divided into the various specialties, the number is reduced considerably.

It should be noted, moreover, that the total population of physicists has a skewed distribution in terms of the scientists' location. University physicists in the leading eighty-six PhD-producing departments number only a little more than two thousand;[12] members of the ten most distinguished departments number about four hundred, or only 2 percent of the total population. In these figures we are not considering the possibility that physicists may use the work of chemists, biologists, and scientists in other related fields. Were we to include this work, the number of potential sources that could be drawn upon by the physicist would be in the hundreds of thousands. Clearly, the physicist makes use of only a minute fraction of the potential material available to him. The question remains: What kinds of material from this mass does he, in fact, select and cite?

Patterns of Influence in Scientific Discoveries

Data on the use of scientific contributions are presented in three sections. First, we briefly consider the general distributions of citations by members of the physics community. Second, we vary characteristics of cited and citing authors in order to discern patterns of use. Finally, we examine only individual papers of varying quality in order to determine what work is used most often by the makers of extraordinary discoveries.[13]

Table 3.1 presents a series of marginal distributions of citations to work by physicists with various social and individual characteristics.[14] The main results of our analysis are foreshadowed clearly in these figures. For example, almost three-quarters of citations in the work of the eighty-four physicists were to scientists at universities. When this figure is compared with the approximately 10 percent of the population within the physics community that is located at universities, the outlines of a sharp stratification in the use of discoveries become clearly etched. Furthermore, among the citations to scientists in academic settings, well over half are to the work by members of the top nine, or most distinguished, physics departments, and four-fifths of the citations are to men holding academic positions in the leading thirty-seven departments. The rest went to physicists in the other twelve hundred universities and colleges in the nation.

The data presented in table 3.1 suggest that the work used by physicists is produced by only a small proportion of those who are active in the field. The problem to be examined presently is the extent to which physicists variously located in the stratification system make use of the research of different types of physicists. For example, do physicists producing work of varying quality rely on the same intellectual antecedents in the stockpile of scientific knowledge?

TABLE 3.1 Marginal Distributions of the Social and Individual Characteristics of the Cited Authors and Comparative Figures for the Entire Field of Physics

Social and Individual Characteristics of Cited Authors		Comparative Populations Statistics
Current Affiliation		
University	72%	
College or Non-academic Research Laboratories	10	43%
Industry	10	34
Government	8	11
	——	——
	100%	88%[a]
	(385)	(26,698)
Rank of Department of Those in Academic Departments		
Distinguished (Top Nine)	60%	21%
Strong and Good	23	42
Lesser Universities and Colleges	17	37
	——	——
	100%	100%[b]
	(299)	(1,308)
Number of Honorific Awards		
Zero	32%	73%
1	18	15
2–3	23	9
4 or More	27	3
	——	——
	100%	100%[b]
	(385)	(1,308)
Quality of Scientific Output (Number of Citations)		
Under 15	25%	67%
15–59	33	25
60 or More	43	8
	——	——
	101%	100%[b]
	(385)	(1,308)

[a] Source: *American Science Manpower 1964: A Report of the National Register of Scientific and Technical Personnel*, National Science Foundation, NSF 66–29.
[b] These figures are drawn from a sample of 1,308 university physicists.

Rank of Department of Cited Author

One of several ways in which the social system of science is stratified is in terms of the quality and prestige of various academic departments. With the exception of a few distinguished scientific laboratories, positions at major universities generally are considered the most prestigious positions in the field of physics.[15] We hypothesize that scientists who themselves are producing superior work will tend disproportionately to cite, or make use of, work by men at the nine distinguished departments of physics. The data presented in table 3.2 confirm this hypothesis. Table 3.2 consists of six sub-tables which illustrate the relationship between characteristics of citers of scientific discoveries and the rank of the physics department of cited authors. In these tables the category "lesser departments" represents the overwhelming majority, including all but the top thirty-seven universities, as well as every college in the United States.

Two consistent and clear results can be seen in these data. First, a relatively small portion of citations are made to the great majority of our nation's physicists. Although most physicists' work is not widely used, it is used to a limited extent; it is not completely ignored. Second, the greater the degree of eminence of the citing physicist, the greater the probability that he will find work by scientists at the most distinguished departments useful, and, correlatively, ignore to a greater degree the work of scientists removed from centers of research.[16]

Take, for example, the Nobel laureates and National Academy members in our sample. As a group, only 9 percent of their citations are to men at the lesser departments; 58 percent are to scientists associated with the top nine. Similarly, citers *at* distinguished departments and those with high visibility also predominantly use the work of men at distinguished departments. Finally, only 10 percent of the citations by physicists who have produced high-quality research go to men located at lesser departments, while nearly 60 percent are to scientists located at the best departments.

These findings clearly suggest that men located in the top strata of academic physics, in terms of their social location and the quality of their life's work, predominantly cite the discoveries produced by other members of the same elite strata. Thus, the data would appear to give credence to the notion held by some scientists that an "in-group" exists in contemporary science, or, an informal mutual admiration society whose members look only to each other's work and disregard the science produced by nonmembers of the elite corps. Furthermore, it is suggested by those holding this view that these relatively few scientists not only neglect scientists not located at the centers of scientific activity but, by controlling the system of communications and rewards, reduce the probability that a scientist located at a lesser institution will be recognized for his discoveries.

TABLE 3.2 Attributes of Citing Authors Related to the Department of the Cited Scientists

Characteristics of Citers	Rank of Department of Cited Author				
	Distinguished	Strong to Good	Lesser	No Academic Affiliation	(N)
Quality of Citer's Work[a]					
High (60 or More)	59%	12%	10%	19%	100% (164)
Medium (15–59)	40	21	14	26	(126)
Low (Under 15)	36	24	18	22	(95)
Total Output in Terms of Scientific Papers[b]					
High (30 or More)	51	13	11	25	(213)
Medium (11–29)	45	24	14	17	(139)
Low (Under 10)	27	27	21	24	(33)
Rank of Department of Citer[c]					
Distinguished	55	13	11	21	(198)
Strong	40	19	18	24	(101)
Lesser	37	28	13	22	(86)
Prestige of Highest Award[d]					
Nobel Prize or National Academy Membership	58	12	9	20	(153)
Medium Prestige	45	18	22	15	(80)
Low or No Award	36	24	12	28	(152)
Total Number of Honorific Awards[e]					
4 or More	69	9	4	18	(45)
1–3	47	19	15	19	(253)
None	36	19	13	31	(87)
Visibility of Citer[f]					
High	60	13	7	20	(162)
Medium	42	20	17	21	(149)
Low	27	24	20	28	(74)

[a] Quality was measured by the number of citations to the scientist in 1965. "High" quality equals more than 60 citations; "medium" quality is 15 to 59 citations; "low" quality is under 15 citations.

[b] The number of papers produced by physicists was obtained from *Physics Abstracts*.

[c] Allan Cartter's list of quality of departments was used for the rank of department.

[d] The prestige of awards was found through an evaluation of a sample of awards by 1,308 university physicists. For an extended discussion of the methodology involved, see Cole and Cole (1968).

[e] All honorific awards and post-doctoral fellowships listed after a scientist's name in the 10th and 11th edition of the *American Men of Science* were counted as awards.

[f] The visibility of scientific research was determined by the percentage of samples of 250 physicists who were familiar with the work of the scientist. An extended discussion of the methodology used in obtaining these scores is also discussed in Cole and Cole (1968). "High" visibility refers to those scientists known to 60 percent of the physics community; "medium" to those known to at least 20–59 percent; and "low" to those known to less than 20 percent.

Moreover, since the big centers of research have the lion's share of scientific resources at their disposal, they can and do publish more and can and do cite one another more frequently. It is proposed that a closed circle of scientists develops, each "plugging" the others' work. The results seen in table 3.2 appear to be consistent with this in-group interpretation. Following this hypothesis, of course Nobel laureates predominantly will cite the work produced by other eminent scientists. These highly honored men are not only their friends, but they are the physicists with whom they are in constant communication and with whom they have the highest rate of social interaction.

As plausible as this interpretation seems, it actually is dealt a serious blow by the data of table 3.2. *These data show that even members of the lower strata of academic physics disproportionately cite the work produced by members of the most distinguished departments, although to a lesser extent than do men who are part of the top stratum.* Sixty percent of the citations made by physicists who produced work of lesser quality are to physicists located at one of the top thirty-seven departments. Thirty-six percent of these citations are to physicists located at the top nine departments, compared to only 18 percent to scientists at lesser departments. In brief, physicists located throughout the stratification system rely heavily on the superior research produced by a relatively few physicists.

Given our recent findings on the visibility of research produced at lesser departments, these data can be expected. The physicists located at one of the lesser departments are no more likely to know about the work of another physicist at a different lower-ranked department than are their colleagues at Harvard. Research produced at lesser universities is uniformly invisible throughout the social system of physics. Therefore, physicists at institutions of lower prestige not only know more about work produced at distinguished settings than at lesser ones, but they also make greater use of the work produced at the top stratum of the social system.

Table 3.2 not only indicates that physicists producing research of limited distinction are less likely than their more renowned colleagues to use the work produced at distinguished departments, it also suggests that there is no relationship between the social location of scientists and the frequency with which they use work produced at lower-ranked departments. If we consider the relationship between the rank of department of citers and the citation rates to physicists at lesser departments, this becomes clear. Thirteen percent of the citations made by physicists at the lower-prestige departments are to others at similarly ranked departments; almost as many citations, 11 percent, in the work of members of the top nine departments are to scientists at departments of lower prestige. These data suggest, then, that physicists at the top departments are not only as aware of the research being done at minor universities as are scientists located at those schools, but, more important, they actually use the work produced there almost as frequently as do the men at those universities.

In a recent paper we showed that although the visibility of scientists' research differed widely, there were only minor differences in the knowledge that scientists differentially located in the social structure had of other scientists' work. There seemed to be no structural impediments to awareness of scientific research: those scientists at the lesser departments were as aware of the work of a Wigner or Weisskopf as were men at distinguished departments (Cole and Cole, 1968). In this sense, the communication system of physics seemed to operate efficiently. But, in that study we were dealing with general familiarity with the life work of physicists. The data here present a significant specification of the difference between two types of knowledge: what William James called "knowledge about" or superficial knowledge and more detailed "acquaintance with" knowledge required for actual use.[17] In the process it presents us with an interpretative problem. If scientists who produce less significant research are as aware of the significant discoveries as are eminent ones, why do they not make as much use of them? It is possible that scientists producing lesser research see this work as significant and are aware of it because of the relatively high level of publicity that it receives, since the major discoveries receive widespread formal recognition. Yet the research may not be relevant to their own work. These physicists also may exhibit poorer taste in choosing sources of information or simply cannot comprehend that much of the great work well enough to incorporate it into their own research. Correlatively, for the eminent physicists working at the research frontier, the high-quality work is more likely to be both significant and relevant to their own research.

Two central variables in this study are the quality of research and the eminence of the scientists whose work is used by research physicists. Table 3.3 presents the relationship between characteristics of citers and the quality of the work they cite. These data once again point to the same general conclusion. If we examine the percent of citations to scientists whose work is cited as heavily as that of the Nobel laureates (sixty or more times), we find consistent results. Almost one-half of the citations made by the producers of high-quality work (the highly productive, the honored, and the highly visible physicists) are to scientists who, at least in terms of citations, are the equals of laureates and Academy members. Of course, many of the cited authors are laureates. If we lower our sights slightly and consider the proportion of citations by various types of scientists to work which is only three times above average, the data are even more striking. Fully 82 percent of the highly visible scientists, 78 percent of the citations of laureates and Academy members, 81 percent of the citations from those who have received more than sixty citations to their own work, and 74 percent of the citations of men at distinguished departments go to scientists who have at least fifteen citations to their work in the 1965 SCI.

The probability of the authenticity of an "in-group" interpretation of these citation patterns is further diminished by the findings revealed in table 3.3.

TABLE 3.3 Attributes of Citing Authors Related to Quality of
Cited Scientist's Research

Characteristics of Citers	Percent of Cited Authors with 60 or More Citations	Percent of Cited Authors with 15 or More Citations	(N)
Quality of Citer's Work			
High (60 or More)	51	81	(164)
Medium (15–59)	39	68	(126)
Low (0–14)	35	60	(95)
Total Output in Terms of Scientific Papers			
High (30 or More)	46	76	(213)
Medium (11–29)	40	67	(139)
Low (0–10)	36	61	(33)
Rank of Department of Citer			
Distinguished	42	74	(198)
Strong	50	75	(101)
Lesser	40	62	(86)
Prestige of Highest Award			
Nobel Prize or National Academy Member	43	78	(153)
Medium	50	76	(80)
Low	40	62	(152)
Total Number of Honorific Awards			
4 or More	49	78	(45)
1–3	43	74	(253)
None	41	62	(87)
Visibility of Citer			
High	46	82	(162)
Medium	42	65	(149)
Low	39	62	(74)

The data do suggest once again, however, that work produced by scientists removed from the research front is used to some extent by physicists located in all strata of the scientific community. After all, about one-fifth of the citations in high-quality work was to scientists whose life work had received fewer than fifteen citations.

When we focus on the patterns of citation to scientists of varying degrees of eminence the same general results are obtained. Comparatively few physicists have received honorific awards. Yet, a disproportionate number of cited physicists are among those who have been honored. And the data in table 3.4 indicate that the more eminent scientists tend to cite other eminent men with greater frequency than do their less decorated colleagues. However, we see that even the less eminent scientists in the social system call upon the work of their honored colleagues with disproportionate frequency. While only 27 percent of our 1,308 university physicists had received any awards, more than half of the authors cited by physicists producing less significant research were those who had been honored.

TABLE 3.4 **Attributes of the Citing Physicist Related to the Number of Honorific Awards Received by Cited Scientists**

| Characteristic of Citer | Number of Honorific Awards of Cited Author | | | |
	None	1	2 or More	(N)
Quality of Citer's Work				
High (60 or More)	25%	12%	63%	(164)
Medium (15–59)	31	24	45	(126)
Low (0–14)	48	22	30	(95)
Total Output in Terms of Scientific Papers				
High (30 or More)	29	12	59	(213)
Medium (11–29)	32	27	42	(139)
Low (0–10)	58	24	18	(33)
Rank of Department of Citer				
Distinguished	31	16	53	(198)
Strong	32	14	54	(101)
Lesser	37	29	34	(86)
Prestige of Highest Award				
Nobel Prize or National Academy Member	28	16	57	(153)
Moderate Prestige	36	18	46	(80)
Low or No Award	36	22	43	(152)
Number of Honorific Awards				
4 or More	22	13	65	(45)
1–3	31	18	50	(253)
None	39	23	38	(87)
Visibility of Citer				
High	25	14	60	(162)
Medium	34	18	48	(149)
Low	45	28	27	(74)

The findings in tables 3.2–3.4 are strikingly uniform; the data move in the same direction with similar magnitude in virtually all of the sub-tables.

A set of more complex relationships on patterns of influence can be generated from the series of two-variable tables. Only one of these more complex patterns is of direct relevance to the hypothesis under consideration. Consider whether the social location of a physicist or the quality of his work is a stronger determinant of who cites him. If the most significant variable in determining whether or not an author's work is cited is its quality, then outstanding research should be cited with equal frequency regardless of its origins. In other words, producers of high-quality work at both distinguished and other departments would be equally cited.

In table 3.5 we see that at least among physicists who produce high-quality work, the social location of the potentially cited author does make for substantial differences in citation patterns. Sixty-four percent of the research used by physicists who produce outstanding work is high-quality work produced by scientists at distinguished departments; twenty-two percent is high-quality work produced at other departments. This pattern is markedly different from the distribution of citations in the work of physicists producing less significant

TABLE 3.5 The Relationship Between the Quality of Research of Citers and the Quality of Work and Rank of Department of the Cited Scientists (Percent of Citations)

Quality of Cited Scientist's Work[a]	Higher		Low		
Rank of Department Cited	Distinguished	Other	Distinguished	Other	Total
Quality of Citer					
High					
(60 or more citations)	64	22	8	6	100
Medium					(133)
(15–59 citations)	42	33	8	17	100
Low					(100)
(0–14 citations)	29	38	14	19	100
					(63)
					$(N = 296)^b$

[a] In this table we have combined "medium" and "high" cited authors into one group. "Higher quality" represents authors with more than 15 citations to their work in 1965. The more refined category of "high quality" (60 or more) yielded the same pattern of results, but the proliferation of categories would have made the table confusing.
[b] The totals do not add up to 385 since men not located at academic departments are not included in this table.

research. While scientists producing high-quality papers still are cited most often, they are located more often at the less distinguished departments.[18]

One possible explanation for these specified results is that the physicists producing less important research papers have greater degrees of association with outstanding work further from the research front and, consequently, use a higher proportion of it in their work. This interpretation would seem to be undermined by our findings that the awareness of work produced at different settings does not differ regardless of the observer's social location. But, given the findings in this paper there may be no contradiction. Work produced at the undistinguished departments which is of high quality may have greater relevance for the work being done by scientists producing lesser-quality research. This assumes that markedly different types of research may be carried on at universities further removed from the research front, which although different from the work at the front, is nonetheless of significance for the advance of science. This research may be allied more closely in both form and content with the work done by the scientists who produce less significant research and, consequently, is of more use to them.

A second explanation of these findings rests on the possibility that physicists who produce high-quality research are more influenced in their definition of what makes for outstanding work by the social context from which a piece of research emerges. It is not impossible that one factor which affects "taste" in science, or the definition by scientists of what is significant, is the source of the research effort. Scientists at top departments might be more prone to define significance in terms of whether the work is produced at a "top" place like Harvard or Brookhaven Laboratory. It is not unlikely that scientists' past experiences with research produced at different settings affects their present judgments of the quality of work produced at various places.

To this point we have examined the relationship between the quality of the life's work of a physicist and the pattern of citations to his work. We now shift our focus slightly. The independent variable will now be the quality of a physicist's "best paper" as gauged in 1965. Obviously, the number of citations to these "best" papers varied greatly. Some papers received only one or two citations; others received twenty to thirty or more. If we divide our sample in terms of the quality of these papers, we can examine the types of work used in *individual pieces of outstanding research.* We would, of course, expect similar results for these papers as for the bulk of a man's life work, since the correlation between citations to physicists' best papers and their total output is high ($r = .73$). A "highly cited" paper was defined operationally as one which received twenty or more citations to it in 1965; "medium cited" were those with ten to twenty citations; "low cited" papers were those receiving fewer than ten. Clearly, even many of the papers here treated as "low" are far above average in terms of their impact on science. Thus, a single paper in our sample which is said to be of medium or high quality has, on the average, more than four times the number of citations to it than an entire life's work of the average scientist cited in the SCI. In fact, of all cited papers listed in the SCI for 1961, 75 percent were cited only once, 12 percent twice, 6 percent three times, and only about 1 percent six times or more.

It turns out that the pattern of findings for individual papers is, in fact, remarkably similar to those described above. The superior papers more often than lesser ones make use of high-quality work produced predominantly at the most distinguished departments. Rather than present a series of tables illustrating this point we have incorporated a number of findings into table 3.6. We see

TABLE 3.6 The Distribution of Citations in Individual Papers of Varying Quality According to the Social and Individual Characteristics of the Cited Scientists

Characteristics of Cited Scientists	Number of Citations to Citer's "Best" Paper		
	High (20 or More Citations)	Medium (10–19 Citations)	Low (0–9 Citations)
Rank of Department			
Distinguished	60	50	36
Strong to Good	14	19	19
Lesser	7	12	18
No Academic Affiliation	19	19	27
	—	—	—
	100	100	100
	(95)	(139)	(151)
Quality of Cited Author			
High (60 or More)	54	48	33
Medium (15–59)	28	30	36
Low (Less than 15)	18	22	31
	100	100	100
	(95)	(139)	(151)
Prestige of Highest Award of Cited Author			
Nobel Prize or National Academy Member	45	32	25
Other Honorific Awards Fellowship*	15	8	12
No Awards	40	60	63
	—	—	—
	100	100	100
	(95)	(139)	(151)

* Fellowships such as the Guggenheim, Sloan, Rockefeller, and Fulbright were considered here as honorific awards distinct from other post-doctoral fellowships.

that a mere 7 percent of the citations in the most outstanding discoveries go to scientists working in the lesser universities and colleges, while 60 percent go to scientists at distinguished departments. Even in the lesser papers the work of men at top universities is cited considerably more frequently than that of men in the lesser settings. As expected, the best papers predominantly cite other significant papers. Fifty-four percent of the citations in the best research go to other superior research. If we consider the citation pattern within lesser papers, we note that there is almost an equal distribution of citations to work of varying quality. Finally, note the extent to which high-quality papers rely on the work of Nobel laureates and National Academy members. Forty-five percent of the citations in these papers go to the work of no more than two hundred men and their collaborators. Although the poorer research does not make use of the work of these "elites" to the same extent, the work of the elites receives proportionately heavy use by them as well.

Throughout this paper we have defined important discoveries simply by the number of citations that they have received. A subjective determination by a well-known physicist of a small group of important contributions to physics in the last ten years could provide the basis for a further test of our results. Accordingly, we asked a well-known academic physicist for a list of the five papers that he believed represented some of the most important contributions to contemporary particle physics. Of course, in many ways this procedure falls short of the rigorous study needed to further test our ideas. For example, we would need a study involving a broad stratified panel of judges evaluating the merits of various pieces of research. It is worth noting, however, that the citation rate to the five selected papers was consonant with our working criteria of "highly-cited" papers. These five papers received an average of sixty-seven citations in the 1965 SCI.[19] We took all the journals and private communication citations in these five papers and then found out where the cited authors were located in the social community of physics. We also obtained citation information on these cited scientists. The five pathfinding articles cited a total of fifty-one articles (not counting single-authored self-citations) involving 126 authors, of which nineteen were located at foreign universities and foreign research laboratories.

The types of scientists who were cited once again corroborate our earlier findings. Of the 107 American scientists cited in these five "pathfinding" papers, *all but one* were located at one of the top nine physics departments in the United States or at such distinguished laboratories as Brookhaven or the Lawrence Radiation Laboratory. All fifty-one articles were produced at one of these top nine departments or laboratories. The average number of citations to the cited authors is equally impressive. This group had a mean number of citations of sixty-nine in the 1965 SCI. This figure includes coauthors of the primary author, that is, a number of younger and not yet widely recognized scientists. If we take only the most cited authors among the coauthors, the mean number of citations to each scientist is 134; 74 percent of these authors had more than sixty citations to their work in 1965.

Since this type of subjective sampling procedure may be methodologically suspect, we decided to perform one final test of the Ortega hypothesis. We replicated the essential aspects of the study design using a set of independent data. We had a complete list of all papers that were cited three or more times in *Physical Review* in 1965.[20] This total exceeded three thousand scientific articles and substantive letters. Of course, some papers were cited often; most received three or four citations. Since we were concerned primarily with the pattern of citation in influential papers, we initially examined the ten papers that were most often cited in *Physical Review*. After identifying these "super" papers, we listed the scientific articles that the authors of these influential papers cited. Finally, we counted the number of citations received in 1965 by authors of papers cited in the "super" articles.

This procedure can be clarified by reference to a specific case. Murray Gell-Mann, recipient of the Nobel Prize in 1969, produced the most heavily cited article on the list. It received a total of forty-nine citations in *Physical Review* in 1965. We took Gell-Mann's paper and listed the citations in it. A total of thirty-three publications, or fifty-five scientists, were cited in the paper. We then noted the number of citations that the life's work of each of these fifty-five scientists had received in the 1965 SCI. This process was followed for all of the scientists cited in the ten "super" papers. Thus, while we were examining only these ten most highly cited papers, we generated a total of 299 authors cited.

The results obtained from this replication not only corroborate our earlier findings but strongly reinforce them. The data in table 3.7 suggest that authors cited in these ten papers were scientists who, on the average, had produced truly influential scientific work. The 299 cited scientists produced research that received an average of 154 citations in 1965, a total more than twice that of the average Nobel Prize winner. Since this figure includes citations received by beginning scientists yet to make their mark, who are collaborating with their more eminent colleagues, the statistic is actually lower than it would be otherwise. In fact, if we take only single authored papers and the most cited author in each collaborative paper and compute the average number of citations to each author's life's work, the mean is increased to 238 citations.

The data in table 3.7 further indicate that if we eliminate the self-citations to the author's own work the average number of citations to cited work remains extremely high. Clearly these data lend added weight to the counterhypothesis that work that is used by the producers of outstanding research is, by and large, itself of the highest rank. The work of the average researcher is rarely the work that is influential in the production of high-impact scientific research.

92 VALUES AND REWARDS IN SCIENCE

TABLE 3.7 Statistical Means for the Quality of Research Produced by
Scientists Cited in Those Ten Scientific Papers That Were Most Heavily Cited
in the 1965 Volume of *Physical Review*

Average number of citations to 299 authors cited in the ten papers	154
Average number of citations considering only major authors[a]	238
Average number of citations not including self-citations to the author's own work	135
Average number of citations to major authors only and excluding self-citations	175

[a] Major author was defined simply as a single author and the author in a collaborative set that received the highest number of citations to his life's work in the 1965 SCI.

A question remains to be tested. What is the quality of research that is cited in the work of the physicists whose papers receive fewer citations than those of the ten "super" papers? Using the same *Physical Review* list we drew a small random sample of papers that had received three to twenty-three citations in 1965. Papers that received twenty-three citations had approximately the same impact as some of the top ten, since the range of citations to the super papers was from twenty-four to forty-nine citations. This small sample consisted of thirty-six papers. Within these papers references were made to 492 communications. We computed the number of citations in 1965 to the 837 physicists who authored these 492 papers. We examined citation rates to cited authors who produced single-authored papers and those in collaborative teams whose work was most often cited.

The data are presented in table 3.8. Two clear results are shown. First, work which is above average quality (i.e., received more than five citations) is

TABLE 3.8 Citation Patterns Found in Papers Cited in
the 1965 *Physical Review*

The Number of Citations to the Paper in the 1965 *Physical Review*	Mean Number of Citations to Life's Work of Major Authors[a] Cited in the Papers
Top ten papers (24–49 citations)	238
	(174)
20–23 citations	229
	(88)
10–17 citations[b]	214
	(215)
5–9 citations	201
	(124)
3–4 citations	115
	(65)

[a] Major author was defined simply as a single author and the author in a collaborative set that received the highest number of citations to his life's work in the 1965 SCI.
[b] There were no papers cited 18 or 19 times in *Physical Review* in 1965.

influenced by other important work and not by minor discoveries. While the top ten papers made use of work produced by physicists who received an average of 238 citations to their life's work, the average quality of cited work found in papers receiving five to nine citations is not appreciably lower. Only when we examine the citation patterns in papers which received three or four citations is the average quality of work cited significantly lower. But even in this group, the average of 115 citations is far in excess of the average number of citations to the work of Nobelists and National Academy members.

These high averages do not conceal the existence of many less influential authors. Forty-one percent of the 837 cited physicists received more than a hundred citations; another 13 percent received sixty to ninety-nine. Thus, a total of 54 percent received more than sixty citations, a figure which is similar to those presented in table 3.3. Only 11 percent of the cited authors received less than five citations to their life's work and 90 percent of the scientists comprising this 11 percent were co-authors on papers for which one of the other authors was cited more heavily. In short, there were virtually no cited authors whose work had received under the average of 5.5 citations to authors in the SCI. These data further reinforce the idea that even the producers of research of limited impact make use of the high-impact research produced by a relatively small number of research physicists.

Conclusions and Interpretation

What, then, are the general conclusions that may be drawn from the findings? Although they deal only with physics, which is perhaps the most highly institutionalized scientific discipline, and therefore must be seen as only suggestive, the data allow us to question the view stated by Ortega and others that large numbers of "pedestrian" scientists contribute substantially to the advance of science through their research. It seems, rather, that a relatively small number of physicists produce work which becomes the base for future discoveries in physics. We have found that physicists throughout the social system of science, regardless of their location, use to a disproportionate degree the work of the eminent scientists. Even physicists removed from centers of research make considerably greater use of the work produced by scientists at the leading departments and by those others who have produced high-quality work than the work produced by their colleagues at lesser universities and colleges. While the data indicate that the research of most physicists is used sparingly, it also suggests that to a limited extent the work of scientists who publish less distinguished research is used by even the most eminent physicists, and eminent physicists are just as likely as non-eminent colleagues to use this less distinguished research.

Although the conclusions of this paper may be reasonably clear, the implications of these data for the structure of scientific activity, at least in physics, need careful consideration. Consider only one problem emerging out of the findings which needs a great deal of further research: the problem of the size of modern science. If future research in other fields of science corroborates our results, what does this say about the number of scientists that are needed to maintain the current rate of advance in science? It is possible that a reduction in the size of Big Science is necessary? Recent behavior on the part of many scientists indicates that some institutional changes may be in order. The proliferation of scientific literature is making it increasingly difficult to locate important work. Many scientists now are communicating their research findings through informal channels. No better example of the real nature of communication in science among today's elite scientists exists than in the testimony of James D. Watson in *The Double Helix*. Most of the new discoveries or new leads that he discusses were communicated privately among members of a small group of researchers.

Moreover, the data indicate that most research is rarely ever used by the bulk of the physics community and even more sparingly used by the most eminent scientists who produce the most significant discoveries. That most articles published in even the leading journals receive few citations was shown in a recent study of citations to articles published in *Physical Review*, the world's leading physics journal (Cole, 1968). It was found that 80 percent of all the articles published in there in 1963 were cited four or fewer times; 47 percent were cited once or never in the 1966 SCI.[21] Clearly, most of the published work in even such an outstanding journal makes little impact on the development of science. Thus, the basic question emerges: Could the same rate of advance of physics be maintained if the number of active research physicists were to be reduced sharply? In a preliminary way, the data in this paper suggest this, in fact, might be possible if some means were found to identify at an early age those scientists who will produce truly significant discoveries.

Finally, what are the implications of these data for the value system of science? Of all social institutions, science turns out to be one of the most highly stratified. There is a substantial gap between the few men at the top and the bulk of scientists. Despite this sharp stratification, science persistently has maintained a value system laced with egalitarianism and a strong emphasis on a sense of community. When the values or the ideology of an institution seem to diverge from reality, we are forced to ask what functions the ideology may be serving. The egalitarian ideology should not be confused with the distribution of rewards on the basis of universalistic standards. Universalism clearly predominates over particularism in contemporary physics. In fact, the egalitarian ideology may be a partial result of the operation of a system which approximates a meritocracy. For how do scientists who have been judged as pedestrian by the system, and who accept the criteria of judgment as legitimate, maintain their commitment to science?

As Parsons repeatedly has pointed out, all social systems must meet both instrumental and expressive functional requirements if they are to be maintained. We suggest that the egalitarian ideology of science serves an expressive pattern maintenance function. The primary instrumental goal of science is the production of significant new discoveries which advance the field. And as we have seen in physics, relatively few physicists actually make significant contributions. Even the top scientists have long periods in which their research does not bear fruit. This indicates that the "business" of science is a highly uncertain one, in which scientists, for the most part, proceed without a high degree of confidence that they will succeed in making important discoveries.

In such a precarious enterprise it may have many positive expressive consequences for maintaining scientists' commitments to the instrumental goals to create and support an egalitarian ideology which holds that the work of all the members of the social system ultimately counts in the historical development of science. Moreover, this ideology may be one functional prerequisite for the production of important science. For, if scientists during the long dry spells before and between new discoveries were to believe that their work, although it presently bore no fruit, did not count in the development of science, they might well choose to drop out of the social system quickly. Further research is obviously needed on the functional consequences of the egalitarian ideology in science.

Notes

1. A recent exposition of this position can be found in Crowther (1968), where Crowther quotes an address to members of the Royal Society by Lord Florey, president of the Royal Society:

> Science is rarely advanced by what is known in current jargon as a "breakthrough," rather does our increasing knowledge depend on the activity of thousands of our colleagues throughout the world who add small points to what will eventually become a splendid picture much in the same way the Pointillistes built up their extremely beautiful canvasses. What I have mentioned here are a few such points which it has taken three hundred years to place in the picture.
>
> For another among many expositions of this position, see Crane Brinton (1960). Interestingly, this position is often implicit in the work of many observers of science (Cf. Linton, 1936). Most often writers simply refer to "scientists" and do not specify what types of scientists are under discussion. We do not intend to convey the idea that the position of Ortega is necessarily the predominant one among observers of science. Certainly, there are those who concern themselves only with the great geniuses in the history of science and who, we can say, profess the "great man theory" of the development of science.

2. Many of the central ideas in this paper are foreshadowed by the perceptive analysis of Price (1963).

96 VALUES AND REWARDS IN SCIENCE

3. Various studies by Crane (1965), Margolis (1967), Merton (1968), Mullins (1969), J. Cole (1969), S. Cole (1968), and Zuckerman (1967a), among others, recently have examined various aspects of social stratification in the scientific community. Watson's *The Double Helix* (1968) provides an insider's look at communication and influence among a leading stratum of scientists. In short, the image of science as a classless society is beginning to fade.

4. For detailed discussion of the relationship between citation counts and the quality of a scientist's work, see Cole and Cole (1967, 1970).

5. The prestige rank of the physics departments was obtained from Cartter's (1966) study of university departments. The number of papers published was based upon *Science Abstracts* and the number of honorific awards on listings in the tenth and eleventh editions of *American Men of Science*. Our original sample consisted of 120 scientists. Some of these never were cited in the SCI; a few had produced only review papers, which were excluded from our sample. Subtracting these nonproducers from our 120 left us with 84 citers for whom sufficient information was available for inclusion in our sample. Since we were interested primarily in the influences upon the pure research of these scientists, and were using citations to measure that influence, we decided for obvious reasons to omit from consideration any paper that was a "review article." In looking up the cited authors within these individual papers, we were faced with the problem of what to do with collaborating authors. We decided to treat a paper as a single unit and locate information on all the collaborators in the research team. To some extent, junior collaborators who often were students of the senior authors dropped out of our sample because no information could be found for them in *American Men of Science*. But it can be seen that the average level of eminence of cited authors is weighted against our hypothesis to some extent since our sample included junior men of distinctly less eminence than their senior collaborators.

6. This sample was chosen for several research problems. Although the sample heavily overrepresents eminent physicists, this is not at all inappropriate for this particular paper. We are interested in examining the influence upon the best work that is produced in contemporary science. Most often this is work which is produced by scientists who have received a fair measure of recognition for their discoveries. Our data does not include articles which were produced by physicists working in industry or government. The sample of eighty-four is restricted to physicists holding academic positions or those associated with superior laboratories.

7. One of the limiting features of this study has to do with the collection of information on the scientists who are being studied. Perhaps the best source for independent information short of a questionnaire sent to a sample of physicists is the *American Men of Science*. The sample only includes men who were listed in *AMS*. Only about one-half of all cited authors appeared in these volumes. We wanted to see whether there were systematic differences in the types of scientists who appear in the *AMS* and those who do not. We found the following biases in *AMS*: foreign scientists, advanced students, and scientists working in industry were underrepresented. The average number of citations to men in *AMS* was approximately 1.5 times that of those not found in it; cited authors in the best work were more often found in *AMS* than those cited in lesser-quality work. Since the work that is not found in these volumes often is cited by men whose work is of poorer quality, the results in one respect lean against our hypothesis.

8. The citations statistics are based upon the work of twenty-four of the twenty-eight living laureates who won the prize in physics. The four living laureates who won the prize more than five years before 1961 were excluded so as not to introduce an age bias.

Included in this computation are the non-American laureates; when they are excluded, the average number of citations to work of American laureates is sixty-eight. The comparison here is between rates of citations to laureates in 1961 and to our sample in 1965. In order to accurately compare the two groups, we would have to multiply the average number of citations to the laureates by one-quarter since, on the average, there is an increase in number of citations of this amount purely due to an increase in the number of journals which have been included in the *Science Citation Index*.

9. These data were obtained from a questionnaire sent to 2,036 physicists located at the eighty-six leading departments producing PhDs. Only 1,308 usable ones were returned.

10. This does not, of course, mean that a scientist whose work is cited only two or three times is producing useless work. The work might have stimulated an important idea. However, on a probabilistic basis, there is not nearly as great a likelihood that his work will have substantial impact.

11. Price (1963) in citing "Lotka's Law" on productivity estimates that the number of people producing "n" papers is proportional to $1/n^2$.

12. These eighty-six departments represent those for which Allan Cartter (1966) obtained quality ratings. They, in turn, represent those departments which granted at least one PhD in each of the years between 1952 and 1962. We took the list of sampled departments and looked up their members as of 1965. For this listing, published by the American Institute of Physics, see *Bulletin of the American Physical Society, Membership Directory* as of June 1, 1965. It turned out that there were approximately two thousand members of these eighty-six PhD-producing departments. This membership number does not include people holding research positions at these departments who are not listed in the catalogs of these schools. Therefore, we may be slightly underestimating the total number of working scientists at those universities.

13. Consider two possible criticisms of the alternative hypothesis to Ortega's. It has been argued that many scientists in the lesser institutions who produce few, if any, scientific papers may not contribute substantially to the development of science through their publications but that their influence is felt through their role as teachers—that is, in terms of their influence on young and potentially brilliant scientists. Here we are dealing only with the research function of physicists. Nonetheless, recent data presented by Zuckerman (1967b) sheds light on the origins of outstanding scientists. Very few of the Nobel laureates had their graduate training in the hinterland. Out of eighty-two American laureates, fifty-eight (or 82 percent) received their doctorates from only sixteen universities and half of them were trained at only four: Harvard, Columbia, Berkeley, and Princeton. Furthermore, of fifty-five laureates interviewed by Zuckerman, thirty-four worked in some capacity, as young men, with a total of forty-six Nobel Prize winners (cf. Zuckerman, 1967b). Putting aside the small special group of laureates we still find that over 40 percent of all physics PhDs are granted by the top ten universities. The relationship between the current level of citation to scientists (or their "quality") and the prestige rank of their PhD department leads to the same general impression. Fifty percent of the scientists who received their PhDs from lesser universities have fewer than ten citations to their life's work in the 1965 *Index*. In contrast, 24 percent of PhDs from distinguished departments have a comparably low rate of citation.

Another problem lies in what might be called the "filtration theory" of utilization of scientific discoveries. It could be argued that although scientists who produce work of the first rank do not directly use the work produced by men at low-prestige universities who make minor contributions, they do use work that does depend upon the less significant research produced by a large body of scientists. Thus, the work produced by the average scientist would filter up to the eminent scientists in a step-wise process involving a number of communications. Since our data deal with a one-step process, we do not take into account this filtration possibility. However, some of our data does implicitly deal with this problem. By examining the "medium" and "lesser" quality research produced by physicists and in particular the work that the less distinguished scientists use, we can see the extent to which this proposition has a valid basis. For this theory says that scientists who produce high-quality work cite a certain proportion of scientists who produce lesser work and, in turn, these medium-quality research pieces depend heavily on pedestrian work produced by a large number of scientists. For an interesting discussion of this filtration idea, see Sarton (1931).

14. This paper relies almost exclusively on tabular analysis. Correlation coefficients with means and standard deviations also could have been used to present the data in this paper.

15. Included in the list of distinguished laboratories are Brookhaven, Lawrence Radiation Lab, Argonne National Labs, Oak Ridge, Institute for Advanced Study at Princeton, Bell Labs, and the Naval Research Lab.

16. There is a tendency to make the high correlation ($r = .73$) between the quality of a piece of research and the total "quality" of a scientist in a one-to-one relationship. This, of course, is avoided where possible. We are actually interested in the use of outstanding research, regardless of its source.

17. For a discussion of various forms of knowledge involving fundamentally different modes of experience, see William James (1970).

18. Two other elaborations of the relationship between the quality of research produced by citers and the quality of work produced by the cited authors were tested. We found that the quality of research of the citers was a stronger determinant of whom he cites than the specialty of the citer. The scientific specialty of the citer did not in any way affect citation patterns independently of the quality of work produced by the citing author. Second, the age of the citing author did have a small but significant effect on the patterns of citations independent of the quality of the work produced by the citer. Younger authors more than their older colleagues used a higher proportion of work produced by scientists whose life work had received at least sixty citations.

19. These five articles included, for example, Lee and Yang's now famous paper on parity conservation. Three of the authors turned out to be Nobel Prize winners; the others, members of the National Academy of Sciences.

20. We would like to thank Dr. Cullin Inman of the American Institute of Physics for supplying us with these data.

21. Stephen Cole provided me with these data on the citation distributions to articles published in *Physical Review*. The only way in which these data differ from those supplied by the AIP is that these data include papers that were cited less than three times, whereas the AIP provided us with data which ranged from three to forty-nine citations.

References

Bayer, Alan E., and John Folger.
1966 "Some Correlates of a Citation Measure of Productivity." *Sociology of Education* 39: 381–90.
Ben-David, Joseph.
1960 "Scientific Productivity and Academic Organization in Nineteenth Century Medicine." *American Sociological Review* 25: 828–43.

Brinton, Crane, et al.

1960 *A History of Civilization*, vol. 1. Englewood, NJ: Prentice-Hall, 654ff.

Cartter, Alan.

1966 *An Assessment of Quality in Graduate Education*. Washington, DC: American Council of Education.

Clark, Kenneth E.

1957 *America's Psychologists: A Survey of a Growing Profession*. Washington, DC: American Psychological Association.

Cole, Jonathan.

1969 *The Social Structure of Science*. Unpublished doctoral dissertation, Columbia University.

Cole, Jonathan R. and Stephen Cole.

1971 "Measuring the Quality of Sociological Research." *American Sociologist* (February).

Cole, Stephen.

1968 "The Reception of Scientific Discoveries: The Operation of the Matthew Effect in Science." Paper presented at the annual meeting of the American Sociological Association (August).

Cole, Stephen, and Jonathan R. Cole.

1967 "Scientific Output and Recognition: A Study in the Operation of the Reward System in Science." *American Sociological Review* 32: 377–90.

1968 "Visibility and the Structural Bases of Awareness of Scientific Research." *American Sociological Review* 33: 398–413.

Crane, Diana.

1965 "Scientists at Major and Minor Universities: A Study of Productivity and Recognition." *American Sociological Review* 30: 699–714.

Crowther, J. G.

1968 *Science and Modern Society*. New York: Schocken, 363.

James, William.

1970 *The Meaning of Truth*. Ann Arbor: University of Michigan Press.

Kuhn, Thomas.

1962 *The Structure of Scientific Revolutions*. Chicago: University of Chicago Press.

Linton, Ralph.

1936 *The Study of Man*. New York: Appleton-Century-Crofts: 300–43.

Margolis, J.

1967 "Citation Indexing and Evaluation of Scientific Papers." *Science* 155: 1213–19.

Merton, Robert K.

1968 "The Matthew Effect in Science." *Science* 159: 56–63.

Mullins, Nicholas C.

1968 "The Distribution of Social and Cultural Properties in Informal Communications Networks among Biological Scientists." *American Sociological Review* 33: 786–97.

Ortega y Gasset, José.

1932 *The Revolt of the Masses*. New York: Norton.

Price, Derek.

1963 *Little Science, Big Science*. New York: Columbia University Press.

Price, Derek, and Donald Beaver.

1966 "Collaboration in an Invisible College." *American Psychologist* 21, no. 11 (November): 1011–18.

Sarton, George.

1931 *History of Science and the New Humanism*. New York: Holt, 34–42.

Sher, Irving H., and Eugene Garfield.

1965 "New Tools for Improving and Evaluating the Effectiveness of Research." Presented at the Second Conference on Research Program Effectiveness, Washington, DC.

Sorokin, P. A.

1962 *Social and Cultural Dynamics*, vol. 2. New York: Bedminster.

Sorokin, P. A., and Robert K. Merton.

1935 "The Course of Arabian Intellectual Development, 700–1300 A.D.: A Study in Method." *Isis* 22: 516–24.

Watson, James D.

1968 *The Double Helix*. New York: Atheneum.

Whitehead, Alfred N.

1925 *Science in the Modern World*. London: Macmillan, 75.

Zuckerman, Harriet A.

1967a "Nobel Laureates in Science: Patterns of Productivity, Collaboration, and Authorship." *American Sociological Review* 32: 391–403.

1967b "The Sociology of the Nobel Prizes." *Scientific American* 217: 25–33.

CHAPTER 4

MEASURING THE QUALITY OF SOCIOLOGICAL RESEARCH

Problems in the Use of the *Science Citation Index* (1971)

JONATHAN R. COLE AND STEPHEN COLE

The problem of assessing the "quality" of scientific publications has long been a major impediment to progress in the sociology of science. Most researchers have typically paid homage to the belief that quantity of output is not the equivalent of quality and have then gone ahead and used publication counts anyway (Coler, 1963; Crane, 1965; Price, 1963; Wilson, 1964).[1] There seemed to be no practicable way to measure the quality of large numbers of papers or the life's work of large numbers of scientists. The invention of the *Science Citation Index* (SCI) in 1964 provided a new and reliable tool to measure the significance of individual scientists' contributions. Starting in 1961, the SCI has listed all bibliographic references appearing in an increasingly large number of journals.[2] The number of citations an individual receives may be tabulated and used as an indicator of the relative scientific significance or "quality" of that individual's publications.

Until now the SCI has not included sociology journals in its files, but director Eugene Garfield informs us that in 1970 the SCI file will include major sociology journals. References made in these journals will also be added to the 1961 and the 1964–1969 files. Thus, we will be able to count the number of references

---

American Sociologist 6, no. 1 (February 1971): 23–29. This study was supported by grant number NSF-GS-2736 from the National Science Foundation to the Program in the Sociology of Science, Columbia University, Robert K. Merton, Director. We thank Professor Merton for his helpful suggestions and for his criticism of an earlier draft of this paper, identified as publication number A-613 of the Bureau of Applied Social Research, Columbia University.

made in recent years to any particular article, book, or sociologist, and we will be able to quickly generate lists of these facts.[3] This should lead to major advances in the sociology of sociology. In practically all studies of the social organization of sociology the quality of work of individual sociologists and of particular institutions is an important variable. With the impending addition of sociological journals to the SCI file, this might be a propitious time to examine certain problems involved in the use of the SCI to measure the quality of work. We have been using this tool in our research in the sociology of science (basically physics) for several years, and we present below a discussion of several problems we have encountered and the solutions we have found for them.

There is some supporting evidence for the assumption that the number of citations a person receives is a roughly valid indicator of the significance or "quality" of his publications. In a thorough study of measures of scientific output, Kenneth E. Clark (1957: chapter 3) asked a panel of experts in psychology to list the psychologists who had made the most significant contributions in their field.[4] He then investigated the correlation between the number of choices received by psychologists and other indices of eminence. The measure most highly correlated with number of choices was the number of journal citations to the man's work ($r = .67$). Clark concluded that the citation count was the best available indicator of the "worth" of research work by psychologists.

Consider another kind of validating evidence for this measure. Recipients of the Nobel Prize are generally regarded as having contributed greatly to advances in physical and biological sciences. Since the number of those prizes is limited, however, there may be other like-sized aggregates of eminent scientists who have contributed as much. Nevertheless, the laureates as a group can be safely assumed to have made outstanding contributions. The average number of citations in the 1961 SCI to the work of Nobel laureates (who won the prize in physics between 1955 and 1965) was fifty-eight, compared to an average of 5.5 citations for other scientists. Only 1.08 percent of the quarter of a million scientists who appear in the 1961 SCI received fifty-eight or more citations (Sher and Garfield, 1965).[5]

We thought it possible that winning the prize might make a scientist more visible and lead to a greater number of post-prize citations than the quality of his work warranted. We therefore divided the laureates into two groups: those who won the prize five or fewer years before 1961 and those who won the prize after. The 1957–1961 laureates were cited an average of forty-two times in the 1961 SCI; the future prize winners (those winning the prize between 1961 and 1965), an average of sixty-two times. Since the prospective laureates were more often cited than the actual laureates, we concluded that the larger number of citations primarily reflects the high quality of work rather than the visibility gained by winning the prize.[6] Here, we have used receipt of the Nobel Prize as an independent measure of the quality of a scientist's work. In recent studies we have found measures of quality based upon citations to be highly correlated with other measures of eminence. For example, we found the quality of work, measured by SCI data, of 120 university physicists to be correlated ($r = .64$) with the number of awards they had received (Cole and Cole, 1967). These data offer further support for the use of number of citations as an indicator of the scientific significance of published work.

Citation counts enable us to distinguish the extent of contributions by various types of scientists. Consider, for example, the citations to one relatively productive and eminent group of scientists: members of university departments of physics in the United States. Among a representative sample of academic physicists totaling 1,308, only 2 percent had sixty or more references to their work in the 1961 SCI, 12 percent had between fifteen and fifty-nine citations, and 86 percent had fewer than fifteen citations. Thus, only a small fraction of university physicists received as many citations as did the average laureate, who received fifty-eight. In short, very few scientists are heavily cited, and there are distinct differences in the number of references to physicists whose quality of research has been validated by other measures of eminence.

Clearly, increasing corroborative evidence suggests that citations provide a good, if rough, indicator of the quality of research output in the natural sciences. With the inclusion of sociological journals in the SCI file we will be able to investigate a number of areas of interest to the sociologist of sociology. For example, we can test whether the correlations between citations and various independent measures of quality of research are comparable to those found in physics, psychology, and other disciplines. Whether stronger or weaker, the correlations should be interesting since they should give clues to the different structures of the fields.

We can also examine the relationship between quality and quantity of research output and determine whether sociologists are rewarded for the bulk or the quality of their publications. Next, we can estimate the extent to which the quality of a sociologist's research affects his visibility to the sociological community. Finally, we can investigate patterns of communication and identify intellectual linkages within sociology. Through an examination of citation patterns we may begin to answer questions about shifting foci of attention and cross-fertilization between subfields of our discipline. With the measure of quality of output obtainable from the SCI, a series of comparative studies will become researchable for the sociologist of sociology. We will be able to make meaningful empirical comparisons of the organization and intellectual structure of different disciplines.

Although citations provide us with an improved method of assessing the quality or impact of a scientist's research, there are, of course, a number of definite problems in the use of this measure. Most of these are basically substantive as opposed to technical, and we shall consider the substantive problems first. In doing so, we shall draw upon data from a number of studies in the sociology of science.

Errors in Evaluation

There may be occasional "errors" in the evaluation of scientific works. The significance of work done by a scientist is not always recognized immediately, for new ideas, especially those that lead to changes in basic scientific paradigms, are sometimes resisted or ignored (Barber, 1962). Some great scientific innovators remain obscure in their own time only to be accorded posthumous recognition. Mendel is a classic example of a scientist whose work was unappreciated by his contemporaries but greatly honored by scientists of later generations. Using citations to measure quality, we may sometimes misclassify work that is currently being "resisted" or that has been judged inadequately. Since history alone will reveal which work has been resisted or misjudged, this flaw in the procedure is inevitable. However, the problem of resistance to significant contributions may be less important in contemporary science than it was in the past. In a recent study of delayed recognition of scientific discoveries we examined citation patterns over time (Cole, 1970). For a sample of 177 papers published in *Physical Review* we found a strong correlation between the number of citations they received one year after publication and the number they received after three years ($r = .72$). When we examined citation patterns to papers published between 1950 and 1961 in different scientific fields we found a similarly high correlation between the number of citations received by the papers in the 1961 SCI and the number received in the 1966 SCI. Although an ideal study of delayed recognition of scientific discoveries would require citation data for a longer period, these data at least suggest that the communication and evaluation systems of modern science work effectively and that relatively few cases of research go unrecognized at the time of publication and then turn out later to be significant.

It is quite possible that our conclusions on the relative unimportance of errors in evaluation will not apply to sociology, for there is probably less consensus in sociology than in physics on the criteria to be used in evaluating work. Nevertheless, with the inclusion of sociology journals in the SCI file we shall be able to compare citation patterns over time in sociology with those in other fields. Are good papers more likely to be ignored in sociology than in the natural sciences?

Critical Citations

Citations may refer to papers that are being criticized and rejected rather than utilized. It is unlikely, however, that work of little value will be deemed significant enough to merit extensive criticism. If a paper presents an error that is important enough to elicit frequent criticism, the paper, though erroneous, is probably a significant contribution. The significance of a paper is not necessarily determined by its correctness. Much work done by the great historical figures of science was in some sense "wrong" or mistaken (Kuhn, 1962). It is unlikely that any work which is wrong without being a "fruitful error" will ever accumulate very many citations. Let us examine this problem more closely. Suppose we had a total of one thousand citations to scientific work and as many as one hundred of these were to work being criticized or rejected. The majority of these "critical" citations are dispersed among a large number of papers, such that most papers cited critically would receive no more than one or two citations. The same dispersion is found for "positive" citations. Let us say that one paper actually receives as many as twenty-five "critical" citations. We suggest that these few pieces of research that stimulate wide criticism have, in fact, stimulated other research. Consequently, it must be considered mistaken but significant; it must be seen as work which has had an impact on future scientific research. In sociology most of us would certainly agree that a paper such as Davis and Moore's "Principles of Stratification" (1945) was a significant contribution even if it has elicited many critical responses.

Treating All Citations as Equal Units

If each citation to a paper is given equal weight, errors in assessing the impact of research may follow. A paper that is cited widely by first-rank scientists should not be equated with a paper cited predominantly by scientists who have made only minor contributions. Since citations do not have equal meaning, should we consider classifying them by the characteristics of the citer? In effect, would a citation count weighted for the quality of research produced by the citer be a superior measure of the quality of scientific papers?

To answer this question, we did a detailed study of the citers of 171 of our 1,308 university physicists. For each physicist in the subsample, we collected data on a random sample of his citers. This enabled us to classify each physicist by the characteristics of his citers. By giving each citer a score that depended upon the number of citations his own works had received, we were able to measure the quality of work of each *subject* physicist by the quality of work of his citers.[7] The correlation between this index and the total number of citations received by the subject physicist was $r = .40$. We believe that this correlation is not higher because a disproportionate number of citations are made by a small group of scientists who publish heavily and are themselves generally highly cited. Consequently, the same men are likely to be found among the citers of high-quality work and low-quality work. Thus, 62 percent of the citers of relatively low-quality work and 70 percent of the citers of relatively high-quality work received ten or more citations to their own work (J. Cole, 1969). Although the two indices

106 VALUES AND REWARDS IN SCIENCE

TABLE 4.1 Correlation Coefficients for Two Indices of Quality of
Scientific Research and Several Characteristics of Physicists

Characteristics of Physicists	Number of Citations to Physicist	Index Based on the Number of Citations to the Citers of the Physicist
Age	.11	.10
Academic rank	.12	.10
Number of awards received	.29	.20
Prestige of highest award	.41	.20
Prestige of PhD department	.11	.13
Rank of present academic department	.19	.22

Note: Statistics based on a sample of 861 citers.

of quality—the total number of citations received and the index based upon the quality of the citers—are not highly correlated, they yield similar correlations with other variables (see table 4.1). On the basis of the correlation coefficients presented in table 4.1 we conclude that a citation index that includes the characteristics of the citing authors would probably yield substantive conclusions similar to those of an index that treats all citations as equal in value.

Quantity and Quality of Research Output

The number of citations a scientist receives may in part depend upon the quantity of his output. A scientist who publishes a large number of papers and receives only a few citations for each may accumulate as many citations as the scientist who publishes only a few papers that are heavily cited. In our sample of 120 university physicists, we found a correlation of .60 between the number of papers a physicist published and the number of citations listed after his name in SCI. However, the correlation between number of papers and number of citations to the three most frequently cited contributions of the scientists (a measure that could not be an artifact of the quantity of publications) is .72 (Cole and Cole, 1967). This is the opposite of what we would find if the total number of citations were primarily a function of the total number of papers. We conclude that the total number of citations could serve as an adequate indicator of quality.

In several of our studies we chose to use more refined measures of citations because they seemed substantively more suitable. For example, in a paper analyzing the relationship between the quantity and quality of the output of physicists, we used the number of citations to the three most frequently cited contributions by each physicist as a measure of quality. This was done in order to eliminate any possible effects of sheer productivity on total citations. Moreover, since a contribution in physics does not typically appear in a single paper but is usually presented in a series of papers, we used citations to the year's output rather than to single papers as the unit of measurement (Cole and Cole, 1967).[8]

A further word is necessary on the relationship between the quantity and quality of a scientist's research. Since in physics these two variables are highly correlated, we may conclude that where citation counts are not readily available (as in historical research) publication counts are roughly adequate indicators of the significance of a scientist's work. As noted above, with the inclusion of sociology journals in the SCI files we shall be able to discuss the extent to which the quantity and quality of research output are correlated in sociology. Where citation counts are available, they should be used, for they have definite advantages over counts of publications. In a recent paper we presented data to show that physicists who were frequently cited, yet had not published numerous papers, received as much institutional recognition as did physicists who were both prolific and widely cited. These "perfectionist" physicists turned out, in fact, to be the most highly recognized scientists in our sample (Cole and Cole, 1967). We hypothesize that quantity of output may be more heavily rewarded in sociology than in physics. When scientists cannot agree upon what high quality is, their concern is likely to be with quantity of output.

Size of Scientific Fields

Differences in the size of various scientific disciplines pose another potential problem in the use of citations as a measure of quality. Comparisons of the works of scientists in different fields must take into account the number of people actively working in those fields. With two writings of equal importance, the one in physics, the other in chemistry, the latter might be expected to receive more citations merely because the field of chemistry is larger than that of physics. The number of citations might be an artifact of the number of chemists and physicists, the number of journals in each field, and the amount of work that is being published. The same applies to specialties within a field. If there are more publications in solid-state physics than in elementary particles, we might expect that of two papers of equal impact in the two fields, the one in solid-state physics might receive the greater number of citations.

Though at first this position seems plausible, under closer scrutiny it does not appear logically sound. While there are more citations being produced in solid-state physics than in elementary particles, there is also more work being done in solid-state physics and, consequently, more literature that is potentially citable. Therefore, the likelihood that a work will receive more citations simply because its field is larger does not logically hold. Furthermore, if size of field were related to number of citations, we would find a positive correlation between these two variables. This would be similarly true for specialties within a given field. Men working in larger specialties would receive more citations than men working in smaller specialties. We have evidence that, at least in physics, there is no relationship between the size of a specialty and the number of citations to the work of men in that specialty. According to the National Science Foundation (*American Science Manpower*, 1966: 183), in 1966 there were 4,593 solid-state physicists and 1,833 elementary-particle physicists. Our data on 1,308 university physicists show no significant differences in the rate of citation to men active in these two specialties. In the 1961, 1964, and 1965 editions of SCI, solid-state physicists had a mean of seventeen citations while elementary-particle physicists had a mean of nineteen citations.

In exceptional cases, of course, there may be a relationship between the size of the field and the number of citations to men working in that field. For example, in a specialty with only a few scientists working in it the number of citations to their works would necessarily be limited by the small total number of citations to work in that specialty.

Contemporaneity of Science

Papers in physics now have a half-life of no more than five years; that is, at least half the citations that appear in a given year are to works published in the five preceding years. The half-life of sociological papers is only slightly longer. We must take this into account in comparing the publications of scientists who made their most important contributions at different times. Two papers which were originally of equal quality may have a different number of citations in the 1961 SCI if one paper was published in 1941 and the other in 1959. This would not matter if the researcher were interested in the *current* significance of both papers. However, he might want a measure of quality that is not time-bound. To handle this problem we developed a technique of weighting citations (Cole and Cole, 1967) wherein older citations were given greater weight than recent citations. For example, since 70 percent of the citations in a particular field (physics) are to works published within the preceding five years and 4 percent are to works published more than twenty years before, we gave a weight of 17 (70 percent divided by 4 percent) to works published twenty or more years before the date of citation. Although weighting citations for their age seems to be substantively necessary, we found a high correlation ($r = .80$) between the total of weighted and unweighted citations. When we compare the number of weighted and unweighted citations to physicists' publications in their three "best" years, we get an even higher correlation (r = .96). These high correlation coefficients once again illustrate the interchangeability of indices (Lazarsfeld, 1958). Since the number of a physicist's weighted citations is correlated so highly with the number of his unweighted citations, substantive conclusions would probably not be affected if the weighting technique were not used.

If weighting is used, one more problem must be considered. Derek Price (1963: 81) has suggested that "although half the literature cited will in general be less than a decade old, it is clear that, roughly speaking, any paper once it is published will have a constant chance of being used at all subsequent dates." In the study of delayed recognition we found that it was in fact empirically correct that papers that were cited in 1961 had on average roughly the same number of citations in 1966 (Cole 1970). Thus, at least for short periods, it is likely that Price was correct. The weighting technique discussed above is not at odds with the model suggested by Price—it is not meant to predict the *number* of citations received by papers in the past but to control for the increasing number of citations. Due to the exponential growth in science, papers presenting important scientific contributions today are receiving many more citations than similar papers did in the past. A paper that receives five citations today is not among the most heavily cited; but a paper published in the nineteenth century that received five citations would probably have been among the most heavily cited of papers. Thus, in comparing works published in different periods we must standardize for the total number of citations made.

Integration of Basic Ideas

Widespread basic ideas are often utilized in papers without explicit citation to their well-known source. Who today cites that paper in the *Annalen der Physik* as the source of $E = MC^2$? For scientists who have achieved eponymy, there may be a decline in the number of formal citations to their work. The "Mössbauer effect" is an example of a recent contribution to science that has been thoroughly integrated into the common body of knowledge and is infrequently given formal citation. Let us examine such cases in terms of our measure. A scientist who makes discoveries that lead to eponymous recognition probably will also produce other research of the first rank that will be heavily cited by the scientific community. Thus, most Mössbauers will be classified as "high quality" despite the fact that one of their outstanding achievements receives relatively few formal citations.[9] It is true, however, that integration of a discovery into the body of scientific knowledge may lead to errors in assessing the quality of that discovery through citations. The use of citations as a measure of quality does involve a degree of error. However, evidence presented here leads to the conclusion that such error, which may be substantial in individual cases, will not be significant in considering the publications of any fair-sized sample of authors.

The problems in using citation counts as measures of the quality of scientific output discussed above were primarily substantive. We now turn to a consideration of two problems that are primarily technical in nature but nonetheless important to those who might use citation indexes.

Citations to Collaborative Papers

Citations to all single-authored papers are recorded in the SCI. Citations to collaborative works are listed only after the name of the first-named author. Since many collaborative papers list authors alphabetically, one might think that collaborators whose names begin with letters late in the alphabet would be misclassified if we counted only citations appearing after their names in the SCI. Data from our research suggest that omission from the SCI of citations to collaborative work does not present a formidable problem. For the sample of 120 physicists we have a full range of citation data for each author: citations to works he produced alone, citations to collaborative works on which his name appeared first, and citations to jointly authored writings where he was not the first-named author. The latter information was obtained by looking up the author's collaborative papers in *Science Abstracts* and then looking up those papers on which he was not first author in the SCI. The correlation is .96 between a straight citation count and total citations (including citations to collaborative work on which the physicist was not first-named author).

We also arrayed the 120 physicists in two ranked lists, one according to a straight citation count and the other according to total citations; the Spearman rank-order correlation coefficient is .85. Although the outcome was to some extent predetermined by the size of the zero-order correlations, we decided to make a final test of the relationship between straight counts and counts that included citations to collaborative work on which the physicist was not the first named author. We divided our sample of 120 physicists into two groups: the first group included physicists whose last names began with A to M; the second group was composed of those whose names started with N to Z. For each group we calculated the percentage of total citations both for single-authored papers and for first-author collaborative papers. By this technique we could estimate the extent to which scientists whose names begin with letters late in the alphabet were "deprived" of their due in terms of citations to work they had actually helped produce. The data indicate little difference between the groups. Sixty-seven percent of the citations in the A to M group and 71 percent of the citations in the N to Z group were to single-authored or first-author collaborative papers. The small difference in the direction opposite from that expected suggests that the omission of citations to collaborative papers on which the author was not listed first does not affect substantive conclusions. For the most part, differences that we did find were among scientists whose work was of the first rank. For example, Murray Gell-Mann had almost six hundred citations to his publications in one volume of the index. When we looked for citations to his collaborative research where he was not first-named author, we found over one hundred additional citations. While these add substantially to Gell-Mann's total of six hundred, they do not affect our classification of the quality of his work. However, when we want to study the quality of specific papers, we must look up collaborative papers under the name of the first author. Also, the researcher must be aware that because of the procedure adopted by the SCI, he may make errors in measuring the quality of work of a particular scientist.

Clerical Problems

Warren Hagstrom (1968) has recently pointed out other technical problems in the use of citation counts. First, he notes that there are clerical errors in the list of citations. Second, the works of two authors may appear under one name. If, for instance, there were two E. McMillans, one the Nobel physicist and the other a relatively unknown sociologist, the citations of both men would appear together under the same name. Although these two problems make for inefficiency in collecting citation data, both can be handled. Clerical errors probably occur randomly throughout the index. Therefore, while the counts may be off slightly, there is no reason to believe that there are patterned errors in the listings. The second problem is more vexing but can be handled by careful compilation of the citation data. The index lists along with the cited author the names of his citers and the title, volume, and page of two journals: the journal that published the article and the one that cited it. Consequently, it is possible to identify the articles produced by the scientists that one is interested in. It is relatively easy to distinguish between scientists working in different fields; it takes more effort to distinguish between two physicists who happen to have exactly the same name.[10] Thus, neither problem materially detracts from the value of the index as a measure of the quality of scientific output.

Conclusion

The data available indicate that straight citation counts are highly correlated with virtually every refined measure of quality. Correlations between straight counts and weighted counts and between straight counts and those that take

112 VALUES AND REWARDS IN SCIENCE

into account citations to collaborative research in which the author is not the first-named author are all greater than .80. Consequently, it is possible to use straight counts of citations with reasonable confidence. In some research situations it may be substantively more appropriate to use weighted counts or to take into account collaborative work, but the use of these refinements is not of methodological necessity.

It is clear that there are problems in using citation counts as indicators of the quality of scientific output. Nevertheless, the value of using them as rough indicators of the quality of a scientist's work should not be overlooked. To interpret small differences in the number of citations as meaningful, however, would be unwise. It would not be accurate, for example, to say that scientists who received six or seven citations to their publications in the 1961 SCI did better work than those who received four or five citations. In other words, citations should not be used as a fine measure of quality. Nevertheless, large differences in the number of citations received by scientists do adequately reflect differences in the quality of their work.

Notes

1. Researchers have had difficulty in estimating the significance of even a small number of papers. Although a panel of judges is often used, problems of standardization of evaluation criteria and individual biases of evaluators are frequently encountered.

2. Compiled under the direction of Eugene Garfield, the SCI in 1961 listed all citations made in 613 journals; the 1962–63 journals have not been indexed. In 1964, 700 journals were covered, and in 1965, 1,147 journals. Virtually all important journals in the natural sciences are included.

3. Even though the SCI files include only references made in journals, they include citations *to* all books or other publications cited in the journals. It is doubtful that the citation patterns in journal articles will differ substantially from the citation patterns in books. We would guess that there would be a high correlation between the number of an author's journal publications and the number of his books.

4. For another study using number of citations as a measure of quality of scientific work, see Bayer and Folger (1966).

5. We thank Dr. Sher for making available some of his unpublished data. By way of emphasizing the difference in the numbers of citations received by laureates and average scientists, we would also point out that many scientists do not appear in the SCI and that the modal number of citations to the work of men who do appear is one.

6. These statistics are based upon the works of twenty-four of the twenty-eight living laureates—those who won the prize in physics as of 1965. The four living laureates who won the prize before 1957 were excluded so as not to introduce an age bias. These computations include non-American laureates; when they are excluded, the average number of citations to the works of United States Nobelists is sixty-eight.

7. Scores were assigned in the following way. Citers whose works had received one hundred or more citations in the 1965 SCI were given a score of 5, citers with fifty to ninety-nine citations were given a score of 4, those with twenty-five to forty-nine citations

received a score of 3, those with ten to twenty-four citations received a score of 2, those with one to nine citations received a score of 1, and those whose works were not cited received a score of 0. These scores were then totaled for all citers, and each subject physicist was classified by the total index score.

8. The year is also an arbitrary unit since physicists do not, of course, arrange their related papers to fit the calendar year. For a more exacting procedure, it would be necessary to identify the series of papers representing an integrated contribution, a requirement extremely difficult to meet in dealing with large numbers of working scientists. Without such detailed information, it would seem preferable to use a period of time as a unit rather than single papers. See also the recent study of scientific productivity by Crane (1965), which treats a series of four papers on the same topic as a "major" publication and single papers as "minor" publications.

9. Although the Mössbauer effect is not as heavily cited as some other major discoveries because of the degree to which it has been integrated into the general fund of knowledge, Mössbauer's other publications received a total of fifty-three citations in the 1965 SCI. Thus, in terms of our measure, his work is rated about equal to the average laureate and National Academy member.

10. This is done by looking up the articles in the journals and using the listed institutional affiliation of the authors to separate the articles.

References

American Science Manpower. 1966. Washington, DC: National Science Foundation, report number NSF 68–7.

Barber, B. 1962. "Resistance by Scientists to Scientific Discovery." In *Sociology of Science*, ed. B. Barber and W. Hirsch, 537–556. New York: Free Press.

Bayer, A. E., and J. Folger. 1966. "Some Correlates of a Citation Measure of Productivity in Science." *Sociology of Education* 39 (Fall): 381–90.

Clark, Kenneth E. 1957. *America's Psychologists: A Survey of a Growing Profession*. Washington, DC: American Psychological Association.

Cole, Jonathan. 1969. "The Social Structure of Science." Unpublished PhD dissertation, Columbia University.

Cole, S. 1970. "Professional Standing and the Reception of Scientific Papers." *American Journal of Sociology* 76 (September): 286–306.

Cole, S., and J. R. Cole. 1967. "Scientific Output and Recognition: A Study in the Operation of the Reward System in Science." *American Sociological Review* 32 (June): 377–90.

——. 1968. "Visibility and the Structural Bases of Awareness of Scientific Research." *American Sociological Review* 33 (June): 397–413.

Coler, Myron B., ed. 1963. *Essays on Creativity in the Sciences*. New York: New York University Press.

Crane, D. 1965. "Scientists at Major and Minor Universities: A Study of Productivity and Recognition." *American Sociological Review* 30 (October): 699–714.

Davis, K., and W. E. Moore. 1945. "Some Principles of Stratification." *American Sociological Review* 10 (April): 242–49.

Hagstrom, W. O. 1968. "Departmental Prestige and Scientific Productivity." Paper read at the annual meeting of the American Sociological Association, Boston.

Kuhn, Thomas S. 1962. *The Structure of Scientific Revolutions*. Chicago: University of Chicago Press.

Lazarsfeld, P. 1958. "Evidence and Inference in Social Research." *Daedalus* 87 (Fall): 91–130.

Price, Derek. 1963. *Little Science, Big Science*. New York: Columbia University Press.

Science Citation Index. (Multiple years). Philadelphia: Institute for Scientific Information.

Sher, I. H., and E. Garfield. 1965. "New Tools for Improving and Evaluating the Effectiveness of Research." Paper presented at the Second Conference on Research Program Effectiveness, Washington, DC.

Wilson, Logan. 1964. *The Academic Man: A Study in the Sociology of a Profession*. New York: Octagon.

CHAPTER 5

THE EMERGENCE OF
A SCIENTIFIC SPECIALTY

The Self-Exemplifying Case of the Sociology of Science (1975)

JONATHAN R. COLE AND HARRIET ZUCKERMAN

Over the course of their careers, all working scientists have the opportunity to observe the growth of new fields of inquiry and the demise of old ones. Yet the emergence of new scientific specialties as cognitive and social entities seems to be a fact of the modern scientific life that is little understood. Physical and biological scientists have understandably been impatient to get on with their own work and few have paused to examine the emergence of one or another special field.[1] Sociologists of knowledge have also not shown much interest in questions about the growth of scientific knowledge. They have occupied themselves primarily with inquiries into the social and existential bases of knowledge. It was not until a few sociologists began to study science as a social institution that more serious inquiry into the growth and differentiation of specialties began. In short, the emergence of scientific specialties became interesting only when a new scientific specialty came into being.

The sociology of science is curiously self-exemplifying. As a scientific specialty, it exhibits many of the social patterns its own practitioners study in other contexts, making it a convenient site for sociological study of emerging

---

Research for this study was supported by a grant from the National Science Foundation (GS 33359X1) to the Columbia Program in the Sociology of Science and by the Center for Advanced Study in the Behavioral Sciences, where the second author was a Fellow in 1973–74. Bernard Barber, Stephen Cole, Yehuda Elkana, Joshua Lederberg, and Arnold Thackray were kind enough to read and comment on early drafts of this paper.

specialties. It is evolving its own system of stratification, its own arrangements for formal and informal communication, its own politics, and its own lines of cognitive and social conflict just as these have become major foci of attention in research by sociologists of science. As participants and observers of these developments, we are in a strategic position to examine how the growth of knowledge in a special field is related to the emergence of its organizational infrastructure.

Tracing the emergence of the sociology of science also provides us with an occasion for examining Robert Merton's contributions to these developments. Since he has shown an almost obsessive fascination with scientific paternity—assiduously cataloguing the fathers of numerous sciences in his analysis of priority and eponymy in science and more recently focusing on George Sarton's efforts to institutionalize the study of the history of science[2]—it seems fitting that he himself now become a subject of inquiry as a father of the sociology of science.[3]

The Sociology of Scientific Specialties

After a long and desultory incubation, the sociology of science now seems to have acquired its own cognitive and professional identity. In the last decade or so, sociological investigations of science have focused primarily on its social organization, in particular, on its reward-and-evaluation systems, its system of communication, and its ethos. Mapping these organizational features of science seemed to many investigators a necessary and congenial set of first steps toward understanding how scientific knowledge grows and becomes codified and institutionalized. By contrast, comparatively little attention has been given to studying the interplay of science and other social institutions.[4]

The research attention of sociologists of science may now be shifting. In the last few years, there has been increasing discussion of the connections between cognitive structures of the sciences and their social structures and some efforts to study them empirically.[5] Sociologists of science are now turning to such problems as the extent to which there is consensus among scientists in different disciplines on theory, method, and substance, and whether there are systematic differences in social organization between sciences which can be linked to their differential degrees of theoretical codification. These studies derive in part from Thomas Kuhn's work on scientific revolutions,[6] Derek Price's studies of the parameters of scientific growth,[7] and from the work of philosophers of science such as Karl Popper and Imré Lakatos.[8] It is in this intellectual context that recent studies of growth and institutionalization of scientific specialties can be located.

Kuhn's model of revolutionary change in the sciences has been especially influential in studies of scientific specialties. He focuses of course on the birth of new theoretical perspectives, or paradigms, on the recruitment of adherents to the new viewpoint, and on the cognitive conflict attendant on the revolution. By working through the implications of the new paradigm and gaining intellectual and social dominance, new recruits ultimately revolutionize the discipline. Even though Kuhn's analysis of paradigms and their role in scientific change has been widely criticized by historians and philosophers of science, sociologists of science have continued to find it useful.[9] As a consequence, sociological studies of the emergence of specialties in the Kuhnian mode have focused especially on cases representing major breaks with disciplinary tradition.[10]

Working from another perspective, Price re-introduced the seventeenth century term "invisible college"—originally used to describe the pioneer members of the group which later became the Royal Society of London—to characterize the informal network of investigators he takes to be the core of any specialty.[11] He links this emphasis on patterns of communication to his other work on rates of growth in scientific manpower and the scientific literature. The impact of both Kuhn and Price is easily discerned in sociological studies of the emergence of specialties.[12] But there are still no agreed-upon problematics for studying the institutionalization of specialties.

Most sociologists of science have assumed that patterns of growth and institutionalization are much the same in different specialties. This assumption ignores the variegated processes that characterize the birth of new specialties. The first step in understanding institutionalization of specialties is more precise description of their cognitive and social development with more systematic attention being given to *variability* in specialty differentiation and its sources. Griffith and Mullins have hinted that the emergence of "elite" specialty groups may differ in important ways from what they call "revolutionary" groups, but the implications of this distinction are not explored.[13]

The cognitive orientation of new specialty groups should be a strong determinant of its rate of institutionalization and its successful establishment. "Cognitively radical" specialty groups which reject the legitimacy of established theoretical and methodological orientations should encounter more intellectual resistance and more difficulty in obtaining resources and recruits and engender more conflict in the process of their development than "cognitively conforming" specialties. The latter base their claims to specialty status on inquiry into new and previously unexamined phenomena or on the use of new research technologies and thus do not challenge prevailing views. The distinction between cognitively radical and cognitively conforming specialties seems central to understanding how scientific specialties emerge. It suggests why the sociology of science, for example—as a specialty with a new subject matter—has encountered relatively little resistance albeit no great enthusiasm from sociological colleagues. We shall take up this matter again when we consider Merton's role in the development of the specialty.

The cognitive standing of a new specialty is only one element affecting its reception. Its goodness of fit with the prevailing structure of the academy should also affect its chances for survival and, if it survives, its pace of institutionalization. Those that grow up in the interstices between disciplines, such as biochemistry, astrophysics, or social psychology, are thwarted by the lack of ready-made academic niches; handicapped by the poor meshing of their intellectual interests with those of deans, journal editors, grant givers, reviewers, and other gatekeepers of resources.[14] These specialties are, in short, structurally atypical regardless of their cognitive content and should encounter different and more challenging functional problems than those, such as nuclear physics or the sociology of science, which have been firmly located within the bounds of established disciplines from their beginnings.

There is, then, some reason to suppose that the development of scientific specialties is highly variable. A paradigm in the early, Mertonian sense is needed that provides for systematic and comparative examination of the cognitive development of specialties and their institutionalization. Such a program would focus attention on at least three sets of problems and processes.

1. *Parameters of growth in personnel and in production of a literature, as they change over time.* Do they grow simultaneously? Do they exhibit patterns of lead and lag? To what extent are they independent of one another or causally connected? How are they linked to the cognitive state of specialties?

2. *Processes of cognitive development of specialties.* To what extent is there consensus or conflict on problematics, methods of inquiry, and on principal contributors at various stages in specialty development? Are ongoing developments built upon a common base or diverse intellectual foundations? How rapidly are new contributions exploited and built upon? How are theory and empirical research linked, if at all, and how do these linkages change over time? When does theoretical codification begin, if ever, and what are its indicators? How do foci of attention in the specialty shift, and are these shifts related to changes in the intellectual orientations of leading authors or influentials? What are the distinctive contributions of the founders to the cognitive development of the specialty? How long do their contributions continue to be used and in what ways?

3. *The development of organizational infrastructures.* How rapidly is the specialty incorporated into the educational curriculum, and how much resistance does it encounter? In what respects do the social and historical contexts of institutionalization affect its pace and success? How rapidly and from what sources are funds and facilities acquired and with what consequences for cognitive development? What provisions are made for routinizing communication

between specialists? In what ways, if at all, is the system of communication among specialists connected to the *general scientific* communication system? How effective have its intellectual founders been as institution builders? Are the tasks of organizing the specialty assumed by its intellectual leaders or is there a division of labor in these activities? And finally, when do specialists develop a sense of professional identity such that they consider themselves, and are considered by others, to be working at a common task?

Having repeatedly referred to cognitive elements in science, we note that the concept of cognitive structure has not yet been defined. In fact, its precise meaning for sociologists of science is still evolving. At present, sociologists of science, cognitive psychologists, and philosophers of science focus on different aspects of cognitive structures. Sociologists regard these structures as multidimensional; they include

1. scientific knowledge as it is reported in theoretical and experimental investigations;
2. the standards by which scientists judge methods, instruments, techniques, and evidence to be acceptable;
3. theoretical orientations which provide criteria for assessing the significance of new problems, new data, and proposed solutions;[15]
4. commonly accepted problematics for further inquiry; and
5. responses to new contributions, particularly the extent and forms of consensus and dissensus.

The cognitive structures of the separate sciences and specialties differ from one another and vary over time.

Analysis of cognitive structures of the sciences from a sociological perspective attempts, for example, to identify those structures' basic theoretical orientations and to determine whether one predominates or whether several compete for scientists' attention. It attempts to determine whether fields or specialties are intellectually fragmented or cohesive; the extent to which theories are interconnected; how intellectual work is organized in terms of substantive problems and theoretical schools; and how theory and experiments are related.

Development and elaboration of the cognitive structure of new specialties appear to depend in part on correlative development of their social structures—on the routinization of an evaluation and reward system, procedures of communication, acquisition of resources, and the socialization of new recruits. In short, the tandem development of both cognitive and social structures of specialties seems central to their institutionalization and establishment as legitimate areas of inquiry. Since institutionalization is more usefully considered as a

120 VALUES AND REWARDS IN SCIENCE

process than as a product, studies of the decline of specialties should be on the agenda for sociologists of science along with studies of their development.[16]

We turn now to the sociology of science as a case study in the emergence of a specialty. Laying claim to the institution of science as a legitimate subject for sociological inquiry, the specialty is cognitively conforming and located firmly within the established disciplinary structure of sociology although it has increasingly elaborate connections to other fields, principally the philosophy and history of science. These cognitive and structural attributes suggest that its growth and institutionalization should be comparatively comforting, uncontested and more rapid than specialties which are cognitively or structurally radical. Had we studied the emergence of plate tectonics or numerical taxonomy, two cognitively radical specialties, quite different patterns of growth might well have been found. Still, as we shall see, the sociological study of science despite its benign character was not enthusiastically embraced by sociologists when it was first introduced.

Parameters of Growth

Although Merton believed that sociological analysis of science was a promising line of inquiry in the mid-1930s, few shared his enthusiasm. He recruited his student and friend Bernard Barber to work on social aspects of science, but their efforts to convert others mostly failed. By 1949, just 1 percent of the members of what was then the American Sociological Society counted the sociology of knowledge among their three fields of competence; no separate tabulation was even made for the sociology of science.[17] A decade later, sociologists had not changed their minds. The *Directory* for 1959 shows that just 1 percent of the membership reported competence in the sociology of knowledge. In 1970, things were still much the same: 1.4 percent of the American Sociological Association's membership who answered the questionnaire on areas of competence—about two-thirds of the total of 13,000—selected the category newly rechristened as the "sociology of knowledge and science" as one of two principal areas of competence.[18] But by 1973, the figure had risen to 2.2 percent, or 301 out of a total of 13,700.

These figures need to be seen in context. First, the most frequently mentioned specialty in 1970 was selected by 9 percent of all sociologists and only four of the thirty-three specialties were selected by more than 5 percent. Second, between 1970 and 1973, the overall membership of the association increased by 5.4 percent, but the numbers reporting competence in the sociology of knowledge rose by a half, suggesting that interest in the composite field grew considerably even though the small numbers involved produced rather high rates of change.

More intriguing than data on increasing numbers are differences between age groups in declarations of competence in the sociology of knowledge and science. Stehr and Larsen find that the specialty is mentioned most often as an area of competence by the youngest members of the American Sociological Association. For sociologists in their 20s, it ranked 17th out of the 33 mentioned, 22nd for those in their 30s, 26th for those in their 40s and 28th for the oldest cohorts who were in their 50s and 60s.[19] These figures imply a growing interest in the field among the younger members and no marked pattern of conversion among the older.

Information on the changing subject matter of doctoral dissertations is consistent with this interpretation. Drawing upon listings in *Dissertation Abstracts* and the new *Comprehensive Dissertation Index, 1861—1972*, we took inventory of dissertations on science and technology written by students in sociology departments in American universities.[20] In all, we identified 105 titles, 65 in the sociology of science, 28 in the sociology of sociology and 12 in the sociology of engineers, engineering, and technology. The two earliest, both submitted in 1929, were sociological studies of sociology.[21] Seven years went by before S. C. Gilfillan, then a student at Columbia, finished his study of the sociology of invention for his sponsor, W. F. Ogburn. The same year, Merton submitted his study of sociological aspects of scientific development in seventeenth-century England to his doctoral committee at Harvard. That committee included the dean of historians of science, George Sarton, the distinguished physiologist and Pareto scholar L. J. Henderson, and the sociologists P. A. Sorokin and Talcott Parsons.

Despite these auspicious beginnings, few graduate students tried their hands at dissertations on the social aspects of science for some time. Just twenty-five dissertations were completed in the thirty years between 1937 and 1967. But things changed markedly after that. In the five years that followed, forty dissertations were turned in, suggesting a sudden increase in interest in the specialty. This number is about 1.4 times as large as would be expected if dissertations in the sociology of science were multiplying at the same rate as they did in sociology as a whole.[22] Students writing dissertations in the sociology of science have been clustered at a few departments: Columbia, Cornell, Chicago, Purdue, and Michigan State, in that order, account for almost half of all dissertations in this specialty. The same universities produced only 28 percent of all PhD's in sociology in roughly the same period.[23] It would seem that a critical mass of students may be developing at a small number of research centers.

Such data only hint at the social processes of specialty growth. Consider the following questions: Do all specialties require an equal number of students to get moving? How rapidly must their numbers grow? What is the ratio between masters and apprentices required for sustained growth? Do new specialties require a higher density of talented newcomers than established ones?

What constitutes a "critical mass" of students and practitioners needed for the development of a cognitive and professional identity? How important for the recruitment of students and the forging of a group identity are the personal characteristics of leaders in the field? How much proselytizing is needed? And what form does missionary work take before the specialty gains legitimacy among potential recruits, young and old?

While a growing number who claim an interest and competence in a specialty would seem to signal its developing professional identity, a growing literature indicates an authentic commitment on the part of practitioners to work in a field. Price and others who work on parameters of scientific growth report almost without exception that the literature of scientific disciplines and scientific specialties first grows slowly and erratically and then exponentially, with a doubling time of ten to fifteen years.[24] In order to examine the growth and texture chiefly of the American literature in the sociology of science, we compiled a bibliography and citation index for publications appearing in nine scholarly journals since 1950.[25] In all, 195 papers published over twenty-four years were identified.[26] In its early years, the specialty had no clear intellectual identity, shared problematics, or techniques of investigation. Early papers were often vague and speculative. It is not surprising that it was far more difficult to decide whether papers published in the 1950s belonged in the bibliography than those published later on.

As in the literature of disciplines and specialties previously examined, the number of papers in the sociology of science published since 1950 has grown exponentially with a doubling time of five to eight years.[27] Taking all 195 publications into account, a third or so (37 percent) were published between 1958 and 1965 and almost half (48 percent) since 1966.[28] In all then, 85 percent of all papers we identified appeared after 1957, the year of Merton's influential paper on priorities in scientific discovery.[29] We also find a marked increase in the relative number of papers reporting quantitative studies in the sociology of science, with 38 percent of the papers published in the 1950s being quantitative in one or another respect and then 52 percent of publications in the 1960s and 56 percent of those in the 1970s. There is little doubt that these data represent an authentic change in the character of research in the sociology of science in recent years, but they clearly do not tell the whole story. Content analysis of books and monographs is needed before firm conclusions can be drawn about changes in the extent of quantification and its relation to other aspects of cognitive development.

Although a growing literature may indicate increased scholarly effort, it is not necessarily evidence for a shared intellectual focus among those at work in the specialty. There are, however, other reasons to think that such a focus was emerging: among them a growing consensus on the usefulness of particular publications, a consolidating research front in which new papers built directly upon those just published, and increased rates of collaborative publication. We want now to consider these and other indications of a developing cognitive structure in the specialty.

Selected Aspects of Cognitive Development

Cognitive Consensus

A growing consensus among specialists on the usefulness of certain publications is a prime indicator that a specialty is developing distinctive problematics and thus a cognitive identity. The extent of convergence of citations to particular papers and to the work of particular authors is a rough measure of such consensus. If there were no common orientation in the specialty, citations would be widely dispersed among cited authors. The emergence of a common orientation, however, should be reflected in increasingly large proportions of citations going to the work of a small group of influential authors whose work is judged useful.[30] Converging citations do not mean that all agree on the significance of cited research or that all highly cited authors have a common orientation but only that the cited work is influential in some respect.

Three measures of convergence of citations or consensus are employed here: the proportion of cited authors receiving two or more citations, the proportion of citations going to the top 10 percent of cited authors, and the overall concentration of citations as measured by Gini-coefficients. The first is a rough gauge of the extent of dispersion of citations among cited authors while the latter two are more sensitive measures of the same variable. Drawing upon references in the 195 papers, these three measures were computed for five successive time periods and the results are presented in tables 5.1A through 5.1C.

Consider first the findings for the sociology of science reported in columns 2 and 3 of table 5.1A. (Comparisons with the sociology of deviance will be made presently.) Only 18 percent of all authors cited in the early 1950s were cited more than once. By the early 1960s, that proportion had increased slightly to 26 percent. A substantial increase in focusing can, however, be observed by the late 1960s. Nearly half of all cited authors were cited two or more times, suggesting growing agreement about the usefulness of work by particular authors. The data reported in columns 2 and 3 of table 5.1B for the sociology of science show an increasing reliance on the work of a few authors. In the early 1950s, there was little consensus in the sociology of science about whose research was useful. The top 10 percent of all cited authors received approximately one-quarter of all citations. Twenty years later, the top 10 percent received 44 percent of all citations. The Gini-concentration ratios reported in column 2 of table 5.1C tell a similar story. In the early years, there is considerable dispersion in the

TABLE 5.1A Comparison of Levels of Cognitive Consensus from 1950–1973 in Sociology of Science and Sociology of Deviance

| | Percent of Cited Authors Receiving Two or More Citations | | | |
| | Sociology of Science | | Sociology of Deviance | |
Period	Percent	Total Authors Cited	Percent	Total Authors Cited
1950–1954	18	(148)	44	(242)
1955–1959	24	(323)	50	(538)
1960–1964	26	(571)	53	(884)
1965–1969	44	(775)	52	(1,347)
1970–1973	45	(899)	45	(723)*

*The literature in the sociology of deviance covers the period 1950–1972 inclusive.

TABLE 5.1B Percent of Citations Received by Top 10 Percent of Cited Authors

| | Specialty | | | |
| | Sociology of Science | | Sociology of Deviance | |
Period	Percent	Total Citations	Percent	Total Citations
1950–1954	24	(198)	36	(1,178)
1955–1959	28	(444)	38	(2,753)
1960–1964	35	(931)	49	(5,560)
1965–1969	43	(1,698)	46	(7,469)
1970–1973	44	(2,145)	36	(3,514)*

*The literature in the sociology of deviance covers the period 1950–1972 inclusive.

TABLE 5.1C Gini-Coefficients for Distribution of Citations in Five Time Periods*

Period	Sociology of Science	Sociology of Deviance
1950–1954	.22	.23
1955–1959	.23	.29
1960–1964	.33	.40
1965–1969	.42	.48
1970–1973	.47	.30

*We are indebted to Stephen Cole for providing us with these data for the sociology of deviance. Extended discussion of these and other measures of intellectual structures can be found in his paper, "The Growth of Scientific Knowledge: Theories of Deviance as a Case Study," in *The Idea of Social Structure: Papers in Honor of Robert K. Merton*, ed. Lewis A. Coser, chap. 9 (New York: Harcourt, 1975). See also O. D. Duncan, "The Measurement of Population Distribution," *Population Studies* 11 (July 1957): 27, for discussion of the Gini coefficient.

distribution of citations. Later on, in the 1960s and 1970s, the data show increasing concentration in the distribution of citations, suggesting the emergence of a cadre of recognized intellectual leaders and convergence in judgments of usefulness among those publishing in the journals examined.[31]

As we noted earlier, the sociology of science derives its claim to specialty status from its focus on phenomena not previously studied by sociologists. In the early 1950s, there was no relevant literature to cite, to respond to critically, to correct, or to elaborate. Each author who published in those early years brought his own highly individualized apperceptive mass to bear on his research. Only as the intellectual identity of the field became fixed did a pattern of increasing consensus on influential authors appear. Once a specialty becomes a recognizable entity, however, the extent of consensus on influential authors need not continue to grow. It may even decline, if rival theoretical or methodological orientations develop. Such is apparently the case in the recent history of the sociology of deviance. (See Stephen Cole's essay in this volume, pages 175–220.)

As tables 5.1A and 5.1C suggest, the larger literature on deviance contains more citations. More authors are cited even though the data are drawn from just four journals, as compared with the nine sampled for the sociology of science. Since "deviance" was the fifth most "popular" on the list of thirty-three areas of sociological competence selected by members of the ASA in 1970 and the sociology of science and knowledge, twenty-fifth, the differences in the size of their journal literatures are not surprising. The significant fact is that this body of literature also contains differences of dispersion of citations over time. The sociology of deviance shows a curvilinear pattern while a linear increase is observed for the sociology of science. Table 5.1A shows that nearly half (44 percent) of all authors cited in the deviance literature between 1950 and 1954 received multiple citations. Ten years later that proportion increased to a high of 53 percent. In the next ten years the rate of multiple citation began to taper off, reaching 45 percent for the current period.

Consider now the more sensitive indicators of consensus. The sociology of deviance shows some convergence in judgments about its most influential authors in the early 1950s: the top 10 percent of authors cited in the journals received 36 percent of all citations in this period. Cognitive consensus in the deviance literature by this measure increased and peaked in the years 1960–1964, when the top 10 percent received almost half of all citations, the same level as has been observed in physics.[32] The Gini-coefficients indicate the same pattern. These years were the high point in the functional analytic study of deviant behavior, as Stephen Cole's essay in this volume reports (pp. 188–205). Toward the end of the 1960s, however, the functional orientation came under attack by ethnomethodologists and symbolic interactionists. The development of these rival perspectives appears to be reflected in the citation data.

126 **VALUES AND REWARDS IN SCIENCE**

Similar developments are not yet discernible in the referencing behavior of sociologists of science. Although in recent years some young English and European sociologists of science have proposed what they consider to be an alternative to the perspective exemplified in Merton's work, their papers are still largely programmatic and critical. They continue to cite Merton and those pursuing similar inquiries and to use their research.[33] Thus the global citation measures of consensus presented here are not an early warning system for cognitive conflict but only provide cues to the emergence of alternative and competing orientations after such orientations are embodied in new self-contained research literature.

The Consolidation of a Research Front

We have observed increasing convergence in citations in the sociology of science and have suggested that this is a signal of developing intellectual coherence and consensus. Such a convergence may indicate a consolidation of new work or mere reiteration of older inquiries. In order to determine whether current work is built on comparatively new contributions rather than on older ones—that is, whether a research front has developed and when—we need to examine the age of cited papers. Derek Price has observed that active research fields rely heavily on recent publications. Citations in active research areas are relatively younger than would be expected on the basis of sheer growth in the literature. This he calls "immediacy."[34] It has also been suggested that rates of citation to recently published work are correlated with the extent of theoretical codification in a science such that the more codified sciences exhibit higher proportions of citations to newly published work. This should be so because codification of theoretical and empirical knowledge makes it possible to identify the connections between new work and old; to gauge the significance of new contributions; and to facilitate their rapid incorporation into ongoing work.[35] Thus, if a specialty were becoming increasingly codified, the age of publications being cited in its literature should decline and "immediacy" measures should rise.

This is precisely what we observe in the literature of the sociology of science. In the first half of the 1960s, when the specialty appeared to "take off" and move toward a high growth rate and institutionalization, cited papers were an average of twelve years old and 48 percent of citations went to papers published in the preceding five years. Between 1965 and 1969, the average age of cited papers dropped to nine years and 56 percent of the references were to papers published no more than five years before. These data on citation of recent publications can be better understood when juxtaposed with figures on production of journal literature in the sociology of science. Although 48 percent of citations in papers published between 1960 and 1964 went to literature published in the same five years, over half the literature (57 percent) then in print appeared during this period. This makes for an immediacy score of 9 percent, a figure that would surely have been larger if literature published before 1950 had been surveyed. Thus, in the years when the sociology of science began to grow rapidly, recent research was not used as quickly as it was produced. In the following five years, things changed markedly. Thirty-six percent of the total literature was published, but 56 percent of citations went to recent work, making for an immediacy score of +20 percent, a figure comparable to rapidly growing specialties in the physical sciences and much higher than that for sociology as a whole.[36]

Although this trend toward citing new publications may be related to the influx of young people into the field and their characteristic interest in new work, it is not confined to their publications. References in Merton's own work reflect this increasing reliance on new research. An avid student of the history of science and obsessed with documenting the filiation of ideas, Merton punctuated his early papers with references to works several centuries old. His recent papers, however, reveal that his perspective has shifted along with those of other sociologists of science. References in his own most-cited papers in the sociology of science, the "Priorities" paper published in 1957 and "The Matthew Effect" published in 1968, exemplify this trend, with the median age of references in the "Priorities" paper being eighteen years and in "The Matthew Effect," eight years.[37] Not surprisingly, references in the same papers also reveal Merton's increased reliance on very recent work: 29 percent of the references in the "Priorities" paper were to publications no more than five years old as against half of the references in "The Matthew Effect." This does not, we think, reflect a conviction that pertinent older references are already well established in the bibliography of the specialty and thus no longer require citation even though they are used. Rather, it is testimony to a genuine shift of attention to newer publications.

The Age Structure of Influentials

Turning from the age of citations to the ages of authors being cited, we see one of the important consequences of the pattern of recruitment among the young that we noted earlier. Not only is the sociology of science more popular among young sociologists, but their intellectual influence on the field has also increased dramatically. Although there are reasons for supposing that the age of contributors of important work to a science declines as that science becomes more codified, there is no reason to think that youthful investigators are primarily responsible for increasing the extent of codification; rather it enables them to deal with important problems in the field.[38] Keeping this observation in mind, we note the decline in average age of the thirty most-cited authors in successive five-year periods. It drops from fifty-three for the years 1950–1964 to forty-six for the years 1965–1972. The principal shift occurs between two five-year periods: 1960–1964, when the mean age of influentials was fifty-four, and 1965–1969, when it fell to forty-four.[39] These data also reveal a slight increase in the age of influentials in the 1970s, a finding easily understood when we note that the young people who first became influential in the late 1960s continue to be so in the 1970s. In fact, the lists of the thirty most influential authors in the late 1960s and early 1970s overlap by 60 percent. Since each author on both lists was growing older, their average ages increased. New entrants among the influentials in the 1970s were not young enough to make up for this trend.

Collaborative Publication

As specialties become organized around a set of problems, the extent of collaboration between specialists increases. This is one outcome of greater numbers simultaneously at work in research centers, increased requirements for specialized skills, and greater agreement on the nature of researchable problems. There is a marked increase in collaborative publication over the twenty-four years covered by our bibliography. In the first eight years, not one of the 29 papers abstracted had more than one author, but in the second, 14 percent, or 10 out of 72, were multiauthored; in the third, 31 percent, or 29 out of 94, were collaborative. Altogether, only 3 percent of the 195 papers had three authors and none more than that. The figures for the last period are slightly lower than that for the journal literature in sociology as a whole.[40]

The Structure of Influence in the Sociology of Science

Thus far, three points emerge from citation analysis of the journal literature of the sociology of science: (1) consensus on the work of particular contributors has been growing; (2) younger authors are increasingly represented among contributors cited most frequently; and (3) recent publications are more often cited now than in earlier years. Nonetheless, Merton has had greater influence on the evolving cognitive identity of the field from its beginnings than any other author. His impact on his colleagues' research is signaled by the extent of citation to his work. With a total of 154 citations in the journal literature surveyed here, his work is used roughly twice as often as that of the second-most-cited author in our list.[41] But apart from Merton and several others, the social and intellectual composition of the most influential group of authors has changed greatly. These changes are cues to marked shifts in foci of attention in the field and in types of work being published. (Given the high correlation in science between intellectual influence and authority, we suspect that they are also cues of changes in the intellectual composition of those occupying

The Emergence of a Scientific Specialty 129

gatekeeping positions.) We increase the resolution of our citation analysis to the micro level by turning now to the question of continuity and change in the influence structure in the sociology of science.

Table 5.2 presents rank-ordered lists of the authors most often cited in the journal literature of the sociology of science in five time periods.[42] We have

TABLE 5.2 Most Cited Authors in the Sociology of Science, 1950–1973 (Self-Citations Excluded)

Period and Rank Order				
1950–1954	1955–1959	1960–1964	1965–1969	1970–1973
Gilfillan, S. C.	Merton, R. K.	Merton, R. K.	Merton, R. K.	Merton, R. K.
Lundberg, G.	Lazarsfeld, P. F.	Crombie, A. C.	Price, Derek	Price, Derek
Dewey, J.	Gaudet, H.	Barber, B.	Garfield, E.	Hagstrom, W. O.
Hart, H.	Wilkening, E. A.	Gillispie, C. C.	Hagstrom, W. O.	Cole, J. R.
Parsons, T.	Wilson, L.	Lazarsfeld, P. F.	Zuckerman, H.	Ben-David, J.
Merton, R. K.	Stimson, D. L.	Kornhauser, W.	Gordon, G.	Cole, S.
Weber, M.	Compton, A. H.	Flexner, A.	Glaser, B.	Zuckerman, H.
Shils, E.	Kellner, A.	Goodrich, H. B.	Garvey, W. D.	Gaston, J. C.
Conant, J. B.	Robertson, T.	Kuhn, T. S.	Kessler, M. M.	Kuhn, T. S.
Leighton, A. H.	Parsons, T.	Caplow, T.	Cartter, A. M.	Crane, D.
Isard, W.	Richards, I. A.	Shepard, H. A.	Ben-David, J.	Barber, B.
Kautsky, K.	Sarton, G.	Shryock, R. H.	Barber, B.	Cartter, A. M.
Lerner, D.	Ryan, B.	Wilson, L.	Pelz, D.	Glaser, B.
Lasswell, H. D.	Kluckhohn, C.	Glaser, B.	Cole, S.	Ogburn, W. F.
Kuhn, T. S.	Gross, N. C.	Gilfillan, S.	Cole, J. R.	McGee, R.
Chase, S.	Berelson, B.	Holland, J. L.	Gamson, W.	Parsons, T.
Durkheim, E.	Shepard, H. A.	Marcson, S.	Kaplan, N.	Polanyi, M.
Corey, L.		McGee, R.	Storer, N. W.	Shils, E.
Goren, G.		Pelz, D.	Lazarsfeld, P. F.	Storer, N. W.
Ogburn, W. F.		Parsons, T.	Kuhn, T. S.	Gouldner, A. W.
Gold, H.		Knapp, R. H.	Berelson, B.	Gordon, G.
Gee, W.		Price, Derek		Caplow, T.
Myrdal, G.				Watson, J. D.
Usher, A. P.				Pelz, D. C.
Sibley, E.				Hirsch, W.
Whitney, V.				Hargens, L. L. Berelson, B.
Number of Citations:				
Range 8–2	Range 12–3	Range 32–5	Range 39–9	Range 67–9

focused on the approximately twenty most-cited authors in each period because patterns are difficult to discern in longer lists. Our more extended enumerations show many of the same attributes as the shorter ones and in no way contradict our general observations.

The list of most-cited authors in the first decade is striking in several respects. Major figures in the wider discipline of sociology—both historical and contemporary—seem to have dominated the literature. Since the sociology of science had not yet developed its own intellectual identity in the form of subject-specific ideas and techniques, specialists applied what they could from the prevailing theoretical and methodological corpus. Thus, Paul Lazarsfeld is among the most highly cited authors in the 1950s. Closer inspection of citations to his work reveals that it is his logic of multivariate analysis that is most often used, not his work in the history of quantification in sociology, which he had not even begun to publish. We can detect no clusters of researchers on the list who worked on related problems and no identifiable similarities in the problems they addressed—in short, no signs of a shared intellectual orientation.

Table 5.2 also shows that few authors frequently cited in the early 1950s are included among those often cited in the later 1950s. In order to convey the extent of continuity from one period to the next, we constructed the transition matrix—or partial "turnover table"—that appears as table 5.3. Two estimates of continuity are presented here. The first, located *above* the main diagonal, shows the proportion of authors in the most-cited decile in one period also in the most-cited decile in subsequent periods. Since the number comprising the most-cited decile grows as the number of cited authors grows, this measure of continuity is less demanding than the second measure, presented *below* the diagonal in table 5.3. This second set of data reports the proportion of the thirty

TABLE 5.3 Continuity Among Influential Contributors to the Sociology of Science (The Most Cited 10 Percent Are Presented Above the Diagonal; the Thirty Most-Cited Individuals, Below the Diagonal)

		Percent Carried Over Between Periods				
		Period				
		1950–1954*	1955–1959	1960–1964	1965–1969	1970–1973
Period	1950–1954	—	13	33	27	40
	1955–1959	8	—	19	22	19
	1960–1964	15	20	—	40	37
	1965–1969	12	17	27	—	47
	1970–1973	19	13	30	60	—

*The base figure for continuity for the period 1950–1954 was 26, since only 26 of the 148 cited authors had work cited two or more times.

most-cited authors for each period who also appear in subsequent periods. Over time, the top thirty make up smaller and smaller proportions of all cited authors and thus represent an increasingly elite group.

Confining ourselves to the extent of continuity of influentials in adjacent periods, we note a sharp increase in continuity in the most-cited decile in the late 1960s, as the data above the main diagonal show. The extent of continuity in the top decile approximately doubled from 19 to 40 percent in the 1960s and then increased slightly to 47 percent in the early 1970s. Data reported below the main diagonal show the same pattern even more sharply. Only two of the thirty authors (Parsons and Merton) whose work was most often used in the early 1950s also appear among those most often cited in the later 1950s. The extent of continuity increases somewhat in the 1960s, but it is not until the early 1970s that a major shift in continuity is discernible. As many as eighteen of the top thirty, or 60 percent, of the most-cited authors appear on both lists. Work in the sociology of science finally was focused on an identifiable set of problems, formally presented in the publications of a limited number of authors. Further, these same authors continued to work in the field and to produce research that was useful to their colleagues instead of moving on to other areas of substantive concern. This marked increase in continuity of influentials between the late 1960s and early 1970s is also reflected in the aging of influentials in these years reported earlier.

For those familiar with the literature of the sociology of science, a glance at the names of the most-cited authors listed in table 5.2 will immediately convey the extent of change in the intellectual interests of the influentials. The lists are dominated after 1965 by sociologists of science and a handful of historians and philosophers of science. Distinguished figures from the physical, biological, and other behavioral sciences disappear from the list of most-cited authors.[43] The sociologists of science whose work is frequently cited are, with notable exceptions, quantitative empirical researchers. Given the increasing number of quantitative studies published and the tendency for authors of such studies to cite prior work of the same kind, this finding is not surprising. It does suggest that citation analysis of journal articles may overestimate the impact of quantitative empirical research and underestimate the role of authors who do not fit this mold. T. S. Kuhn is of course the most conspicuous example of this group.

These lists contain cues to a second kind of continuity among influentials: not simply across adjacent time periods but through more extended periods of time. Although Merton alone turns up in each of the five lists covering twenty-three years of literature, Thomas Kuhn, Derek Price, Bernard Barber, and Donald Pelz all appear on the last three lists.[44] Thus, at least a small number of leaders have been working at the research front for some time. Mullins's contention that leaders of research specialties often turn to new problems before their specialty is fully institutionalized is not borne out in this case.[45] There are no signs that Merton, Price, or Kuhn, among the other leading influentials, are turning away from their interest in the sociology of science.

132 VALUES AND REWARDS IN SCIENCE

Thus, we return once again to the question we raised earlier in other contexts: How much variability is there in the emergence of new specialties? Under what conditions do research leaders lose interest in specialties they helped to establish? Are founders of "cognitively radical" specialties, and those which are structurally atypical, more apt to move on to entirely new problems than leaders of conforming specialties? Or does the exodus of founders depend on the extent to which the specialty remains fertile ground for studying fundamental issues?

"Insiders" in the sociology of science will also notice a third kind of continuity among the most-cited authors listed in the fourth and fifth columns in table 5.2. For the first time, a significant number of students of influentials appear along with those who trained them. Of the thirty-three different authors who were cited most between 1965 and 1973, seven are Merton's former students.[46] Insiders will also note that a four-generation chain of masters and apprentices appears here, a chain much like those observed in the physical and biological sciences.[47] Starting with Talcott Parsons, one of Robert Merton's teachers at Harvard, we can trace an intellectual lineage through Norman Kaplan, who studied with Merton at Columbia and went on to supervise Norman Storer's doctoral work at Cornell. Had we set our cutoff point in citations just one step lower, the 1970–1972 list would also have included Nicholas Mullins, who did his doctoral work with Storer at Harvard. These master-apprentice chains reflect the twin processes of self-selection and selective recruitment among different generations of influentials. By providing structural supports for continuing social relations between members of the same intellectual family, they make for a degree of cohesion between academic generations.

The next steps in citation analysis of specialty formation are clear. First, analysis of citations in books must be undertaken. Second, these lists of most-cited authors contain names of the most influential scholars in the specialty, but the intellectual linkages between them is not explored. Quantitative and qualitative analysis of such linkages between papers and influentials is needed. This will enable us to learn whether some authors are influential for all their colleagues, whether schools actually exist or are in formation, how the cognitive texture of the specialty has changed, and the extent to which the boundaries of groups identified are permeable. This analysis is in process.[48]

Robert K. Merton: Teacher, Founder and Influential

So much for the intellectual concerns and social composition of leading authors in the sociology of science. We turn now to Merton's impact on the specialty.

Unlike his own teacher, George Sarton, Merton has had some success in recruiting students to the discipline. In his concern to establish the history of science as a respectable scholarly enterprise, Sarton made demands on students so severe as to be self-defeating. Not many learned the classical and Asian languages whose mastery, along with five or six major modern languages, Sarton deemed necessary. And still fewer obtained the equivalent of advanced degrees in both the physical and the biological sciences he also considered necessary for historians of science. He also failed to develop a coherent formulation of principal problems in the field and a set of usable research techniques. Although Sarton developed a distinctive perspective on the history of science it was not one that could be readily adopted by potential recruits. It is not surprising then that few historians of science count themselves among Sarton's students.[49] Although Merton shares some of Sarton's perfectionism in his demands on students (a characteristic not all of them find endearing), his work has brought many students into the specialty. It lays out a series of problems in the sociology of science and provides an orientation to sociological work in general. This becomes evident when we look closely at the uses made of Merton's work by different intellectual constituencies.

Although Merton continued his studies in the sociology of science and published more than twenty papers in this area in the two decades after he completed his dissertation, these efforts were not immediately recognized by sociologists, historians of science, or, for that matter, anyone else. This was so in spite of the considerable attention paid his theoretical work and his studies in the sociology of organizations, professions, and mass communications.[50] It now seems obvious that Merton's early papers (and, we think, those of any founder) had limited initial impact because they had no audience specifically attuned to publications on this subject. Merton's "Priorities in Scientific Discovery," his presidential address to the American Sociological Society, was warmly received.[51] Sociologists clearly knew about this paper and may even have read it. Yet it had little immediate impact even in the literature of the sociology of science. Of all the citations it has received—which are more numerous than for any other paper in the literature—fewer than a third came in the eight years following its publication and only half in the first ten. This is most unusual considering the general pattern of intensive citation immediately following publication and a gradual tailing off afterward.[52] It also turns out that the average time elapsed—mean and median—between the publication of this paper and its use is ten years. Increasing citation of the "Priorities" paper in recent years no doubt reflects growth in the literature of the sociology of science. But over and above this artifactual element, the pattern of use observed here suggests how little influence ideas will have until a core of professional researchers with a common orientation are around to use them.

Although Merton's work overall has been increasingly influential in recent years, this is not uniformly true of all his papers. His publications appearing since 1959 seem to have had disproportionately great impact on the specialty.

134 VALUES AND REWARDS IN SCIENCE

Comprising fewer than half of the papers he has published on science, they have received 65 percent of all citations to his work. With the exception of his studies of the normative structure of science, his early papers on the compatibility of Puritanism and the scientific ethos, on the sociology of knowledge, and on science and totalitarian politics are far less often used by sociologists of science than his later inquiries into competition for priority,[53] multiple discovery,[54] the Matthew Effect,[55] the evaluation system in science,[56] and Insiders and Outsiders.[57] Thus the active interest taken in his later papers does not appear to have triggered renewed interest in his earlier ones.

The reasons for the differential influence of these earlier and later papers are more complex than it would first appear. As we noted, new publications are more apt to be cited in the growing literature than older ones. In no small part, this is so because older publications are "dated" or their content has already been incorporated into the cognitive structure of the field. Neither of these explanations appears to fit here. Instead, sociologists of science found in Merton's later work on priorities and the stratification in science and its reward system greater "potential for elaboration" and a reasonably clear program of research.[58] It focused attention on the operation of the evaluation and reward systems, their efficacy in extending scientific knowledge, and the role of recognition in scientists' motivations to continue their work. Close inspection of the papers by newcomers to the field who appear on the list of most-cited authors in the 1960s and 1970s shows that much of their empirical work begins with a problem posed in one or another of Merton's later papers.

Moreover the "potential for elaboration" of his recent work has been much enhanced by three developments beyond the confines of the sociology of science: establishment of the *Science Citation Index*, publication of the American Council of Education's appraisals of graduate education, and growing interest in empirical studies of social stratification.

Developments in quantitative analysis of social stratification fed concurrent studies of stratification in science and, it would appear, gave research in the esoteric specialty a more general appeal. At the same time, sociologists have made little progress in studying structures of norms and values or the relations between social institutions, the most important issues Merton addressed in his earlier work in the sociology of science. Their potential for elaboration is largely untested by sociologists of science.[59]

Historians of science have had different concerns. Had our bibliographic search and citation index been extended to recent publications in the history of science, greater interest in Merton's studies of Puritanism and the rise of science would have been registered. This would be particularly so among younger historians of science working on the social contexts of scientific development.[60] The influence of Merton's work on seventeenth-century science was invisible for some time, as Henry Guerlac tells us, since it ran counter to the dominant trend toward "internalist" or "idealist" history of science.[61] Merton's influence on historians of science has surfaced in the last decade or so. Although we find only one other sociologist even mentioned in papers dealing with the history of science in *Past and Present* and in *History of Science*, references to the Merton thesis have multiplied considerably in recent years. And such established "internalist" historians of science as I. Bernard Cohen,[62] A. Rupert Hall,[63] Charles Gillispie,[64] and Guerlac appear to be increasingly sympathetic to considerations of the social and ideological contexts of science, Guerlac calling Merton's dissertation a "landmark" in the literature of the history of science,[65] and Gillispie remarking, "Not many a thesis furnishes fuel for a controversy lasting as long as its author's career, much less bidding (as this one is beginning to do) for immortality."[66]

There are, however, a few signs that Merton's sociological perspective was being incorporated into the history of science. Gillispie reports that on first reading the "Priorities" paper it appeared "a bit trivial. I don't believe I also said 'unworthy' but recollect that such a dark thought was in my mind." He goes on to say that he and his colleagues in the history of science did not really understand what it was all about: "Only a few years later, when I began to study and teach materials in the social and institutional as well as the more traditional internal and intellectual history of science, did I come to take the full thrust of what he [Merton] had in fact said, and said clearly and convincingly."[67] Nevertheless, of the entire corpus of Merton's writings in the sociology of science it is the Puritan roots of English science which still preoccupy most historians of science. Merton's recent work that sociologists of science have found so "puzzle-producing" has not yet found its way into the thinking of the historians. But, nonetheless, he is an intellectual presence for them. Commenting in a book review on an error in indexing, John Murdoch writes, "Robert Merton . . . has been classified under 'Merton College.' (This kind of transformation of Merton into an institution is presumably a bit premature!)"[68]

Returning to the impact of Merton's work on sociologists of science, we note that citation analysis cannot take us very far toward understanding how authors are influenced by publications they cite even though it does tell us how often they do so in our limited sample of journals. Procedures for analyzing types of citations and their frequency are not well developed. In fact, sociologists of science have not even settled on a standardized classification of citations.[69] Our own efforts at a content analysis of citations to Merton's work are, as a consequence, rather tentative. Altogether fifty-eight different authors writing in the nine journals sampled have used thirty-four of Merton's papers. Although his later publications have been disproportionately cited in recent research in the sociology of science, the kinds of citations these papers receive are not much different from those given his earlier publications at comparable times in their own life histories. By and large, Merton's work is cited for two purposes: to confer authority on statements authors make and to identify the source of problems. There are few ceremonial or perfunctory citations and a few disparaging citations. In the absence of statistical norms on the relative frequency of different kinds of citations in the sociological literature it is not possible to interpret the distribution observed here. Authors who draw on Merton's work as a point of departure for their own research typically do so by developing ideas he originated or crystallized rather than using it as a source for specific hypotheses along the same lines as Stephen Cole's study of the "Matthew Effect."[70] Clearly sociologists of science are now taking the first steps toward understanding the ways in which knowledge develops and how research by different generations of scholars is linked together as these are reflected in citations.

Thus, citation analysis at its present stage of development can only take us so far toward understanding the character of intellectual influence. Serious content analysis of manuscripts is required to fill out the skeletal facts provided by citations. The same is true when we attempt an account of networks of influence. The gross connections between publications can of course be mapped. But citations do not even hint at the decisive if only partly visible effects scholars have on one another via informal discussions and critical readings of manuscripts and grant applications. Painstaking documentary research and interviewing are needed to get at these informal linkages and their diverse consequences. Not only would studies of authors' acknowledgments be germane here but so would notes on seminars, research memoranda, letters, diaries, and commentaries on texts. Consider this fragmentary exchange in which Merton serves as facilitator and reference individual. In June 1959, while at the Center for Advanced Study in the Behavioral Sciences, Thomas Kuhn wrote to Merton thanking him for his close reading of the paper Kuhn had sent him on measurement in science:

> I am sending you . . . a much-revised draft of my first chapter on "Scientific Revolutions." If you have a chance to look at it, I shall be very grateful for your reactions. Meanwhile, or at least until you call me off, I shall continue to pester you with pieces in this vein as they become available.[71]

Merton was deeply impressed by the manuscript. When Kuhn was ready to send *The Structure of Scientific Revolutions* to the University of Chicago Press, he wrote to Merton asking that he intercede with the Press if it proved reluctant to publish the volume independently of the *International Encyclopedia of Unified Science*. Merton replied:

> Of course, I'll be glad to write to the Chicago Press along the lines you suggest. After all, I've read the earlier draft and this alone is enough to justify a strong recommendation to the Press that they proceed as you would have them do.[72]

Kuhn's apprehensions proved groundless. The Press agreed to publish the book as Kuhn requested and it appeared at the end of 1962. Merton wrote to Kuhn:

> I have just this day received a copy of your new book. . . . Having read this version in its entirety, I must say that it is merely brilliant. More than any other historian of science I know, you combine a penetrating sense of scientists at work, of patterns of historical development, and of sociological processes in that development.[73]

Kuhn replied:

> I think you know how much your good opinion of the sort of work I have tried to do in the book means to me. . . . Of course I'll inscribe your copy. . . . I always hate that particular task, but it will be a small price to pay for the chance to talk the whole area over with you.[74]

Bernard Barber, in his essay on L. J. Henderson, and Nicholas Mullins in his analysis of specialty formation, both conclude that the people who fill the role of the "trusted assessor" critically affect scientific development not only by setting standards but also by providing social and psychological support for those working in new areas.[75] The physicist and knowledgeable observer of science John Ziman goes even farther and asserts that "the creation of a group of reliable experts who can be trusted to give fair consideration to all new work . . . [is] one of the difficulties in the establishment of an entirely new science."[76]

And, finally, crude quantitative analysis of citations is not fine-grained enough to detect the subtle and often unacknowledged influences of general theoretical and methodological orientations. In the process of "obliteration by incorporation," which is especially marked in the case of general ideas, original sources are lost to view.[77] We want now to consider how some of Merton's generic sociological concepts have found their way into the sociology of science and, in doing so, illustrate the interplay between the cognitive structures of specialties and those of the larger disciplines.

Merton's analysis of science and scientists did not develop independently of his other theoretical efforts. On the contrary, the problems he has selected and the mode of attack he has used are clearly related to his general interest in applying structural and functional analysis to social patterns in various institutional spheres. Four theoretical themes will serve the purpose:

1. Anomie and deviant behavior
2. Multiple consequences and manifest and latent functions
3. Over-conformity and maladaptation
4. Self-reinforcing social processes

Anomie and deviant behavior. Merton's analysis of competition for priority in science closely parallels his widely known work on deviant behavior. The origins of deviant behavior in science and in the larger culture are located in a disjunction between goals and normatively prescribed means. And its incidence depends in large part on the structure of opportunities to conform to the norms.[78] He suggests that the strong emphasis in science on the extension of certified knowledge and thus on original discovery has much the same effect as the comparably strong emphasis in American culture on financial success. Although scientists are enjoined to humility and to disinterestedness, they are, at the same time, driven to seek recognition of their originality since that is the only way they can be sure they have truly made contributions to science. As a consequence, they are under great pressure to stake claims to what they take to be their scientific property, to assert their priority of discovery, and, however uncontentious they might be personally, to engage in priority disputes. Rarely, however, do these efforts shade over into thoroughly deviant acts including plagiarism, data manipulation, and slander of competitors for priority. Merton's analysis of competition for priority is formally similar to his treatment of other kinds of patterned nonconformity in two important respects: he is unwilling to accept psychological accounts of what he takes to be socially structured behavior, and he looks for explanations of its origins and frequency in particular in normative ambivalence (norms and counter norms), in cultural inconsistencies, and in socially structured opportunities. Storer,[79] Gaston,[80] and Hagstrom,[81] among others, have used this analysis in studies of the sources of competition in science, its incidence, and the forms it assumes in various social contexts.

* * *

Multiple consequences and manifest and latent functions. Merton's analysis of the Matthew Effect in science draws upon the perspective laid out in his "Paradigm for Functional Analysis."[82] He begins with the observation that scientific recognition tends to accrue to those who already have it. Without deliberate intent, the Matthew Effect penalizes the young and the unknown and, in the process, reinforces the already unequal distribution of rewards.[83] But this is not all. Merton characteristically takes a step back from the phenomenon he has been examining from the perspective of individuals and asks how things look for the system as a whole and finds multiple and diverse consequences. He argues that the misallocation of credit which results from the Matthew Effect is unjust and exacts a high emotional cost from individuals. But, at the same time, it has the surprising effect of making the communication system more efficient; it thus contributes to the extension of certified scientific knowledge. For one thing, the Matthew Effect calls attention to the work of proven scientists and thereby increases the probability that work of value will get noticed and read.

It also calls attention to different parts of the output of distinguished scientists, increasing the visibility of reappearing themes in their work that might otherwise go unnoticed. Merton's technique of shifting the angle of theoretical vision from social consequences for individuals to those for systems and from manifest to latent outcomes has been fruitfully used by sociologists of science. Two diverse examples are Menzel's[84] analysis of the functions of scientific communication and our own studies on the intended and unintended effects of rewards on scientists' subsequent productivity.[85]

✳✳✳

Overconformity and maladaptive behavior. The notion that the same social pattern is likely to have multiple consequences for different units of a system is closely allied to the view that the same social patterns have different consequences over time and in diverse circumstances. For more than thirty years, Merton has been concerned with the idea that conforming behavior that is adaptive in some circumstances readily becomes maladaptive when circumstances change. One aspect of the analysis of "Bureaucratic Structure and Personality" illustrates this idea. For their effective operation, bureaucracies require their members to be punctual and methodical. However, characteristics which have positive consequences under one set of conditions can have negative consequences under other conditions. When temporal and procedural flexibility are required, excessive concern with temporal and formal rules makes for bureaucratic obsessiveness, red tape, and the development of bureaucratic virtuosity. Merton continually reminds us of the troubles created by too much of a good thing. In science, he observes, an excess of commitment to the norm of disinterestedness—to science for its own sake—has social consequences that make for public alienation from science. Similarly, overconformity to the norms requiring recognition of all participants in scientific research has made for increasingly large author sets and, in some fields, great difficulty in identifying those responsible for the research who should be credited or tarred for it.

✳✳✳

Self-reinforcing social processes. Among the self-reinforcing processes Merton has examined, the accumulation of advantage has special interest for sociologists of science.[86] Introduced in his 1942 essay on normative structure, the idea suggests that even evaluation which is impersonal and universalistic and perceived as fair can lead to the accumulation of differential advantage among certain segments of a population.[87] This idea has been taken up and elaborated by Crane, the Coles, Zuckerman, and by Merton himself in the Matthew Effect.[88] The process begins with an initial definition of some individuals as promising. These individuals are then given more resources and facilities for their work than those who are initially defined as less talented. Assuming that those who are thus advantaged are sufficiently competent to use the tools placed at their disposal effectively, it is hardly surprising that they often produce better research than others who are less advantaged. Once in operation, differences in performance between the "haves" and the "have nots" increase as the "haves" are, on universalistic grounds, consistently given more resources for their research and more recognition for it. This process and its components are difficult to study empirically because it is not evident how the performance effects of differential access to resources can be distinguished from effects attributable to differences in capacity. It is clear, however, that systems of evaluation operating in this fashion tend to be self-confirming. However effective they may in fact be in allocating resources to those who can best use them, their effectiveness cannot be judged by simple comparisons of performance between the advantaged and disadvantaged.

This brief review suggests how the sociology of science has drawn upon generic sociological ideas and how they have been adapted to its distinctive purposes. As special fields of inquiry develop, they typically elaborate ideas and techniques that are subject specific. At the same time, the problems they take to be central as well as the concepts and procedures used to study these problems are often drawn from the larger discipline of which they are a part. How the relative proportions of these two aspects of cognitive structure, generic ideas and procedures, and those that are subject specific, vary among types of specialties and over time is still another unanswered question in the cognitive and social evolution of science.

The Development of an Organizational Infrastructure

New specialties, radical and conforming, require more than a developing cognitive identity for institutionalization. We do not yet know the extent to which cognitive development in special fields and the development of their organizational infrastructures are interdependent or how this interdependence varies at different stages of specialty development. It is clear, however, that mundane problems such as training students, arranging for jobs, obtaining funds for research, and finding outlets for publication have to be dealt with in routinized ways for specialties to develop.

Specialties can grow without provision for regular training if there are new recruits sufficiently interested and willing to teach themselves. However, chances for recruitment are greatly improved if undergraduate and graduate students have access to systematic coursework in the special field. On this assumption, we surveyed the catalogues of the twenty-one departments of sociology receiving the highest rank in the 1969 American Council on Education study of the quality of graduate faculties for courses being offered in the sociology of science in each of three academic years: 1960–61, 1964–65, and 1973–74.[89] At the beginning of the 1960s, Norman Kaplan, then at Cornell, offered the only course in the sociology of science in the United States. Four years later, six courses were offered by four departments. At the last reading, fourteen of the twenty-one ranking departments offered eighteen courses in the sociology of science.

Some implications of these data are obvious. Not only are there more students but there are also faculty members of leading departments willing to teach and presumably qualified by work in the field to do so. These signal that the specialty has acquired a certain legitimacy among sociologists—a legitimacy related to its compatibility with the fundamental theoretical and methodological commitments of most American sociologists. The same developments are also related to the historical fact that the takeoff points in interest in the sociology of science coincided with a period of expansion and affluence in American science and in universities. The increased funds available in the 1960s led to substantial increases in the size of university faculties. Consequently, sociology departments were willing to hire people trained in new specialties such as the sociology of science. Growing interest among sociologists in science and increasing resources for it coincided. Both seem to be outcomes of the same underlying fact. Scientific development had become both a great national asset and a global problem.

Nonetheless, there still is little undergraduate teaching of the sociology of science. No general textbook has been published, Joseph Ben-David's *The Scientist's Role in Society* being more monographic than didactic.[90] In fact, few general introductory texts deal systematically with science.

Research Support

One of the distinctive features of American science is pluralism in funding. Consequently, information about who has gotten research money, how much, and for what purpose is scattered. The records of the National Science Foundation, the principal source of support for work in sociology of science, indicate that a steady increase in funding paralleled growth in publications, dissertations, and recruitment.[91] Less than one percent of NSF expenditures in sociology went to the sociology of science between 1957 and fiscal year 1964–65, 2.3 percent between 1965–66 and 1969–70, and 5.7 percent between 1970–71 and 1971–72. In this sixteen-year period, NSF grants to the sociology of science increased seven times compared to an increase of 1.75 times for sociology as a whole. Since sociological research is supported

142 VALUES AND REWARDS IN SCIENCE

by multiple government agencies and by a variety of private foundations, NSF figures for the discipline underestimate total expenditures on sociological research. Nonetheless, whatever share of funds the sociology of science has received, it has been supported with increasing generosity. Since the NSF employs the procedure of peer review in allocating its resources, the increase in funds registers growing confidence among sociologists (and government administrators) that research in the sociology of science is worth doing.

Formal Communication and Formal Organization

There is other evidence that the specialty was becoming interesting to sociologists and gaining a measure of legitimacy among them. An increasing number of the papers in the sociology of science appearing in sociology journals were published in the two principal journals in the field, the *American Sociological Review* and the *American Journal of Sociology.* Before 1957, too few papers were published to permit meaningful comparisons, but afterward things changed markedly. The number of papers published in the *ASR* and the *A JS* more than doubled from the ten appearing between 1957 and 1965 to the twenty-two appearing between 1965 and 1972. During the same period, the average number of articles published in these journals remained roughly constant. Since the prime journals have by far the greatest circulation among sociologists, the specialty was clearly gaining visibility. This is of no small importance to young people interested in working in the specialty who want their papers to be read by sociological colleagues. That one could publish in one of the important journals in the field (and, as we have seen, have a position in a leading department) means that doing the sociology of science could now be a career as well as a labor of love.

Unlike other new and growing specialties, the sociology of science has not yet acquired its own journal. The willingness of the principal sociological journals to publish papers in the sociology of science, the limited size of the specialty, and the presence of functional alternatives in the form of other journals addressed to the wider audience of historians, philosophers, political scientists, and sociologists of science—*Minerva* (founded in 1962) and *Science Studies* (founded in 1970)—probably mean that a specialty journal will not be needed for some time to come.

Still another sign of its growing legitimacy among sociologists is the appearance of sociologists of science on the programs of the various regional, national, and international meetings. Since the early 1960s almost every national meeting of the American Sociological Association has had a session devoted to the specialty, and since 1966, there has been a Research Committee in the Sociology of Science in the International Sociological Association. Merton's influence and efforts to build an organizational infrastructure are highly visible in this domain. He encouraged the scheduling of sessions at ASA meetings in the early 1960s by agreeing to chair them or to prepare papers and was one of the chief organizers of the ISA committee. But he does not find these activities congenial. He does not like to organize things or to run them. Unlike his teacher, George Sarton, who avidly devoted himself to establishing an elaborate organizational infrastructure for the history of science, Merton has set about most of these tasks reluctantly and has been far less effective than Sarton. But among physical and biological scientists his standing has helped him to call attention to the sociology of science in quarters such as the National Academy of Sciences and the National Science Foundation.

Professional Identity

Patterns of consensus in citation practices noted so far are outcomes of unselfconscious behavior of sociologists of science. For the specialty to develop a full-fledged cognitive and social identity, however, specialists, particularly influential ones, must self-consciously define themselves as having a common task. We use the citation data once again to see if a growing proportion of influentials specifically identify themselves as sociologists of science. Limiting our survey to two overlapping groups of cited authors, the thirty most often cited and the top 10 percent of cited authors in successive five-year periods, we tabulated their self-described areas of specialization as they appeared in standard directories of the ASA and *American Men (and Women) of Science.*[92] Table 5.4 shows the sharp increase in the proportion of the thirty most

TABLE 5.4 Self-Definitions of Influential Contributors to the Sociology of Science, 1950–1972

| | *Percentage Who Define Themselves as Sociologists of Science* | | | | | |
| | Top Thirty Authors | | | Top 10 Percent of All Cited Authors* | | |
Period	1st choice	1st and 2nd choice	(N)	1st choice	1st and 2nd choice	(N)
1950–1954	7**	7**	(15)**	7	7	(15)
1955–1959	7	10	(30)	7	9	(32)
1960–1964	10	17	(30)	7	11	(57)
1965–1969	36	40	(30)	16	18	(78)
1970–1972	40	47	(30)	18	22	(90)

*The specialties of two authors on the 1955–1959 list could not be identified along with one on the 1960–1964 list, four on the 1965–1969 list, and five on the 1970–1972 list. For totals, see table 5.1A.

**Only fifteen authors had more than a single citation in this period.

influential authors who consider themselves sociologists of science. It rose from 10 percent in the early 1960s to 36 percent in the late 1960s and to 40 percent in the 1970s. While there are also significant increases in the extent of identification with the specialty among the top decile of influentials, the figures suggest that the major shift has occurred among the thirty most-cited authors. Such changes in self-definitions among influentials register a growing consolidation of the specialty and new commitment to it. This heightens its visibility and helps accord it legitimacy.

More than twenty years ago, in his foreword to Bernard Barber's *Science and the Social Order*, Merton asked why "the sociology of science is still a largely unfulfilled promise rather than a highly developed special field of knowledge, cultivated jointly by social, physical and biological scientists?"[93] His inventory of neglect was dismaying indeed. Few courses were offered in the sociology of science and no standard textbook took notice of it; little empirical research was in process and what was being done was divorced from theory; publications were speculative, relying more on historical examples than on systematic historical evidence. Merton's own content analysis of Barber's bibliography of the field showed that half of all the referenced works were by "practicing physical and life scientists or by scientists who have turned to administration; more than a quarter by historians and philosophers of science and only the remaining fraction by sociologists."[94] In short, the field showed no signs then of impending institutionalization.

Things have changed. Taken together, the evidence we have examined suggests that the sociology of science is emerging as a special field of interest with a distinctive cognitive and professional identity. Having passed through an initial takeoff point in the middle 1960s, the field is still growing. It is too soon yet to tell how damaging hard times will be to further consolidation of the field. But it is not too soon to report that the sociological analysis of science is no longer the province of amateurs.

The next steps in studying the development of research specialties are reasonably clear. Continuing work will have to take account of the larger context of the growth of scientific knowledge and the relations between cognitive and social structures of science. Merton himself has now shifted his attention to developing an historical sociology of scientific knowledge. The growing emphasis among sociologists of science on the interplay of substantive and social aspects of science signals the beginnings of a new phase in the specialty.

Notes

1. See Cornelis B. Van Niel, "The Microbe as a Whole," in *Perspectives in Microbiology*, ed. S. A. Waksman (New Brunswick, NJ: Rutgers University Press, 1955), 3–12; Gerald

Holton, "Models for Understanding the Growth and Excellence of Scientific Research," in *Excellence and Leadership in a Democracy*, ed. S. R. Graubard and G. Holton (New York: Columbia University Press, 1962), 94–131; D. R. Stoddart, "Growth and Structure of Geography," transactions and papers, Institute of British Geographers, 1967, no. 41: 1–19; and J. S. Hey, *The Evolution of Radio Astronomy* (London: Paul Elek, 1973).

2. Robert K. Merton, "Priorities in Scientific Discovery," in *The Sociology of Science: Theoretical and Empirical Investigations*, ed. Norman Storer (Chicago: University of Chicago Press, 1973), 286–324. First published in 1957; Arnold Thackray and Robert K. Merton, "On Discipline Building: The Paradoxes of George Sarton," *Isis* 63 (1972): 473–95.

3. This is not to say, of course, that Merton was the first sociologist to study science and invention. Durkheim, Marx, Mannheim, Scheler, Znaniecki, Sorokin and others had addressed questions about the social determination of scientific knowledge long before him, as did W. F. Ogburn and Dorothy Thomas in their studies of the role of cultural accumulation in scientific and technical innovation. See their "Are Inventions Inevitable? A Note on Social Evolution," *Political Science Quarterly* 37 (1922): 83–98. Rather, Merton took the lead in the systematic study of science as a social institution and provided models for how this might be extended. Sociologists of science generally agree that Merton established the field as an intellectual and social activity. For examples, see Michael Mulkay, "Some Aspects of Cultural Growth in the Natural Sciences," *Social Research* 36 (1969): 22–53; Kenneth Downey, "Sociology and the Modern Scientific Revolution," *Sociological Quarterly* 8 (1967): 239–54; and Barry Barnes, ed., *Sociology of Science: Selected Readings* (Harmondsworth, UK: Penguin, 1972).

4. Joseph Ben-David, "Scientific Productivity and Academic Organization in Nineteenth-Century Medicine," *American Sociological Review* 25 (1960): 828–43, and Joseph Ben-David and Awraham Zloczower, "Universities and Academic Systems in Modern Societies," *European Journal of Sociology* 3 (1962): 45–84; and Joseph Ben-David, *The Scientist's Role in Society: A Comparative Study* (Englewood Cliffs, NJ: Prentice-Hall, 1971).

5. For examples of Merton's early interest in cognitive as well as social aspects of scientific knowledge, see his collaborative papers with Sorokin in Pitirim A. Sorokin, *Social and Cultural Dynamics*, 4 vols. (New York: American Book, 1937), 125–80, 439–76; chapters 7–11 of his *Science, Technology and Society in Seventeenth-Century England* (Bruges, Belgium: St. Catherine, 1938); and reprinted with new introduction (New York: Howard Fertig and Harper & Row, 1970). For recent examples of empirical and theoretical sociological studies of cognitive aspects of science, see chapter 7 in this volume and John Law, "The Development of Specialties in Science: The Case of X-ray Protein Crystallography," *Science Studies* 3 (July 1973): 275–303; and Michael J. Mulkay and David O. Edge, "Cognitive, Technical and Social Factors in the Growth of Radio Astronomy," *Social Science Information* 13 (1974): 25–61.

6. Thomas S. Kuhn, *The Structure of Scientific Revolutions* (Chicago: University of Chicago Press, 1962).

7. Derek J. de S. Price, *Science since Babylon* (New Haven, CT: Yale University Press, 1961); *Little Science, Big Science* (New York: Columbia University Press, 1963); "Networks of Scientific Papers," *Science* 149 (1965): 510–15; and "Citation Measures of Hard Science and Soft Science, Technology and Non-Science," in *Communication Among Scientists and Engineers*, ed. Carnot E. Nelson and Donald K. Pollak (Lexington, MA: Heath, 1970); and Derek J. de S. Price and Donald Beaver, "Collaboration in an Invisible College," *American Psychologist* 21 (November 1966): 1011–18.

8. See for example, Karl R. Popper, *Logic of Scientific Discovery* (New York: Basic Books, 1959), a translation of *Logik der Forschung* (1935); *Conjectures and Refutations* (London:

146 VALUES AND REWARDS IN SCIENCE

Routledge, 1963); and *Objective Knowledge* (Oxford: Clarendon, 1972). Much of Imre Lakatos's seminal work is now being prepared for posthumous publication. His influential publications include "History of Science and Its Rational Reconstructs," in R. C. Buck and R. S. Cohen, eds., *Boston Studies in the Philosophy of Science* 8 (1971): 91–136, 174–82; "Falsification and the Methodology of Scientific Research Programmes," in *Criticism and the Growth of Knowledge*, ed. I. Lakatos and A. Musgrave (Cambridge: Cambridge University Press, 1970), 91–195; "Popper on Demarcation and Induction" in the two-volume collection of critical essays on the Popperian tradition, *The Philosophy of Karl Popper*, ed. P. A. Schilpp (LaSalle, IL: Open Court, 1974), 241–73; and the paper which he earmarked as best epitomizing his concept of "research programme," Imre Lakatos and Elie Zahar, "Why Did Copernicus's Programme Supersede Ptolemy's?" presented at the Quincentenary Symposium on Copernicus of the British Society for the History of Science. London, January 5, 1973.

9. See in addition to Lakatos and Musgrave, *Criticism and the Growth of Knowledge*, Israel Scheffler, *Science and Subjectivity* (Minneapolis, MN: Bobbs-Merrill, 1967), and Dudley Shapere, "The Structure of Scientific Revolutions," *Philosophical Review* 73 (1964): 383–94.

10. See Nicholas C. Mullins, "The Development of a Scientific Specialty," *Minerva* 10 (January 1972): 51–82; "The Development of Specialties in Social Science: The Case of Ethnomethodology," *Science Studies* 3 (1973): 245–73; and Belver C. Griffith and Nicholas C. Mullins, "Coherent Social Groups in Scientific Change," *Science* 177 (September 1972): 959–64.

11. Price, *Little Science, Big Science*, and chapter 3 of Crane, "Invisible Colleges . . ."

12. Among those who have discussed the growth and change of scientific specialties are Warren Hagstrom, *The Scientific Community* (New York: Basic Books, 1965); Joseph Ben-David and Randall Collins, "Social Factors in the Origins of a New Science: The Case of Psychology," *American Sociological Review* 31 (August 1966): 451–65; Charles S. Fisher, "The Death of a Mathematical Theory: A Study in the Sociology of Knowledge," *Archive for the History of Exact Sciences* 3 (1966): 137–59; Terry N. Clark, "Emile Durkheim and the Institutionalization of Sociology in the French University System," *European Journal of Sociology* 9 (1968): 37–91; Diana Crane, "Social Structure in a Group of Scientists: A Test of the 'Invisible College' Hypothesis," *American Sociological Review* 34 (1969): 335–52; Diana Crane, *Invisible Colleges* (Chicago: University of Chicago Press, 1972); Nicholas C. Mullins, "The Development of a Scientific Specialty," "The Development of Specialties in Social Science," and *Theories and Theory Groups in Contemporary American Sociology* (New York: Harper & Row, 1973); Griffith and Mullins, "Coherent Social Groups in Scientific Change"; M. J. Mulkay and D. O. Edge, "Cognitive, Technical and Social Factors"; John Law, "The Development of Specialties in Science"; Thackray and Merton, "On Discipline Building"; Griffith and Mullins, "Coherent Social Groups in Scientific Change," 960; and Michael J. Apter, "Cybernetics: A Case Study of a Scientific Subject-Complex," in P. Halmos, ed. *The Sociological Review Monograph* 18 (September 1972): 93–115.

13. Mullins, whose work *Theories and Theory Groups in Contemporary American Sociology*, is the most ambitious to date, proposes a four-stage model of specialty development. It emphasizes the structure of communication between specialty group members rather than the content of scientific innovations. The principal components are the roles of intellectual and social leaders, the role of programmatic statements, the diffusion of group members from centers of activity to the periphery, and the thickening of communication nets with growth. The model involves problems of defining boundaries of

groups and their orientations and requires revision to apply to cognitively dissident specialties. Further investigation is needed to specify the structural and intellectual conditions of movement from the "normal" to the "network" to the "cluster" and finally to the "specialty" stages of development. See also Paul D. Allison, "Social Aspects of Scientific Innovation: The Cases of Parapsychology," master's thesis, University of Wisconsin, 1973.

14. For two pertinent case studies, see Aaron J. Ihde, "An Inquiry into the Origins of Hybrid Sciences: Astrophysics and Biochemistry," *Journal of Chemical Education* 46 (April 1969): 193–96, and Robert E. Kohler, "The Enzyme Theory and the Origin of Biochemistry," *Isis* 64 (1973): 181–96.

15. Harriet Zuckerman and Robert K. Merton, "Age, Aging and Age Structure in Science," in in *The Sociology of Science*, ed. Robert Merton (Chicago: University of Chicago Press, 1979), 497–559.

16. Fisher, "The Death of a Mathematical Theory."

17. Matilda W. Riley, "Membership of the American Sociological Association, 1950–1959," *American Sociological Review* 25 (1960): 914–26.

18. Nico Stehr and Lyle E. Larson, "The Rise and Decline of Areas of Specialization," *American Sociologist* 7 (August 1972): 5.

19. Stehr and Larson, "The Rise and Decline of Areas of Specialization," 6.

20. Five colleagues in the sociology of science, Warren Hagstrom, Walter Hirsch, Norman Kaplan, Janice Lodahl, and Norman Storer, reviewed entries in the inventory, suggested additional titles, and identified sponsors of dissertations.

21. See Theodore F. Abel, "Analysis of Attempts to Establish Sociology as an Independent Science," PhD dissertation, Columbia University, 1929; and Wilfred Binnewies, "A History and Evaluation of the Quantitative Trend in Sociological Analysis," PhD dissertation, University of Nebraska, 1929.

22. Tabulated from *Doctoral Dissertations Accepted by American Universities* (Washington, DC: U.S. Office of Education) and *Earned Degrees Conferred* and *Doctoral Records File* (Washington, DC: National Research Council, National Academy of Sciences).

23. Calculated from Lindsey R. Harmon and Herbert Soldz, comps., *Doctorate Production in United States Universities, 1920–1962* (Washington, DC: National Research Council, National Academy of Sciences, 1963), no. 1142, Appendix 3, 74–85.

24. Price, *Little Science, Big Science;* Crane, *Invisible Colleges;* Henry Menard, *Science, Growth and Change* (Cambridge, MA: Harvard University Press, 1971); and David L. Krantz, "Research Activity in 'Normal' and 'Anomalous' Areas," *Journal of the History of the Behavioral Sciences* 1 (January 1965): 39–41.

25. As well as Price, *Little Science, Big Science*, see Crane, *Invisible Colleges;* Henry Menard, *Science, Growth and Change* (Cambridge, MA: Harvard University Press, 1971); and David L. Krantz, "Research Activity in 'Normal' and 'Anomalous' Areas," *Journal of the History of the Behavioral Sciences* 1 (January 1965): 39–41.

   Holton has proposed several models for interpreting comparable data. See chapter 12 in Gerald Holton, *Thematic Origins of Scientific Thought: Kepler to Einstein* (Cambridge, MA: Harvard University Press, 1973).

26. Six of these nine journals are wholly sociological: *American Sociological Review, American Journal of Sociology, Social Forces, Social Problems, Sociology of Education*, and the *British Journal of Sociology*. The others include *Minerva*, first published in 1962, which focuses on various aspects of the history, politics, philosophy, and sociology of science, the *American Behavioral Scientist*, which has devoted issues to problems in the sociology of science, and *Science*, the official publication of the American Association for

the Advancement of Science, which has included papers on the sociology of science. A preliminary check of the very small number of papers published in other books or journals shows that their age and references do not differ systematically from those we have analyzed. Nevertheless, this sample underestimates growth in the literature since 1971 since it does not include newer journals such as *Science Studies* and *Research Policy*.

There is prima facie evidence that few papers were published before 1950. Bibliographies of work in the field compiled by sociologists show few entries before that date: Bernard Barber and Robert K. Merton, "Brief Bibliography for the Sociology of Science," *Proceedings of the American Academy of Arts and Sciences* 80 (1952): 140–54; Norman Kaplan, "Science and Society: An Introduction," in *Science and Society*, ed. Norman Kaplan (Chicago: Rand-McNally, 1965), 1–8; and Bernard Barber and Walter Hirsch, eds., *The Sociology of Science* (New York: Free Press, 1962).

27. Of the total 195 papers, 165 were published in journals in print for the full twenty-four-year period under examination. Analysis of trends in publication is confined to these 165 papers. All papers were included that considered scientists and the institution of science from the sociological perspective. Papers on science were excluded if they were wholly historical or philosophical or if they focused on questions of science policy. Two lists of candidate papers were compiled independently and discrepancies between them resolved by two judges.

The citation index was constructed by compiling information on the citing authors, place and year of publication, and authors and dates of publications cited. Since citations and pairs of citations are the unit of analysis, multiauthored papers produced a multiplication of citations. When a publication is cited by collaborating authors, every author is counted as having been influenced by the cited publication. This gives extra weight to papers cited in empirical studies since they are more often multiauthored. Self-citations were excluded. An additional procedure was used when Merton was cited. In addition to registering his name and the year of the publication, its title and page number were recorded. This permitted a detailed content analysis of citations to his work and enabled us to identify precisely which of his ideas had been influential.

28. The two journals established in the 1960s were excluded to avoid the biasing effects of increased publication outlets.

29. The slope of a scatterplot of publications over time approximates an exponential curve. We have not included the figure in the text, since it looks similar to other exponential growth curves. We suggest that in the future authors consider presenting temporal data on productivity in the form of a log-normal transformation. For illustrations of growth curves in science, see D. J. de S. Price, *Little Science, Big Science*; Diana Crane, *Invisible Colleges*.

30. Reprinted in Merton, *The Sociology of Science*, 286–324. For observations on the significance of this paper, see Norman W. Storer, "Prefatory Note: The Reward-System of Science," *The Sociology of Science*, 281–85; and Joseph Ben-David, "The Sociology of Science," *New York Times Book Review*, November 11, 1973, 32.

The "Priorities" paper does not constitute a formal paradigm in the Mertonian or Kuhnian sense or a research program in the Lakatos sense. It does propose a general orientation for the sociological study of science and directs attention to certain central problems.

31. On the use of citations to measure scientific influence and significance, see M. M. Kessler, "The M.I.T. Technical Information Project," *Physics Today* 18 (March 1965): 28–36; Eugene Garfield, "Citation Indexing for Studying Science," *Nature* 227 (1970): 669–71; and "Citation Indexing Historico-Bibliography and the Sociology of Science," E. Davis and W. D. Sweeny, eds., *Proceedings of the Third International Congress of Medical Librarianship* (Amsterdam: Exerpta Medica, 1970), 187–204; Jonathan R. Cole and

Stephen Cole, "Measuring the Quality of Sociological Research: Problems in the Use of the Science Citation Index," *American Sociologist* 6 (February 1971): 23–29, and *Social Stratification in Science* (Chicago: University of Chicago Press, 1973).

32. For other examples, see Stephen Cole, "Scientific Reward Systems: A Comparative Analysis," in *Research in Sociology of Knowledge, Sciences and Art*, ed. R. A. Jones, 167–90 (Greenwich, CT: JAI Press, 1978).

33. Cole and Cole, *Social Stratification in Science*.

34. For a recent collection of papers adopting this point of view, see Richard D. Whitley, ed., *Social Processes of Scientific Development* (London: Routledge, 1974), 69–95; Mulkay, "Some Aspects of Cultural Growth in the Natural Sciences"; S. B. Barnes and R. G. A. Dolby, "The Scientific Ethos: A Deviant Viewpoint," *European Journal of Sociology* 11, no. 1 (1970): 3–25; and Barnes, *Sociology of Science*.

Since journal articles are subjected to refereeing, they may be less polemical than books. Content analysis of citations drawn exclusively from journals may underestimate "critical" and overestimate adulatory comments. In order to determine whether this is so, we examined citations to Merton and other authors in the indexes of several books recently published by Merton's critics. Using Barnes, *Sociology of Science*, Leslie Sklair, *Organized Knowledge* (London: Hart-Davis, 1973), and Whitley, *Social Processes of Scientific Development*, as a crude sample, we found evidence consistent with our hypothesis. We also found fewer references to empirical literature.

35. Price, "Citation Measures of Hard Science and Soft Science,"; see also Price, "Networks of Scientific Papers," and J. Margolis, "Citation Indexing and the Evaluation of Scientific Papers," *Science* 185 (March 1967): 1213–19.

36. See Zuckerman and Merton, "Age, Aging and Age Structure in Science," 506ff.

37. Stephen Cole, Jonathan R. Cole, and Lorraine Dietrich, "Measuring Consensus in Scientific Research Areas," in *Toward a Metric of Science: Thoughts Occasioned by the Advent of 'Science Indicators'*, ed. Y. Elkana, J. Lederberg, R. K. Merton, A. Thackray, and H. Zuckerman (New York: Wiley-Interscience, in press). For comparable data, see also Price, "Citation Measures of Hard and Soft Science."

38. Both reprinted in Merton, *The Sociology of Science*.

39. Zuckerman and Merton, "Age, Aging and Age Structure in Science," 510–19.

40. Average ages for the extended periods 1950–1964 and 1965–1972 were calculated from the ages of the thirty most-cited authors at the midpoint of each of five periods.

41. For changing rates of collaboration in sociology, see Narsi Patel, "Quantitative and Collaborative Trends in American Sociological Research," *The American Sociologist* 7 (1972): 5–6.

42. Merton is, of course, heavily cited in the literature of general sociology and in its neighboring disciplines of anthropology and psychology. Recent studies of the impact of leading sociologists' work show that he is the most cited author in the current literature and is cited more often than all authors but Durkheim in current textbooks. Mark Oromaner, "The Most Cited Sociologists," *The American Sociologist* 3 (May 1968): 124–26; "The Structure of Influence in Contemporary Academic Sociology," *The American Sociologist* 7 (May 1970): 11–13; "Comparison of Influentials in Contemporary American and British Sociology," *British Journal of Sociology* 13 (1970): 324–32; Frank R. Westie, "Academic Expectations for Professional Immortality: A Study of Legitimation," *The American Sociologist* 8 (February 1973): 19–32. See also Howard M. Bahr, T. J. Johnson, and M. R. Seitz, "Influential Scholars and Works in the Sociology of Race and Minority Relations, 1944–68," *The American Sociologist* 8 (November 1971): 296–98; and William H. Swatos Jr., and Priscilla L. Swatos, "Name Citations in Introductory Sociology Texts," *The American Sociologist* 9 (November 1974): 225–28.

150 **VALUES AND REWARDS IN SCIENCE**

43. Citation counts for individuals are omitted because the addition of journals other than those surveyed here would have changed the absolute numbers, if not the approximate relative positions. Moreover, we are less concerned with individual scores than with changing patterns of citation over time.

44. We note that just two authors (Barber and Merton) in the recent lists have also worked in the sociology of knowledge. To some extent, the sociology of science and of knowledge have developed independently. By way of illustration, a compendium of recent work in the sociology of knowledge includes just one paper in the sociology of science and one in the sociology of sociology. The remaining twenty-five papers, classified as "Later Perspectives," contain no reference to any work in the sociology of science except the several in Merton's "Paradigm for the Sociology of Knowledge." See James E. Curtis and John W. Petras, eds., *The Sociology of Knowledge: A Reader* (New York: Praeger, 1970).

45. Kuhn's work on revolutions in science, Price's studies of patterns of scientific growth, Barber's comprehensive analysis of the institution of science, and Pelz's investigation of organizational climates of research have continued to be widely used since their publication.

46. Mullins, *Theories and Theory Groups in Contemporary American Sociology.*

47. The seven are Bernard Barber, Stephen Cole, Barney Glaser, Alvin Gouldner, Norman Kaplan, Jonathan Cole, and Harriet Zuckerman. An eighth, Diana Crane, studied with Merton but did not do her doctoral work under his supervision.

48. For master-apprentice links among psychologists, see Joseph Ben-David and Randall Collins, "Social Factors in the Origins of a New Science: The Case of Psychology," *American Sociological Review* 31 (1966): 451–65; and among Nobel laureates, see Harriet Zuckerman, "Nobel Laureates in Science: Patterns of Productivity, Collaboration and Authorship," *American Sociological Review* 32 (1967): 391–403, and *Scientific Elites: Nobel Laureates in the United States* (Chicago: University of Chicago Press, 1975).

49. The Institute for Scientific Information, the organizational home of the *Science Citation Index*, is now undertaking different but related cluster analyses in order to identify groups of linked and frequently cited papers believed to presage the development of specialties. See Eugene Garfield, Morton V. Malin, and Henry Small, "Citation Data as Indicators of Scientific Activity" in Elkana et al., *Toward a Metric of Science.*

50. Thackray and Merton, "On Discipline Building."

51. Oromaner, "The Structure of Influence in Contemporary Academic Sociology."

52. Merton, "Priorities in Scientific Discovery."

53. See P. E. Burton and R. W. Keebler, " 'Half-Life' of Some Scientific and Technical Literature," *American Documentation* 11 (1960): 18–22.

54. Merton, "Priorities in Scientific Discovery."

55. Robert K. Merton, "Singletons and Multiples in Scientific Discovery," in *Proceedings of the American Philosophical Society* 105 (October 1961): 470–86.

56. Merton, "The Matthew Effect," in *The Sociology of Science*, 439–59.

57. Harriet Zuckerman and Robert K. Merton, "Patterns of Evaluation in Science: Institutionalization, Structure and Function of the Referee System," *Minerva* 9 (January 1971): 66–100.

58. Robert K. Merton, "Insiders and Outsiders: A Chapter in the Sociology of Knowledge," *American Journal of Sociology* 78 (July 1972): 9–47.

59. Allison, "Social Aspects of Scientific Innovation: The Case of Parapsychology."

60. Robert K. Merton, "The Normative Structure of Science," in *The Sociology of Science*, 267–78. Ben-David has argued that Merton's analysis of the normative structure of science "gave a static and idealized picture of science as a social system and did not reveal how the system actually worked" ("The Sociology of Science," *New York Times*

Book Review, 32). We would add that the later discussion of ambivalence in the norms ("The Ambivalence of Scientists," in *The Sociology of Science*, 383–412) is not only sociologically instructive but also the point of departure for intriguing empirical studies of ambivalence by Ian Mitroff ("Norms and Counter-Norms in a Select Group of Apollo Moon Scientists: A Case Study of the Ambivalence of Scientists," *American Sociological Review* 38 [June 1974]: 579–95; and *The Subjective Side of Science: A Philosophical Inquiry into the Psychology of the Apollo Moon Scientists* [Amsterdam: Elsevier, 1974]). The time has surely come for thorough empirical investigation of the distribution of norms among members of the scientific community and their conformity to them (see Marian Blissett, *Politics in Science* [Boston: Little, Brown, 1972], for one effort) and for an end to speculation about whether scientists actually are conforming or deviant.

61. Until recently, the "Merton Thesis" has had greater influence on historians of science than any other part of his work. But the fact that Thomas Kuhn devotes practically one-third of his review of the "History of Science" in the *International Encyclopedia of the Social Sciences* (1968, 14: 74–83) to that thesis may say as much about what Kuhn finds interesting as about actual foci of attention in the field itself. Not surprisingly, Marxist historians of science find Merton's sociological analysis of science more congenial than do their colleagues doing traditional internalist history of science. The Marxists argue, however, that Merton does not confront the central issue of scientific knowledge as being "objective or value neutral." See Robert Young, "The Historiographic and Ideological Contexts of the Nineteenth-Century Debate on Man's Place in Nature," in *Changing Perspectives in the History of Science: Essays in Honour of Joseph Needham*, ed. M. Teich and R. Young (London: Heinemann, 1973), 69–95.

62. Henry Guerlac, "History of Science: The Landmarks of the Literature," *The Times Literary Supplement*, April 26, 1974, 450.

63. I. Bernard Cohen, "Science, Technology and Society in Seventeenth-Century England," *Scientific American* 228 (1974): 117–20.

64. A. Rupert Hall, "History of Science: Microscopic Analyses and the General Picture," *The Times Literary Supplement*, April 26, 1974, 437–38.

65. Charles C. Gillispie, "Mertonian Theses," *Science* 184 (May 1974): 656–60.

66. Guerlac, "History of Science," 450. In the last fifteen years, a parallel concern with the history and the sociology of science has developed among some philosophers of science, especially Lakatos and Feyerabend. See R. N. Griere, "History and Philosophy of Science: Intimate Relationship or Marriage of Convenience," *British Journal for the Philosophy of Science* 24 (September 1973): 282–97.

67. Gillispie, "Mertonian Theses," 658.

68. Gillispie, "Mertonian Theses," 656.

69. John Murdoch, "Review of M. Witrow, *Isis Cumulative Bibliography*," *British Journal for the Philosophy of Science* 25 (March 1974): 89–91.

70. Norman Kaplan, "The Norms of Citation Behavior: Prolegomena to the Footnote," *American Documentation* 16 (July 1962): 179–84.

71. Stephen Cole, "Professional Standing and the Reception of Scientific Discoveries," *American Journal of Sociology* 76 (September 1970): 286–306.

72. Professor Kuhn has kindly given us permission to quote from his letters. Letter, June 1959.

73. Professor Merton has kindly given us access to his files of manuscripts and correspondence dealing with the sociology of science. Letter, May 4, 1961.

74. Letter, December 13, 1962.

75. Letter, January 21, 1963.

76. "Introduction" in *L. J. Henderson, On the Social System*, ed. Bernard Barber (Chicago: University of Chicago Press, 1970), 42–43; Mullins, *Theory and Theory Groups in Contemporary American Sociology*.

77. John Ziman, "Science Is Social," *The Listener* (August 18, 1960), 251.

78. On the process of obliteration by incorporation, see Robert K. Merton, *On Theoretical Sociology* (New York: Free Press, 1967), 27–35.

79. Robert K. Merton, "Social Structure and Anomie," *American Sociological Review* 3, no. 5 (October 1938): 672–82.

80. Norman Storer, *The Social System of Science* (New York: Holt, 1966).

81. Jerry C. Gaston, *Originality and Competition in Science* (Chicago: University of Chicago Press, 1973).

82. Warren Hagstrom, *The Scientific Community* (New York: Basic Books, 1965), and "Competition in Science," *American Sociological Review* 39 (February 1974): 1–18.

83. Merton, "The Matthew Effect," *Science* 159, no. 380: 56–63; Robert K. Merton, "Manifest and Latent Functions," in *Social Theory and Social* Structure (New York: Free Press, 1949), 73–138.

84. In the words of St. Matthew, "For unto everyone that hath shall be given, and he shall have abundance: but from him that hath not shall be taken away even that which he hath."

85. Herbert Menzel, "Scientific Communication: Five Themes from Social Research," *American Psychologist* 21 (1966): 999–1004; Cole and Cole, *Social Stratification in Science*, 113–15; and Zuckerman, "Nobel Laureates in Science," 398–403.

86. Briefly introduced in Merton's 1942 paper on the ethos of science, the notion of "accumulation of advantage" and of disadvantage in systems of social stratification, which relates to the notions of the self-fulfilling prophecy and the Matthew Effect, has been developed in a series of investigations: Merton, *The Sociology of Science*, 273, 416, 439–59, and 497–559; Harriet Zuckerman, "Stratification in American Science," *Sociological Inquiry* 40 (Spring 1970): 235–57; Zuckerman and Merton, "Age, Aging and Age Structure in Science"; Cole and Cole, *Social Stratification in Science*, 237–47; Paul D. Allison and John A. Stewart, "Productivity Differences Among Scientists: Evidence for Accumulative Advantage," *American Sociological Review* 39 (August 1974): 596–606; and Zuckerman, *Scientific Elites*, chapter 3.

87. Merton, "The Normative Structure of Science," in *The Sociology of Science*.

88. Diana Crane, "The Academic Marketplace Revisited: A Study of Faculty Mobility Using Cartter Ratings," *American Journal of Sociology* 75 (May 1970): 953–64.

89. K. D. Roose and Charles J. Andersen, *A Rating of Graduate Programs* (Washington, DC: American Council of Education, 1970). Although the ranks of some departments change slightly, the intercorrelations between readings taken in 1957, 1964, and 1969 are above 0.95. The twenty-one departments receiving the highest ratings granted 77 percent of all doctorates in sociology in 1968–69.

90. Ben-David, *The Scientist's Role in Society*.

91. These data were drawn from National Science Foundation, *Annual Reports*, fiscal years 1950–51 to 1971–72, and from its *Grants and Awards* series, first published in fiscal year 1963–64. Information for 1972–73 was generously provided by Dr. Donald Ploch of the National Science Foundation.

92. The self-designations of sociologists were tabulated but not those of information scientists, historians, or philosophers of science. These data refer only to the increasing tendency for sociologists to identify themselves as sociologists of science.

93. Robert K. Merton, "Foreword," in Bernard Barber, *Science and the Social Order* (Glencoe, IL: Free Press, 1952).

94. Merton, *Sociology of Science*, 212.

CHAPTER 6

THE REPUTATIONS OF AMERICAN MEDICAL SCHOOLS

(1977)

JONATHAN R. COLE AND JAMES LIPTON

Abstract

This study is one of a larger inquiry into organizational stratification. About U.S. medical schools it asks: How does a sample of full-time clinical and basic science medical school faculty rank 94 medical schools as to quality of faculty and effectiveness of instruction? And: What are the structural correlates of such rankings? The resulting rank order takes on significance as it affects recruitment and placement of students and faculty. Measures of aggrandizement (inflated estimates of worth by insiders) are estimated. Characteristics of medical schools that correlate with perceived quality are: research and publication, eminence of faculty, training and research grants available, size of full-time faculty, and perceived effectiveness of training. While the data support the view that reputation stems from functionally appropriate performance, there is some evidence of a ceiling effect (Harvard) and a halo effect for schools affiliated with universities having national reputations. Regional location is positively associated with perceived reputation in the North and West, negatively in the South. Caveats are entered about interpreting the data.

Social Forces 55, no. 3 (March 1977): 662–84. This study was supported by a grant from the National Science Foundation to the Columbia University Program in the Sociology of Science, NSF-SOC72-95326; by the Center for Advanced Study in the Behavioral Sciences, where the first author was a Fellow in 1975–76; and by a Guggenheim Fellowship, held by the first author in 1975–76. We thank Bernard Barber, Peter M. Blau, Stephen Cole, and Robert K. Merton for their helpful comments on an earlier draft of this paper.

154 VALUES AND REWARDS IN SCIENCE

There is now a history spanning almost fifty years of studies that attempt to measure the quality of graduate departments of Arts and Sciences in the United States. The earliest assessment was by Raymond M. Hughes; three decades later a study was made by Hayward Keniston, followed by two American Council on Education studies: one by Allan M. Cartter and the other by Kenneth D. Roose and Charles J. Andersen.[1] The last two studies reported rankings for engineering departments but not for other professional schools. Very little previous work has been done on the reputations, or comparative perceived quality, of professional schools generally and medical schools specifically.[2] This paper reports findings of a study designed to examine the reputational standings of American medical schools as assessed by the community of physicians and scientists who work in them.[3]

While reputation should not be equated with quality, it also should not be dismissed as an insignificant part of the social reality of the medical community.[4] Reputation makes a difference because it has multiple consequences for students, for faculty members, and for medical schools. Students are concerned with the reputation of medical schools when they elect to apply to some rather than others. They are aware that the reputation of their alma mater has an impact on their subsequent career mobility; they perceive the medical school as a critical first step in their medical career, opening or closing future opportunities. Further, the reputation of their school influences students' self-esteem and affects perceptions of their ability within their significant reference groups. Faculty members are interested in reputations of schools when they consider appointments and promotions, not only because these reputations affect their own visibility and perceived ability in the larger medical community but also more basically because they hinder or enhance opportunities to obtain resources and facilities necessary for research. Correlatively, medical schools are concerned with their reputation because it affects their success in recruiting able faculty and outstanding students and in obtaining resources to carry out basic and clinical research. In short, general reputation has much to do with the actual quality of a medical school.

Previous work in the sociology of science has demonstrated that the reputations of scientists depend largely on meritocratic or universalistic bases. The quality of research performance and the honors received by scientists explain a large portion of the total variance on reputational standing. Sheer quantity of research has little independent effect on reputations (Cole and Cole, 1973; Gaston). The process of accumulating advantage also operates to reinforce and enhance the reputations of scientists who make important discoveries early in their careers (Merton, 1972; Zuckerman). The difference in reputation between the "rewarded" and the "unrewarded" increases over time because resources for future research are disproportionately placed in the hands of scientists who have made discoveries in the past. Most of this earlier work focused on individuals. In this paper, we examine schools and ask whether the basic stratification processes that affect reputations of scientists are similar to those affecting the standing of medical schools.

The data reported in this paper are, then, part of a larger ongoing study of stratification processes within a variety of social institutions. They are intended to promote further inquiries by students of the social stratification system within the medical profession. More immediately, we center on two questions: What is the distribution of American medical schools in terms of their perceived quality and their visibility among a stratified random sample of full-time faculty members on the medical and basic science staffs of these schools? What are the organizational and structural correlates of these rankings?

Method

A short questionnaire was sent to full-time medical school faculty members within all clinical and basic science departments in eighty-seven American medical schools. All schools in the United States approved by the American Medical Association as of 1971–72 were included in the population of schools from which the sample of respondents was drawn. There were ninety-four such schools in all, but seven were eliminated from the sample: four because available catalogs did not include the names of full-time faculty members after 1970 and three because they did not differentiate their faculties by academic rank.[5] The sample was stratified by three academic ranks—full professor, associate professor, and assistant professor—and included every twentieth faculty member within each academic rank. The questionnaire was designed to obtain an appraisal of the faculty quality and the effectiveness of training in all ninety-four fully approved institutions.

The questionnaire has two parts. Part I asked the medical faculty to rate medical schools on two dimensions of reputation: the perceived quality of its medical faculty and the perceived quality or the effectiveness of its medical training program. This questionnaire format is comparable to that used in the American Council on Education (ACE) studies of Graduate Arts and Sciences departments.[6] Faculty members were asked: "Please circle the number under the term that corresponds most closely to: A. Your judgment of the quality of the medical faculty at each institution. B. Your rating of the effectiveness of the medical training program at each institution." Seven response categories for A were presented to the physicians and scientists: Distinguished (6); Strong (5); Good (4); Adequate (3); Marginal (2); Poor (1); Insufficient Information (0).[7] Five response categories were provided for B: Extremely Attractive (4); Attractive (3); Acceptable (2); Not Acceptable (1); Insufficient Information (0).[8]

Part I contained an additional question: "What factors did you consider important when evaluating the quality of the medical faculty—and the effectiveness of medical training—for the schools listed?" Answers to this question were used to compare the subjective criteria of evaluation with quantitative correlates of perceived quality. Part II of the questionnaire requested information about the social background of respondents, such as their age, academic rank, and medical or scientific specialty.

It was not practical to ask each faculty member to rate all of the schools. Instead, three forms of the questionnaire were designed. Each contained the names of forty medical schools arranged alphabetically by the state in which they are located. Thirteen randomly selected schools appeared on all three forms to test for comparability of rankings by faculty members responding to the various forms. We received 186 usable questionnaires for Form I, 193 for Form II, and 204 for Form III. Rankings on the different forms are highly consistent. Johns Hopkins and UCLA medical schools are two typical examples drawn from the thirteen. The quality-of-faculty scores (sometimes referred to as "perceived quality scores") for Johns Hopkins were 5.11, 5.13, and 5.11, respectively, on each of the three forms; for UCLA they were 4.66, 4.55, 4.62, respectively. In short, the respondents to all three forms of the questionnaire answer in roughly the same way.

In all, 2,049 physicians and scientists were sent questionnaires; 30.3 percent responded. This rate of response and its distribution requires examination. There were 583 usable questionnaires; 19 could not be used, and 61 were returned because the faculty member was no longer affiliated with the medical school. The response to the questionnaire by faculty members of different ranks was not fully representative of the population. Full professors represented 29 percent of the respondents, but only a fifth of the sample; associate professors, 36 percent of the respondents but only a quarter of the sample; and assistant professors, 35 percent but roughly half the sample. There are, of course, several plausible reasons for this unrepresentative response rate. It may be that full and associate professors, with longer experience in the medical science community, are more likely to have extensive information about programs at other medical schools. They are apt to feel an obligation to know what is going on at other schools; it is a role-appropriate response. Perhaps younger scientists and physicians, more engrossed in surviving within the competitive world of American medical schools, are less likely to allocate time for answering questionnaires. Or, age and academic rank may be inversely related to skepticism about the value of social science research focusing on medical schools. Other factors may be operative as well.

In any case, though different ranking scientists and physicians respond in disproportion to their numbers in the population, this question is critical: Do they rate medical schools differently? As it happens, the differing response rate among the variously ranked faculty members creates no special problem in measuring the reputation of medical schools. The pattern of evaluations is almost exactly the same for all academic ranks. The rank-order correlation between the rankings of schools by full professors and associate professors is $r = .99$; between full and assistant professors, $r = .96$; and between associate and assistant professors, $r = .97$.[9] The extraordinarily high correlations indicate that there is no significant bias in terms of professional rank; rank simply does not influence the rating of the schools.

Of course, there are many possible bases of differential awareness and evaluation of medical schools. We examined several of these, including those that might result from differences in age, quality of PhD departments among scientists in the schools, and scientific or clinical specialty. Here no significant rank-order differences were obtained.[10] There was, however, one factor that did significantly influence evaluations: there were self-aggrandizement effects in rating medical schools.

To examine self-aggrandizement, the rating of Columbia, for example, by physicians and scientists currently affiliated with Columbia would be compared ideally with ratings by non-Columbia faculty. Since there were not enough respondents from each school to make individual comparisons, we aggregated respondents affiliated with similarly ranked schools. The procedure followed requires review. For each school in a group, the average rating given to a school by all those faculty affiliated with it was obtained. These ratings were summed over all raters in a group, and a group mean was computed. This mean is the average evaluation given by insiders to their own schools. For instance, forty respondents were affiliated with medical schools ranked among the first ten in quality of faculty. The sum of the ratings they gave to their own school, 216, was divided by forty, giving an average rank of 5.40. This insider score was then compared to the average rating of the same schools as judged by faculty members not affiliated with them (outsiders). Of course, in comparing a group of, say, ten schools, a faculty member may hypothetically be an insider only once but an outsider as many as nine times.[11] The grand mean for affiliated and non-affiliated respondents was computed for each group. The difference between the mean perceived quality scores of insiders and outsiders is taken as a rough estimate of self-aggrandizement resulting from current affiliation.[12] A second type of self-aggrandizement, that resulting from receiving medical training at a school, was estimated. The results of this analysis may be seen by comparing the columns of table 6.1.

There are two consistent patterns to the data. First, faculty members rate their own schools significantly higher than do others in the medical community. The average difference in means between insiders and outsiders is .67. Among the top ten medical schools, for example, insiders rated their schools at 5.40 compared to 4.93 among outsiders. Self-aggrandizement is particularly

TABLE 6.1 Estimates of Self-Aggrandizement Among Physicians and
Scientists Located in American Medical Schools

Rank of Medical School	Mean Perceived Quality Ratings of Medical Schools			
	Affiliated with the School*	Not Affiliated with the School	Attended the School*	Did Not Attend
1–10	5.40 (40)	4.93 (2,246)	5.60 (93)	5.08 (2,426)
11–20	5.37 (49)	4.50 (2,087)	5.13 (30)	4.53 (1,900)
21–40	4.60 (47)	3.92 (2,204)	4.76 (33)	3.94 (2,023)
41–60	3.97 (31)	3.51 (2,062)	4.20 (25)	3.54 (1,859)
61–94	3.86 (43)	3.00 (3,687)	4.06 (17)	3.01 (3,440)

*The number of faculty included under the affiliated and attended columns does not equal the total
sample size because there were three forms to the questionnaire, and in some cases respondents returned
questionnaires rating schools other than the one at which they were currently located.

strong among faculty members affiliated with lower-ranked medical schools,
from sixty-first to ninety-fourth: insiders rated these schools 3.86; outsiders
rated them 3.00. These are statistically significant differences.

Second, there is more self-aggrandizement in rating alma mater than in
rating current affiliation. Stratifying the faculty members by the medical school
from which they received their MD training produced average differences
between insiders and outsiders of .73. Faculty members who received their
medical education from the lower-ranked group of schools are most apt to
overrate their alma maters; correlatively, products of the highest-rated schools
are least likely to overrate them.[13] Although we find self-aggrandizement in
scores, it does not significantly distort the overall rank ordering of medical
schools. The rank-order correlation between perceived quality scores for insiders and outsiders is .84; between the rankings of schools by the entire sample
and outsiders it is .998. In sum, there appears to be an extraordinary degree of
consensus within the medical school community about the relative standing of
the faculty at the ninety-four schools.

These results are consistent with the Cartter study findings on self-aggrandizement. But they run counter to self-aggrandizement findings reported in
occupational prestige studies. At every level of prestige in the occupational
structure, people rate their own occupations almost exactly as others rate
them.[14] How can we account for the different patterns?

Perhaps the absence of significant self-aggrandizement in estimating occupational prestige results from continual reinforcement of the actual prestige
position of an occupation. Attempts by individuals to inflate the prestige of their
own occupation are continually negatively sanctioned in interactions. There
are fine-tuned social thermostats which constantly feed back to incumbents

the relative prestige of their occupation. Fine-tuned feedback mechanisms may not exist in the medical school community. Physicians and scientists in variously ranked medical schools tend to associate predominantly with others in the medical community who are in similarly ranked schools.[15] This pattern of differential association may reinforce tendencies toward self-aggrandizement.

These faculty quality estimates assume that those who respond represent the population. We know that this was not so in the medical school study, but the slight unrepresentativeness of the sample among different ranks apparently makes little difference in the rankings obtained. It does remain possible, however, that all of the respondents, regardless of academic rank and other social characteristics, differ significantly in their assessments from nonrespondents.[16]

The unit of analysis is the medical school, not the individual faculty member. There are two dependent variables in this study: the perceived quality and the visibility of medical schools.[17] Perceived quality will be discussed in detail but visibility only cursorily. A school's visibility score is taken as the percentage of all respondents who felt that they had sufficient information to rate it. It is obtained by dividing the total number of assessments (6 to 1), regardless of the quality ascribed, by the total number of respondents who returned the questionnaire. Perceived quality of faculties of medical schools is measured by considering only the evaluations of respondents who actually assessed the quality of a school's faculty. Among those who made judgments, the perceived quality score is the mean rating of the respondents. It has been noted that these scores vary from a high of 6 ("Distinguished") to a low of 1 ("Poor"). The number of cases on which perceived quality scores are based varies among schools of differing degrees of visibility.

We collected data on thirty-seven characteristics of each medical school. Some of these variables were computed by aggregating information obtained for each of the faculty members within the school.[18] For example, data on the number of papers published in scientific journals in 1972 by members of the clinical and basic science faculty of each medical school were collected from the Source Index of the *Science Citation Index* (SCI).[19] The aggregated number of publications was used as a characteristic of the medical school. Similarly, the eminence of the faculty of each school was estimated by aggregating the number of faculty members elected to the Association of American Physicians (AAP), "an organization limited to individuals who have made distinguished contributions to medical science." A second aggregated indicator of eminence was the total number of chairmen and deans of medical schools that any school had produced. Other characteristics of each school, not based on aggregation procedures, included the total funds it received in 1969 from the National Institutes of Health and from the Department of Health, Education and Welfare; the total funds received from these sources for research and development; and the total for training graduate as well as postgraduate students and fellows.

Data were obtained on faculty size, on social and individual characteristics of students, such as the sex composition and the applicant-to-acceptance ratio, as well as on the achievements of the school's graduates. All of these data were obtained from available sources (American Association Medical Colleges; Association of American Physicians; Council on Medical Education; Dube; Giza and Burns; Institute for Scientific Information; Singletary; Theodore).

Findings

A basic hypothesis of ours is that the reputations, or perceived quality, of medical schools will depend largely, in fact almost entirely, on their performance as scientific and research organizations. This assumes, of course, that reputation is largely a consequence of rational processes and of the application of performance criteria of evaluation. To test this hypothesis we will want to consider several questions. How much do the reputations of medical schools reflect their functional performance in research and their contributions to the growth of knowledge in the clinical and basic sciences? How strongly is reputation influenced by the level of federal funding of research? How closely is reputation related to the number of eminent stars on a school's faculty? Do irrational, or functionally irrelevant, characteristics of a school, such as its geographical location, or its sex and racial composition, influence its reputation? Finally, what social processes contribute to the persistence or change in reputations of American medical schools?

To address these questions, we turn to the results obtained in the survey. Table 6.2 presents the medical schools in rank order of perceived quality of their faculty. The itemized list is presented in detail both for its intrinsic interest and for its possible use in future research. The analysis in this paper, however, deals rather with those characteristics of schools which are associated with the distributions of these rankings than with the absolute values of assessments for particular schools.

The perceived quality scores range from a high of 5.71 to a low of 2.23; the mean score is 3.68 with a standard deviation of .70.[20] To guide interpreting differences in perceived quality scores, we computed the standard error of the estimate for each score.[21] These errors varied very little from one score to another. The mean standard error for the entire set of rankings is .082. It would be a mistake to attribute any statistical significance, therefore, to scores differing by less than twice the standard error, that is, by .164.[22] For example, it would be an obvious mistake to take the difference in scores between, say, Stanford (5.11) and Johns Hopkins (5.11), on the one hand, and Yale (5.00) on the other as either statistically or substantively significant. However, differences between school ratings of 5.00 and those of, say, 4.00 would be statistically significant.

TABLE 6.2 Perceived Quality and Visibility Scores of American Medical Schools

Medical School	Perceived Quality Score	Std. Dev.	Visibility Score	Number of Raters
Harvard	5.71	0.54	87.3	509*
Johns Hopkins	5.11	0.92	84.7	494*
Stanford	5.11	0.84	81.2	151
California, San Francisco	5.01	0.76	75.1	145
Yale	5.00	0.79	82.0	478*
Columbia	4.93	0.86	79.2	462*
Duke	4.77	0.82	82.4	159
Michigan	4.74	0.82	76.2	147
Cornell	4.71	0.80	76.9	143
Washington, St. Louis	4.68	1.00	80.3	155
University of Pennsylvania	4.66	0.84	75.6	146
Minnesota	4.62	0.82	69.0	402*
UCLA	4.61	0.81	74.4	434*
Albert Einstein	4.60	0.94	70.1	143
University of Chicago, Pritzker	4.52	1.06	57.0	110
University of Washington, Seattle	4.52	0.81	69.5	405*
Case Western Reserve	4.41	0.86	76.7	148
Rochester	4.37	0.79	69.4	134
Colorado	4.36	0.86	71.5	133
California, San Diego	4.27	1.01	60.3	123
Mount Sinai	4.22	0.94	67.9	131
New York University	4.18	1.03	60.8	113
Texas, Southwestern	4.11	1.00	48.2	93
Vanderbilt	4.10	0.90	61.8	126
North Carolina	4.07	0.76	59.8	122
Baylor	4.06	0.95	64.7	132
Tufts	4.05	0.91	68.6	140
University of Wisconsin	4.02	0.85	59.3	121
Northwestern	3.98	0.87	67.7	126
Emory	3.95	0.97	63.2	129
Boston University	3.95	0.94	65.1	121
Iowa	3.93	0.82	63.7	123
University of Virginia	3.89	0.85	58.0	112
Ohio State	3.79	0.84	64.0	119
Alabama	3.78	1.00	48.9	91
University of Florida, Gainesville	3.77	0.88	60.1	116
Dartmouth[†]	3.73	0.91	59.5	347*

(continued)

TABLE 6.2 (*continued*)

Medical School	Perceived Quality Score	Std. Dev.	Visibility Score	Number of Raters
Illinois	3.70	0.89	59.7	111
Tulane	3.68	1.00	62.7	128
Georgetown	3.68	0.96	75.1	145
Utah	3.68	0.82	50.0	93
Cincinnati	3.68	0.87	55.9	114
California, Davis	3.66	0.85	50.5	94
Penn State	3.64	1.03	46.8	87
Pittsburgh	3.64	0.80	59.7	111
Vermont	3.62	0.80	50.0	102
Virginia Medical College	3.62	0.85	60.2	112
Oregon	3.58	0.85	48.2	93
State University of New York, Syracuse (Upstate)	3.57	0.89	58.6	109
Michigan State	3.54	1.20	49.5	101
Indiana	3.54	0.87	52.0	106
Buffalo	3.53	0.84	59.1	114
Texas, Galveston	3.48	0.94	48.9	91
St. Louis	3.48	1.19	60.8	124
Temple	3.46	0.79	61.3	125
Miami	3.43	0.92	65.6	122
Medical College of Wisconsin	3.43	1.05	52.8	102
Maryland	3.41	0.82	64.2	124
Kansas	3.40	0.84	52.2	97
Albany	3.37	0.75	56.4	115
Bowman Gray	3.37	0.87	59.7	98
Arizona	3.34	0.95	37.7	77
Missouri	3.33	0.87	50.5	94
California, Irvine	3.32	1.04	45.6	93
George Washington	3.31	0.95	67.2	125
State University of New York, Brooklyn (Downstate)	3.31	0.84	57.4	117
Texas, San Antonio	3.25	0.75	34.7	67
Wayne State	3.24	0.89	54.8	102
Chicago Medical School	3.21	1.21	55.2	322*
Oklahoma	3.20	0.76	43.9	256*
Kentucky	3.20	0.82	52.9	97
Jefferson	3.19	1.02	61.1	118

New Mexico	3.17	0.80	50.0	93
Georgia Medical College	3.12	0.81	42.0	81
Tennessee	3.12	0.92	45.2	84
Louisiana State	3.05	1.07	45.7	85
Arkansas	3.05	0.78	39.9	77
Connecticut	2.96	0.95	44.6	91
Louisville	2.94	0.90	52.8	102
Medical College of Pennsylvania	2.94	1.07	47.8	89
Hahnemann	2.94	0.96	52.5	107
Loma Linda	2.93	1.00	46.6	90
West Virginia	2.92	0.81	36.7	214*
Nebraska	2.92	0.86	38.7	79
New York Medical College	2.92	1.00	59.0	344*
South Carolina	2.91	0.83	36.3	70
Mississippi	2.86	0.95	41.5	80
Ohio, Toledo	2.84	1.02	29.9	61
Howard	2.69	0.93	52.9	108
Loyola	2.64	1.02	42.2	86
Creighton	2.48	1.04	45.7	85
New Jersey	2.40	0.90	47.7	92
Puerto Rico	2.29	0.75	17.2	35
Meharry	2.23	0.85	47.8	239*

*These thirteen medical schools appeared on all three forms of the questionnaire. This accounts for the larger number of raters.

† Dartmouth, of course, provided education only in the basic sciences at the time of survey.

Of course, even scores that differ by more than twice the standard error may not be substantively significant.

Medical schools vary significantly in terms of their visibility to others within the medical school community, that is, in the percentage of faculty from other schools who feel they know enough about the designated school to appraise its faculty. The mean visibility score is 57.7; the standard deviation, 13.3.[23]

We come then to the central question: What characteristics of medical schools are correlated with their perceived quality, and what variables have the strongest independent effect in predicting these scores?

We hypothesized that the stratification of medical schools' perceived quality was largely a result of their functional performance in research.[24] Schools with the greatest resources in support of research, with faculty producing the most research, and with faculty recognized and honored for their research performance, would be rated most highly.[25] If the hypothesis has any face validity there must be fairly strong correlations between indicators of research performance and the perceived quality of schools. How much support is there, in fact, for this hypothesis?

In arriving at their appraisals, medical faculty gave quality of research far greater importance than teaching and other features of schools. Coding criteria given for the assessments of faculty research performance (180 mentions) and eminence of faculty (203 mentions) ranked just behind personal knowledge of or acquaintance with faculty members at other schools (214 mentions) as the most frequent bases given for evaluations. Teaching performance, listed next most frequently, was mentioned only 61 times. Are these bases for appraisals consistent with the correlation between perceived quality scores and objective measures of research output and eminence of faculty?

Table 6.3 presents a set of zero-order correlations of selected independent variables with the perceived quality of medical schools.[26] The data suggest a strong association between the publication of scientific research and the perception of quality of the schools at which the work originated. The primary indicator of productivity, as noted above, is the aggregated number of papers listed in the 1972 SCI, published in scientific journals by both the clinical and the basic science faculties of each medical school.[27] The correlation between perceived quality and this indicator of scientific output is .87, thus itself accounting for 75 percent of the variance on perceived quality.

The association between the number of research publications and perceived quality scores for the ninety-four medical schools is represented in the scattergram of figure 6.1. This association corroborates the criterion that the physicians and scientists reported having used in arriving at evaluations. If the unit of analysis had been the individual physician or scientist, rather than the medical school, the correlation between perceived quality and productivity would have been lower than .87 because there is, of course, significant variability in the number of scientific papers published by individual faculty members in schools of varying perceived quality.

While productivity is in general strongly associated with the perceived quality of medical schools, the various indicators of research productivity presented in table 6.3 yield differing correlations. As noted, the correlation for the faculty as a whole is .87. But if we standardize for size of faculty, thus producing average productivity per faculty, the same correlation is .58. This reduced correlation reflects the association between faculty size and perceived quality, as well as the skewed distribution of productivity within most medical schools. Indeed, it points to the possibility that the reputation of schools is more a function of the performance of a few stars located at the school than of the performance of most of its faculty.

TABLE 6.3 Correlation Coefficients Between Sets of Independent Variables and the Perceived Quality of Medical Schools

| | Dependent Variable |
Independent Variables	Perceived Quality
Productivity	
Total papers (clinical + basic science faculty – 1972)	.87
Papers/FTF (clinical + basic science)	.58
Papers (clinical faculty only – 1972)	.36
Eminence	
AAP members (total – 1972)	.79
Elites (deans and chairmen in 1972)	.60
Resources (1969)	
NIH total funds	.61
NIH graduate training and fellowship funds	.84
NIH research and development funds	.85
DHEW total funds	.63
DHEW graduate training and fellowship funds	.82
DHEW research and development funds	.84
Size (1971–72)	
FTF: total full-time faculty, clinical and basic science	.59
Total number of applicants	.23
Total number of students enrolled	.15
Percentage of female students	.11
Total number of graduates	.17
Percentage of first-year students from same state as medical school	−.34
Effectiveness of Training	.99

Still further, the data suggest what we might expect: there are different rates of scientific publication in the pre-clinical basic sciences and the clinical departments. Considering only the publications of clinical faculty members, less given than their basic science colleagues to work on research leading to publication, the correlation between scientific output and perceived quality is reduced to .36. This datum suggests that the perceived quality of medical schools is influenced more by the productivity of basic science than by clinical faculty.[28] In fact, the structural organization of medical schools may influence, in part, their perceived quality. Although most universities locate pre-clinical

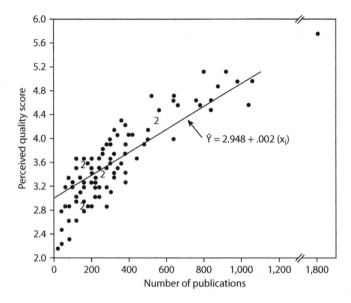

6.1 Scattergram of productivity of medical school faculty by perceived quality.

departments within the structure of the medical schools, some do not. Schools that do have affiliated pre-clinical departments benefit from the esteem associated with work produced by faculty members in those departments. Thus, it may seem questionable to include publications produced by pre-clinical faculty in the school totals, since some schools do not have well-funded basic science departments to contribute to their total. But the presence or absence of these departments in the medical school is part of the social reality of that school as viewed by the medical science community, and it is this social reality that we are trying to capture.

If the functional hypothesis is supported by the data on research productivity, it is given additional weight by data on the honorific achievements of a school's faculty. Total membership in the Association of American Physicians, one indicator of faculty eminence, is highly correlated with perceived quality, $r = .79$. A second indicator of faculty eminence was obtained by finding the number of medical school deans, and chairmen of departments of psychiatry, medicine, and surgery, in the academic year 1972–73, who had graduated from the school. The positions of department chairman and dean carry more prestige in medical schools than they do in Arts and Sciences departments. Schools were, therefore, classified by their record in producing chairmen and deans. This indicator of eminence is correlated .60 with perceived quality. These associations between faculty eminence and perceived quality suggest that it is the leading figures, the relatively few eminent men and women on the basic science and clinical faculties, who are largely responsible for a medical school's general reputation. As we shall see, however, the presence of a significant number of eminent faculty is itself correlated with other attributes of medical schools.

Given the strong correlations between the perceived quality of schools and the research performance and recognition of their faculties, we should also find a strong association between the financial resources available for research at medical schools and their ratings.

Indeed, this is the case. We collected data on the amount of NIH and Department of Health, Education, and Welfare (HEW) funds available for research and development and for educating doctoral and postdoctoral scientists at each medical school in 1969. These totals are strongly associated with perceived quality. Total NIH funding, for example, is correlated .61 and total NIH research and development funds, .85. Although we do not present the figures in the correlation matrix (table 6.4) for total HEW support of the various medical schools, the correlations between the HEW figures and perceived quality do not differ significantly from the perceived quality and NIH correlations. To cite only one example, the zero-order correlation between HEW research and development funds and perceived quality is .84.

These then are the basic zero-order correlations that lend support for the functional performance hypothesis. Is there any evidence to support an alternative hypothesis that irrational or functionally irrelevant criteria influence the reputations of medical schools?

Let us return to the data on faculty research performance and examine them more closely. The linear regression equation for the relationship between the production of scientific papers and perceived quality, illustrated in figure 6.1, is

$$\hat{Y} = 2.948 + .002\ X_i$$

where $\hat{Y}$ = predicted perceived quality score, and
X_i = scientific productivity of the medical faculty

This equation estimates how well faculty productivity predicts the perceived quality of a school. The high zero-order correlation means, of course, that it predicts perceived quality quite well. But there are some notable deviations from the least-squares regression line, a finding that gives us a clue to a functionally irrelevant basis of evaluation. It turns out that only four of the ninety-four schools had residuals greater or less than twice the standard error. Therefore, we lowered the criterion for identifying deviant cases, examining the list of schools that had estimated scores, or residuals, greater than or less than one standard error of the estimate, or ±.35 from their actual scores. Clearly, we are searching for substantively interesting clues to behavior rather than statistically significant patterns.

On the basis of faculty productivity, twenty-seven schools had predicted scores higher than their actual scores; twenty-two had scores lower than predicted. Harvard is the most notable school among those whose actual perceived quality is less than predicted by scientific productivity. Although Harvard ranked first among all medical schools, its faculty is so extraordinarily prolific that we would predict its perceived quality to be 1.06 points higher than it in fact is—a score that would exceed the upper bound of the scale. Only one other medical school associated with a major university, UCLA, had a perceived quality score substantially underestimated by its faculty productivity.

Of the schools with scores lower than predicted by faculty productivity, eight are located in the South. None of the eight is associated with better-known southern universities. Consider a concrete example. On the basis of faculty productivity alone, the University of Florida, Gainesville, should have a perceived quality score .36 higher than it actually received. Similar discrepancies are obtained for the Medical College of Georgia, LSU, University of Mississippi, South Carolina, Tennessee, and West Virginia. A number of southern schools with national reputations, including the University of North Carolina, Duke, University of Virginia, Emory, and Vanderbilt, had scores which were higher than we would predict on the basis of their faculty productivity.

That the predictions are higher than the actual ratings does not mean, of course, that the faculty at these southern schools are as prolific as those at higher-ranked ones.[29] The residuals do suggest that halo effects influence the perceived quality of medical schools. Those schools that are part of universities with national reputations have actual scores which either are predicted well or are somewhat higher than predicted by faculty productivity (Harvard and UCLA excepted). Correlatively, schools affiliated with universities without national standing tend to be victimized by their association—their actual scores are lower than predicted. Since these data do not control for the quality or impact of papers published by faculty members, the deviations from predicted values might well be lower if these factors were incorporated in the model.

The residuals suggest, then, a significant relationship between the geographical location of schools and their perceived quality. Regression analysis, in which region was recoded into a set of dummy variables, indicates that medical schools in the Northeast and the West have higher perceived quality scores, on average, than schools in other regions.[30] Location in the Northeast is correlated .19 and in the West, .17, with perceived quality. Correlatively, location in the South was negatively correlated with perceived quality scores, $r = -.20$. These effects of region, especially in the South, are not significantly reduced by the addition of performance variables to the regression equations.

Thus, we have at least one indication that the reputations of medical schools are not based solely on functional research performance. Are there other bases of evaluation which depart from the rational ideal? The data we have collected suggest that variables that could be taken as indicators of functionally irrelevant bases of evaluation have only minor effects. For example, the sex composition of the student population, which varies very little, is almost uncorrelated with perceived quality scores ($r = .11$). Similarly, privately endowed schools tend to have slightly higher scores than those that are state-supported ($r = .20$). Older schools, with longer traditions in medical education, have somewhat higher scores than younger schools ($r = .16$).[31] These correlations are quite weak, and in fact, since each of these independent variables is significantly correlated with the performance variables, their independent influence on perceived quality after controlling for faculty performance is minimal. In sum, geographic location of schools is the only factor that had an independent effect on perceived quality after controlling for research performance of faculty. The zero-order correlations suggest, then, that there is strong support for the research performance hypothesis.

We turn from the efficacy of the functional performance hypothesis to the relative predictive effects that variables associated with performance have on perceived quality scores. To estimate relative effects, we regressed perceived quality on six predictors: faculty productivity; faculty eminence; faculty size; total NIH graduate training and fellowship funds; total NIH research and development funds; and total postdoctoral students, fellows, and graduate students in clinical departments. The correlation matrix (table 6.4) indicates that faculty productivity is strongly correlated with other performance-related variables. In fact, the addition of the five predictor variables increases R^2 from .75—that is, the total variance explained by faculty productivity alone—to .80.[32] The relative weights of these six variables are strongly influenced by the high intercorrelations between the predictors, since there is strong multicollinearity in these data. For example, the regression coefficient in standard form for productivity is .38, but for faculty eminence, which is correlated .79 with perceived quality, it is only .05. The level of NIH funding, however, does have some independent predictive effect. The beta coefficient for total research and development funds received by schools is .23; it is .16 for total NIH training and fellowship funds. The relative effects of these six predictor variables are presented in table 6.5. It should be emphasized that given the high level of multicollinearity in these data, other regression equations produce roughly the same amount of total explained variance. But the estimates of R^2 are not influenced significantly by the restricted range in the dependent variable. Probit regression estimates yielded similar results.

We have avoided causal language, since it is in fact difficult to determine the causal sequence of the functional performance variables. The social processes that determine perceived quality are undoubtedly mutually interactive and self-reinforcing. By and large the same processes that determine quality also tend to maintain it. Although there may be notable shifts for a few universities that make extraordinary efforts at structural change and at faculty recruitment, there is probably much stability in these scores over time.

Consider how the current sharp differentiation in facilities, resources, and production of original scientific discoveries tends to maintain existing reputation distinctions. Past outstanding research performance by medical faculty increases the probability of financial support for current and future projects. But a high level of funding also makes high productivity and the production of original ideas feasible, especially in an age when basic science and medical research depend so heavily on adequate facilities and resources. It is this reciprocal combination—resources and performance—that attracts physicians and scientists of the first rank, as well as quality medical students. Since these institutional features of medical schools also result in added prestige, schools which already are superior are more apt to remain so.

TABLE 6.4 Correlation Matrix of Selected Characteristics of American Medical Schools

		Q	V	P	N	M	F	E	S	G
Q	Perceived quality									
V	Visibility	.846								
P	Productivity – total[1]	.866	.749							
N	NIH – total funds[2]	.608	.582	.681						
M	NIH – research and development funds[3]	.846	.745	.880	.773					
F	Faculty size[4]	.594	.610	.612	.573	.643				
E	AAP members – eminence of faculty[5]	.790	.641	.879	.642	.805	.555			
S	Number of postdoctoral students in clinical departments[6]	.683	.607	.706	.490	.614	.470	.622		
G	NIH – graduate training and fellowship funds[7]	.843	.716	.861	.765	.936	.632	.827	.684	
Mean		3.68	57.7	346.6	6,017[8]	3,203[8]	314.7	3.7	60.9	1,244
Standard deviation		0.70	13.3	288.4	5,339	2,952	207.2	4.9	94.2	1,231

[1-7] For a list of the sources of these variables, see note 26.

[8] In thousands of dollars.

Discussion

The value of the earlier Cartter and Roose-Andersen studies of the quality of graduate departments is often obscured by the negative reactions to some of the gross measures used in their design. Those studies have been properly faulted for not specifying and measuring the multiple dimensions of the comparative quality of departments and also for depending largely on mere estimates of prestige (Elton and Rodgers; Knudsen and Vaughan; Lewis; Magoun). The rough measures encouraged readers to assume that the small differences in scores could be equated to actual differences in quality. The numerical rankings tended to become reified.

With all their limitations these studies assessing the quality of departments had, to use Paul Allison's felicitous phrase, "potential for elaboration," and in fact contributed greatly to studies in the social organization of science. For instance, studies of social mobility among academic scientists would have been difficult to do without even rough estimates of the quality of academic affiliations, in as much as these studies measured a form of recognition associated with prestigious academic affiliation (Cole and Cole, 1967; Crane; Gaston). The Cartter ratings made it possible to study in detail this feature of the reward system of science. Further, the Cartter and Roose-Andersen rankings provided

TABLE 6.5 Correlations and Standardized Regression Coefficients of the Determinants of the Reputations of American Medical Schools

Predictors	Dependent Variable: Perceived Quality of Medical School	
	Zero-Order Correlation	Regression Coefficient
Productivity—total	.866	.38
NIH graduate training and fellowship funds	.843	.16
Size (number of fellows, graduate and postdoctoral students in clinical science departments)	.683	.12
NIH research and development funds	.846	.23
Size of full-time faculty	.594	.03
Eminence of faculty (number of members in AAP honorific society)	.790	.05
Multiple R		.89
R squared		.80
Residual		.45

an opportunity to investigate the social correlates of the evaluations and to uncover the variables that are associated with the rise and fall of academic departments (Hagstrom). Testimony to the impact of these ranking studies may be found in the number of citations to them in the *Social Science Citation Index* (SSCI). In 1974, the Cartter study, published eight years earlier, received a total of twenty-five citations. When this total is compared to the average number of citations received by any cited paper, which is just over one, the relative impact of the report should be clear.[33] Further, of these citations, twenty-three came from different authors publishing in sociological, economic, and political science journals. The twenty-five citations do not include, of course, the many citations to papers that made extensive use of the ratings.

Nonetheless, studies assessing the quality of schools—the present study of medical schools, the Cartter study, and others—are subject to several limitations. Let us note the limits to the data reported here. First, we do not investigate the *quality* of medical schools per se but only their *perceived quality* (Cole and Cole, 1973). This is not a trivial distinction. The perceptions of these schools by the physicians and scientists may be based on inadequate or false information; their images may be distorted or dated. The perceived quality may reflect a halo effect generated by the reputation of the larger university with which the medical school is associated. When future research on the quality of medical schools, using different indicators of quality, is accomplished, it will be useful to compare the several measures.

Second, the rankings presented in this paper are not fine measurements. Differences in rank between any adjacent ranking schools are not necessarily substantively significant. In short, let us not reify these numbers. Third, the data are the product of a single measuring instrument at one point in time. Other measurements might change some of the obtained rankings. Moreover, slightly different wording of the questions to which the medical faculty responded might yield somewhat different results. Fourth, our study does not deal with the problem of intersubjectivity. We did ask physicians and scientists to indicate the criteria they used to arrive at their judgments. But they may have used differing subjective criteria and levels of actual knowledge to reach the same judgment that the quality of a particular school is, for example, distinguished. We have no way of telling the relative weights assigned to evaluation criteria by respondents reporting their judgments.[34] Finally, the questions on which our rankings are based ask only for evaluations of the quality of the faculty of a medical school as a whole.

Clinical and basic science departments within each institution are not assessed separately. This makes for rough overall estimates, since there is, of course, extensive variability in both the actual and the perceived quality of different departments within the same school. In large part, physicians and scientists base their evaluations on knowledge of their own specialties or related ones in other schools rather than on actual knowledge of the entire faculty.

The measures of perceived quality and of visibility are for the medical schools as a whole, not for specific specialties.

What then can we tentatively conclude from the data reported in this paper? Medical schools in contemporary American society vary widely in terms of their reputations, that is, in terms of their perceived quality and visibility. The characteristics of medical schools that predict these ratings, and are likely to determine them, are associated with basic science and clinical research performance: faculty productivity, eminence of the leading members of the faculty, and resources available for research. In sum, the hypothesis that functionally relevant criteria of evaluation of medical schools predominate in the formation of assessments of their quality is strongly supported by the data. There is some limited evidence, however, that functionally irrelevant characteristics of schools and of raters influence evaluations. Medical schools located in different geographical regions receive somewhat varied evaluations, and ratings by alumni of their alma maters and by current faculty of their present affiliation produce some self-aggrandizement.

Notes

1. Actually, estimates of quality date back to 1911 when the Bureau of Education prepared a rating of 344 institutions at the request of the Association of American Universities. The study by Hughes in 1925 was repeated by the American Council on Education in 1934. The results of the 1934 study were reported by Logan Wilson.
2. Margulies and Blau and Blau and Margulies report evaluations of American professional schools, which were based on questionnaire responses by deans of the various schools. Each dean was asked to list the top five institutions in the profession. Of course, for each discipline only a limited number of schools were mentioned. For medicine, only five schools were listed, and these ratings were based on a sample of eight deans responding to an initial survey. Based upon a follow-up survey, eleven schools, rated by fifty-one deans, were listed in rank order.
3. The absence of systematic estimates of the rank of medical schools has created multiple research problems for sociologists of medicine. Some of these are discussed in Barber et al.
4. Extended discussions of the significance of reputational rewards in science are found in Merton (1960).
5. The available catalogs for Loma Linda, Loyola, Harvard, and Michigan State were dated prior to 1970; Howard, New Mexico, and Texas-San Antonio did not distinguish the ranks of faculty members in their bulletins.
6. The ACE format was adopted in order to compare the scale scores for the medical schools and those for Arts and Sciences departments.
7. The actual questionnaire used a scale range of 7 to 1, with "Insufficient Information" being scored "1." The category "Insufficient Information" is here scored "0," and the scale has been transformed into a six-point scale.
8. It turns out that the correlation between the faculty scores and the training scores is .99. In effect, there are no differences in the ratings of medical schools along these two dimensions. Therefore, the training scores need not be discussed.

9. These rank-order correlations are for the responses to Form II. The same pattern is found for the other two forms.

10. On differential awareness of the performance of scientists, see Cole and Cole (1968).

11. For those schools in the top ten, there were a total of 2,246 responses, since each respondent could theoretically rate at least nine institutions, and ten if he or she was not affiliated with a school in this group. When the sum of the ratings was divided by 2,246, the overall mean was computed as 4.93 for the ten schools.

12. The insider and outsider perspectives are developed by Merton (1972).

13. These results are not attributable to a "ceiling effect"; the average scores do not approach the scale's upper bound.

14. Religious groups tend to aggrandize their own reputations. D. J. Treiman reported these unpublished data to me in private communication.

15. The idea of differential association must remain speculative until more research is done on the interaction patterns of physicians and scientists at different schools.

16. An attempt to check on nonresponding faculty by a limited telephone inquiry found that faculty members who did not answer in the first place would not respond to further queries. Consequently, bias resulting from differences between those who responded and those who did not remains a possibility.

17. Many interesting questions about the variations in individual responses to the set of medical schools will not be addressed here. For example, are the perceptions of schools influenced by the eminence of the observer? For discussions of problems similar to these in the larger scientific community, see Cole and Cole (1968).

18. On types of aggregation as a method of constructing new variables, see Lazarsfeld.

19. Publication counts are limited to the journals abstracted and listed by the Institute for Scientific Information in the Source Index of the SCI. More than 2,500 journals are abstracted annually.

20. The omission of the University of Southern California medical school was the result of a clerical error in printing the questionnaire. Its perceived quality and visibility scores are not presented in table 6.2. Data on the characteristics of Southern California and its faculty were collected, however, and on the basis of a prediction model discussed below, which explains more than 90 percent of the variance on perceived quality, Southern California had an estimated perceived quality score 3.96. A small follow-up study could be designed to assess the accuracy of this prediction.

21. The standard error is, of course, easily obtained by dividing the standard deviation of the ratings for a school by the square root of the number of raters, i.e., $s / \sqrt{n}$.

22. When examining scores in terms of their standard errors, the difference of two standard errors is the conventional rule for establishing statistical significance. Thus, scores which do not differ by more than this should be viewed cautiously in terms of their statistical significance.

23. There will be no further detailed discussion of visibility scores and their correlates. The determinants of visibility are much the same as those for perceived quality.

24. For the functional principle from which this hypothesis derives, see Barber; Cole and Cole (1973); and Merton (1968).

25. The relationship between scientific productivity and scientific recognition has been a major focus of attention in the sociology of science for the past ten years. In particular see Barber et al. and Cole and Cole (1973) for a discussion of this relationship for faculty members in two types of hospitals.

26. There are only two significant differences in the way visibility and perceived quality scores correlate with the independent variables presented in tables 6.3 and 6.4.

Indicators of research productivity, faculty eminence, and federal resources are consistently, but only slightly, more highly correlated with perceived quality than with visibility; correlatively, indicators of size of schools, student populations, and numbers of graduates are more strongly correlated with visibility than with perceived quality.

27. Citation indicators of quality of output were not collected, since that would have meant collecting data for each faculty member at every medical school. Publication counts could be more conveniently obtained because the Source Index of SCI lists all publications by the affiliation of the authors as well as by authors' last names. Thus, the count for productivity of schools was easily obtained. Adequate controls for the impact of scientific papers produced by faculty at the medical schools would require the count of citations for individual faculty members and the aggregation of these individual scores. On the correlation between productivity counts and citation counts, see Cole and Cole (1973).

28. Productivity counts for clinical faculty vary greatly among and within the different schools.

29. If the actual scores of these southern schools were closer to those predicted by their faculty productivity, the span between highest and lowest scores would be further reduced. A comparison of medicine and other professions in terms of the distance between higher- and lower-ranked schools should be done.

30. Five dummy variables were constructed, in which a region was given a value of 1 and all other regions were scored 0. In the regression equations we omit one dummy variable in each equation. Cartter's (Appendix E) categorization of a school's geographic region was used.

31. The age of medical schools is to some extent related to the eminence of its graduates. Clearly, medical schools that only recently opened could not have many graduates who are deans or in AAP.

32. The maximum amount of variance explained on perceived quality by a subset of the thirty-seven school characteristics is .947, with a total of thirteen independent variables; it is .999 with sixteen variables in a stepwise regression equation. These prediction equations did not include either the school's visibility or its rated effectiveness in training students. Of course, as the explained variance approaches 1.0, the beta weights become unstable. The variables included in the sixteen-variable regression equation are publications; total NIH funds; year of founding of the school; number of department chairmen and deans; percent of graduates who are not a member of any specialty board; percent of first-year class completing four years of college; total HEW funds; number of Fellows, postdoctoral and graduate students enrolled in the clinical sciences; total number of graduates; total number of graduates alive as of December 31, 1967; number of "elites" per graduate; percent of graduates who are not federally employed; total number of full-time faculty; HEW research and development funds; total number of students; geographic region: East; and geographic region: not East.

Since we have a small number of cases we corrected the total R_2 explained for degrees of freedom. This correction is obtained with the following equation (McNemar).

$$R = 1 - [(1 - R^2)(N - 1)] / (N - n)$$

where N = number of cases, and
n = number of variables, independent and dependent.

The sample size does not significantly affect the R_2. For the six-variable regression discussed in the text, the adjusted R_2 is .79; for the sixteen-variable equation, it is .9988.

33. For impact of the evaluations as primary sources, we examine works that refer to the authors of the studies in the *Science Citation Index*; for a more extensive measure of impact we would count citations to works that have made extensive use of the ratings.

34. Standardized criteria for evaluation plainly require more research. In fact, consensus on standardized evaluation criteria is a problem faced in many settings, but it is particularly significant in the process of refereeing journal articles and in peer review decisions.

References

American Association of Medical Colleges. 1972. *Directory of American Medical Educators, 1972–73.* Washington, DC

Association of American Physicians. 1972. *Transactions* 95. 85th Session.

Barber, B. 1975. "The Limits of Equality: Social Stratification in Complex Societies." Paper presented for the More Equality as a Goal of Public Policy conference. Princeton, NJ: Institute for Advanced Study.

Barber, B., J. J. Lally, J. L. Makarushka, and D. Sullivan. 1973. *Research on Human Subjects.* New York: Russell Sage.

Blau, P. M., and R. Z. Margulies. 1974. "The Reputation of American Professional Schools." *Change* 6 (December): 42–47.

Cartter, A. 1966. *An Assessment of Quality in Graduate Education.* Washington, DC: American Council on Education.

Cole, S., and J. R. Cole. 1967a. "Scientific Output and Recognition; A Study in the Operation of the Reward System in Science." *American Sociological Review* 32 (June): 377–90.

——. 1968b. "Visibility and the Structural Bases of Awareness of Scientific Research." *American Sociological Review* 33 (June): 397–413.

Cole, J. R., and S. Cole. 1973. *Social Stratification in Science.* Chicago: University of Chicago Press.

Council on Medical Education, American Medical Association. 1972. "Medical Education in the United States, 1971–1972." *Journal of American Medical Association* (November): 921–1081.

Crane, D. 1965. "Scientists at Major and Minor Universities: A Study of Productivity and Recognition." *American Sociological Review* 30 (October): 699–714.

Dube, W. 1973. "Study of U.S. Medical School Applications, 1971–1972." *Journal Medical Education* 48: 395–403.

Elton, C. F., and S. A. Rodgers. 1971. "Physics Department Ratings: Another Evaluation." *Science* 174 (November): 565–68.

Gaston, J. 1973. *Originality and Competition in Science.* Chicago: University of Chicago Press.

Giza, R., and C. Burns. 1971. "DHEW Obligations to Medical Schools, Fiscal Years 1967–1969." Washington, DC: Department of Health, Education and Welfare.

Hagstrom, W. 1971. "Inputs, Outputs and the Prestige of the American University Science Departments." *Sociology of Education* 44 (Fall): 375–97.

Hughes, R. M. 1925. *A Study of the Graduate Schools of America.* Oxford, OH: Miami University.

Institute for Scientific Information. 1973. *Science Citation Index, 1972.*

Keniston, H. 1959. "Appendix: Standing of American Graduate Departments in the Arts and Sciences." In *Graduate Study in the Arts and Sciences at the University of Pennsylvania.* Philadelphia: University of Pennsylvania.

Knudsen, D. D., and T. R. Vaughan. 1969. "Quality in Graduate Education: A Reevaluation of the Rankings of Sociology Departments in the Cartter Report." *American Sociologist* 4 (February): 12–19.

Lazarsfeld, P. F. 1958. "Evidence and Inference in Social Research." *Daedalus* 87: 99–130.

Lewis, L. S. 1968. "On Subjective and Objective Rankings of Sociology Departments." *American Sociologist* 3 (May): 129–31.

Magoun, H. W. 1966. "The Cartter Report on Quality in Graduate Education." *Journal of Higher Education* 37 (December): 481–92.

Margulies, R. Z., and P. M. Blau. 1973. "America's Leading Professional Schools." *Change* 5 (November): 21–27.

McNemar, Q. 1969. *Psychological Statistics*, 4th ed. New York: Wiley.

Merton, R. K. 1960. "'Recognition' and 'Excellence': Instructive Ambiguities." In *The Sociology of Science: Theoretical and Empirical Investigations*, ed. N. Storer, *The Sociology of Science: Theoretical and Empirical Investigations*. Chicago: University of Chicago Press, 1973.

——. 1968. *Social Theory and Social Structure*. New York: Macmillan.

——. 1972. "The Perspectives of Insiders and Outsiders." In *The Sociology of Science*. Chicago: University of Chicago Press, 1973.

Roose, K. D., and C. J. Andersen. 1970. *A Rating of Graduate Programs*. Washington, DC: American Council on Education.

Singletary, O. (ed.). 1972. *American Universities and Colleges*, 11th ed.

Theodore, C. 1968. *Medical School Alumni, 1967.* Chicago: American Medical Association.

Treiman, D. J. Forthcoming. *Occupational Prestige in Comparative Perspective*. New York: Academic Press.

Wilson, L. 1942. *The Academic Man: A Study in the Sociology of a Profession*. New York: Oxford University Press.

Zuckerman, H. Forthcoming. *The Scientific Elite*. New York: Free Press.

CHAPTER 7

AGE AND SCIENTIFIC PERFORMANCE

(1979)

STEPHEN COLE

The long-standing belief that age is negatively associated with scientific productivity and creativity is shown to be based upon incorrect analysis of data. Analysis of data from a cross-section of academic scientists in six different fields indicates that age has a slight curvilinear relationship with both quality and quantity of scientific output. These results are supported by an analysis of a cohort of mathematicians who received their PhDs between 1947 and 1950. There was no decline in the quality of work produced by these mathematicians as they progressed through their careers. Both the slight increase in productivity through the thirties and the slight decrease in productivity over the age of fifty are explained by the operation of the scientific reward system. By encouraging those scientists who produce the most favorably received work and discouraging those who produce work that is not favorably received, the reward system works to reduce the number of scientists who are actively publishing. Those who continue to publish throughout their careers are a "residue" composed of the best members of their cohort. Increases in productivity through the thirties and into the forties are shown to be a result of command over the resources necessary to be highly productive. In the last part of the paper I examine the hypothesis that it should be easier for young scientists to make important discoveries in the more highly developed or codified sciences than in the less highly developed ones. The data do not support this hypothesis but, rather, suggest that scientists in all six fields are approximately equally likely to make important discoveries shortly after they receive their PhDs.

Einstein was twenty-six when he developed his special theory of relativity; Newton was twenty-four when he began his work on universal gravitation, calculus, and the theory of colors; and Darwin was twenty-nine when he developed his theory of natural selection (Zuckerman and Merton 1973). Because of the well-known fact that many important discoveries have been made by very young scientists, it is commonly believed by scientists and laymen alike that age is negatively correlated with scientific productivity and creativity. It is assumed that significant research output declines with advancing age. Perhaps the most frequently cited source of support for this assumption is the now classic work of H. C. Lehman, *Age and Achievement*, published in 1953. Although Lehman's methodology has been criticized (Dennis 1956a, 1956b, 1956c, 1966), his conclusions are still widely accepted. It is the purpose of the first section of this paper to reexamine the relationship between age and scientific output. Is it really true that older scientists produce fewer and less significant papers than younger scientists and, if so, why? The answers to these questions are perhaps more important today than in the past. The exponential growth of the scientific community is leveling off. This means that in the next twenty years the average age of American scientists will probably increase, as there will be relatively few new positions for young people to fill. Will the increase in age of our scientists reduce our scientific capacity? To answer this question, we must know whether the commonly held belief about the relationship between age and scientific performance is correct.

Let us begin by reviewing Lehman's work. His basic source of data is histories of science. He takes all the discoveries listed in prominent histories of science and then constructs a chart showing the number made in each five-year age period. To handle the problem of differences in life span, he uses the average number of discoveries per living scientist in his sample. His charts show that more discoveries are made by young scientists than by old ones. He does not, however, take into consideration the number of scientists alive in each age group in the population. Since science has been growing exponentially for the past few centuries, at any time the population of scientists is likely to be made up disproportionately of young people. Therefore, the conclusions based on Lehman's method are questionable.

What Lehman really has are data on only a small portion of the population of scientists—those who have made important discoveries. Instead of asking what proportion of scientists in different age groups make important discoveries, he has asked what proportion of important discoveries were made by scientists of different ages. Thus, Lehman's results are not necessarily contradictory with the conclusion that age has no causal influence on the process of discovery.

Table 7.1 presents a hypothetical example to illustrate the point. Suppose that at any given time we were to make a list of all the discoveries that had ever been

180 VALUES AND REWARDS IN SCIENCE

TABLE 7.1 Hypothetical Figures Illustrating Lehman's Error

A. Hypothetical Data on Population of Scientists

Age of Scientist	N Scientists Making Important Discoveries	N Scientists Not Making Important Discoveries	Total N Scientists
Under 30	500	4,500	5,000
30–39	400	3,600	4,000
40–49	200	1,800	2,000
50–59	100	900	1,000
60 or over	50	450	500
Total	1,250	11,250	12,500

B. Lehman's Method of Percentaging

Age of Scientist	% Scientists Making Important Discoveries
Under 30	40
30–39	32
40–49	16
50–59	8
60 or over	4
Total ($N = 1,250$)	100

C. Percentaged Correctly

Age of Scientist	% Scientists Making Important Discoveries	N Scientists
Under 30	10	5,000
30–39	10	4,000
40–49	10	2,000
50–59	10	1,000
60 or over	10	500
Total		12,500

made and were to classify them by the age of the discoverer and their importance. Most of the discoveries, important and unimportant, would have been made by young people, since science at any time has had many more young practitioners than old ones. This procedure might yield figures such as those in table 7.1A, except, of course, that the real figures would be many times larger. What Lehman has done is to examine only the first column of figures, scientists making important discoveries, and percentage in the manner illustrated in part

B. From these percentages he would conclude that the ability to make important discoveries declines with age. A correct percentaging of the table (shown in part C) would indicate that age has no influence at all on making important discoveries. Among the scientists in each age group, 10 percent have made such discoveries. The proper conclusion would be that the bulk of important discoveries is indeed made by young scientists because most scientists are young, not because age has a causal influence on scientific creativity.

Age and Scientific Output

Two more accurate ways than that chosen by Lehman to examine the influence of age on scientific performance would be to analyze the research output of a sample of scientists of varying ages at one time and to examine the productivity of a cohort of scientists as they move through their careers. I have done both but will begin by reporting the results of the cross-sectional analysis. The data consist of the research output of random samples of scientists working in PhD-granting institutions. Depending on the size of the field, the samples are drawn from either all departments or every other department included in the 1969 American Council on Education (ACE) study of departmental prestige (Roose and Andersen 1970). Within each department every other member was sampled.[1] The six fields selected for study were chemistry, geology, mathematics, physics, psychology, and sociology. For each field except mathematics the sample was drawn from scientists employed in 1970 in PhD-granting institutions rated by the ACE in 1969; the sample of mathematicians was drawn from those employed in ACE-rated schools in 1975.[2]

For each scientist, data on the quality and quantity of his or her current scientific output were collected. "Current output" is defined as papers published from 1965 to 1969.[3] Data on the number of papers published by each scientist were obtained from the appropriate abstracting journal. For example, the name of each chemist was looked up in *Chemical Abstracts* and a count was made of the number of papers published from 1965 to 1969.[4]

To measure the quality of publications, the 1971 volume of the *Science Citation* Index (SCI) was used. Counts were made only of citations to work published between 1965 and 1969 in order to arrive at a measure of the quality of current publications (Cole and Cole 1973, chap. 2).[5] By now there is a large body of literature which suggests that citations are a roughly adequate indicator of the quality of work, as long as "quality" is defined as the opinion that the scientific community currently holds of the work. The number of citations to a scientist's work has been shown to be highly correlated with the way colleagues evaluate the significance of that work on questionnaires (Cole 1978), with election to the National Academy of Sciences (Stern 1978), with the likelihood of receiving

182 VALUES AND REWARDS IN SCIENCE

a Nobel Prize (Cole and Cole 1973, chap. 2), and with scientists' reports of who has exerted the most significant influence on their work (Mullins et al. 1977). As long as we are examining groups of scientists, as opposed to a specific individual scientist, there can be little doubt that those scientists who, on the average, have received a high number of citations to their work have produced work which most of their colleagues would deem more significant than those scientists who, on the average, have received relatively few or no citations.[6]

In table 7.2 data are presented showing the relationship between age and current productivity for the six fields, both separately and in combination. It turns out that, in general, age is curvilinearly related to productivity. Productivity rates rise, peaking either in the late thirties or the forties and then drop off. The same curvilinear relationship was observed in five of the six fields, with the point of inflection varying slightly.[7] Similar results were obtained by Bayer and Dutton (1977), who looked at the number of self-reported publications in the two previous years by samples of scientists in seven different fields. Using regression techniques, they attempt to compute the best-fit function for each field. They conclude: "Application of linear models are especially weak in

TABLE 7.2 Age and Scientific Productivity, 1965–1969, Six Fields: Mean Number of Published Papers

| Field | Age | | | | | | |
	Under 35	35–39	40–44	45–49	50–59	60+	Total
Chemistry	11.5	11.4	16.0	16.3	13.9	10.3	13.0
	(115)	(62)	(55)	(34)	(61)	(29)	(356)
Geology	7.3	7.6	8.1	7.0	3.8	3.4	6.3
	(45)	(60)	(60)	(36)	(49)	(41)	(291)
Mathematics[a].	5.1	7.3	6.2	3.5	5.2	6.1	5.6
	(101)	(96)	(67)	(63)	(73)	(35)	(435)
Physics	4.5	5.3	6.2	5.6	4.4	3.4	5.1
	(138)	(153)	(111)	(84)	(61)	(45)	(592)
Psychology	5.6	6.4	6.4	4.9	3.3	4.4	5.4
	(151)	(101)	(92)	(94)	(79)	(27)	(544)
Sociology	1.8	3.8	5.2	6.3	5.1	4.4	4.1
	(60)	(41)	(40)	(33)	(39)	(29)	(242)
Six fields combined	6.1	6.8	7.7	6.3	5.9	4.6	6.4
	(610)	(513)	(425)	(344)	(362)	(206)	(2.460)

Note: Ns are in parentheses.
[a] For mathematicians, 1970–1974 is the period used.

explaining variance in professional activity with career age. However, even the 'best-fit' a priori model for most fields, represented by the fourth-degree polynomial, explains less than seven percent of the variance even in the case of the field (experimental psychology) with the highest obtained R for prediction of publication output over the past two years" (1977, p. 274). Although Bayer and Dutton observe some differences in the best-fit models for the seven fields, all but chemical engineering show a curvilinear relationship between age and productivity. Since computations made on the data presented here, as well as those made by Bayer and Dutton, indicate that age does not explain much variance in productivity, I have concentrated on explaining why productivity gradually increases in the earlier years of a scientist's career and drops off at the end.

Since we are not at this point interested in field differences, the data from all six fields are combined in the last row of table 7.2. In general, the data support Bayer and Dutton's conclusion that age explains very little variance in productivity. There is a difference of only 3.06 papers between the most productive age group (40–44) and the least productive (60+). Productivity gradually increases to about the age of 45 and then gradually decreases. In most of the fields studied the scientists over the age of 60 were not much less productive than those under 35.

Although young scientists on the average may publish fewer papers than colleagues in their forties, the papers they do publish could be more significant. Perhaps when we consider the relationship between quality of work and age we will find the expected relationship. The truly important discoveries may be disproportionately the work of young scientists. Using the number of citations received by recent work as a measure of the significance of publications a rough test of this hypothesis can be made.[8] Table 7.3 presents data on citations to recent work in the six fields. Again, we find a basically curvilinear relationship, although the curve rises less steeply than it does for productivity. Combining data from all six fields yields the figures in the last row of the table. Basically, age has a slight curvilinear effect on the quality of published work by scientists between the ages of 30 and 50. Scientists over the age of 45 are slightly less likely to publish high-quality research than those under 45. It might be objected that the operational definition of high-quality work is not exacting enough. Perhaps the highest-quality discoveries are made disproportionately by young scientists. For the data presented in table 7.3, no matter how we classified "high-quality" work the result remained unchanged.

It has frequently been pointed out that longitudinal data are superior to cross-sectional data in measuring the effects of age on scientific productivity (Allison and Stewart 1974). Cross-sectional data do not allow us to disentangle the effects of age from cohort effects. Therefore, longitudinal data were collected for a cohort of mathematicians. Mathematics was chosen for this purpose because youth is believed to be more significantly correlated with creative productivity in math than in any other science (Lehman 1953). Thus, in a cohort study of mathematicians we would presumably have the greatest chance

184 VALUES AND REWARDS IN SCIENCE

TABLE 7.3 Age and Citations to Work Published, 1965–1969, Six Fields:
Mean Number of Citations in 1971 SCI

Field	Age						Total
	Under 35	35–39	40–44	45–49	50–59	60+	
Chemistry	14.4	11.4	20.2	18.4	12.1	7.4	14.2
	(115)	(62)	(55)	(34)	(61)	(29)	(356)
Geology	5.6	6.5	7.2	5.7	2.7	1.2	5.0
	(45)	(60)	(60)	(36)	(49)	(41)	(291)
Mathematics[a]	2.7	3.8	5.8	3.4	5.6	5.1	4.2
	(101)	(96)	(67)	(63)	(73)	(35)	(435)
Physics	11.2	15.1	10.8	6.8	7.4	15.9	11.5
	(138)	(153)	(111)	(84)	(61)	(45)	(592)
Psychology	5.2	6.6	6.8	5.1	3.3	3.3	5.3
	(151)	(101)	(92)	(94)	(79)	(27)	(544)
Sociology	.8	1.6	2.4	3.6	1.8	1.5	1.8
	(60)	(41)	(40)	(33)	(39)	(29)	(242)
Six fields combined	7.5	8.8	9.1	6.4	5.7	6.3	7.5
	(610)	(513)	(425)	(344)	(362)	(206)	(2,460)

Note: Ns are in parentheses.
[a] For mathematicians, 1970–1974 and the 1975 SCI were used.

to find the anticipated effect of age not shown by the cross-sectional data. For all 497 people who earned a PhD in mathematics in American universities between 1947 and 1950, data were collected on the number of papers published in each of twenty-five years and the number of citations to these papers.[9]

Productivity and citation data for this cohort are presented in table 7.4. As the first column indicates, the mean productivity of the members of the cohort did not change significantly during that period. In the first five-year period after the cohort had completed its dissertations, the average number of papers published was 2.4. In the last period for which we have data, 1970–1974, ending twenty-five years after all members of the cohort had received their PhD's, the average number of papers published was 2.6. Thus, the data on a cohort of mathematicians support the conclusion based upon a cross-section analysis that productivity does not differ significantly with age. The second column of table 7.4 shows the mean number of citations in 1975 to papers published by members of the cohort in each of the five-year periods. The means range only from 0.84 to 1.4 for work produced in the first and fourth periods, respectively. For the fifth period the mean is 1.1. Therefore, if we used citations in 1975 as

TABLE 7.4 Mean Numbers of Papers Published and Citations to Them for Cohort of Mathematicians Receiving Ph.D. between 1947 and 1950

Date of Publication	Mean N Papers Published	Mean N Citations Listed in 1975 SCI	Mean N Citations Listed in Volume of SCI Closest to Time of Publication
1950–1954	2.4	.84	.33 (1961 SCI)
1955–1959	2.8	1.2	.78 (1961 SCI)
1960–1964	2.3	1.1	.96 (1965 SCI)
1965–1969	2.8	1.4	1.4 (1970 SCI)
1970–1974	2.6	1.1	1.1 (1975 SCI)

the only indicator of the quality of work, we might conclude that there have been no significant changes in the quality of work produced by this cohort throughout the twenty-five years. The last column of table 7.4 shows the mean number of citations received by work published by members of the cohort in the different periods when we use the edition of the SCI which is closest in time to the publication date. For example, for 1950–1954 the closest volume of the SCI would be 1961. When we use this measure we see that there has been a gradual increase in the average number of citations received to work published up until the late 1960s and then a slight decline in the early 1970s. If we were to use these citation counts as indicators of quality, we would have to conclude that on the average the members of the cohort produced more significant work in their later years than they did in the years immediately following the receipt of their PhDs.

The citations received by a given paper are a function of three different variables: the quality of the paper, the change in its relevance as time progresses, and the increase in the size of the SCI file. One would presume that, as a paper gets older, on average its relevance will decrease because the significance of the problem it addressed may decline or it may be surpassed by other work. However, since science has been growing rapidly in the period that we are studying, and the SCI file increased significantly in size between 1961 and 1975, the probability of a given paper being cited has been increasing as a result of this increase in the size of the file.[10] The data we have indicate that the increase in the size of the file has a far more significant effect on the number of citations received than does the possible decline in relevance of earlier work. Therefore, the least biasing measure of quality for testing the hypothesis of this paper is the use of citations in 1975 to work published throughout the twenty-five-year period.[11]

In order to trace further what happened to members of the cohort as they progressed through their careers, a typology of productivity was developed. Each mathematician in each period was placed in one of three categories: the nonproductive, weak publishers, or strong publishers. Nonproductive mathematicians were those who published no papers at all in a given time period; weak publishers were those who published at least one paper in the five-year period but received at most one citation to their work; strong publishers were those who had published at least one paper in a five-year period and received two or more citations to the work published in that period. The data presented in table 7.5 tell an interesting story. In the first five-year period after receipt of the PhD, almost two-thirds of the members of the cohort published at least one paper. However, only 17 percent published papers which were cited twice or more. Throughout the time periods under consideration the proportion of the cohort which was classified as strong publishers remained almost constant, varying between 17 percent for the first period and 14 percent for the last period. However, as time progressed there was a gradual dropping off of the proportion of weak publishers and an increase in the proportion of the nonproductive. Thus, by 1960–1964, fifteen years after the cohort had received its PhDs, 58 percent of them were nonproductive, 26 percent were weak publishers, and 16 percent were strong publishers. The distribution of the cohort among these types remained constant throughout the remaining periods. Of primary concern for the hypothesis under consideration is the fact that the proportion of strong publishers remained nearly constant throughout the time period. As a group these mathematicians were not likely to be more creative when they were young than they have been in their older years.

Thus far, I have shown only that the proportion of strong publishers remains constant throughout the time period. I have not addressed the question of whether the mathematicians who began their careers as highly productive scholars remain highly productive or whether there is movement back and forth between the group of productive scholars and nonproductive ones.

TABLE 7.5 **Typology of Productivity for Cohort of Mathematicians Receiving PhD between 1947 and 1950 (%)**

	Time Period				
Productivity Type	**1950–1954**	**1955–1959**	**1960–1964**	**1965–1969**	**1970–1974**
Nonproductive	38	50	58	60	61
Weak publishers	45	35	26	25	24
Strong publishers	17	15	16	15	14
Total	100	100	100	100	99

The Pearsonian correlation coefficients presented in table 7.6 indicate that there is a relatively high level of stability in productivity of individual members of the cohort. Since the correlation coefficients for the number of papers published by each member of the cohort in different periods vary between a low of .61 and a high of .79, we may conclude that those members of the cohort who were productive at one time were the most likely to be productive at any other time. Using the typology of nonpublishers, weak publishers, and strong publishers, I computed a five-variable table showing the location of each member of the cohort during all five time periods. I was interested in the extent to which members of the cohort would change from one category to another. Of particular interest was a switch of more than one place, that is, nonpublishers who became strong publishers or strong publishers who became nonpublishers. Only forty-one mathematicians made a two-position move during all periods under consideration. Thus, less than 10 percent of the cohort made a significant change in their productivity patterns during the course of their career. Two hundred and sixteen members of the cohort did not change their position at all, remaining nonpublishers, weak publishers, or strong publishers consistently. Two hundred and forty members of the cohort made changes of one position up or down; for example, a weak publisher would become a nonpublisher. Finally, a simple cross-tabulation of the mathematicians' index score at time 1, 1950–1954, and time 5, 1970–1974, shows that of the seventy-one mathematicians who were strong publishers at time 5, only one was a nonpublisher at time 1, and thirty-four were weak publishers at time 1. Of the 304 who were nonpublishers at time 5, only fourteen were strong publishers at time 1, and 123 were weak publishers at time 1. On the basis of these data and those already presented we may conclude not only that the proportion of productive and creative scholars in the cohort remains relatively constant over time but also that the individuals within the cohort are in general not likely to make significant changes in their productivity patterns throughout their careers. Those who start by being highly productive and creative tend to remain so as they grow older. Those who are not productive and creative early in their careers very infrequently become more productive.

For the typology used in table 7.5 a relatively low cutting point was employed for dividing the publishers into "weak" and "strong" ones. To be classified as a strong publisher one needed to receive a total of only two citations to all papers published during the period. It might be argued that age will have an effect on creativity only if we look at the most significant discoveries made. This hypothesis was examined by pulling out from the cohort all those members who at any time during their career had published a paper which received five or more citations in 1975. If mathematicians are indeed more creative when they are young than when they are old, we would expect the resulting sample of the forty-seven best papers published by members of this cohort to have been

188 VALUES AND REWARDS IN SCIENCE

TABLE 7.6 Correlation Matrix for Number of Papers Published by
Mathematicians during Different Time Periods

	1950–1954	1955–1959	1960–1964	1965–1969	1970–1974
1950–1954		.68	.62	.61	.63
1955–1959			.72	.74	.71
1960–1964				.74	.73
1965–1969					.79
1970–1974					

published predominantly during their earlier years. In fact, the data offer no
support for this hypothesis. Of the forty-seven mathematicians who had published at least one paper receiving five or more citations, 22 percent published
their most cited paper in the first five-year period, 21 percent in the second,
21 percent in the third, 23 percent in the fourth, and 13 percent in the last. It
should be pointed out that some of the papers published in the last five-year
period may have been published too close to 1975 to receive their full comple-
ment of citations. Thus, the data based upon the most significant discoveries
made by the cohort of mathematicians lead once again to the conclusion that
mathematicians are no more creative in their early years than in their later
years.[12] We can at least tentatively reject the commonly held belief that the creativity of scientists declines after the age of thirty-five. In fact, age seems to have
very little influence on the quality and quantity of work produced by contemporary academic scientists.

Scientific Output and the Reward System

How can we explain the curvilinear relationship of age and productivity and
the slight decline in the average quality of work produced by scientists over the
age of fifty? I believe that these patterns are influenced by the scientific reward
system. Most scientists are trained in major departments. About 65 percent
of those studied received their doctorates from departments ranked in the top
twenty in the ACE survey. Students in these major departments internalize
the norm that they are expected to publish. Thus, most recent PhDs in science
seek recognition by publishing. As they continue to publish, some find their
work rewarded and go on to publish more. But what about those who are not
rewarded? I hypothesize that those who are not rewarded are less likely to continue publishing. Thus, as a cohort of scientists advance in age the number of

prolific publishers is likely to decline. Most people will not continue an activity as arduous as scientific research unless they are rewarded for it.

We can, therefore, hypothesize that since the members of an age cohort who continue to be prolific publishers after the age of fifty are the scientists who have in the past produced significant work and been rewarded, their current research will be of relatively high quality.

Data from another study can be used to demonstrate that the productivity of scientists whose early work goes unrecognized drops off. In a sampling of male and female biologists and chemists who received their PhDs in 1957 and 1958, data were collected on yearly productivity and the number of citations received in 1961. Here citations to early work are used as a measure of recognition. The data are presented in table 7.7. About half of the scientists started their careers by being fairly productive—publishing at least three papers in the four years after receipt of the doctorate. The work of some of them was recognized as significant (i.e., was highly cited); that of the others went unrecognized. The table clearly demonstrates that scientists whose early research efforts are cited are more likely to continue to be productive. Among the initially productive scientists, those whose early work was never cited published an average of 3.6 papers in 1966–1969, whereas those whose early work received ten or more citations published an average of 9.7 papers in 1966–1969.

Data from the math cohort confirm the conclusions reached from the data presented in table 7.7.[13] The cohort was divided into members who published one or more papers in the first five years after receipt of the PhD and those who published no papers in that period. The first group was then subdivided into those who had received citations in 1961 to their early work and those who had

TABLE 7.7 Mean Number of Papers Published in 1966–1969, by Early Recognition and Early Productivity

Early Recognition	Early Productivity	
N Citations Received in 1961	Wrote 3 or More Papers in 4 Years after PhD	Wrote Fewer Than 3 Papers in 4 Years after PhD
0	3.6	1.3
	(56)	(166)
1–9	5.8	1.9
	(125)	(89)
10 or more	9.7	3.2
	(48)	(5)

Note: The results were exactly the same when proportion producing four or more papers in 1966–1969 was used as the measure of the dependent variable; figures in parentheses are *N* scientists.

not. The mean number of papers published by each group in the period 1960–1964 was calculated. Mathematicians who had received no citations to their earlier work published an average of 2.35 papers during the period 1960–1964. But those who had received one or more citations to their early work published an average of 6.55 papers during that period. Thus, once again the data show that scientists whose early work is recognized are more likely to continue to be highly productive than are others.

I was able to make one additional test of the reward system hypothesis. As a cohort of scientists ages a decreasing proportion will be highly productive, but those who remain productive should be a "residue" of the most talented members of the cohort. Therefore, we should find that the work of older scientists who are still productive should be equal or superior in quality to that of younger scientists. In order to test this hypothesis I used data from the cohort of mathematicians. In 1950–1954, the first five-year period after the cohort had received its PhDs, 307 of the 497 mathematicians published at least one paper. On the average, the 307 productive scientists received 1.37 citations to their work in the 1975 SCI.[14] In 1970–1974 only 193 of the 497 members of the cohort were still publishing papers. On the average, these 193 members of the cohort received 2.94 citations in 1975 to work published in 1970–1975. Thus, we may conclude that the productive members of the cohort in the last period were publishing on average higher-quality work as older scientists than were the productive members of the cohort in the early period.

The data just presented are useful in explaining the decline in average productivity and quality of work among older scientists. We must now consider why in most fields very young scientists do not publish as many papers as scientists in their forties. This too may be explained by the operation of the reward system. The production of scientific research is to some extent dependent upon the control of resources. Most research requires at least some funding. Certainly, it is easier to conduct research leading to many publications if one has research grants, graduate students, assistants, and technical help. Because the reward system does not generally make large amounts of such resources available to scientists until they have established themselves, scientists in their late thirties and in their forties are likely to control more resources than their younger colleagues. The age-graded differential access to resources could possibly explain differences in publication rates. Although data with which to perform a direct test of this hypothesis are not available, certain data for the fields of geology and chemistry allow an indirect test. For these two fields we can examine data on the number of single-authored and multi-authored papers published by each scientist.

Scientists who work with teams of graduate students and junior collaborators are more likely to publish multi-authored papers. Therefore, I hypothesized that production of single-authored papers should be less likely to increase with age than that of multi-authored papers. The data presented in table 7.8 offer support for this hypothesis. In geology there is a slight negative association between age and the production of single-authored papers. In chemistry both very young scientists and those over sixty publish slightly more single-authored papers than their middle-aged colleagues. In both fields, when we look at multi-authored papers we find the same curvilinear relationship as we observed in table 7.2. We may conclude that the effect of age on productivity is a result of the reward system "cooling out" unsuccessful scientists and making resources available to successful ones.

Field Differences in Age Productivity Patterns

Thus far we have found that in general age is curvilinearly related to both the quantity and the quality of published papers and that this results from the operation of the scientific reward system. We have not yet examined the differences among the six scientific fields considered in this paper. Age may have a more important effect on scientific performance in some fields than in others.

The cognitive structures of scientific fields differ. Some fields, like physics, for example, have what Kuhn (1962) calls a highly developed paradigm, while others, like some of the social sciences, are in a preparadigmatic phase or are divided into competing schools. Some sciences are highly theoretical and mathematical, while others are descriptive. Recently Zuckerman and Merton (1973) have characterized scientific fields by the extent to which knowledge is codified. "Codification refers to the consolidation of empirical knowledge into succinct and interdependent theoretical formulations" (p. 507). In highly

TABLE 7.8 Mean Number of Single-Authored and Multi-Authored Papers Published in 1965–1969 by Geologists and Chemists of Different Ages

	Age					
Field	Under 35	35–39	40–44	45–49	50–59	60+
Geology:						
Single-authored	3.2	2.3	2.2	2.0	.9	.9
Multi-authored	4.1	5.3	6.1	4.9	2.9	2.6
N	45	60	60	36	49	41
Chemistry:						
Single-authored	2.3	1.5	1.7	1.6	1.2	2.8
Multi-authored	9.3	10.0	14.3	14.7	17.8	7.6
N	115	62	55	34	61	29

codified fields knowledge is compacted into a relatively small number of theories which can usually be expressed in elegant mathematical language. Physics and chemistry are examples of highly codified fields, botany and zoology are less codified, and the social sciences are the least codified.

Zuckerman and Merton hypothesize that in the more codified fields like physics it should be easier for young scientists to make significant discoveries than in the less codified fields like sociology. There are two reasons for this. First is the method of gaining competence in fields of varying degrees of codification. Because in the highly codified fields knowledge is compacted, graduate students can learn quickly the current state of their field from textbooks and while still graduate students begin work on the research frontier (Kuhn 1962). In the less codified fields knowledge is not as compacted and far greater experience is needed to gain competence. "In these [the less codified fields], scientists must get command of a mass of descriptive facts and of low-level theories whose implications are not well understood" (Zuckerman and Merton 1973, p. 507). Zuckerman and Merton conclude that "codification facilitates mastery of a field by linking basic ideas in a theoretical framework and by reducing the volume of factual information that is required to do significant research. This should lead scientists in the more codified fields to qualify earlier for work at the research front." (p. 510).

The second reason it is easier for young scientists to make significant discoveries in the highly codified fields is variation in the social process through which discoveries are identified as significant. In the highly codified sciences there is presumably greater consensus, and it may therefore be easier to identify new discoveries as important or unimportant. In the less codified fields the identification of a discovery as significant is more dependent upon the reputation of its author. "In these less codified disciplines, the personal and social attributes of scientists are more likely to influence the visibility of their ideas and the reception accorded them. As a result, work by younger scientists who, on the average, are less widely known in the field, will have less chance of being noticed in the less codified sciences" (p. 516). For these two reasons, the ease of moving quickly to the research front and the ease of having an important contribution identified as such, younger scientists should be more likely to make important discoveries in the more codified sciences than their age peers in the less codified disciplines.

The data presented in table 7.3 are relevant for a rough test of the codification hypothesis. Of the six fields included, physics and mathematics are the most highly codified, followed by chemistry, geology, psychology, and sociology. If the codification hypothesis is correct, we should find a higher proportion of young scientists making important discoveries in physics, mathematics, and chemistry than in the other fields. Since the distribution of citations to scientists working in the various fields differs greatly, in order to compare scientists in different fields we must standardize the data separately for each field. All scientists with standard scores greater than zero can be classified as having made relatively high-quality contributions to their fields. The proportions of scientists under the age of thirty-five whose work published between 1965 and 1969 received a relatively high number of citations (a positive standard score) were as follows: chemistry (29 percent), geology (38 percent), mathematics (11 percent), physics (25 percent), psychology (24 percent), and sociology (13 percent). If we examine these proportions, we find little support for the codification hypothesis. Geology is the field in which the highest proportion of young scientists has made significant discoveries and mathematics is the field in which the lowest proportion of young scientists has made significant discoveries. Furthermore, young psychologists are just about as likely to make important discoveries as young physicists and chemists.

Let us consider some possible flaws in the empirical test we have performed. It might be argued that the codification hypothesis is meant to apply only to discoveries of high import and that in the more codified fields it will be easier for younger scientists to make truly outstanding contributions. However, for the data in table 7.3, no matter how high we set the cutoff point for classification of important discoveries, we find no systematic differences among fields. I decided, however, to perform an additional test of the codification hypothesis and examine the ages at which the most eminent scientists in various fields make their first significant discovery.

Two sets of data were used. The first comes from questionnaires sent to random samples of university scientists in physics, chemistry, biochemistry, psychology, and sociology. The respondents were asked to name the five scientists who they felt had contributed most to their field in the past two decades. I have examined the ages at which the ten most frequently mentioned scientists in each field made their first significant discovery. In this procedure it is important that the same criteria of what an important discovery is be used in each field. The first approximation to a more definitive test was to record the age at which each scientist published his first paper to receive five or more citations in the 1971 SCI. As an example, the earliest paper by C. N. Yang, the Nobel laureate in physics, receiving five or more citations in 1971 was published in 1951, when Yang was twenty-nine. There are several problems in the use of this procedure. The SCI contains many more chemistry, physics, and biochemistry journals in its file than it does psychology and sociology journals. It is, therefore, easier for a paper to receive a large number of citations in chemistry, for example, than in sociology.[15] To compensate for the heavier representation of hard science journals, one can use different cutoff points as to what is a significant discovery. The cutoff point for the three natural sciences was ten citations; for the two social sciences, five citations. However, as the data presented in table 7.9 indicate, even if the cutoff point of five citations had been used for all the fields the results would have been essentially the same.

These data offer slightly more support for the codification hypothesis. Physicists do make their first important discoveries at slightly younger ages than

194 VALUES AND REWARDS IN SCIENCE

TABLE 7.9 Mean Age at First Important Discovery for Five Fields

Field	Mean Age on Publishing First 5-Citation Article	Mean Age on Publishing First 10-Citation Article	N[a]
Physics	27	28	10
Chemistry	30	34	10
Biochemistry	35	36	11
Experimental psychology	34		11
Clinical psychology	34		10
Sociology	34		10

[a]In biochemistry and experimental psychology two scientists were tied for tenth place. Both were included—giving 11 scientists for these two fields.

other scientists. However, in the relatively highly codified field of biochemistry scientists take just as long on the average to make their first important discovery as in sociology. Also, in experimental psychology, which is presumably more codified than clinical psychology, scientists are no younger when they make their first significant contribution.

I replicated this test on another set of data: lists of the most frequently cited physicists in the 1965 volume of *Physical Review* and the most frequently cited sociologists in the 1970 volumes of *American Sociological Review* and *Social Forces* (Oromaner 1972).[16] These lists can be considered to be samples of the scientists currently producing the most useful work in each of the fields. At what age did each scientist make his or her first significant contribution? Using the cutoff point of five citations I found that the mean age at which the physicists made their first important contribution was thirty-one, as compared with a mean age of thirty-four for the sociologists (see table 7.10).[17]

It is possible, however, that the results might be influenced by the difference in age at which physicists and sociologists earn their doctorates. The mean age at which the physicists earned their doctorates is twenty-six, as compared with twenty-eight for the sociologists. Although it might be claimed that this merely reflects the shorter time required to learn the more codified field, it also might be a result of the different requirements for the PhD. In physics doctoral dissertations generally concern shorter, more discrete, problems than dissertations in sociology, which are usually more equivalent to books than articles. At any rate, scientists in both fields generally have little opportunity to publish prior to receipt of their PhD. Since physicists are younger at receipt of their doctorate, to make the two fields more comparable we should examine the professional age (years since earning the doctorate) at which scientists in

TABLE 7.10 **Mean Age at Making First Significant Discovery for Most-Cited Physicists and Sociologists**

	Physics	Sociology
Mean age at publication of 5-citation discovery	31	34
N	50	31
Mean professional age	5.0	6.0
N	50	31
Mean professional age: scientists under 60 in 1971	4.9	4.8
N	48	24
Mean professional age: scientists under 50 in 1971	3.9	3.4
N	42	17

the two fields made their first important contribution.[18] It took the physicists on average 5 years from receipt of their doctorate and the sociologists 6 years to make their first significant contribution. Since citations made in 1971 may not be a very reliable indicator of the quality of very old work, we can omit from the sample all scientists sixty years old or older in 1971. The mean professional age at which the physicists under sixty made their first important contribution is 4.9, as compared with 4.8 for the sociologists. When scientists fifty or older in 1971 are excluded, we get a mean professional age of 3.9 for physicists and 3.4 for sociologists. No matter which comparisons we make, we must at least tentatively conclude that there is no significant difference in the age at which scientists in a highly codified field like physics and a relatively uncodified one like sociology make their first significant contribution.

Conclusion

The studies reported in this paper suggest that the relationship between age and scientific performance is influenced by the operation of the reward system. In other research I have found that the different scientific fields have very similar reward systems. The reward systems of highly codified fields like physics and chemistry do not differ significantly from those of less codified fields like psychology and sociology. The homogeneity of reward systems is influenced by the fact that the central locus for the various scientific fields is the university. Both physicists and sociologists must be evaluated for promotion and tenure within the same time period (usually five to seven years). Thus, all university-based scientists must make a significant

VALUES AND REWARDS IN SCIENCE

contribution to their discipline relatively quickly if they are to maintain their positions.

A comparison of the ages at which scientists in different fields make their first important discovery leads to the suggestion that the Zuckerman and Merton (1973) codification hypothesis must be modified to take into account the type of discovery made. In some fields, for example, it may be easier for young scientists to make significant theoretical contributions. Also, the age at which synthesizing contributions are made may vary with the level of codification. In testing the modified codification hypothesis we should be careful to look at the distribution of various types of discoveries as well as the ages at which they are made. If, for example, the proportion of theoretical discoveries made in physics is greater than that made in sociology, it would not be surprising to find more young physicists making theoretical discoveries than young sociologists. A careful analysis of the age distribution of different types of discoveries should be the next step in the study of the influence of cognitive content on the relationship between age and scientific performance.

All the data reported in this paper lead to the conclusion that, at least in the six fields we have studied, age has only a minor influence on scientific performance. Those scientists who begin their careers by publishing and are recognized for making important contributions generally continue to publish high-quality work. It is unlikely that an increase in the mean age of our scientists will in and of itself bring about a meaningful decline in our scientific capacity.

Notes

I thank Pamela Summey for aid in data collection, Lorraine Dietrich for help with the computer, and Kenneth Bryson for statistical advice. Nancy Stern collected and analyzed the cross-sectional data on mathematics. The analysis was made easier by conversations with Ann H. Cole, Jonathan R. Cole, Lorraine Dietrich, Scott Feld, Robert K. Merton, Eugene A. Weinstein, and Harriet A. Zuckerman. This research was supported by a National Science Foundation grant (NSF-SOC 72-95 326) to the Columbia University Program in the Sociology of Science.

1. The names of department members were obtained from either directories of graduate programs or university catalogs. Emeritus professors were excluded to limit the study to active research scientists. Scientists without PhDs, most of whom were graduate students, were also excluded in order to make the sampling procedures among departments and fields as uniform as possible.

2. Since not all scientists who begin their academic careers at PhD-granting institutions will get tenure there, many scientists begin their careers in these institutions and continue them at different types of institutions. This might introduce a possible source of bias, as the older scientists in the sample had already been deemed worthy of receiving tenure, whereas the younger ones, many of whom were still assistant professors, had not yet been screened. On the other hand, it is possible that criteria for granting tenure in the major institutions have become more stringent since the end of World War

II. The fact that the results from the following cohort data to be reported on are the same as those obtained from the cross-sectional data leads me to believe that any bias introduced by using a sample of scientists employed in PhD-granting institutions is not significant.

3. For mathematics, current output was defined as work published in the period 1970–1974.

4. All papers were included, whether they were written by the subject alone or in collaboration.

5. Citations to the work of mathematicians published between 1970 and 1974 were looked up in the 1975 SCI. Past research has suggested that the failure of the SCI to include citation data on collaborative papers of which the subject is not first author has no effect on substantive findings.

6. For a complete bibliography of studies using citations, see any one of the annual guides published by the Institute for Scientific Information to accompany the SCI.

7. For mathematics we find no systematic effect of age. Mathematicians in the age group 45–49 have an unusually low publication rate. The reasons for this are not known.

8. Bayer and Dutton (1977) use the total number of works cited without regard to when they were published or how many citations they received. Therefore, their data on citations are not relevant to the hypothesis we are interested in.

9. The number of papers was counted in *Math Reviews;* citations were counted in the 1961, 1965, 1970, and 1975 editions of the SCI. One problem in studying a cohort is that some people drop out as a result of death or a career change. Of the 497 people in the cohort, fifty-two could not be located in 1975. Some of these were probably women who married after receipt of the PhD and then starting using their married names. Of the people about whom we had information, 84 percent were employed in academic math departments, 7 percent in other academic departments, and 10 percent in industry. Despite the fact that some people clearly were no longer mathematicians, we did not exclude anyone from the cohort. Since people who are no longer mathematicians cannot be making scientific contributions in the later years, any bias introduced by counting all 497 people as part of the cohort goes against the null hypothesis that creative productivity is not related to age.

10. I am currently working out some mathematical estimates of the effects of both these variables on the citation rates over time.

11. In fact, all the analyses of the cohort used as a measure of quality of work published throughout the period both 1975 citations and the citation index closest in time to the date of the papers' publication. All the analyses yielded virtually identical results, differing by only a percentage point or two.

12. It might still be argued that even papers receiving five citations or more, although produced by only forty-seven people, or less than 10 percent of the cohort, still do not represent the truly significant breakthroughs. When the cutoff point for a significant discovery is placed at a very high level it becomes virtually impossible to study such breakthroughs statistically. In fact, when we refer to the highest level of contributions, those which introduce new paradigms, we are probably dealing with somewhat idiosyncratic contributions which are not representative of the general patterns within a scientific discipline.

13. The conclusions drawn from table 7.7 are also supported by a study conducted by Allison and Stewart (1974). Examining the publications of a large sample of biologists, mathematicians, chemists, and physicists in the previous five years, they found that the dispersion of the productivity distribution was much higher for older scientists than younger ones. As age increases, a smaller proportion of the scientists are responsible for a larger

proportion of the total publications. Using the Gini index as a measure of inequality, they found a correlation of $r2 = .98$ between the size of the Gini index and the number of years since the PhD was obtained. This supports the conclusion that as a cohort of scientists ages there is a gradual decrease in the proportion who remain productive.

14. If we use the 1961 SCI as an indicator of the quality of this work, we find an average of 0.53 citations to the work of the 307 productive mathematicians.

15. For example, in table 7.3 the mean number of citations to the recent work of sociologists is two, and the mean for the recent work of chemists is fourteen.

16. The data for *Physical Review* were provided by Cullen Inman of the American Institute of Physics.

17. Using the same cutoff point for each field should compensate for any possibility that work in physics might become obsolete more quickly than work in sociology.

18. If a contribution was made prior to receipt of the doctorate, it was counted as a negative number in computing the averages. Only five of the physicists and one of the sociologists published a five-citation discovery prior to receiving their doctorate.

References

Allison, Paul D., and John A. Stewart. 1974. "Productivity Differences among Scientists: Evidence for Accumulative Advantage." *American Sociological Review* 39 (August): 596–606.

Bayer, Alan E., and Jeffrey E. Dutton. 1977. "Career Age and Research—Professional Activities of Academic Scientists: Tests of Alternative Non-linear Models and Some Implications for Higher Education Faculty Policies." *Journal of Higher Education* 48, no. 3 (May/June): 259–82.

Cole, Jonathan R., and Stephen Cole. 1973. *Social Stratification in Science.* Chicago: University of Chicago Press.

Cole, Stephen. 1978. "Scientific Reward Systems: A Comparative Analysis." In *Research in Sociology of Knowledge, Sciences and Art,* ed. R. A. Jones, 167–90. Greenwich, CT: JAI.

Dennis, Wayne. 1956a. "Age and Achievement: A Critique." *Journal of Gerontology* 11, no. 2 (July): 331–37.

——. 1956b. "Age and Productivity among Scientists." *Science* 123 (April 27): 724–25.

——. 1956c. "The Age Decrement in Outstanding Scientific Contributions." *American Psychologist* 13 (August): 457–60.

——. 1966. "Creative Productivity between the Ages of 20 and 80." *Journal of Gerontology* 21, no. 1 (January): 1–8.

Kuhn, Thomas S. 1962. *The Structure of Scientific Revolutions.* Chicago: University of Chicago Press.

Lehman, Harvey C. 1953. *Age and Achievement.* Princeton, NJ: Princeton University Press.

Mullins, Nicholas C., Lowell L. Hargens, Pamela K. Hecht, and Edward L. Kick. 1977. "The Group Structure of Cocitation Clusters." *American Sociological Review* 42 (August): 552–62.

Oromaner, Mark. 1972. "The Structure of Influence in Contemporary Academic Sociology." *American Sociologist* 7 (May): 11–13.

Roose, Kenneth D., and Charles J. Andersen. 1970. *A Rating of Graduate Programs.* Washington, DC: American Council on Education.

Stern, Nancy. 1978. "Age and Achievement in Mathematics." *Social Studies of Science* 8 (February): 127–40.

Zuckerman, Harriet A., and Robert K. Merton. 1973. "Age, Aging and Age Structure in Science." In *The Sociology of Science: Theoretical and Empirical Investigations,* ed. Norman Storer, 497–559. Chicago: University of Chicago Press.

CHAPTER 8

BALANCING ACTS

Dilemmas of Choice Facing Research Universities (1993)

JONATHAN R. COLE

Prognostications of the decline of American higher education, and particularly of our research universities, are not infrequent. Analyses of the "current crisis" by critics and friends surface only somewhat less frequently than the seven-year locust.[1] The list of diseases and etiologies leading to the imminent decline of these institutions is long and includes such familiar items as claims of administrative waste, fraud, and abuse, and of the corruption of fundamental academic values and standards; a retreat from the undergraduate classroom; the perversion of the academic reward system; the end of meritocracy; the triumph of corporate, bureaucratic models of governance over the more congenial ecclesiastical style of shared decision-making through consensus formation; the ceding by faculty of academic authority and responsibility for producing a rigorous and sound curriculum that defines the shape and scope of the educated person; the erosion of public trust; and among many more, the absence of visionary academic leaders who can articulate the mission of research universities, who can write the brief for them, and who can argue the case persuasively before critical attentive audiences—in short, the lamentable absence of the voices that represented the academy from the time of Eliot and Hutchins to Conant and on down until only a generation or two ago. Yet, none of these perceived problems is particularly new.[2]

Simultaneously, many shrewd, knowledgeable veterans of higher education point out that the American research university continues to be the jewel in the

---

Daedalus 122, no. 4, *The American Research University* (Fall 1993): 1–36.

higher education crown, that it remains the envy of the world, the set of universities with the highest prestige and distinction in the nation, the institutions that hold the most sought after positions for talented faculty and students, the continuing producer of more Nobel Prize–quality science than any other type of educational or research institution in the world, and one of the few remaining American "industries" with a favorable balance of trade. None of these defenses has been recently copyrighted.

What is new in the current debates about the state of research universities? The problems in generic form are not particularly new. They have existed above or below the surface over an extended period of time quite simply because they are linked fundamentally to the basic social and organizational structure of research universities. When these complex social systems experience disequilibrium, the problems surface. They tend to become open to substantial discourse only periodically—usually when the economies of research universities are constrained or when the conflicts within the university echo broad and fundamental conflicts within American society. But when they do surface anew, they are more often than not brought forward by a new set of critics who are unfamiliar with the history of research universities and the earlier appearances of these problems. In short, problems of governance, leadership, the foci of faculty and student attention, and the relationship of universities to external social systems of government, industry, and the general culture always exist. Nonetheless, our concerns today with the state of research universities do reveal some important new substantive variations on older themes. These new variations are the foci of this issue of *Dædalus*.

The contemporary problems involve dilemmas of choice that have become more pressing over the past several decades. I want to outline a number of these dilemmas and suggest that while the research university as an institution is not about to disappear or to lose its fundamental character and basic strengths, those universities that successfully deal with the dilemmas of choice will have important strategic advantages over their peers in the decades to come. Before discussing a number of these dilemmas, I want to indicate how patterns of growth and change in higher education over the past fifty years have created tensions within the academy and between the academy and some of its traditional partners who support research.

Patterns of Growth and Change

We tend today to think of the major research universities—Berkeley, Chicago, Columbia, Harvard, and Stanford—much as we did in the past. In important respects, there are great similarities in the pasts and presents of these universities, perhaps most notably in their basic commitment to teaching and research at a very high level of excellence. They are, after all, elite institutions whose reputations have been largely intact for the better part of the century.

Moreover, there has been relatively little change over the past half century in the number of schools within the universities or in their basic organizational structure.[3] But in ways that are not entirely appreciated, the Harvard or Columbia of today is a very different institution from what it was in 1945. The fundamental difference is in its size and complexity and in its responses to the exponential rate of growth of knowledge. Research universities, our principal incubators of new discoveries and ideas, reflect the pattern of exponential growth first described by the historian and sociologist of science Derek J. de Solla Price.[4]

While Columbia, for example, is in many ways the same university that Nicholas Murray Butler left behind in the academic year 1944–45 (his last as president), it is in basic ways an entirely different enterprise.

Even in 1944, Columbia was concerned about a balanced budget: it faced a $1.6 million deficit, which it adroitly turned into a $65,000 surplus by June 1945. This was all on an operating budget that totaled about $11 million. Today, still struggling to balance our books, Columbia will have an operating budget in 1993–94 that is estimated to be roughly $1.1 billion—a budget one hundred times greater than at the point of Butler's departure. Even a cursory glance at the intervening decades reveals dramatic growth in the university's expenses, signaling growth in the number and size of academic programs: from $57 million in 1959–60 to $170 million in 1969–70, to $317 million by 1979–80, to about $800 million in 1989–90. We have witnessed at Columbia more than a doubling in budgetary size about every ten years, with our annual expenses increasing at a compound rate of close to 10 percent for the past forty-five years.[5] Even allowing for the substantial inflation in portions of that period, this is an enormous rate of real growth. The same pattern of growth has been sustained by most of the other major research universities.[6]

Patterns of budgetary growth simply reflect patterns of expanding research and teaching opportunities. These changes are the underlying causes of the dilemmas I will consider. Change has come as a consequence of the salience of relatively new concerns of these universities about such matters as equal access to education (which has led to need blind/full need admissions and financial aid policies), affirmative action, and greatly increased support services for students. The positive results of the civil rights movement have led to increased diversity of the university population, and with increased diversity have come new conflicts over the curriculum, university hiring and promotion policies, and admissions and financial aid standards and practices at university colleges. Now that research universities reflect more closely the socioeconomic composition of the larger society, it is inevitable that they will experience more conflicts that were avoided when they had too little of a good thing.

The increased dependence of research universities on federal government financing of research and student financial aid has changed academic and financial relationships within universities. This has led to the enormous growth of health science divisions at research universities, has altered the relative size of health science compared with the arts and sciences and other professional schools, and has produced uncertainty about the future of scientific research at universities dependent on continued government support.

Add to these changes the growth of claims for new scholarly disciplines, the expanded number of PhD programs competing for resources, the emergence of philosophical relativism, the increasing imbalance between research and teaching, and the transformation produced by the information revolution. When this is mixed in with a set of externally imposed constraints caused by a national economy that is not expanding at a rate comparable to that experienced in the 1980s, you have the conditions for dynamic change that will require research universities to confront many difficult dilemmas of choice. When resources contract even as the legitimate demands for sustaining academic excellence expand, universities will face dilemmas of choice, as they do now.

Dilemma One: Governance

How do research universities define their priorities? Who decides what to build, what to favor, what to contract, and what to eliminate? What gives the process legitimacy?

Of critical importance to the research university is the exponential growth of knowledge during the postwar period. Much of this growth was fueled by a massive increase in the federal government's investment in scientific research at the major universities. As knowledge expanded at this pace, there emerged a plethora of new claims for resources to fund new areas of knowledge. Even the great research universities began to experience a gap between the expanding knowledge base and the capacity to offer programs of high quality in *all* of these new areas—while also retaining excellence in the programs that had been sustained for generations.

During periods of rapid expansion in resources—which occurred to a substantial degree through the 1950s, 1960s, and 1980s—the research universities were able to live with the illusion that they could remain "full service universities" without having to make many difficult choices about which new areas of knowledge would take programmatic form and would be supported at a level needed to achieve true distinction; which currently supported areas would have to be phased out; and which areas of knowledge would go uncovered, left for others to develop, thus creating a true division of intellectual labor in higher education.[7]

It is rapidly becoming accepted that the 1990s will not allow for the expansion of the research university at the rate achieved in the 1980s and before. We have seen, accordingly, a spate of articles, generally authored by university or college presidents, former presidents, or those who make it their business to monitor the economics of higher education, that call for "doing more with less," "growing through substitution," or "making difficult choices between competing goods." There is widespread recognition that it is no longer possible for research universities to afford excellence in all areas of knowledge, including those supported currently and those required to cover the most important areas of new knowledge. Most leaders of America's great research universities recognize that they have to make choices and that failure to do so bespeaks implicit choice in any event. Nonetheless, there has been far more talk about the need to make critical choices than a willingness to engage directly the problems associated with choice: to make creative, strategic decisions and then to implement them within a reasonable time.

The fundamental problem of choice at research universities has more to do with basic ambiguity over governance than with the ability to articulate alternatives. Who has the authority, beyond the formal authority registered in the statutes or the table of organization, to make such choices? Who has the power to "veto" the choices made? What are the processes by which the choices of the decision-makers are legitimated within the university community? What is the role of faculty, students, administrative leaders, trustees, and alumni in making such choices? Traditional business organizations have little problem assigning responsibility for decisions, while universities have failed to do so. The structure of universities impedes decisions from being made, creates suspicion among schools and departments about the explicitness and fairness of criteria for dividing up scarce resources, and reduces the flexibility institutions require to respond imaginatively and reasonably to new academic needs and priorities.[8]

Of course, research universities are not, cannot, and should not be organized in imitation of corporations. The process of decision-making is going to take longer than in the hierarchical culture of the corporate world. The goal is not to imitate the business community but to take some lessons from it (especially in the administrative and business side of research universities). We must recognize that the rhythms of the external world have changed and that these changes directly affect the internal life of universities. The faster pace and the rapid growth of the institution requires more rapid, year-round responses and initiatives. This new environment requires a clearer process of decision-making so that universities can make meaningful changes and adaptations in a timely way.

The problem that universities have in reaching difficult decisions is not simply a matter of speed but of certain structural features that produce difficulty in reaching conclusions.[9] First, they tend to be organized around a "company of equals" pattern. Second, they rely heavily on peer judgments of academic quality, which has great value but is not noted for producing high levels of consensus or for unambiguous judgments. Third, there is a high level of motivated unwillingness of any academic unit to criticize any other—at least when the stakes are as high as reductions in size or possible program elimination. Fourth, the pattern of economic commitments and rigidities associated with tenure can place a significant drag on movements to create changes in the composition of academic units and subunits. Finally, since most academic deans anticipate rejoining the faculty, they are reluctant to burn bridges behind them—a likely outcome if they make difficult decisions that cannot possibly please everyone.

Given these constraints, should such matters of choice be left principally, if not entirely, in the hands of the faculty? Can faculties with highly diverse and often competing interests lead each other to consensus? Is it the mark of outstanding academic leaders that they define priorities and build coalitions within the faculty to support a strategic plan for change—one that involves elimination of some programs and expansion of others? Will popular academic leaders lose their luster at the precise moment that they propose substantial cuts in some academic programs in order to focus resources on other areas of comparative strength? Should presidents and provosts of universities articulate the academic mission and vision for the university and then consult with the faculty about proposed changes?[10] What forms of consultation are appropriate for decisions involving reallocation of resources? To what extent should a limited number of active faculty be permitted to forestall proposed change? Answers to these questions are hardly self-evident since research universities do not have constitutions to govern this decision-making process and the "common law" at universities remains quite ambiguous about how and where decision-making authority resides.

It is, after all, one thing to say that universities will thrive if they have leaders who can build faculty coalitions supportive of difficult choices; it is another matter to articulate how that gets done. The difficulty derives, in part, from the strong value placed on faculty governance, when the vast majority of faculty focus appropriately on their teaching and research and know little if anything about the economics of the university.[11] It is also far easier to argue that there should be "competition" for resources among academic programs— followed by faculty discussion of the relative merits of these programs, which in turn would lead to faculty consensus on choices—than to operationalize a structure for this competition and achieve faculty consensus. Admirable efforts at consensus building have been known to break down at the first mention of eliminating a department, reallocating faculty billets from one department to another, or reassigning laboratory space from one research program to another.

Efforts at making difficult choices have led to tense times at research universities. Many faculty tend to be opposed to any significant program change—any shift in academic priorities that is accompanied by shifting resource allocations—because they believe it is the slippery slope that could end with reconsideration of their own department's allocations. History will likely show that where substantial changes have occurred as a result of "choices," they have been unsystematic in their development, have involved small units and large expenditures of effort, and have been only tangentially related to any well-defined effort to shape the future direction of the university.

My recent experience at Columbia provides three exquisite, if not entirely admirable, examples. Over the past seven years, while there was substantial growth of academic programs at the university, two departments were closed—Geography and Linguistics—and also the School of Library Service. Each of these fields is important but was deemed not to be central to the future mission of the university. In the case of the School of Library Service, it took two years of intensive work by faculty and administrative committees, senate reviews, and responses to hundreds of individual and many group protests before the decision was implemented. That is the success story. It took seventeen years between the decision and the actual closing of the Linguistics Department; it took nearly as long to do the same with the Geography Department.

The closing of the School of Library Service is particularly instructive because of the implicit criteria of choice that the faculty and administration articulated in the process and debated at some length before the final decision was made. The fact that it was a small unit (four tenured faculty) probably contributed to the eventual outcome, but it was not a central factor. The framework for choice included the following elements: 1) an effort to establish a balance between core activities of the university and those that are peripheral (if enriching) activities; 2) academic priorities that juxtaposed the cost of maintaining and enhancing a preeminent school against the resources required for higher priority arts and sciences needs and the necessity to invest in other new programs; 3) an evaluation of whether the School was critical to the educational and research missions of *other* schools of the university; 4) the opportunity costs associated with over twenty-five-thousand square feet of space (in a space-poor campus) that might otherwise be used for renovation and expansion of Columbia's main library; 5) an evaluation of whether the School would move decisively into information science, a goal that had been set five years earlier; 6) the possibility of students interested in traditional forms of library service obtaining a quality education in the discipline at other universities in the nation; and 7) the impact on the university's larger reputation of closing a school in an area in which we had been pioneers.

These criteria were used, often without explicit articulation, throughout the discussion of the School's future. There was never any disagreement about the

quality of the past contributions of the School or of several of its current programs. There was substantial disagreement, however, in evaluations on some of
the criteria and the weight that individuals place on the various elements in the
framework. In the end, the decision was not one that called for weighing dollars
against academic purposes but one which confronted academic priorities in
weighing the merits of competing academic needs.

The resistance to closing such academic units highlights not only the dis-
position of faculty, students, staff, and loyal alumni to protect everyone's turf
lest their own become vulnerable but also the distorted conception of the "life
cycle" of academic departments, specialties, institutes, and centers at research
universities. We have a marvelous sense of fertilization; we are experts at ges-
tation and early development; we know about maturation and full expansion;
but we refuse to confront dying and death. The academic way of death is traditionally through atrophy at a Darwinian pace. We rarely consider the idea of
a full life course—of what should be associated not only with a beginning but
with an end. And this is so because we have neither the rules that permit for
orderly governance of choice nor the conceptual frameworks to guide those
choices. Moreover, without clear, agreed-upon criteria, many academic leaders,
looking at the consequences of "boldness" among some of their brethren, see,
quite accurately, that making significant changes in the face of limited faculty
opposition often leads to larger-scale faculty opposition and potentially to a
loss of personal authority and legitimacy.

Establishing informal criteria of choice in the case of the Library School
constituted the beginning of a framework to guide difficult choices of this
type.[12] The point is simply that discussions of choices would best be held within
an agreed-upon framework of evaluation. In fact, we have only rare examples
of faculties and their leaders engaged in thoughtful discussion of principles to
guide choice. Research universities must consider what internal processes will
increase the chances that choices will be viewed as legitimate by the university
community.

If research universities can no longer cover all areas of knowledge, then
each university will have to determine those areas in which it has comparative
advantages in developing and maintaining true distinction. It will also have to
judge which are the "core" areas of knowledge, the areas of such importance
to the future of knowledge that any great research university, to be defined as
such, will have to demonstrate excellence in them.

Finally, research universities will have to develop mechanisms that will
enable them, despite their substantial fixed costs, to gain greater control over
the resources needed to support new areas of knowledge.[13] Perhaps the greatest limit on flexibility is the tendency to allocate "permanent" tenure billets to
departments. Mechanisms need to be developed at many universities for redis-
tributing faculty lines and for developing full resource models for departments

that treat resources as more fungible assets that can be distributed to support faculty, student fellowships, scientific facilities, and support services of academic departments.

What, then, is in order? It is time that these universities articulate a division of primary responsibility and authority in decision-making. It has often been repeated that the university *is* the faculty. But in the contemporary world of universities, faculty governance must be shared in an effective way with administrative leaders. Administrative leaders are drawn almost always from the faculty and do not renounce their faculty citizenship when assuming the office of president, provost, or chancellor. The false dichotomy between the faculty and administration ought to be replaced by a more sociologically appropriate view that some members of the faculty change their roles and their role obligations during their tenure as administrators. As administrative leaders, their interests may no longer be entirely consistent with the interest of their "home" department or school, but they continue to embody the core values and interests of the faculty. Nonetheless, primary control and responsibility for curricular decisions, faculty appointments, and promotions should reside, as they do now, in the active teaching and research faculty. The development of academic priorities should be a collaborative enterprise, with the faculty working with academic leaders. The academic vision and institutional priorities should be articulated by the university's president, provost, and deans—with the president as the voice of the university. The business of translating goals into achievements must be delegated to the executive arm of the university, backed explicitly by the trustees.

Difficult decisions will be better understood within the community if they are consistent with a well-defined, visible set of academic priorities for the university. But, ultimately, there must be clear, final authority over the allocation of resources and the changing foci of attention at the university that is vested in its academic leaders. Consultation with faculty, students, and alumni about the bases for choices is essential, but there cannot be inordinate delays in decisions to mollify everyone. Academic leaders should present the faculty and students with clear explanations for their decisions. And, once the choices have been made, there ought to be open reporting of the outcomes that will permit the university community to evaluate the academic and financial consequences of the actions taken.

Dilemma Two: Who Owns the Null?

Research universities are facing a set of challenges and choices of a wholly different kind from those associated with the allocation of scarce resources. One of these is represented by a significant attack on the prevailing organizational axioms, or presuppositions, on which research universities have been built. A second is represented by a fundamental challenge to what John Searle calls "the Western Rationalistic Tradition" in his essay in this volume of *Dædalus*. This attack is leveled against the presuppositions of rationality, of objectivity, of truth, of "there being a there out there," among other basic epistemological and metaphysical presuppositions that have guided discourse throughout most of Western history, and certainly since the seventeenth century. These challenges to the university's organizational principles and to its philosophical presuppositions are interrelated. They involve conflicting views of the basic principles and what is required to prove that one or another organizational principle is right or wrong.

I shall call this conflict, which involves fundamental choices, "a conflict over who owns the null."[14] As users of statistical analysis will know, hypothesis testing involves setting up a "null hypothesis" and trying to overturn it. The null hypothesis states the hypothesis of zero difference or equality. This can be contrasted with the research hypothesis, which involves a statement of expected differences. For example, suppose that I believe that science treats women unfairly in hiring, promotions, salaries, and peer recognition. That belief can be framed as my research hypothesis. To test that hypothesis, I set up the null hypothesis: science treats women fairly, that is, there are no differences in these forms of recognition between men and women. As a researcher, in order to "prove" that there is unfair treatment, I try to overturn the null by collecting sufficient evidence to demonstrate that the null hypothesis of equality must be rejected, that it is not true. I can use various tests, but generally in order to make the research hypothesis credible, I must minimally show that the pattern of difference between men and women would not have occurred by chance more often than five out of one hundred times.

As any empirical social scientist can attest, it is extremely difficult to accumulate enough acceptable evidence to reject, or overturn, the null hypothesis, given the limited power of social science theory and our inability to identify adequate methods and techniques that can be applied to complex social situations. Therefore, whoever controls, or "owns," the definition of the null is apt to preserve it against attacks based on existing evidence. The formulation of the null also determines who bears the burden of proof. It makes a great deal of difference if the null hypothesis is "University X is a meritocratic institution without racism" rather than "University X is fundamentally a racist institution." Since overturning the null is difficult, the individual or group that "owns" it has a good chance of controlling the conclusion reached. This is particularly true because owning the null gives the owner control over the standards and practices in establishing "truth." It gives the owner the power to establish the methodology that is acceptable in trying to overturn the null—and that is what can make it doubly hard to overturn.

Consider first the challenge to the basic core values of the research university: meritocracy, rationality, organized skepticism, which enjoins members of the community to test ideas against appropriate evidence, and an open society which supports a free marketplace of ideas. Universities have been organized particularly around the value of meritocracy, which defines and requires the use of universalistic rather than particularistic standards of judgment. This value holds that the admission of students, the hiring and promotion of faculty, and the allocation of other forms of rewards and recognition will be based upon the quality of performance, not on the personal, ascribed characteristics of the individual. To be sure, universities have too often not fully approximated this ideal, but the value has been deeply ingrained in the institution and its self-definition.[15] Those who have governed the elite research universities and who have taught at them strongly believed in these core values. They have thus far controlled the definition of the null hypothesis.

At colleges and research universities, there is a substantial political drama unfolding over who owns the null—who gets to define the "truth" that must be falsified. Interestingly, the current attack on the existing "null" comes from both the cultural Right and the Left. From the Left, in its crudest form, comes the attack that the research universities as institutions are basically repressive, corrupt, racist, sexist, homophobic, biased in favor of Western cultural history and its literary forms, particularistic and nonmeritocratic, and are organized to perpetuate these values. Part of this attack is associated with aspects of "political correctness": efforts to limit "offensive," hurtful speech on campus through the introduction of speech codes; to review teaching and course materials for their content; and to review the content of presentations in the classroom for their offensive character.

Somewhat more subtle are the claims that a reward system that depends upon peer-reviewed publications and peer-reviewed assessments of quality undermines opportunities for "outsiders" to become "insiders." Under the prevailing system, the defender of the old order (read "old null") purports to make evaluations and decisions about admissions, appointments, and tenure without regard for the personal characteristics of individuals. In fact, the claim is made that these personal characteristics have always been relevant, but not acknowledged, and should now be made an explicit part of the decision-making process. These personal characteristics are relevant in at least two ways—even if *all parties* to the transaction are entirely unaware of their relevance. First, the person or group that owns the null cannot help judging the world from that point of view; and second, there are intrinsic differences between the way in which men and women, Blacks or whites, do their work and, accordingly, in the criteria by which their work should be judged.

The attack on the existing presuppositions from the cultural Right assumes that the transfer of control of the null has been all but completed. The presumption is that the cultural Left has won, that the leaders of research universities have capitulated and officially endorsed various forms of limitations on free speech, have supported the creation of academic programs for political rather than substantive reasons, have adopted quota systems in admissions, financial aid, and academic appointments (or at least have adopted a set of different standards that are applied to groups rather than individuals), and have failed to defend faculty and students against unfounded, stigmatizing attacks. Faculty now live with great apprehension that they can be labeled a "racist" or a "sexist" without substantial support from their colleagues or university leaders and the burden of proof now lies with faculty to demonstrate that such allegations are false. Universities are attacked for capitulating to pressure from the Left to increase diversity and multiculturalism, for adopting the principles of group justice while abandoning the concept of individual opportunities without guaranteed outcomes. It is further claimed that entitlement has replaced meritocracy and opportunity as the governing principle in university decision-making.

Whether and how to formulate the null hypothesis is not a trivial decision. On the one hand, if we hold to the presupposition that the university is meritocratic, and that the university defines what "meritocratic" means, the burden of proof remains on those who believe otherwise. If, on the other hand, the null is framed as "the university is fundamentally racist, sexist, and homophobic," then the burden of proof lies with those who want to prove that this is not the case. Since disproving the null is difficult, the ownership of the null corresponds to a set of important consequences in the formation of university policy.

Because it is difficult to prove or disprove complex phenomena like discrimination or racism, especially when the conflict is in part over the methods of proof, it is not surprising that the conflict takes on an ad hominem character. Assertions and counterassertions substitute for evidence, in part because the methods of establishing facts are at the heart of the dispute. In Neil J. Smelser's essay in this issue of *Dædalus*, Smelser suggests that academic leaders and faculty at research universities today have an easier time defending themselves against attacks from the Right than from the Left. He notes that universities are traditionally liberal institutions and leaders have had more experience defending against the cultural Right—and they feel more comfortable doing so. But liberal academic administrators and members of the faculty are ambivalent about defending themselves against attacks from the cultural Left, since they share a commitment to many of the goals associated with the Left, such as increased diversity on campus. Nonetheless, they have substantial difficulty with the means used to achieve those goals and do not share the basic goal of redefining the null hypothesis. Thus, administrators and faculty leaders tread lightly in turning back assertions from the Left that are not supported by evidence. Smelser poses the dilemma faced by these leaders:

Liberal academic administrators and faculty generally applaud and welcome "diversity" if it is carried out within the confines of meritocracy and the preservation of the values of the academy. When those values themselves come under attack, however, and when the attacks on them appear to be made in the context of antimeritocratic demands for entitlement, liberals are cast in an uncomfortable role, in which they experience a dissociation of—indeed a conflict between—meritocracy and egalitarianism. Their role now becomes one of a conservative elite, jealously guarding those values of universalism that were invented and best suited to challenge conservative elites.[16]

Ironically, liberal administrators feel reluctant to take a liberal stand for fear of not appearing liberal enough. A good example of this reluctance can be found in the recent debates over free speech and speech codes on campus. The prevailing null is that the university campus should be a free marketplace of ideas, with no limits on speech except, perhaps, in those rare events when the physical safety of the community is at risk. But that position, which would be held by liberal advocates of First Amendment protection, is under attack. The position to the Left suggests that the protection of disadvantaged groups and the creation of a civil society on campus call for some judicious limitation on speech when speech takes the form of hate speech or displays of "offensive" symbols, such as sexually explicit photographs, a Confederate flag, or a swastika on the outside of a dormitory wall or inside a student's room. This attack on the null is cautiously, and ambiguously, resisted—often because of apprehension that those who defend free speech will become its victims, a result of stigmatizing labels. The liberal administrator fails to use speech in defense of his position for fear of being labeled racist or sexist and as a result suggests an absence of commitment to the null.

What makes the current dilemma particularly interesting at universities is that the conflict over who owns the null hypothesis is a struggle for political power between groups within departments, centers, student bodies, faculties, institutes, administrations, and professional associations and is being influenced by the changing social and ideological composition of these groups. The ultimate "fate" of the null may not be the result of any single choice but of a series of choices, each having only a limited effect on the final outcome. This is the way social change often comes about.

Control of the null is no less important in the contemporary debates over the content and methods of scholarly work within many of the humanities and social science disciplines. It is not clear that the debates are carried on in terms of standard criteria of scholarship—or should I say traditional forms of scholarly discourse—since in some sense the criteria themselves are the subject of the conflict.

With increasing frequency, scholars in the humanities and social sciences at research universities are extending the older attack on positivism but often without much knowledge or understanding of the deep philosophical and sociological questions that are involved in the challenge. The challenge is for control of the content of scholarship, and in some cases that means basic intellectual control over the core journals, the disciplines, and departments. In its current form, this challenge asserts that the fundamental tenets of Western philosophy, those on which modern science and social science have been built, are misguided. The challenge is to the basic concept of rationality. The constructivist argument is that there is no objective reality, that scientific knowledge—indeed all knowledge—is subjective and socially constructed, and that facts cannot exist independently of the attributes of their producer.[17]

Many of the critics of rationality, reality, objectivity, and the correspondence theory of truth associate that epistemology and metaphysic with a repressive social organization of the research universities—if not larger communities. The critics and their followers have multiple objectives. For some members of the professoriate who have thought deeply on these issues, it is to overturn the cognitive null because they believe that the older Western metaphysic and epistemology is wrong or no longer has positive heuristic value. For some within the professoriate and student body, it is to further a political agenda that has little to do with philosophy or scholarship. In some departments at research universities, the critique of the principles of reality, rationality, and objectivity has become the "politically correct" position and those scholars who fail to accept the critique are apt to find promotion and peer recognition increasingly difficult to acquire. It is not clear that members of university communities are fully aware of the implications of the attack on the cognitive null. In the meantime, scholarship at universities is changing without many members of the scholarly community coming to grips with the implications of these trends. For better or worse, control of the null is being relinquished in many departments at research universities without a serious discussion of the consequences of the transformation for scholarship and the training of students.

The unwitting abandonment of ownership of the null involves verbal transactions that are interesting enough to have attracted the attention of playwright David Mamet.[18] His play, *Oleanna*, shows us three meetings between John, a professor, and Carol, his student, in which the ownership of the null passes progressively from his hands to hers. At their first meeting, Carol has come to see him because she has written an unsatisfactory paper and is failing his course. Carol is, as she states early on, from a lower social and economic class than the faculty and many of the other students. She has worked hard and sacrificed to come to college and is diligent and earnest. She finds that despite her hard work she does not understand most of what transpires in her classes: her determination to succeed makes her aggressive about her failure to understand. Mamet's audience laughs when John, in an attempt to show Carol that her paper is gibberish, reads it aloud: "'I think that the ideas contained in this work express the author's feelings in a way that he intended, based on his results.'" John then asks her, "What can that mean? Do you see?" Mamet's audience are also people who own the null and although they may find John a bit pompous, they share with him the judgment that Carol's words do not mean anything. Carol, although she does not understand why her words lack meaning to John, believes entirely in John's evaluation of her work and in the absolute and eternal correctness of the ideals and standards which give him the right to judge. John owns the null, the power to define the vocabulary and syntax of the classroom, to define the kinds of logic and reasoning that are legitimate, to define who will receive a college degree and go on to reap the social and financial rewards it confers and who will not. John is clearly the beneficiary of this system: he has just received word he will be granted tenure and he is preparing to buy a house for his wife and child.

Mamet begins the first interview by giving John an innocently pompous speech which gives nothing away to Carol but shows the audience the way in which the routine speech of academic daily life contains within itself the seeds of its own destruction. Carol asks, "What is a 'term of art'?" John's answer is innocently filled with academic terms of art:

> What is a "term of art"? It seems to mean a *term*, which has come, through its use, to mean something *more specific* than the words would, to someone *not acquainted* with them . . . indicate. That, I believe, is what a "term of art" would mean.[19]

His definition of a "term of art" reveals to the audience, if not to Carol, that language is not a universally clear medium. At the same time, his use of the rhetorical question, the conditional tense, and the insincerely self-deprecating "seems to mean" and "I believe," which reveal him as an academic insider to the audience, are taken literally by Carol, who asks, "You don't know what it means?" The literalness of Carol's response confirms that she is not privy to these academic terms of art. The stock diffidence, self-examination, and self-deprecation of academic speech of which John is a master and which mark him as an insider will turn out to belie real diffidence, real introspection, and real self-deprecation. The desire to *sound* open-minded and thoughtful does indeed reveal an ambivalent and fatal desire to be open-minded.

In words that should be uncomfortably familiar to everyone in academics, John goes on in the remainder of their first meeting to give away both his ownership and his claim to ownership of the null. In the face of Carol's persistence and lack of understanding, John attempts to mollify her and end the endless interview by means of a series of partially hypocritical, partially truthful

214 VALUES AND REWARDS IN SCIENCE

statements of self-deprecation. First, John tries to put Carol off by insincerely suggesting to her that she is very bright but angry. Then he almost saves himself by beginning to suggest to Carol that her failure is her own fault. When Carol protests, he does not finish his thought. Almost immediately, John makes one more pass at upholding his standards and then he flounders:

> *John*: What do you want me to do? We are two people, all right? Both of whom have subscribed to . . .
> *Carol*: No, no . . .
> *John*: . . . certain arbitrary . . .[20]

He reminds Carol that she is failing according to criteria to which she sub-scribed in a disinterested manner before she could know the outcomes of her subscription, but then in a moment of honesty, doubt, and weakness, he char-acterizes these standards as "arbitrary." Contrary to John's intentions, this only upsets Carol further, and she presses on, insisting that there are standards that he must teach her, that she must understand. Finally, John succumbs to Carol, to his own self-doubt, to the complexity of the issue, and to open-minded diffidence and self-scrutiny which his discipline has held up to him as good. He suggests to Carol that she did not understand his book because "perhaps it's not well written," that the distinction between teacher and student is an "Artificial *Stric-ture*," that the tests which students take in school and in college are "nonsense," designed "for idiots . . . *by* idiots," that he would not employ the people on his ten-ure committee to wax his car, that she will get an "A" in his course because they will "break the rules" and "start over," and that they can do this because "What is the Class but you and me?" Carol is, of course, shocked. She has been told that there is absolute meaning in the world and in words and that her professors will teach her to understand. She has been told, as John says, and holds it as "an article of faith, that higher education is an unassailable good." In this interview, John has revealed something that he does not entirely believe—that he owns the null by power rather than by right. The possibility that this is true will transform both of them and redefine the entire discourse of the world of the play.

At their second meeting, Carol has brought John up on charges of elitism, sexism, and sexual harassment. Carol has quoted him accurately to the com-mittee, and he finds in their report his own words. As he realizes that he no lon-ger controls the definition of the vocabulary, John begins to find it difficult to make sense—or to understand. On the contrary, Carol, whose speech has been rather minimal, begins to speak in longer, more sustained and impassioned phrases. Of the charges, she says,

> You think, you think you can deny that these things happened; or, if they *did*, if they *did*, that they meant what you *said* they meant. Don't you see? You drag

me in here, you drag us, to listen to you "go on"; and "go on" about this, or that, or we don't "express" ourselves very well. We don't say what we mean. Don't we? Don't we? We *do* say what we mean. And you say that "I don't understand you.[21]

Now, the words and the charges mean what Carol says they mean and it is John who does not understand. When John tries to deny that he intended to harass or intimidate her, Carol eloquently tells him what his own words meant to her:

> How can you *deny it*, You did it to me. *Here*. You *did* . . . You *confess*. You love the Power. To *deviate*. To *invent*, to transgress . . . to *transgress* whatever norms have been established for us. And you think it's charming to "question" in yourself this taste to mock and destroy. But you should question it. Professor. And you pick those things which you feel *advance* you: publication, *tenure*, and the steps to get them you call "harmless rituals." And you perform those steps. Although you say it is hypocrisy. But the aspirations of your students. Of *hardworking students*, who come here, who *slave* to come here—you have no idea what it cost me to come to this school—you *mock* us.[22]

At their third meeting, Carol, and her "group," has seized the null not only within the university but in the society at large: John has been charged with rape, the statutory definition of which, according to the authorities, matches his actions. The audience knows that according to the statutes and definitions of the old null John neither intended nor committed attempted rape. To prevent their sympathies from going over to him and also to reveal the weakness of his unexamined principles, Mamet has John try to take refuge in his belief in "freedom of thought." Carol understands perfectly that from her point of view "freedom of thought" is the last refuge of professorial scoundrels:

> Then why do you question, for one moment, the committee's decision refusing you tenure? . . . You believe in what *you call* freedom of thought. Then, fine. *You* believe in freedom-of-thought *and* a home, and, *and* prerogatives for your kid *and* tenure. And I'm going to tell you. You believe *not* in "freedom of thought," but in an elitist, in, in a protected hierarchy which rewards you. And for whom you are the clown. And you mock and exploit the system which pays your rent. You're wrong. I'm not wrong.[23]

Carol brings the point home both to John and to the audience that "any atmosphere of free discussion is impossible" when one of the discussants has power over the other. As long as there is a null, someone will own it, and as long as someone owns the null, speech can never be free. Mamet does not explore the implication that without the null, or with a uniformly and universally held null, speech might be said to be free—except that it would be meaningless. Mamet does suggest that he sees no net gain or loss in the transferring of the null we have just witnessed, simply a transfer of power and a shift in terms.

Oleanna has created quite a bit of controversy and has been disliked particularly by people who feel that the play portrays Carol's feminist awakening unsympathetically. These audiences are, in fact, people who believe in John's version of the null so strongly that they cannot imagine it ever being justly changed. They, like John, want to enlarge the ownership of the null without changing its content. History and Mamet both teach us that this is unlikely.

Why does the current struggle for ownership of the null pose a dilemma of choice for university faculty and administrators? Because our own ideal of free inquiry and our own pride of intellect requires us to acknowledge the ways in which universities are not meritocratic or open to free inquiry and speech. Simultaneously, many of us believe in the ideals of meritocracy, organized skepticism, rationality, objectivity, and truth and wish to preserve them. To agree that there have been violations of the basic principles does not require their abandonment. To agree, however, that such principles are unattainable ideals may perhaps reduce them to empty words which allow one group to exercise power over others. The dilemma that universities face is how to deal effectively with demonstrable abridgments of the principles of meritocracy, objectivity, and rationality without fatally undermining them. Unless it rests on these principles, control of the null at the university is simply another sort of political power and social coercion.

It is hardly a new idea that knowledge is in some measure situationally based. However, does accepting the fundamental idea that scientists and the sciences are affected by their social characteristics and location mean that we must abandon the idea that there is an objective reality that is being increasingly approximated with additional knowledge? If owners of the null acknowledge that knowledge is to some extent socially constructed, that there are limits to objectivity, must they give up the ideal of objectivity and the correspondence theory of truth?

When we examine social and intellectual changes in scholarly disciplines and in the character of the university itself, the seeds of the transformation may prove to be sowed by the groups that currently claim ownership of the null. Those who relinquish the null may contribute to their own loss because they see some "truth" in the criticism and because they are committed to rationality and objectivity. Ownership can change as an unanticipated consequence of a commitment to ideals that are under attack; ideals that will be abandoned. If leaders go further, as in Mamet's play, and agree that the basic principles are empty and that they represent nothing other than expressions of prevailing power relationships, the stage is set for the overturn of the null. The leaders of university administrations and departments are being challenged to defend the core principles on a philosophical and sociological basis. If they choose not to respond to the challenge, transformations or stability in the null is apt to hinge on the efficacy of local political maneuvering and tests of power.

Two propositions are worth considering. First, overturning the null comes easily when the owners are not deeply committed to the principles that underlie the ownership. Second, with the shift in ownership of the null comes a shift in control of moral authority. For these reasons, among others, the stakes of the challenge are high at the research universities.

Dilemma Three: Striking a Proper Balance between the Demands of Scientific and Scholarly Research and of Teaching

Research universities continue to face the dilemma of how to fulfill their dual mission of excellence in teaching as well as in research. Is it possible in the highly competitive world of research universities, where academic free agency flourishes, for universities to produce faculty members who are among the most distinguished in the world in terms of research productivity and who will devote sufficient time and energy to teaching, particularly of undergraduates?

The legitimacy of the research university is at stake in being able to demonstrate that the answer to this question is yes. The dilemma is how to maximize productivity on both fronts so that these universities can reinforce their claims to preeminence in research with those who support and evaluate it and demonstrate teaching excellence and commitment by senior faculty with a public that is beginning to demand it.

If academic leaders feel that there is currently an imbalance between the time allocated to research and to teaching, it is of their own making. The current state of affairs results from research universities being in a highly competitive environment where the goal is to be "the best" (among the top five to ten ranked departments or professional schools) and to be *perceived* as among the best. Such perceptions will not come from hiring and promoting those who have extraordinary track records as teachers without equivalent research records. To be recognized as the best, research universities try to monopolize the talent market. This is even more difficult today than fifty years ago, but that is the goal: to bring in as many truly distinguished faculty as budgets and persuasion will permit—both younger and more established eminences, whose research publications are envied by others and who have won recognition from institutions that confer recognition and rewards for research achievements. That is the script for legitimating the strength of a department, a school, or a research university. It is the basis on which universities make claims for their unique quality and preeminence. It is how research universities gain legitimacy and increased resources in the competitive world of research funding and in the competition for the best students and faculty. It is the principal basis for their reputational standing and prestige. Moreover, it is the basis for prestige for individual members of faculties—even those who gain recognition not for their own achievements but through their association with a distinguished department or school. Thus, personal and institutional legitimacy is obtained predominantly through research achievements. That is what academic leaders have coveted as much, if not more, than the faculty. Indeed, to a significant degree, enhancing research excellence is a measure of an academic leader's performance in office.

Thus, the dilemma is often incorrectly cast in terms of individual faculty members trying to avoid teaching while academic administrators seek to steer them back to the classroom. I know very few faculty who are not interested in teaching bright students and very few academic leaders who do not spend time recruiting scholars and scientists who are known for their research rather than for their teaching. In fact, academic leaders have consistently applied strong pressure and provided large incentives for faculty to pursue their research interests with almost singular devotion. And for good reason. This not only reinforces what most faculty find exciting and enjoy doing and leads to a national and international reputation, but research excellence legitimates the university's claim to greatness. Greatness, as currently defined, depends almost exclusively on the quality and quantity of research produced by the faculty and on knowledge within important reference groups of that quality. Academic leaders recruit and support scientists and scholars who have made or are apt to make seminal discoveries—those who define fields and specialties.

The real puzzle is how to reshape a reward system, which has been created by the competition for quality and prestige in research and which has upset the balance between teaching and research, so that the scales are rebalanced and research is unimpaired. Is it possible to achieve very high levels of commitment to and excellence in teaching among the most prominent scholars at research universities without damaging the quality of the research enterprise? What price needs to be paid and will the outcome prove worthwhile?

Some years ago, former Yale University president A. Bartlett Giamatti, in one of his lyrical essays about the "real world of the university," enjoined research universities to increase their commitment to teaching. "All the research we want to do, all the obligations we must carry as faculty are in some sense nurtured by and are versions of that first calling, which is to teach our students. We want always to do more, but we can never do less."[24] Many university presidents have followed Giamatti in calling for increased attention to teaching, and particularly to undergraduates. There is, of course, much virtue in these statements of mission. As lyrical and appealing as Giamatti's prose may be, his rhetoric fails to capture the structural tension that exists at research universities between these two dimensions of the mission, and the language surely fails to recognize fully the set of fundamental cross pressures and structurally induced ambivalences felt by many faculty who aim simultaneously to be "the best" in the laboratory and in the classroom. The cross pressures result from being encouraged to apply for and to obtain as many research grants as possible; to support expensive research programs and laboratories, including support for graduate and post-doctoral students; and to publish research that brings renown to the university, while being pushed to devote time to graduate and undergraduate classroom teaching at a level of commitment and performance equal to that displayed in running a research program. Not only do these normative prescriptions create substantial time-budget problems, but they often lead to uncertainty among faculty about how they are expected to spend their time. Under these stressful conditions, most faculty members look to the reward system for guidance. Until the reward system changes and the incentive structures shift, there will continue to be a preponderance of effort directed toward research.

It is not impossible to address this dilemma and to make an effort to rebalance the scales. Academic leaders can do more to shift the balance through their actions than their rhetoric. Consider some of the things that could be done at research universities to gain greater support for undergraduate and graduate classroom teaching and ultimately to place greater emphasis on the lasting contributions scholars can make through the achievements of their students. First, as former Harvard University president Derek Bok has suggested, research universities should not compete for faculty by negotiating reduced teaching loads or unusually generous paid leave arrangements.[25] During the 1980s, the bidding wars for academic stars often led renowned professors away from students. This created a "class" structure within the ranks of the professoriate and reinforced the perceptions that classroom teaching was not prized. The legitimacy of these universities began to be undermined as the public became aware of escalating tuition costs coupled with the retreat from the classroom.

Second, research universities must try to create a culture that explicitly honors excellence in teaching as well as in research. We must not only insist on good teaching, but we must demonstrate that it will be rewarded. Again, Bok, among others, has suggested that we create "teaching portfolios" that will not only be used in promotion and tenure decisions but in helping young scholars become outstanding teachers.[26] We must not simply demand better teaching, as if the demand will be sufficient to create the supply of distinguished teachers. We should invest in programs that help young researchers become outstanding classroom teachers and begin to develop better indices of the quality of their teaching performance—ones that measure different types of teaching in different types of settings.

Third, this desirable cultural change is more apt to happen if research university leaders insist that quality of research dominate quantity. Promotion and tenure decisions must focus on the best that a scholar or scientist is able to produce, not on the sheer volume. Limiting the number of publications that could be submitted for review by a candidate for tenure would reinforce the effort to limit output for its own sake. It might also permit greater concentration on teaching roles and the interaction between research and teaching.

Finally, we must clarify the problem itself. Complaints about undergraduate teaching at American universities have occurred regularly at least since Benjamin Franklin ridiculed the instruction offered to Harvard undergraduates in 1772. The problem today, as two hundred years ago, is not one of the quantity of teaching as much as of the content and the quality. The dilemmas associated with a rebalancing of the roles of teaching and research will not be solved even if we manage the difficult assignment of changing the reward system. It is not simply a matter of substituting full professors for advanced graduate students in undergraduate classrooms or of faculty offering more courses. The absence of full professors from classrooms may be symptomatic of the problem, but it fails to confront the major issues of the quality of teaching and the lack of coherence in the curriculum that we offer students—that is, unstructured curriculums that do not represent the books and materials that the faculty believes college students should engage but a grab bag of courses that capitulate to market forces and current fashion. A serious examination of classroom teaching will undoubtedly reveal that some advanced graduate students are brilliant teachers who will become the great, full professor teachers of tomorrow, while some of the "giants" of today who are absent from the classroom were the poor teachers of the past. The problem is not really the professorial rank of instructors (although professors of all ranks should be active teachers of undergraduate and graduate students); it is the absence of institutional interest in understanding the bases for a productive advanced learning experience and an unwillingness of many research universities to commit the resources necessary to improve teaching performance.

Until now, research universities have failed miserably in teaching young scholars and scientists about the art of teaching. We tend to scoff at pedagogy, are unwilling to take seriously the idea that young scholars can acquire skills as teachers, and we do not prepare them for one of their two fundamental roles as professors. This does not mean that universities foster poor teaching. They do not, but the quality of teaching that exists is a function of individual endowments and effort, largely made in isolation, and there is little being done to help young scholars become better teachers—and to have them consider new, nontraditional modes of acquiring and transmitting knowledge. We would never contemplate a similar approach to the research training of graduate students. This set of attitudes needs to be changed if we are to improve the quality of the teaching offered to both undergraduate and graduate students at research universities.

The real challenge then for research universities is not to lower research standards in appointment, promotion, and tenure decisions in order to accommodate "better teaching" but to recognize and facilitate demonstrated quality in teaching performance among brilliant researchers. The message sent by academic leaders to the faculty must be unambiguous; the actions that follow must demonstrate that the words in the message are not empty. It would be a significant mistake, and unnecessary, for research universities to lower the threshold on research quality required for recognition and tenure. Research universities need to increase expectations and rewards for teaching excellence—and to require that all members of the permanent faculty demonstrate their capability as teachers. They need to systematically evaluate teaching performance in every hiring and promotion decision; they need to increase the visibility of extraordinary teaching in the university community; and they need to initiate programs that will help brilliant young scientists and scholars become outstanding teachers. The research university should become the place where it is once and for all demonstrated that it is a myth that excellence in research and teaching performance are fundamentally incompatible.

Dilemma Four: The Partnership between Research Universities and the Federal Government

How can the partnership between research universities and the federal government be redefined and new sources of research support be acquired without entering into Faustian bargains? The 1940s Vannevar Bush paradigm, which defined the partnership between the federal government and the research university, is rapidly changing.[27] It is ironic, of course, that this is the case since this partnership has resulted in American preeminence in science in the postwar period. When all is said and done about changes in the Bush paradigm, the federal government must and will continue to be the principal supporter of basic research in the nation and at universities. But it is not apt to invest on the same terms that existed during the period of extraordinary growth in knowledge over the past fifty years.

Consequently, research universities face increasingly important dilemmas about the support of basic science and technology. 1) What role should the research universities play (indeed what role can they play) in modifying or replacing the older Bush paradigm with a new framework that maintains American preeminence in science and preserves the research university's role as the principal incubator of scientific ideas and talent? 2) How can research universities retain commitments from the federal government while simultaneously developing new sources of research support that do not exacerbate existing tensions between the government and the universities? 3) Can and should university scientists redefine their scientific goals and reorient themselves toward new types of scientific and technological problems that have the potential for short-term practical results? 4) Can research universities adapt successfully to changing research conditions by increasing the number of interuniversity collaborations and consortial research efforts? 5) Can research universities increasingly collaborate with international partners without undermining national economic interests and American support for their research efforts? 6) How can research universities develop new research relationships with the industrial and corporate world without entering into a Faustian bargain?

The dilemmas facing research universities are nothing less than how to sustain the world's most creative science and technology enterprise without the rate of increases in federal support that would appear to be needed to do so. But these dilemmas are not simply about obtaining new research resources. They are about the types of changes that the university scientific community will have to undergo and the bargains it will have to strike in the effort to preserve and expand the research enterprise while ensuring its continued quality. The drama in the situation lies in the nature of the bargains: What is being given up, at what cost, to achieve what goals?

Within the past five years, it has become increasingly clear that the rate of increase in government investments in science and technology, which has been doubling every decade or so, and which marked most of the postwar period, will no longer be sustained. Given the nation's economic problems and the current efforts at deficit reduction, real growth during the 1990s is apt to diminish. Moreover, the increased cost of conducting pioneering scientific research will increase, intensifying still further the existing competition for federal dollars among research universities.

Unfortunately, the reduced rate of investment in science has exacerbated tensions between Congress and the research universities. The points of view in these recent debates between Congress and the research community have been extensively covered in the media. In the end, leaders of the research universities fear that the recent polemics and congressional actions have further undermined the special status that research universities have enjoyed in American society over the past half century: higher education has become just one more competitor for a piece of the federal budget.

In the post–Cold War era, the military rationale for government investments in science (which was in fact more central to the Bush paradigm than most observers acknowledge) will have to be replaced by a new rationale—one that builds on the social and economic benefits for the continued investments in American science and technology. As I have noted, there can be no substitute for federal support of science if American leadership in science and technology is to be maintained. But there are possibilities for new sources of substantial, supplemental capital that could fuel the next phase of scientific advance by releasing creative scientific energy that thus far has remained fettered.

One such source of financial and human capital can be found in a closer partnership between the federal government, American industry, and the research university—with the American public as potential beneficiaries. The building blocks for that partnership are, in fact, already in place. They have been developed as a direct consequence of the prescient Dole-Bayh Bill of 1981, which authorized universities to hold the patent and licensing rights to discoveries that were produced with federal funding. A new, expanding partnership with the industrial and corporate world holds great promise for new sources of capital that can produce important scientific and technological discoveries, but entering into that new partnership is fraught with its own dilemmas and difficult choices.

Recognizing the potential for support of biomedical and other scientific activities, university leaders have developed new offices of science and technology that examine research discoveries for their potential practical applications. While these offices are only in their blueprint phase of development, the patents and licenses that result from their work are linking the university research community with biotechnology incubator companies and with more established firms, as well as with new computer software businesses. Some universities are developing new high-technology "parks" that are introducing new industries into urban centers desperately in need of economic development. The income from the patents and licenses is bringing substantial new resources to the universities that can be used for internal reinvestments in their scientific and engineering activities. This new capital allows universities to seed innovative, high-risk, high-payoff programs; to invest in novel ideas that cannot initially obtain government funding. These investments can, however, be leveraged into research programs that are highly attractive in the longer run to government funding agencies.

The resources that could be made available for investments in new scientific efforts at the research universities are not trivial. At Columbia, for example, annual revenues from patents and licenses have risen from roughly $4 million to $24 million over the past five years. Other major research universities have experienced similar growth. It is noteworthy that this represents the annual return on an endowment of about $480 million, given a 5 percent spending rule. Moreover, it is widely believed that we are seeing just the tip of the iceberg. Over the next decade, we could see these figures grow to as much as $75 million a year.

But there is a potentially darker side to these bright possibilities. What price, if any, will have to be paid for these new partnerships, and therefore, what balancing acts must be considered by research universities? Consider six problems that already exist for those that have taken the lead in developing these new relationships.

First, industrial support has, of course, its own uncertainties. Motivated more by the bottom line than universities, businesses that invest in university-based research can and will make rapid decisions to cut support when and if they feel it lacks profitability. Reducing dependence on federal grants and contracts through partnerships with industry has its own set of built-in uncertainties that can affect university capital investments as well as hiring and promotion decisions. Second, universities will have to balance investments in high economic payoff research against sustained effort in more basic and intellectually challenging research. It is not, in fact, known whether or not these efforts truly compete with each other, or whether the efforts are additive or complementary. Third, research universities will have to examine increasingly the allocation of effort by the faculty when some patentable, but less significant, research may lead to large personal gains for the faculty. There is a real possibility that the normative code of scientific research will be modified as a result of the terms of the new partnership with industry. In the past, individual scientists sought recognition for their discoveries but eschewed direct economic gain. That has now changed. Many scientists with extraordinary capabilities are now direct beneficiaries of the patents and licenses produced by universities. This is a matter of university policy. An increasing number of university faculty members are stakeholders in incubator biotechnology or computer software companies. Indeed, many universities are becoming holders of equity in these new companies. These new relationships between economics and science pose a set of dilemmas for universities that are just beginning to be addressed. In fact, the possible changes in the normative structure of science are related to a fourth problem: universities must balance their dedication to a neutral position regarding the outcome of scientific experiments against their efforts to support the entrepreneurial efforts of their talented faculty. This may not seem like a thorny issue, but anyone who has served recently on internal science and technology review committees can tell you that universities are increasingly facing ethical and moral issues that previously they rarely had to confront. Conflict of interest policies are being redrafted with an eye toward maintaining the norm of organized skepticism while reinforcing creative faculty research energies. Fifth, universities must deal with new problems regarding the training of their graduate students. They will have to be concerned about how scientists who stand to gain from patented discoveries mentor students. When there is a potential conflict, do faculty continue to steer students toward the most intellectually interesting and challenging projects rather than those with the greatest potential for personal profit? Finally, research universities will have to examine closely how their commitment to open science is affected by their relationships with both foreign and domestic businesses. Each of these problems represents new policy questions and choices to be made by universities as they seek to fulfill their research mission.

Some observers of research universities foresee major structural changes over the next decade. They envision the end of the nineteenth-century Germanic model of departmental and school boundaries. I doubt that we will see these types of structural reorganizations, and we surely will not see them artificially imposed on the current structure of departments, centers, and interdepartmental research institutes. The best of the research universities will continue to be sources of national pride—an American institution that remains superior to its counterparts around the world. Nonetheless, by the turn of the century, we will probably return to many of the same problems and dilemmas that we have discussed in this volume. The themes will be the same; the variations will have changed. When we think about which research universities during the 1990s made significant gains in their relative quality and reputational standing, we will, I believe, focus on those that dealt effectively with the dilemmas of choice discussed in this volume.

Acknowledgment

Special thanks to Dr. Elinor Barber for her critical role in developing this volume and for her insightful comments on this paper.

Notes

1. Among other works, see Martin Anderson, *Impostors in the Temple: American Intellectuals Are Destroying Our Universities and Cheating Our Students* (Englewood Cliffs, NJ: Simon & Schuster, 1992); Jacques Barzun, "We Need Leaders Who Can Make Our Institutions Companies of Scholars, Not Corporations with Employees and Customers," *Chronicle of Higher Education*, March 20, 1991; Allan Bloom, *The Closing of the American Mind* (Englewood Cliffs, NJ: Simon & Schuster, 1987); Derek Bok, *Universities and the Future of America* (Durham, NC: Duke University Press, 1990); David Bromwich, *Politics by Other Means: Higher Education and Group Thinking* (New Haven, CT: Yale University Press, 1992); Dinesh D'Souza, *Liberal Education: The Politics of Race and Sex on Campus* (New York: Free Press, 1991); A. Bartlett Giamatti, *A Free and Ordered Space: The Real World of the University* (New York: Norton, 1988); Roger Kimball, *Tenured Radicals: How Politics Has Corrupted Our Higher Education* (New York: Harper and Row, 1989); Henry Rosovsky, *The University: An Owner's Manual* (New York: Norton, 1990); John Searle, "The Storm Over the University," *The New York Review of Books* XVII (19) (December 6, 1990): 34–42; and Charles J. Sykes, *ProfScam: Professors and the Demise of Higher Education* (Washington, DC: Regnery Gateway, 1988).
2. See, for example, "American Higher Education: Toward an Uncertain Future," vols. 1 and 2, *Dædalus* 103, no. 4 (Fall 1974) and 104, no. 1 (Winter 1975). Eliot and Conant were, of course, distinguished presidents of Harvard; Hutchins established the preeminence of the University of Chicago.
3. I will not attempt to define "the research university" beyond the obvious: its core mission is both teaching and research in the form of contributions to new knowledge through original scientific and scholarly discoveries and interpretations. Such a broad definition takes us only so far. Clearly, even the universities classified as research universities by

the Carnegie Commission differ in the number of their professional schools and in their coverage of liberal arts subjects. The question remains whether some components of the research university are essential to its identity while others are not.

4. Derek J. de Solla Price, who charted exponential growth in science, noted, for example, that fully half of all the scientists who have lived since the seventeenth century are alive today and that the intellectual half-life of the scientific literature is rapidly decreasing because of the exponential growth of those literatures. For example, in a specialty like high-energy particle physics, the half-life is a mere five years.

5. During the same forty-five-year period, gifts to the university have grown from $1.4 million in 1945 to over $120 million in 1989–90; faculty size has grown less rapidly, from 362 full professors in 1945 to about 750 today. Perhaps the anomaly in the overall pattern is found in the growth in the student population. This has varied widely by school, with the college almost tripling in size, while the graduate faculty of arts and sciences has grown by less than 20 percent.

6. While I do not have precisely parallel data, budget materials obtained for Harvard, for example, suggest a budget growth from about $217 million in 1972–73 to $1.2 billion in 1991–92; and growth from $174 million in 1972 to $653 million in 1992 at the University of Chicago. Faculty size and student populations have not grown nearly so rapidly at these universities.

7. The problem of dealing with a gap between the bases of knowledge and the resources to cover an expanding area has been the subject of concern in earlier periods as well. It is noteworthy that a number of the great research universities, such as Princeton, MIT, and Cal Tech, never defined their mission in terms of "full service." Nonetheless, the great private and public research universities have tried, by and large, to sustain substantial, if not full, coverage.

8. I have focused here almost entirely on choices involving academic programs. Although universities do not find it easy to reach agreement about administrative cuts, and they too often do not link these cuts with an ordering of academic priorities that require some services more than others, making administrative cuts is much easier than making hard choices about academic programs.

9. Dr. Elinor Barber's comments were particularly helpful.

10. In one of his first communications, President Richard Levin suggests one method, building on the work of a predecessor at Yale: "On July 1, 1978, A. Bartlett Giamatti issued the first memorandum of his Presidency: 'In order to repair what Milton called the ruin of our grandparents, I wish to announce that henceforth, as a matter of University policy, evil is abolished and paradise is restored. I trust all of us will do whatever possible to achieve this policy objective.' I have appointed a committee, chaired by the University Chaplain, to investigate why the Giamatti Proclamation failed to produce the intended result. I have asked the committee to study the feasibility of abolishing evil and to develop a strategic plan for the restoration of paradise. The committee will present its findings to the University Budget Committee, which will determine whether paradise can be restored without further cuts in academic programs and support services. Before any action is taken, I assure you that there will be opportunity for full discussion by the appropriate faculties, the Yale College Council, the Graduate and Professional Student Senate, the Association of Yale Alumni, Locals 34 and 35, and the New York Times. I expect to transmit recommendations to the Yale Corporation before the end of the millennium." E-mail communication from Richard Levin to members of the Yale community, July 1, 1993, 11:27 A.M.

11. The strong value placed on faculty governance today is often mistakenly believed to have originated with the inception of the research university. In fact, Edward Shils has argued that there has been a de facto shift in authority from presidents and trustees to faculties since the early part of this century. Before the turn of the century, presidents were autocrats with complete backing from their trustees. By 1940, the faculty had gained control over appointments, promotions, degree requirements, and new courses of study. See W. Allen Wallis, "Unity in the University," *Dædalus* 104, no. 1 (Winter 1975): 72.

12. A recent joint faculty and administration effort at Columbia attempted to identify some criteria that might be used in establishing academic priorities. Consider the nine criteria that we thought should be considered when making choices among "competing goods": 1) centrality of the field to the university's mission and goals; 2) current state of the field, discipline, or specialty; 3) current academic excellence of the field at the university—whether its organizational shape is department, school, institute, or center; 4) projected vitality of the field over the next several decades; 5) relevance and contribution of the field to the undergraduate curriculum and to the training of graduate and professional students; 6) contribution to other fields, disciplines, and schools at the university; 7) additional investment required to improve significantly the quality of the department, school, or organizational structure; 8) sense in which work in the field meets important social needs; and 9) reversibility of the required commitment, such that the investment can be terminated or redirected if it yields less advancement of knowledge than anticipated. A host of questions could be raised about any framework such as this. For instance, how do we define and determine "excellence" or the current or projected future state of a discipline? Who decides such matters? Plainly, this particular set of principles is not definitive; many others could be developed. The appropriateness of these or other values will vary, of course, at different universities. No one would be foolish enough to claim that we could, or should, strive at this time to develop an algorithm for choices.

13. A high proportion of fixed costs are associated with commitments to tenured faculty and maintenance of a physical plant. While I emphasize here the lack of flexibility in resource allocation, there is some value in moving so slowly. Institutions are less apt to shift significant resources to currently fashionable cognitive areas that prove of little lasting educational value. This functional consequence of a dysfunctional structure should not be lost on us or minimized.

14. I first encountered this usage of the null hypothesis concept in a review essay by Harrison White of Jonathan R. Cole, *Fair Science: Women in the Scientific Community* (New York: Free Press, 1979), which appeared in the *American Sociological Review* 87, no. 4 (January 1982): 951–56.

15. Limited opportunities for members of various religious, racial, and ethnic groups at research universities are well documented, along with limitations placed on women. Substantial changes have occurred in opportunities offered to minorities and women at these universities in the recent past.

16. Neil J. Smelser, "The Politics of Ambivalence: Diversity in the Research Universities," *Dædalus* 122, no. 4 (Fall 1993): 40–41.

17. Richard Rorty, among others, presents an alternative view to Searle's. See Rorty's "Science as Solidarity," in Richard Rorty, *Objectivity, Relativism, and Truth* (Cambridge: Cambridge University Press, 1991), 35–45. There is an extended debate on these issues in the recent literature in philosophy, history, and sociology of science. For an important presentation of the alternative positions, which suggests both the value and limitations

228 VALUES AND REWARDS IN SCIENCE

to both the traditional positivist and recent social constructivist points of view, see Stephen Cole, *Making Science Between Nature and Society* (Cambridge, MA: Harvard University Press, 1992), chaps. 1–3. Cole stakes out a middle ground that has potential for theoretical and empirical elaboration.

18. David Mamet, *Oleanna* (New York: Vintage, 1992). All quotations from the play are drawn from this edition. This discussion of Mamet's play and its relationship to ownership of the null has benefited from extensive discussion with Joanna Lewis Cole.

19. Mamet, *Oleanna*, 3.

20. Mamet, *Oleanna*, 10. The quotations in the following paragraph are drawn from pages 10–33.

21. Mamet, *Oleanna*, 48–49.

22. Mamet, *Oleanna*, 52.

23. Mamet, *Oleanna*, 67–68.

24. A. Bartlett Giamatti, *A Free and Ordered Space*, 56–57.

25. Derek Bok, "The Improvement of Teaching," *Teachers College Record* 93, no. 2 (Winter 1991): 236–51.

26. A number of universities have begun a more systematic study of teaching performance and effectiveness. Among the more interesting recent efforts is one at Harvard University led by Richard J. Light. See the reports beginning in 1990 of *The Harvard Assessment Seminars*. Richard J. Light, "Explorations with Students and Faculty about Teaching, Learning, and Student Life," Cambridge, MA: Harvard University, Graduate School of Education and Kennedy School of Government, First Report 1990.

27. Vannevar Bush, *Science—The Endless Frontier* (Washington, DC: National Science Foundation, 1960). The Bush Report was first published in 1945 and transmitted to President Harry Truman. It has probably been more heavily cited in the past several years than in its first forty years.

CHAPTER 9

ROBERT K. MERTON, 1910–2003

(2004)

JONATHAN R. COLE

Robert K. Merton was a giant among twentieth-century social scientists. Both in print and in person, he had a profound influence on modern social science, and particularly on the study of bureaucracy, deviance, mass communications, social stratification, and sociology of knowledge and of science. He created an American style of sociological inquiry—"theories of the middle range"—that linked theories closely with empirical testing. He was the consummate essayist and lover of language. And he used it with such grace, tact, and analytic precision that many of the terms he coined have a deceptively self-evident clarity and have passed into common usage. I am thinking, of course, of "the focus group," "the Matthew Effect," the perspectives of the "insider" and "outsider," and "the self-fulfilling prophecy".

The magic of Robert K. Merton reached audiences beyond sociology. Although many of his students, such as James S. Coleman, Peter Blau, Seymour Martin Lipset, Lewis Coser, Rose Coser, Alvin Gouldner, Alice Rossi, and Stephen Cole would become leading sociologists, Bob Merton's influence was felt not only by historians of science, economists, political theorists, and anthropologists but also made its way across disciplinary boundaries into the worlds of the sciences, humanities, and law. The range and depth of his reach was recognized especially in his early election to the National Academy of Sciences and in his award of the National Medal of Science.

Scientometrics 60, no. 1 (2004): 37–40.

In fact, Merton's influence has been so widespread and sustained that I find evidence of it wherever I turn. Just a few weeks ago, Nick Lemann, the remarkable journalist and Columbia's new Journalism School dean, told me that he had ventured into the "classroom" chez Merton at 450 Riverside Drive while working on his book *The Big Test*. There he joined the select group of "informal students" that Merton taught throughout his life. Linked to many disparate groups, Merton delighted in bringing them together and in creating a more connected and "smaller world."

Despite his seminal contributions to many areas of sociological knowledge, Robert Merton's first scholarly love was the historical and sociological study of science. He was the father of the discipline. His *Science, Technology and Society in Seventeenth-Century England*, which was the 1938 published version of his Harvard doctoral dissertation, is almost certainly the most important sociological study of science ever produced. In August 1957, after twenty years spent cultivating other fields, Merton announced his return to the study of science in his renowned presidential address to the American Sociological Association—a talk entitled "Priorities of Scientific Discovery." Many subsequent studies of the social system of science, of the ethos of science, and of the theoretical basis for conflicts over scientific priority followed from the ideas presented in that extraordinary address. For the next thirty years, Merton and his students developed and legitimated the study of science as a social institution.

Although Merton's work touched on many disparate subjects, common theoretical elements and strategies knit them together into a highly coherent body of work. Whether focusing on the political machine or on the priority dispute between Newton and Leibniz over the invention of the calculus, Merton examined the way in which social structures constrain and direct the choices of individuals and the way in which, in turn, their choices had institutional consequences. Why was it so difficult to root out the political machine? How did the emphasis on originality in science lead to deviance among scientists? Merton often found solutions by going at the problem from an odd angle, by turning away from received wisdom and from the language in which it was enshrined. He looked for the unintended consequences of action—such as the way in which the ethos of Puritanism led to the growth of science in seventeenth-century England.

Merton naturally gravitated toward irony and paradox. In his famous paper "Social Structure and Anomie" he shows how an American virtue—the desire to succeed, to get ahead—creates an American vice—deviant behavior. He analyzed the way in which political bosses do more good with less justice than the just government does when it fails to fulfill the social needs of people who are unable to navigate complex government bureaucracies. In the ironies of the unintended consequences of patterned behavior, Merton found the magic—the material that was there but could not be seen immediately by most observers—an explanatory magic that was not revealed to his readers until Bob was ready to pull the rabbit out of the hat.

To begin to understand Bob you have to accept the idea that the power of personality can be almost as influential as the power of intellect. Merton was not only a transcendent mind but also a life force. He was a mesmerizing classroom teacher, a magician in front of would-be prestidigitators. I came to know him as my teacher in the mid-1960s, first as an undergraduate auditor of one of his most famous courses, "Analysis of Social Structures," which was offered annually to over 150 students. The content of the course was often original from year to year. In it, Merton taught Mertonian theory—that is, he taught "himself." The dramatic tension in RKM's classes was palpable. We witnessed the genesis, the evolution, and sometimes the extinction of ideas. Concepts of social structure were chiseled out, smoothed over, and used as building blocks for larger theories of social structure. Little of this material could be found in his or others' published work. There were dramatic reversals and dénouements, disasters and triumphs, as students took stock and tried to establish their own identities through critical analyses of these Mertonisms. We watched someone with the skill of a Bobby Fisher toying with a Boris Spassky. We tried to follow along and anticipate the next six moves. But most often, Merton seemed magisterial to his students. Whether they were his own students or those who had drifted in from other social science disciplines, they were in awe of him and often scared of disappointing him.

A few years later, when I was his teaching assistant, it occurred to me and to his other students that Merton seemed larger than life. Consistent with my training, I tested that hypothesis in a survey of students in that course, Analysis of Social Structures. Over 150 responded to the question: How tall is Robert K. Merton? There was little variance in opinion. The class average had Merton at 6 feet 3-and-a-half inches in height—a full two inches taller than he actually was. It was true. Merton was, in fact, larger than life.

In the 1950s and 1960s, the Columbia Department of Sociology was full of energy and hope—energy that came from the presence of Merton and his closest colleague, Paul Lazarsfeld, and from the other extraordinarily talented faculty who were there at the time—each generating a belief in the potential of sociology as a discipline. Merton, of course, led the way. He was its most renowned and most influential member.

Merton recruited me to graduate school in 1964 and early on I was fortunate to work with him as his research and teaching assistant. He had an extraordinary critical eye and sensibility, and he was a meticulous editor of manuscripts. Nothing got past him. I once misused the word "stipend" in a draft of a paper I was writing. I experienced, then, what a generation of leading sociologists had before me—the Merton red editorial pen. Sociology was fortunate that it had its Robert Merton who understood so well that clear, elegant thought goes hand in hand with clear elegant language—and who led by example in his essays, as well as in his brilliant digressive satire of learning, *On the Shoulders of Giants*. This was especially so in mid-century, when sociological jargon appeared regularly for ridicule in magazines like the *New Yorker*. When Bob found my "stipend" error—and it surely was not the only one that he identified over the years—he gave me a hand-written memo many pages long on the etymology and various correct uses of the word "stipend." When Bob's papers become available in the rare books and manuscripts collection of the Columbia library, researchers will find out how much Merton's editing helped to shape the arguments of many of the best works of sociology produced by his students and others over several generations.

If Bob was tough on his students, he was tougher on himself. Nothing left his hands for publication until his own high standards were met. Often, years would pass before Bob released papers for publication—sometimes reducing rather than enhancing the impact of these papers.

Of course, Bob was not without foibles. He had his moods that seemed to be related to how well his writing was going, writing that he began religiously at about 6 a.m. each day. After class, conversations about his ideas or mine were always intense, but they were more or less rewarding depending on how Bob was feeling. After a while, I discovered that Bob had—as poker players say—a "tell." In those days, Bob smoked both a pipe and cigars. After many experimental trials, I learned that when Bob was smoking a pipe, he was accessible. I could knock on his door and wind up spending a wonderful hour talking with him. However, if Bob were smoking a cigar, I would never knock on his inner door but instead do a fast about-face and walk quickly and quietly out of 415 Fayerweather.

As my relationship with Bob evolved into friendship, he continued to teach me. His instruction was multidimensional. When Joanna and I were young marrieds, he advised us not to put our scarce dollars into works of current fiction but instead to buy reference books. Over the years, books have come and gone in our house, but, as Bob predicted, the multi-volumed OED, the DNB, Strong's Biblical Concordance, and many other reference works have happily remained—thanks to that good advice from a good friend. Bob's teaching in other areas was equally wise. It was he, after all, who taught me the inestimable and transcendent value of very good single malt Scotch. And nothing of interest escaped Bob's critical eye. When Joanna and I became part of his extended family, we would watch Knicks games together. Bob would apply his analytic skills to breaking down opponent defenses and would occasionally jump out of his seat in frustration at a particularly stupid play by Walt "Clyde" Frazier or Dick Barnett—two great Knicks backcourt players. Over all these years, in ways too numerous to count and too various to enumerate, Bob—and Harriet—have always been ready, willing, and supremely able to give good and disinterested counsel. It is a debt that cannot be repaid.

The last time Bob and Harriet came over for a family dinner, my 24-year-old daughter, Nonnie, a doctoral student in art history at Columbia, joined us. Bob and Nonnie were old friends and had corresponded a bit when she was at college and needed some advice about independent multiple discoveries. That evening, Bob, at 92, still standing tall and straight, had lost some of his physical energy but not one iota of mental acuity, and, quite frankly, the presence of a beautiful young woman seemed to energize him. With Bob in the lead, he and Nonnie began talking about the possibility of analyzing the history of art in terms of the social organization of artists. Bob delineated three lines of inquiry that might prove fruitful. He expounded on each of them and took Nonnie and the rest of us on a scholarly journey replete with lengthy detours. Afterward, Nonnie told us that she was always amazed and delighted when Bob, after long excursions and parentheses, finally wove the various strands of his argument coherently together. And so, yet another generation had fallen under the spell of that Mertonian magic.

The death of Robert K. Merton brings to a close the extraordinary work of twentieth-century sociology. It was my great good fortune that for the last four decades of that century, Bob was my teacher, my close colleague, my adviser and my friend. His death—as Harriet said in a recent conversation—has left behind a great silence.

CHAPTER 10

WHY ELITE-COLLEGE ADMISSIONS NEED AN OVERHAUL

(2016)

JONATHAN R. COLE

The current system for gaining entry to elite colleges discourages unique passions and deems many talented students ineligible.

March madness is almost here. No, I'm not referring to the college-basketball playoffs; I'm alluding to the anxious waiting of young people and their families of word about their fate from the highly selective colleges of America. And I'm talking as well about those who are about to venture forth on the ritualistic campus tours to determine where they will apply next fall. What few of these families realize is how broken the admission system is at these selective colleges.

At these institutions of higher learning, the goal is to "shape a class," which involves trying to admit qualified and diverse students who will learn from each other as well as from their experiences in the classroom. These are the students who have the greatest potential to use their education in productive ways and to contribute to their own well-being and to the needs of the larger society. Diversity is not defined here as solely pertaining to race, ethnicity, or gender, although that weighs on decisions, but also on a range of interests and talents that students can develop and share with others during their college years. These are high-minded goals.

Undergraduate admissions decisions rest in the hands of a staff of well-trained and highly motivated young people: the often-dreaded admissions officers. They travel around the country touting the virtues of their school, train students to give campus tours, and provide professional videos of what life is like at their institution. A director of admissions, usually significantly older and more experienced, oversees their work. Faculty members, however, rarely have any input in these undergraduate admissions decisions. In fact, at most elite colleges and universities, the faculty have almost nothing to say about admissions policies or what criteria should be emphasized in admitting students. Even at the Ivy League schools, there is rarely a discussion with the faculty about how the admissions office defines a "success" or a "failure" in a past admissions decision.

Despite their considerable abilities, many admissions officers are arguably not as talented or as interesting as the thousands of students who are applying to these schools—lots of whom will be rejected. The smartest, most imaginative, and most creative administrators are seldom located in the office of admissions, despite the office's dedication and determination to admit the most qualified students.

There is a superabundance of applicants who are extraordinary by almost any of the standard numerical indicators: GPA, SAT, and ACT. But as much as applicants would like to think that there is some inherent rank of quality applicants from, say, 1 to 36,000 at Columbia, there is not. Instead, diversity guidelines are set, including race, ethnicity, gender, and geographic distribution. It is not simply by chance that the proportion of students in each of these categories rarely varies much from year to year. These may not be quotas, but they certainly represent goals or targets.

Beyond demographic and geographic criteria, there are also athletic teams that have to be filled, bands that need a trombone player, alumni children that need a break, and talented students in a variety of disciplines that need to be recruited. For colleges and universities that don't have deep financial-aid pockets, the ability to afford the education may also be a factor that is considered along with the student's record. Contrary to the opinion of some secondary-school guidance counselors, these colleges are looking for a well-rounded class as much as for well-rounded individuals. And yet it seems that the nation's elite colleges rarely take chances when it comes to filling each freshman class—they are too often guided by what the final result will look like in numerical terms compared with their competition, and how that might play out in U.S. News & World Report rankings.

An empirical study conducted in 1996 by Patricia Conley, a political scientist at Northwestern University, of three hundred admissions applications at a highly selective college demonstrated that although admissions officers talk about having a good deal of discretion in making decisions that circumvent customary attributes, College Board scores, GPA, race, gender, ethnicity, legacies—the standard factors that can be easily examined in an application—are by far the most significant determinants of the admissions decision.

If you are not a kid who has gone down the straight-and-narrow path for your entire high-school career, doing exceptionally well in everything and racking up impressive scores, you are rarely advised to apply to one of these highly selective colleges—unless you fall into some category (e.g., a star athlete) where it is well-known that lower standards are typically applied in terms of many academic credentials.

Within the group of high achievers whose SATs and GPAs are already off the charts, youngsters are often pushed by their parents and secondary-school teachers to differentiate themselves from the thousands of others by doing something special in extracurricular activities. So they may, say, enroll in a volunteer program, not necessarily because of true passion but for the record.

The brilliant poet, distinguished novelist, or political cartoonist of the future who just did not care about that physics course in his or her sophomore year (and received a grade that showed it) is told that he or she doesn't have a prayer of getting into one of the selective schools. So is the kid who starts out entertaining tourists on the street but who will eventually do extraordinary work as a performance artist. There is an appreciation for diverse talents but only if they go hand-in-hand with great College Board scores and uniformly high GPAs.

But that should not be the way the world works. If Columbia can produce a poet of Allen Ginsberg's quality, who cares if he was lousy in mathematics? And if the university can produce a physicist as brilliant as the eventual Nobel laureate Julian Schwinger, does it matter if he had no interest in high-school European history?

By gauging the achievement of secondary-school students according to current admissions standards, many of the top schools seem to have taken the quirkiness out of the student body—and the rebelliousness of intellect, style, and thought that is often critical to doing something important in fields other than law or medicine. And in my experience, it shows. I've noticed that students today are rarely willing to challenge their teachers in class. College becomes an instrumental bridge between high school and graduate school—or to a good job. The admissions process currently used by many highly selective colleges leaves behind some of the most talented kids. Of course, many of these youngsters go to other places and thrive—and have wonderfully productive careers pursuing their interests. It is unfortunate, however, that the admissions criteria used by the some of the most selective schools end up classifying the exceptionally talented but "one-sided" youngster as "not eligible" for admission.

The schools' proclivity to "do everything right" may be limiting students' impulses toward the rebellion and inquisitiveness that could lead to greater skepticism and creativity.

Is a transformation of the admissions system doable? For many colleges and universities, it may prove too costly in terms of both faculty time and dollars spent. But at the top schools in the nation—with their relatively large endowments—perhaps resources can be transferred and properly used in the admissions process. The quality of the institution depends almost as much on the interesting nature of its student body as it does on its faculty members. Consequently, student admissions ought to be one of the main functions of the university and its faculty. The job is too important to leave to well-meaning young administrators. It requires intervention and guidance—indeed, decision-making—by faculty dedicated to admitting students with what the Harvard developmental psychologist Howard Gardner calls multiple intelligences and providing them with an exciting and creative learning environment.

There could be a standing committee of experienced and judicious faculty members who work on shaping the class, the "truffle dogs," with the ability to sniff out talented individuals who may have gone against the grain but who have exceptional potential. Being part of such a committee could be as prestigious as being part of a tenure-review committee. Admissions-committee faculty members could serve three-to-five-year terms, rotate on and off, and be given a reduced teaching schedule (or substantial additional compensation) for agreeing to be part of the process.

Their first order of business would be to define the types of students they seek and how those students will fit into the education offered at their institution. Obviously, the criteria for admissions are apt to be quite different at a place like MIT than they would be at Amherst. The admissions staff, composed of people with experience identifying talent and potential in the various domains of intelligence, could make the first big cut. When the number of students has been winnowed from, say, 35,000 to 3,500, the faculty committee could then discuss these applicants. Of the 3,500, perhaps 2,000 would be interviewed by two of the faculty on the admissions committee—as is done in the final phase of admissions to medical schools.

The goal is to create an environment where those interested in writing, computer science, physics, cognitive psychology, anthropology, earth sciences, and economics can join together and use their abilities in an optimally creative way. Scott E. Page, a professor at the University of Michigan, has found that diversity in forms of skills and intelligence leads to better group decisions, more productive firms and schools, and ultimately to a more creative society.

For students who have exceptional artistic talent, the great universities could adopt a system of final selection similar to that used at the Juilliard School or Cooper Union. The Committee on Admissions at Juilliard, for example, selects students on the basis of their performance at competitive auditions, which are evaluated by members of the faculty.

The current high-stakes-testing mania in schools also plays a role in these trends. From its inception, the testing movement has been an effective way of classifying people as either "gifted," "average," or "challenged" ("challenged" is the current jargon that replaces such terms as "dumb," "stupid," or "not bright"). These labels, often applied to students at an early age, can stick with

238 VALUES AND REWARDS IN SCIENCE

people for the rest of their lives, affecting their own definition of self as well as how they are viewed by others. Those who are currently in positions of power and wealth control the content of the IQ, ERB, SAT, ACT, and other tests administered to youngsters before they enter college; and these tests can be skewed in any number of ways. Moreover, profit-making organizations like U.S. News and World Report overly rely on the derivatives of the standardized test.

These examinations not only classify people into various groups but also offer vastly different opportunities depending on how one is classified. The American reward system is built by those who control the examination and creates the illusion that those who do only moderately well on these exams are not as worthy of recognition as those who do well. Moreover, the quality of schools and teachers in the public schools is being measured disproportionately by how students do on standardized tests. Teachers teach to the tests, for their own careers depend on how their students do on them. As one prominent educator at the Bank Street College of Education said of the Common Core: "I have only one problem with the Common Core. It forgot about the children."

At a more fundamental level, the exams are also deeply problematic as predictors of talent. Although these tests try to measure one form of intelligence, even in that role they turn out not to be very good predictors. They measure the speed at which a person can get what the examiner defines as the "correct" answer; they are slightly correlated with first-year college grades and with how well the test-taker does on subsequent examinations of the same kind—such as examinations for entrance to professional schools. Basically, if you're good on these kinds of tests, you'll be good on these particular tests—but how good are the tests themselves in measuring quality of mind, creativity, and potential?

Maybe it is time to return to some of the fundamentals that John Dewey discussed in his philosophy of education more than a hundred years ago. Dewey espoused the pragmatist's idea of "learning by doing"—of learning from engagement with life and experiences that are actually interesting and that lead students to formulate new questions for which there may not be answers at the back of some textbook. Some primary and secondary schools, like the Bank Street School for Children in New York, actually stimulate the creativity of children to work as individuals and as part of groups to have fun solving problems.

Students ought to understand that they learn from failure. It is not evidence of their lack of ability; it is testimony to their need to practice more and work harder at gaining skills. Great teachers give students confidence to fail because they know they can succeed. Finally, educators and schools need to instill in students a thirst to pursue their own interests, even if it is at the expense of creating the "well-rounded" student. Colleges want to produce young people with enormous curiosity who have developed talents and skills in a finite set of areas that conform to their type of intelligence—youngsters who are at least as much interested in generating provocative, unanswered questions as in producing answers to existing questions. This is what all institutions of higher learning ought to be looking for as they consider applicants for admission to their schools.

This piece is adapted from Jonathan R. Cole's *Toward a More Perfect University*.

CHAPTER 11

THE PILLAGING OF AMERICA'S STATE UNIVERSITIES

(2016)

JONATHAN R. COLE

America's great public research universities, which produce path-breaking discoveries and train some of the country's most talented young students, are under siege. The result may be a significant weakening of the nation's preeminence in higher education. Dramatic cuts in public spending for state flagship universities seem to be at odds with widespread public sentiment. Americans say they strongly believe in exceptional educational systems; they want their kids to attend excellent and selective colleges and to get good, well-paying, prestigious jobs. They also support university research. After fifteen years of surveys, Research! America found in 2015 that 70 percent of American adults supported government-sponsored basic scientific research like that produced by public universities, while a significant plurality (44 percent) supported paying higher taxes for medical research designed to cure diseases like cancer or Alzheimer's. Nonetheless, many state legislators seem to be ignoring public opinion as they essentially starve some of the best universities—those that educate about two-thirds of American college students.

According to the American Academy of Arts and Sciences's recently completed Lincoln Project report, between 2008 and 2013 states reduced financial support to top public research universities by close to 30 percent. At the same time, these states increased support of prisons by more than 130 percent. New York City's budget office reported in 2013 that incarcerating a person in a state prison cost the city roughly $168,000 a year. California apparently does it on the cheap: It costs roughly $64,000 annually for each prisoner—a bit more than the cost of a year at an Ivy League university (average tuition is $50,000) and far more than at the University of California, Berkeley ($13,000) or at CUNY ($8,000).

The withdrawal of state funds is often one of the direct causes of increased college tuition—not necessarily an increase in faculty size, spending on construction, or administrative costs. Yet, many state policymakers attribute the increased tuition to wasteful spending by the universities. To fill the financial hole, state universities are going national and international—admitting many more out-of-state and foreign students, who sometimes pay as much as three times the tuition of state residents.

Today, families pay more than half of the cost of a public-university degree. In 1970—a period during which middle-class wages stagnated—they paid about one-third. The tuition at the University of Michigan, Ann Arbor, this year is roughly $7,000 for the first two years; $9,000 for the second two, according to school officials (not the $55,000 commonly attributed to Ivy League universities). And these figures represent the sticker price, not the discounted price after taking into account financial aid. At the two largest public systems of higher learning (the City University of New York and the California system), the average tuition runs from about $8,000 to $13,000 a year—the price tag at UC Berkeley. For example, fully 60 percent of City University undergraduates pay no tuition after taking into account financial aid, such as Pell grants and state support.

And again, as student tuition and fees rise, the United States spends between $60 billion and $80 billion a year on its incarceration system. Meanwhile, it charges interest on student loans that yields roughly $66 billion off of six years of federal student loans—$51 billion in 2013 alone, according to the Massachusetts Senator Elizabeth Warren. Some argue that college-student debt could be reduced dramatically if the government charged students the same interest on their loans as it charges banks to obtain money. And if the United States reduced or eliminated the unnecessary hundreds of thousands of people (predominantly minorities) who wallow in jails for multiple non-violent misdemeanors, perhaps it could invest in the higher education of a great many more of its young people—and thus create many more productive members of society.

A type of delusional thinking seems to convince American policymakers that excellent public colleges and universities can continue to be great without serious investment. As the former Secretary of State and Stanford University provost Condoleezza Rice and Joel Klein, the former chancellor for New York City's Department of Education, wrote in a Council of Foreign Relations report, higher-education investments are a form of national security at least as important as direct investments in bombers, military drones, missiles, or warships. In other words, these education investments have a very high payoff for states, the nation, and the larger world.

242 VALUES AND REWARDS IN SCIENCE

All this amounts, arguably, to a pillaging of the country's greatest state universities. And that pillaging is not a matter of necessity, as many elected officials would insist—it's a matter of choice. If Wisconsin's governor and legislature succeed in eliminating or emasculating tenure for faculty members at the University of Wisconsin, Madison, they can say goodbye to the greatness of that institution of higher learning. If Florida's governor asks students in the humanities or arts to pay higher tuition than those who major in business or STEM subjects, Florida's universities are apt to deteriorate in quality. And just so it doesn't seem like I'm cherry picking, consider what North Carolina's governor said not long ago: "If you want to take gender studies, that's fine, go to a private school and take it. But I don't want to subsidize that if that's not going to get someone a job." The consequence of such policy choices, it seems, is that tuition will go up and access for kids from poorer families will go down.

But such outcomes can be prevented. Those in the voting public who believe that they can get something for nothing or that quality will simply materialize out of the ether can revisit their assumptions. Governments can increase the marginal tax rates on substantial incomes so that those who have benefited most from the nation's prosperity pay a fair share of taxes that enables both access and educational opportunity for talented young people. The United States currently has one of the lowest marginal tax rates in the industrial world. Transferred resources from the very rich (less than 1 percent of nation's population controls more than 25 percent of its wealth), corporations, and from lower-priority institutions could build a more robust educational system in our country. There are important positive consequences in economic growth from such investments at the state and local level, as has been demonstrated in studies of Silicon Valley and in the area surrounding Boston.

In short, today essentially everyone who attends Berkeley pays a maximum of around $13,500 a year—even if his or her parents are billionaires. At Stanford or at Ivy League universities that same student would pay (and could afford to pay) the full sticker price of tuition (around $50,000 a year), but the youngster from a poorer economic background might well go free. There is not enough differentiation in tuition pricing between those who come from very wealthy background and those whose parents can barely make ends meet.

Higher education also needs more political leaders who are willing to risk their jobs by committing to ostensibly risky investments in K-12 and higher education. Every elected politician should have quitting issues—in short, be willing to resign or risk reelection when others undermine his or her core values. Take the former Connecticut Governor and Congressman Lowell Weicker, who was willing to sacrifice his leadership positions in defense of his tax policies. It's time for the states and the federal government to step up to the need to invest creatively in education and raise public awareness about the wisdom of these investments. And it's time for more members of the voting public to support candidates who meet this description.

Finally, it's up to the federal government to become an even more active player in supporting the quality of and access to higher public education, just as it did in 1862 when during the Civil War it passed the great Morrill Act that established the land-grant universities—and as it did after World War II when it passed the G.I. Bill. As Benjamin Franklin, who gave a good deal of thought to higher learning and founded the University of Pennsylvania, once said, "An investment in knowledge pays the best dividend." This investment is a choice that the country must make if the United States wants to maintain a leadership role among the community of nations in the twenty-first century.

PART II

FREEDOM AND UNFREEDOM

THE CASE OF WOMEN IN SCIENCE

CHAPTER 12

WOMEN IN AMERICAN SCIENCE

(1975)

JONATHAN R. COLE AND HARRIET ZUCKERMAN

The small numbers of women in the sciences, physical and biological, and in the other learned professions in the United States result from early and cumulative discrepancies in the extent and character of educational attainment. Women are less likely to attend university or college than men. About a fifth of all women of university age were at university in 1969, compared to more than a third of the males of the same age.[1] Women who attend university or college are slightly less apt than their male classmates to continue to completion of their course of study.[2] Male undergraduates are about three times as likely as women to specialize in science and engineering, thus providing a largely male pool of potential scientists.[3] Finally, men who specialize in science as undergraduates are also more apt than women to go on to graduate studies; about twice the proportion do so.[4]

This successive filtering means, of course, that a far smaller proportion of women than men emerge from graduate schools in the United States with PhDs in the physical and biological sciences. Only ten out of every hundred doctorates in the physical and biological sciences and six out of every thousand in engineering were granted to women in 1972.[5] This comes as no surprise. But what is surprising is that no greater share of all PhDs awarded in the sciences in the United States go to women now than in the 1920s. Women received about 11 percent of all doctorates in the physical and biological sciences between 1920 and 1929, compared with 10 percent, in 1972.[6]

Minerva 13, no. 1 (Spring 1975): 82–102.

Although their proportionate representation is much the same, absolute numbers of women receiving PhDs in the sciences have increased since the 1920s, when an average of about fifty women a year received degrees.[7] In 1972, the figure ran to 1,096.[8] The proportion of doctorates in the physical and biological sciences awarded to women has, however, fluctuated from its highest level of about 11 percent for the 1920s to its lowest level of 5 percent for the 1950s.[9] The present rate of 10 percent is a composite of the comparatively low rate of 7 percent for the physical sciences and the comparatively higher rate of 15 percent for the biomedical sciences.[10]

The more recent picture is somewhat different. Between 1960 and 1972, the annual number of doctorates conferred in the physical and biological sciences increased 2.7 times, but the number of women receiving these degrees rose 4.7 times. Thus, the number of women receiving doctorates in the sciences rose faster than the total number of recipients. At the same time, the proportion of women among all scientific and technical employees has dropped slightly during the last twenty years, a period of rapid growth in the scientific population and of the increasing integration of women into the labor force.[11]

The Principle of the Triple Penalty

What social processes produce these distributions? How has the number of female scientists and engineers come to be so restricted? And what happens to the comparatively few who do embark on scientific careers?

We suggest that women encounter three barriers to becoming productive scientists. First, science is culturally defined as an inappropriate career for women; the number of women recruited to science is thereby reduced below the level which would be obtained were this definition not prevalent. Second, those women who have surmounted the first barrier and have become scientists continue to be hampered by the belief that women are less competent than men. Whatever the validity of this belief, it contributes to women's ambivalence toward their work and thereby reduces their motivation and commitment to scientific careers. And third, as we shall see, there is some evidence for actual discrimination against women in the scientific community. To the extent that female scientists suffer from these disadvantages, they are victims of one or more components of the triple penalty.

Recruitment to Science

It is widely believed that women are neither fit for scientific and technical careers nor interested in them.[12] The evidence for these beliefs is equivocal. For one thing, women seem to be no less interested in science than are men. For example, the nationwide survey of university and college graduates conducted by the National Opinion Research Center in 1964 indicates that about half of all men and women express interest in a career in scientific research. This is not the case for engineering, which is preferred overwhelmingly by males.[13] The same survey also shows considerable differences between the sexes in self-appraisals of scientific ability. More university women than men believe they do not qualify for careers in science and engineering.[14] Women, more often than men, believe that careers in science, however interesting, require too great an investment.[15] Many female university and college students believe that they are unfit for scientific careers; as a result, they are less often willing to undertake the rigorous training required for these careers, choosing instead occupations which they believe are more in line with their self-defined abilities. However, this evidence is now a decade old and thus does not convey changes in women's attitudes which may have occurred in the years intervening.

Female students in universities and colleges believe that scientific careers are open only to exceptional women.[16] Apparently, a considerable number of women reject careers in science long before the point at which they might be rejected by those who control admission, appointment, and promotion. Some of the women believe themselves to be unqualified and others believe that a scientific career, in conjunction with marriage, would be excessively demanding. For these reasons, it is believed to be an unrealistic occupational choice.[17]

The belief that a career in science, together with marriage and motherhood, form a combination too difficult for most women to handle is consistent with the attributes of the female scientists and engineers young women know or know about. The few who teach in universities or whose names and faces appear in the press and on television are apt to be extraordinarily competent, and a large proportion have not taken on the conventional female commitments of marriage and family. As exceptional female scientists, they may seem too remote to be effective models.[18] "Average" male scientists are more familiar to young persons than "average" female scientists, both because they are more numerous and because they are more apt to have regular posts in universities and industry. As a consequence of the small numbers of women in science, many female undergraduates never encounter a female scientist at all. Although Madame Curie may serve as an inspiring model for some young women, for others she is all too remote and formidable a figure to emulate.[19]

Thus, few women even consider science as an occupation. It is not known whether women who have the potential to become scientists choose other occupations considered more appropriate for women, or whether they choose not to work at all and are thereby wasted resources.[20] For women, then, the selection of occupations is, in part, the outcome of a self-fulfilling prophecy.[21] Widely defined by themselves and others as inappropriate candidates for scientific careers, women are less apt to go into science and more likely to choose occupations more in line with prevailing expectations. In doing so, they confirm beliefs about their inclinations and apparently about their capacities as well.

That self-fulfilling prophecies contribute to differences between the sexes in the choice of scientific careers becomes more plausible when we remember that one of the most important determinants of differences in occupational attainment, social class, does not pertain here. Women, after all, share the same social origins as men; in matters of status they share characteristics, and they should, other things being equal, appear in the scientific professions in the same numbers as men. But other things are not equal. As we have noted, women are less apt to be educated at university than men of the same ages.[22] This is the case in each social and economic stratum, at each level of academic aptitude.[23] In general, families have been less willing to invest in educating daughters than sons, a practice which is quite consistent with the expectation that the chief roles of women are to serve as mothers and wives rather than to move into the occupations requiring academic training.[24] Perhaps this explains why female scientists consistently come from families of higher social status than do male scientists.[25]

The ambivalence of many women about professional careers—with their demanding schedules, high levels of competition, and potentials for conflict with family life—and about careers in the domain of science, which are also considered to be masculine, is reflected perhaps in women delaying the decision to go to graduate school longer than men. Just over half of the female graduates of superior undergraduate colleges, contrasted with a third of the men, postpone their decision to go into graduate study until their senior year or later; this is the case even though both groups report that they become interested in science at about the same time.[26] These decisions are supported by the widespread belief that occupational achievement is unfeminine; this is especially true of achievement in science and engineering.[27]

Graduate Study

Many academics believe that female students are as able as men, but women are still uniformly discriminated against in admission to graduate school, and those who are admitted tend not to be granted financial assistance.

Although intelligence tests are generally designed to eliminate sex differences in test scores, those women who obtain a doctorate in scientific subjects consistently do somewhat better than men in these standardized tests.[28] Cole's study of matched samples of women and men who have received doctorates shows that in each of five fields and at every level of prestige of university graduate departments, women had consistently higher intelligence scores than men although the differences between them were rarely significant. These consistent differences, although not large, deserve further investigation. Women's higher scores may result from the tendency of all but the very best female graduates not to pursue graduate studies and from the relatively more rigorous selection which occurs in the course of their graduate training. At the very least, the evidence indicates that women who get doctoral degrees in the sciences are no less able than their male colleagues in the intellectual capacities measured by intelligence tests.

Although it is not known whether there are systematic differences in the criteria graduate schools use in admitting men and women, such evidence as is available suggests that members of both sexes are admitted in the same proportions as they apply. Since women apply in such small numbers, they ultimately comprise a small proportion of all graduate students.[29] Female candidates do receive slightly higher scores on tests of verbal facility incorporated into the nation-wide Graduate Records Examination, but men do considerably better than women on tests of quantitative skills.[30] The qualifications of applicants, male and female, would have to be known in much greater detail in order to discover whether discrimination has occurred, and since we lack this information it is impossible to say with authority whether or not women are discriminated against in admission. As for the allocation of financial assistance to women, once they are in graduate school the record is clear: about the same proportion of women are granted financial assistance as men—at least this was the case in the 1950s and 1960s when graduate students in the sciences received much financial support in the United States.[31] How the women's movement and the scarcity of resources will affect these findings is not known.

It is also widely believed by faculty members that more women are poor risks as graduate students. They are said to be less likely to complete their postgraduate degrees, and when they do, to take longer than men.

These beliefs appear to be essentially correct. Women in general are considerably more likely than men to withdraw from graduate studies before completion. At twenty-four American universities, 54 percent of the female graduate students and 36 per cent of the male students failed to complete their studies.[32] The rate of attrition has been even higher among the recipients of Woodrow Wilson fellowships, which are awarded nationally according to very exacting standards to graduates who intend to follow academic careers. Six to eight years after having begun graduate studies, 44 percent of those men (and 64 percent of the women) had neither completed their degrees nor were actively pursuing them. The differences in the rates of attrition between men and women was greatest among graduate students in the sciences: only 26 percent of the men had discontinued their studies, while 54 percent, of the women had done so.[33]

High rates of attrition may be attributable in part to the negative attitudes of teachers toward female students. Thirty-five percent of faculty in the sciences questioned by the Carnegie Commission in the survey of American university teachers said that they believed that female students were not as dedicated as men—a considerably higher proportion than among teachers of history, English, political science, and sociology.[34] But since attrition rates are higher in these fields than in the sciences, this seems an unlikely explanation.

Some have suggested that female graduate students have particular difficulty establishing themselves in relationships of master and apprentice and that they are thereby deprived of an important element in training for a scientific career.[35] In general, they are less apt than men to identify themselves as apprentices or colleagues of their research supervisors.[36] In the sciences, however, this is not consistently the case. In physics and chemistry, there are no significant differences between male and female graduate students in the proportion reporting as being apprenticed to faculty members, although there are such differences between those students in the biological sciences and mathematics. In these latter sciences, men indicated more often than women that they were regarded by faculty members as apprentices or colleagues.[37]

Women obtain postgraduate degrees, on average, two years later than men.[38] This discrepancy is apparently not attributable to having begun their graduate studies later. It is partly the outcome of the greater propensity of women—married or unmarried—to study part-time.[39] This delay is as characteristic of the outstanding young women who win Woodrow Wilson fellowships as of the others.[40] Still, in part it is marriage and motherhood which slows women down. Married women, especially those with children, have therefore been poorer risks than their unmarried or divorced colleagues.[41]

Many believe that women are less likely to have been trained at the most distinguished universities, since they are discriminated against in admissions and because their range of alternatives in choosing graduate schools are more limited than those of men. But women are just as likely to have received doctoral degrees from leading universities as men, with 51 percent of the one and 52 percent of the other being trained at these institutions. This holds in the sciences as well as in other fields.[42] Such information as there is, as we have already observed, does not strongly point to any pattern of discrimination in admission to graduate study. It seems that if the choices of women are constrained by the requirements of their husbands' careers, these constraints are not evident in the distributions of doctoral origins of male and female scientists.[43]

Patterns of Employment

Not only is it believed that training women to be scientists is a poor investment but also that if they get their degrees, they marry, have children, and stop working. Among those few who return to work, skills and knowledge, which were once fully up to date, have become obsolete.

This view is contradicted by the best evidence available. Nearly nine out of ten women with PhDs in the sciences were at work seven years after they took their degrees—presumably the period in which most would have had children if they were to have them at all—and three-fourths were working full-time.[44] Women with PhDs rarely stopped working—eight out of ten worked at their professions without interruption after receiving their degrees.[45] The 18 percent who suspended their professional work—to have children or to accommodate changes in their husband's employment—were absent for an average of fourteen months, less time than the once usual absence of unexempted males in the military.[46] Another study of over forty-two thousand teachers in American colleges and universities in 1972 showed that approximately a fifth of all female teachers had interrupted their professional careers for at least a year, but one-fourth of the male teachers had done the same, for military or familial reasons.[47] Of course, marriage and motherhood are related to patterns of employment. Married female scientists with children more often give up employment than married women who are childless, and they in turn are slightly more often not at work than single women in science.[48]

The employment histories of women holding doctorates in science are remarkably stable. There was no difference between male and female PhD holders in mean numbers of employers over their entire careers, and when years spent with current employers are included in that total, women with doctorates show more continuous employment with one employer than do men.[49] Those employers who are reluctant to appoint women because they are allegedly less dependable than men would do well, as one letter-writer to *Science* recently observed, "to adopt the more generally corrective principle of exacting the penalty after rather than before the crime is committed."[50] If they do not, they will bring still another self-fulfilling prophecy into being.

Female holders of the doctorate may work as much as men, but they do not work for the same employers nor do they hold the same sorts of post. Female scientists are disproportionately under-represented in industry and over-represented in academic institutions.[51] Women are not relegated to appointments in colleges and minor universities as is commonly supposed. And this was so even before the official policy of "affirmative action" requiring universities to increase the proportions of female faculty went into effect. The academic affiliations of women with doctorates are about the same as those of men, but academic affiliation is one thing and academic rank quite another.[52] Women with doctorates in science more often teach than men, and two to three times the proportion teach full-time.[53] But the differences are not great: 39 percent of female faculty members in colleges and universities teach as many as thirteen hours a week against 30 percent of male faculty. However, while teaching more, women are less apt than men to report having teaching assistants: 30 percent do against 41 percent of the men.[54]

Scientific Performance

It is commonly believed that female scientists individually, and of course collectively, do not contribute as much to science as men; they are simply less productive.

Female scientists are indeed less productive than their male counterparts, according to one standard measure of productivity—the publication of scientific papers. They publish at a lower rate than men.[55] One interpretation of such differences in productivity is that they publish less because they are affiliated more often than men to academic institutions which neither expect research and publication nor reward them for it. But the facts are not consistent with this contention. Men in academic posts publish more than their female colleagues in the same institutions.[56] Thus, differences in productivity are not a consequence of differences arising from the differences in institutional affiliation of male and female scientists.

The other conventional explanation of women's lower rates of publication, namely, the responsibilities of marriage and motherhood, also does not fit the facts. Unmarried female scientists and those who are married and have no children publish slightly more than half as many papers as men with the same marital and parental status.[57] Among scientists who are parents, motherhood, more than fatherhood, results in reduced publication.[58] The more children a female scientist has, the fewer papers she publishes.

The correlation between early and later productivity is substantial for men but considerably weaker for women.[59] Women who are productive early in their careers are less apt than men to continue to be productive, and they are more apt to disappoint those who might have tried to facilitate their work. In the process, women reinforce the belief that even the best of them do not realize their promise.

The observed differences between the sexes in scientific productivity need not be attributed to discrimination. In fact, a smaller proportion of women than men are interested primarily in research, and the difference is most marked in universities: 39 percent of men and 15 percent of women who hold academic appointments in universities count themselves as being primarily interested in research.[60] In part, interest in research probably determines rate of publication and may be reinforced by it.

Turning from quantity of publication to its quality, once again we find that women do less well than men. It is difficult to assess the quality of published scientific work. Frequency of citation in the scientific literature is, however, considered a rough but valid measure.[61] Differences in the frequency of citation to the publications of male scientists compared to those of female ones are considerable but not quite as great as the differences in rate of publication.

Those women publish less than the men, and what they do publish appears to have less impact on their field.[62] The reasons for differences in rates of citation are not self-evident. It may be that women work in comparatively less popular specialties and thus have fewer resources to permit them to work at the forefront of their fields, or it may be that they have less sense than men of "good problems."[63]

There are some grounds for thinking that men are better integrated into the network of informal scientific communications. Women are less apt than men in the same fields and at the same rank to receive preprints from other scientists or to send their own work out for comment. Women who go to scientific meetings report fewer productive conversations about their work. However, about the same proportion of female faculty members as men report being in frequent communication with others in their specialties.[64] Although we do not know what effects, if any, informal communication has on scientific work, we do know that the reverse holds. Highly productive scientists are more often in close touch with other scientists than are unproductive ones.[65]

Although women apparently are as often associated with leading universities as men, there are indications that they spend less time on research, have more limited access to resources for it, and are less fully integrated into the scientific community within their fields of specialization, thus reducing the probability of carrying on useful scientific inquiry.

Rewards and Rank

Regardless of differences between the sexes in types of employment, they hold ranks lower than men.[66] This is the case both in industry and in universities, in women's colleges as well as in co-educational institutions. Women are usually concentrated in non-supervisory positions and in positions which do not provide permanent tenure.[67] Male and female scientists begin their careers at the same rank, but men generally hold higher rank seven years later, and this is so even when the quality of their research output and seniority are held constant.[68]

Thus far, we have reported scant evidence of discrimination against women in science. Although the processes affecting their recruitment to scientific occupations are external to the community of scientists, they nonetheless affect women's aspirations, access to higher education, and willingness to embark upon demanding careers. Yet there is little to suggest the existence of discrimination once women have begun postgraduate study. They are admitted to the same universities and seem to have an equal chance of financial support. When women move into the occupational world, however, they are discriminated against in the matter of rank. The extent to which this is so varies in universities and industrial corporations and among sciences. Within universities, women in physics do better than women in the biological sciences, and both are ranked higher than women in the social sciences and humanities.[69]

The differences between men and women in academic rank need not be attributed to discrimination alone. They also reflect some measure of self-conscious choice by women. The tendency to interrupt their careers and to work part-time clearly increases sex differences in rank.[70] Such choices reflect women's accommodation to domestic responsibilities. Nevertheless, since furloughs from work and part-time employment come just at the time in the scientific career when others are in the process of moving from the status of junior research worker to senior investigator and are preparing for positions of greater responsibility, these accommodations have important consequences for the women's scientific careers. They fall behind at the very time that men are moving ahead. Even though the costs of marriage and motherhood to their careers are self-imposed, they are costs nevertheless—costs which do not have to be met by men of science.

These observations on married women need to be put into context. Only half of all women with doctorates in science are married and living with their spouses, as compared with something like 90 percent of the men.[71] The combination of marriage and a scientific career does present difficulties—in budgeting time and in dealing with conflicting obligations—as female college graduates often anticipate. Yet about half of female scientists—those who are unmarried at any given time—do not face these difficulties. But whatever the reasons for their frequently unmarried state, female scientists are "damned if they do and damned if they don't." Universities and business firms are reluctant to appoint married women whose domestic responsibilities are assumed to interfere with their work, and they are also reluctant to appoint unmarried women who, they say, will marry and acquire these responsibilities. This contributes to the ambivalence about work which women experience. Such beliefs about female employees make for self-fulfilling prophecies. As a consequence, young women are apt to conform to the expectations others have of them. The social processes involved in self-fulfilling prophecies thus have strong implications for recruitment to science, for actual access to opportunity, and for motivation as well.

If one outcome of the obstacles encountered by female scientists is that they are less productive scientifically, another is that they are less well rewarded, in salary and in honor. Almost every inquiry—in industry and the universities—shows that women are paid less than men for the same work.[72] Data for the sciences alone suggest that differences in salary between men and women increase with ascent in rank. This is only partly the result of the wider range in salaries paid at higher ranks. There are also differences between the various sciences; among full professors of physics, women earn about 90 percent as much as men, and in microbiology, about 75 percent. These figures, however, do not take into account variations associated with institutional affiliation, nor do they adjust for individual differences in productivity.[73] To the extent that female scientists are less productive and less motivated to do research because of the obstacles they encounter, they also receive fewer offers of employment and thus are less able to bargain for higher salaries. However, women on average are paid less than men after taking into account a variety of individual and institutional attributes, including education, class of university or college, rank, length of tenure, and productivity. When the same variables which account for differences in salary among men are applied to women, it seems that women were underpaid in 1969 by about $1,040 annually.[74]

It would not be correct to infer that all difference in remuneration is attributable to discrimination or to lower levels of performance. In an unknown proportion of cases, women choose types of work which pay less, or they leave the field for a time, thus reducing their seniority.[75] Women, in general, and female scientists in particular, put greater emphasis on the intrinsic satisfactions of the work and less on remuneration.[76] This is as much true now as it was a decade ago. They sometimes avoid more remunerative appointments in favor of those they find more congenial or do not conflict with their domestic obligations. And since women are less often under financial pressure than men, they are free to refuse promotion to managerial posts which paradoxically entail renouncing research for administration. Greater fidelity to research may be a luxury women are better able to afford than men. While it might appear that decisions of this kind would give women more opportunity for research than comparable men, family and home life usually claim their time. Whatever their reasons, some women prefer positions which pay smaller salaries. Yet more women who have attained the doctorate level feel discriminated against in the matter of salaries than in any other aspect of their careers.[77] If some women voluntarily refuse promotions involving increased administrative responsibility, they do not reject academic promotion. Women nonetheless are promoted less rapidly than comparable men.[78]

Promotion and Recognition

There is a widespread belief that women scientists are consistently under-rewarded with regard to promotion, the class of university to which they are appointed, and recognition, regardless of the quality of their work. The relation between the performance of female scientists and the rewards they receive is, however, more complex than is often supposed. The question is, of course: Under what conditions are men and women treated differently and under what conditions are they treated the same? Do male and female scientists who are highly productive and whose work is widely esteemed hold the same ranks in higher educational institutions of approximately the same standing? In other words, when there is little question about competence in research, are those scientists by and large treated as equals? They are equally likely to be affiliated to major universities, but women are less highly ranked than men, even when their different rates of productivity are taken into account. Among scientists of both sexes who have never published a paper and thus for whom there is scant evidence on which to appraise their research performance, women do less well than men in the rank they attain and the class of institution in which they find appointment.[79] In short, gender, in principle a functionally irrelevant characteristic, affects the allocation of rewards particularly strongly when there is no relevant criterion by which capacity or achievement can be judged.

Two recent psychological experiments support this finding. In one, descriptions of psychologists, matched in all respects but sex (implied by first names) were sent to 155 department heads who were asked to indicate the appropriate level of appointment for each candidate and to appraise the psychologist's desirability as a member of the department. Women were just as often judged desirable members as men, but they were significantly more often considered for posts at the rank of assistant professor or lower than were men.[80] The chairmen who replied to the questionnaire may well have followed the economic principle of considering applicants—in this case, women—for lower rank who, experience told them, would accept appointment at a lower rank. Whatever the chairmen's motives, this suggests that, other things being equal, the sex of candidates may affect the ranks of appointment for which they are being considered.

In the other study, chairmen of 179 science departments were asked to evaluate pairs of curricula vitae of "candidates" for appointment as associate professors. One pair included sets of qualifications of two average candidates, differing only in sex; the other set included a male candidate with mediocre qualifications and a female candidate with superior qualifications. The departmental chairmen preferred the superior woman scientist to the mediocre man, but the same chairmen also preferred the average man to the average woman when the absence of outstanding achievement provided no basis for distinguishing between them.[81] If this is so, then we would expect even greater differences in rewards between men and women of similarly undistinguished quality than between men and women whose achievements are clearly superior.

Few women have been elected to the National Academy of Sciences in the United States or have won major scientific awards, including the Nobel Prize. Just over 1 per cent, or 13 of the 986 members of the Academy in 1974, are women, although women comprise something like 6 percent of the PhDs in the comparable age strata from which members of the Academy are drawn.[82] It may be more to the point to observe that the few women who are now members of the Academy received their doctoral degrees at 27, which was the same age as all members of the Academy, but they were promoted to the rank of full professor at 48, which was ten years later on average than the rest. In the aggregate, they were elected to the Academy at the age of 59, nine years after the average male counterpart. This difference cannot be explained by variations in the age of election of physical and biological scientists. We do not know, however, whether these female scientists produced important research at ages comparable to male academicians.

Among the 294 scientists who have won Nobel Prizes, five are women, and three of them shared the award with their husband—Marie Curie, Irene Joliot Curie, and Gerty Cori. Dorothy Hodgkin and Maria Mayer, both married, became Nobel laureates independently. Cori was not promoted to a full professorship until the year she received the Nobel Prize. The other four laureates, although promoted before they won their prizes, became professors later than their husbands and the male laureates.

What, then, can be said about the allocation of rewards among men and women of science? In some measure, the difference in the distribution of rewards reflects lower levels of performance by women and the choices women make in decisions about careers. And, in some measure, as we have noted, discrimination contributes to these differences, for it is not only that women hold lower ranks than men but also that they are deprived of the perquisites that accompany rank. In academic life, low rank is usually accompanied by limited clerical help and little free time for research. Almost three times the proportion of male compared to female faculty members spend a minimum of thirteen hours per week on research.[83] Low rank also means heavy teaching loads—it should be remembered, however, that women, more often than men, report a preference for teaching—and often exclusion from applying for research funds as principal investigator.[84] In short, these concomitants of low rank reduce women's opportunities to do scientific research, lower their chances to perform as well as men of approximately similar scientific capacity, and thus restrict their access to honorific awards. Given the complex interaction between discrimination, motivation, and level of performance, and the paucity of systematic data on these, it is not at present possible to estimate their relative contribution to differences in rewards between the sexes.

This brings us to more subtle aspects of the problem: to the relations between scientific colleagues. Since so few women have elected to become scientists their presence in the world of science is not taken for granted. They are apt to be treated with excessive courtesy or with excessive hostility. As a consequence, women tend to be uncertain about the response they will receive and are themselves either unduly reticent or unduly defensive. Discrimination against women has become so salient an issue among academics that it often elicits strong and inappropriate responses from both sexes. Beyond this, men have not been quick to recognize the implications of their attempts to preserve the "men's club" character of academic and corporate life. The scheduling of meetings where women

Overcoming the Triple Penalty

The principle of the triple penalty, as we have observed, asserts that female scientists are triply handicapped: first, by having to overcome barriers to their entering science, second, by the psychic consequences of perceived discrimination—limited aspiration—and, third, by actual discrimination in the allocation of opportunities and rewards. Discrimination often reduces motivation to perform; those subjected to it come to feel, regardless of their competence, that nothing they do will make any difference to their ultimate attainments. This motivational component of the triple penalty is usually combined with restricted access to the resources necessary for outstanding accomplishment. Judged on the grounds of performance in research alone, that is, in extending scientific knowledge, women do not in the aggregate perform as well as men. In consequence, female scientists have tended over the course of their careers to be deprived of resources needed to do good work, a condition which widens the gap between their scientific achievements and those of men and makes for cumulative disadvantage.[85] Such cumulative differences in scientific achievement tend to confirm and reinforce beliefs that women are less competent and less motivated than men and in their turn maintain the conditions, which first made for different levels of performance.

There is another feature of the dilemma of female scientists. Those persons who are characterized by an unusual combination of roles, such as Black physicians, youthful presidents of corporations, or female scientists, never quite know which of their roles is being responded to in social intercourse. Are they being appraised as Blacks or as physicians, as a person who is young or as one who heads the corporation, as a woman or as a scientist? They are affected by what Professor Robert Merton has called "the haunting presence of functionally irrelevant statuses." The "haunting presence" leads persons who possess these unusual combinations of characteristics to believe that positive and negative evaluations are responses to irrelevant characteristics rather than to those which are appropriate to a particular function or task. Such a person is often diverted from the obligations of his principal status and role to unproductive ruminations. Female scientists are apt to blame their failures and their successes on their being women rather than on their actual performance and to feel unjustly deprived or unjustly rewarded as a consequence. At the very least, many feel that they do not receive an appropriate response to their work and thus are deprived of the opportunities which men have of having their research criticized and corrected.[86]

The principle of the triple penalty does not apply to female scientists in every aspect of their careers. The small number who earn doctoral degrees are no less likely to have received first-class training than men nor is their distribution among academic institutions, later on, appreciably different. But they seem to do less well in rank, remuneration, and perquisites. Self-selection by women as well as discrimination contribute to these differences. The difficulties faced by female scientists—like those of other women—are not altogether remediable by reforms which would extend the allocation of rewards and resources in science. Even if it were possible to eliminate discrimination and its secondary effects, the problems associated with being a woman in American society would remain. It is in this respect that sex can be said to influence women's performance of the scientific role.

Accommodations of some sort have to be made to the responsibilities of parenthood, but women need not be the only ones to make them. Until recently, of course, women who wished to have serious professional careers were expected to forgo marriage and motherhood or to accept a compromise by seeking full-time help and part-time work. Neither of the latter alternatives, it now seems, is practicable, the former being increasingly scarce and the latter carrying its own penalties by impeding serious work and the advancement of career.[87]

If the institution of government policies designed to increase the proportion of women in science makes more pressing the need to deal effectively with the extra-professional responsibilities of women, the institution of the same policies raises important questions for science and its practitioners. First, what will be the effects of the contracting market for scientific skills on the success of the efforts to alter the sexual composition of scientific occupations? Second, what impact will an increase in the proportion of women have on the prestige of science and thus on its capacities? Third, and most important, how will such increases affect the quality of scientific research?

It is self-evident that the number of men entering science would have to be reduced substantially in order to keep the size of the number of scientists in line with demand and at the same time increase the proportions of women. The present labor market in science, with its restricted employment and opportunities for promotion, provides an inhospitable environment for elevating women into posts which they have not had previously. In some fields, there is now a surplus rather than a shortage of trained scientists. Although the evidence on the rates of unemployment of scientists is fragmentary, it does appear that the general contraction of the market for PhDs in scientific subjects has hit women harder than men.[88]

When opportunities for employment are restricted, efforts by government and women's groups to alter the sex composition of the scientific professions are apt to run into difficulty.

262 FREEDOM AND UNFREEDOM

Fears about a reduction in the prestige of science as female scientists become more numerous rest on two assumptions: one, that the prestige of occupations is affected by the proportion of women in them; the other, that increasing the number of women will discourage competent young men from choosing that occupation. These assumptions may have some basis in fact. By and large, the greater the proportion of males in an occupation, the greater its prestige and, by available measures, the better the quality of its recruits. (Although the correlation is well established, its causes are not clear.) Compare, for example, the standing of medicine, law, and science with that of teaching, social work, and nursing.[89] If the sciences are pressed harder than other occupations to recruit and to elevate women, their comparative prestige may decline. Since all professions in America are now being subjected to pressures for equality of the sexes, increasing the representation of women in science in particular should not alter its relative prestige or its ability to recruit able prospects.

Finally, we come to the troublesome question of the quality of achievement. Fear of the consequences of greatly increasing the number of women scientists in universities and in industry is partly based on the assumption that the pool of qualified women is smaller than the demand for them. The size of that pool and its character is an empirical question about which we know far too little. We do not, for example, know whether the supply of qualified women scientists is now large enough to meet the changed requirements of universities and industry. It is at least possible that the supply of comparably qualified women is not equal to the rapidly increasing, effective demand for female scientists.

How many female scientists have actually overcome the triple penalty so that their scientific performance is equivalent to that of male scientists? The answer is not known. For the present it is probably not enough. In the long run, with an enlarged pool of talent on which to draw, the quality of the scientific population is not apt to deteriorate. But what will probably be good for science will still have its costs for female practitioners who try to meet the obligations of the scientific profession and of family simultaneously.

Notes

1. United States Bureau of the Census, *Statistical Abstract of the United States*, 91st ed. (Washington, DC: U.S. Government Printing Office, 1970), 107. Although the proportion of all Americans of university and college age who have actually been attending such institutions has been increasing, the increase has been more rapid in recent years among women than among men. See Abbot L. Ferris, *Indicators of Trends in the Status of American Women* (New York: Russell Sage Foundation, 1971), 320.
2. Computed from Ferris, *Indicators of Trends in the Status of American Women*, 307, and United States Bureau of the Census, *Statistical Abstract of the United States*, 131. Four years later, 41 percent of all bachelor degrees were awarded to women (131). Bernard

reports consistently higher attrition rates for female university and college students than for male students: Jessie L. Bernard, *Academic Women* (University Park: Pennsylvania State University Press, 1964), 70.

3. Computed from the United States Bureau of the Census, *Statistical Abstract of the United States*, 131. Twenty-three percent of male college graduates specialized in science or engineering in 1968, compared with 7 percent of female graduates. The difference between the sexes, however, is largely due to the differences in their representation in engineering: 10 percent of all male graduates but less than 1 percent of female graduates received degrees in engineering.

4. Laure M. Sharp, *Education and Employment: The Early Careers of College Graduates* (Baltimore, MD: The Johns Hopkins University Press, 1970), 9. Four years after graduation, 60 percent of the male graduates of 1958, against 29 percent of female graduates who specialized in science, had gone on to graduate studies. Among engineers, 31 percent of the men went to graduate school compared to 24 percent of the women. Even among those who specialized in science at the leading liberal arts colleges, women are half as likely to take doctorates as men; see W. Rodman Snelling, Robert F. Boruch, and Nancy B. Boruch, "Science Graduates of Private and Selective Liberal Arts Colleges," *College and University* 46 (Spring 1971): 231–44.

5. Office of Scientific Personnel-National Research Council, *Doctorate Recipients from U.S. Universities: Summary Report 1972* (Washington, DC: National Research Council, May 1973), 1. Although the following data are not for the same cohort of recipients of bachelor's and doctoral degrees, the third column shows the extent to which selected sciences differ in the amount of filtering out of women between college and the PhD.

Percentage of Women among All Recipients of Bachelor's Degrees and PhDs in 1969, by Scientific Field

	% Bachelor's (a)	% PhD (b)	Ratio of (b to a)
Biological Sciences	28.0	15.4	.55
Chemistry	18.3	7.5	.41
Physics	5.8	2.4	.41
Geology	10.4	3.2	.31
Mathematics	37.4	6.2	.17

Source: United States Bureau of Census, *op. cit.*, 1971, p. 130.

6. Lindsey R. Harmon and Herbert Soldz, *Doctorate Production in United States Universities: 1920–1962* (Washington, DC: National Academy of Sciences-National Research Council, 1963), Publication no. 1142, 52, for statistics relating to 1920s; United States Office of Scientific Personnel-National Research Council, *Doctorate Recipients from U.S. Universities*, 1, for statistics relating to 1972.

7. Harmon and Soldz, *Doctorate Production in United States Universities*, 50.

8. Office of Scientific Personnel-National Research Council, *Doctorate Recipients from U.S. Universities*, 1.

9. Office of Scientific Personnel-National Research Council, *Doctorate Recipients from U.S. Universities*, 1; and Ferris, *Indicators of Trends in the Status of American Women*, 321. Ferris reports that, after reaching a low point in the late 1950s, the proportion of women receiving doctorates in all fields began to increase and has continued at an irregularly increasing rate. According to the United States Bureau of the Census, *Statistical Abstract*

of the United States, 65, 227, about 7 percent of all physicians were women in 1960 and again a decade later. According to Margaret Law (ed.), *Goals for Women in Science and Engineering* (Boston: Women in Science and Engineering, 1972), 23, the percentage of women among entering medical students is now increasing sharply and was estimated at 16.7 percent for 1972.

10. Office of Scientific Personnel-National Research Council, *Doctorate Recipients from U.S. Universities*, 1.

11. United States Bureau of the Census, *op. cit.*, 1965, pp. 233–234, and 1970, 525. In 1950, women comprised 11.4 percent of all scientific and technical workers, as compared with 9.4 percent in 1970. According to United States Bureau of the Census, *op. cit.*, 515, the proportion of women in this occupational group dropped to its lowest point about 1964 and is now on the increase.

12. Paul H. Mussen, "Early Sex-Role Development," in D. A. Goslin (ed.), *Handbook of Socialization Theory and Research* (Chicago: Rand-McNally, 1969), 707–32. We shall not examine here the complex evidence on sexual differences in intelligence and personality or the differences which may exist in childhood training; nor will we compare the status of American women in science with their counterparts in other countries. For a comprehensive review of research on sexual differences, see Eleanor Maccoby (ed.), *The Development of Sex Differences* (Stanford, CA: Stanford University Press, 1966).

13. Joseph H. Fichter, *Graduates of Predominantly Negro Colleges: Class of 1964* (Washington, DC: U.S. Government Printing Office, 1967), Public Health Service Publication no. 1571, 91, 141. Fichter's study draws upon data collected by the National Opinion Research Center.

14. Fichter, *Graduates of Predominantly Negro Colleges*, 97, 138.

15. Fichter, *Graduates of Predominantly Negro Colleges*, 143.

16. Fichter, *Graduates of Predominantly Negro Colleges*, 160.

17. Alice S. Rossi, "Barriers to the Career Choice of Engineering, Medicine, or Science Among American Women," in *Women and the Scientific Professions*, ed. J. A. Mattfeld and C. A. Van Aken (Cambridge: Massachusetts Institute of Technology Press, 1965), 92, 94–96.

18. Patricia Albjerg Graham, "Women in Academe," *Science* CLXIX, no. 395 (September 1970): 1284–90. Although many conclude that the presence of appropriate models is an important determinant of differences between the sexes in occupational choice, we know of no evidence that this is so.

19. Gloria Lubkin, "Women in Physics," *Physics Today* XXIV, no. 4 (April 1971): 23–27.

20. See Cynthia F. Epstein, *Woman's Place: Options and Limits in Professional Careers* (Berkeley: University of California Press, 1970), 152; and Rose Laub Coser and Gerald Rokoff, "Women in the Occupational World: Social Disruption and Conflict," *Social Problems* XVIII, no. 4 (Spring 1971), 542ff., for further discussion of the implications of the association of certain occupations with one sex or the other.

21. Robert K. Merton, "The Self-Fullfilling Prophecy," in *Social Theory and Social Structure* (New York: Free Press, 1968; originally published 1948), 475–92.

22. Ferris, *Indicators of Trends in the Status of American Women*, 27.

23. John K. Folger, Helen S. Astin, and Alan F. Bayer, *Human Resources and Higher Education* (New York: Russell Sage Foundation, 1970), 310; and William Sewell and V. P. Shah, "Socio-Economic Status, Intelligence, and the Attainment of Higher Education," *Sociology of Education* XL, no. 1 (Winter 1967): 1–23.

24. Melvin L. Kohn, "Social Class and Parental Values," *American Journal of Sociology* LXIV, no. 4 (January 1959): 337–51; and William H. Sewell, "Inequality of Opportunity for Higher Education," *American Sociological Review* XXXVI, no. 5 (1971): 793–809.

25. Deborah David, "Career Patterns and Values: A Study of Men and Women in Science and Engineering," unpublished doctoral dissertation (Department of Sociology, Columbia University, 1971), table 3–9; Helen S. Astin, *The Woman Doctorate in America* (New York: Russell Sage Foundation, 1969), 25; and Lindsey R. Harmon, *Profiles of Ph.D.s. in the Sciences* (Washington, DC: National Academy of Sciences-National Research Council, 1965), Publication no. 1293, 40.

26. W. Rodman Snelling and Robert Boruch, "Factors Influencing Student Choice of College and Course of Study," *Journal of Chemical Education* XLVII, no. 5 (May 1970): 327–28.

27. Rossi, "Barriers to the Career Choice of Engineering, Medicine, or Science Among American Women," p. 95.

28. Jonathan R. Cole, *Woman's Place in the Scientific Community* (New York: Wiley, forthcoming); Bernard, *Academic Women*, 78. The fields referred to in Cole's study are physics, mathematics, chemistry, biology, psychology and sociology. Astin, *The Woman Doctorate in America*, is used as a source for information on women, while Cole provides a matched sample of men who took degrees from the same universities. Unlike Astin's data, his data follow some women for twelve to thirteen years after they received their doctorates.

29. Pearce, R. H., et. al., *Women in the Graduate Academic Sector of the University of California: Report of an Ad Hoc Committee of the Coordinating Committee on Graduate Affairs* (Los Angeles, CA, 1972); *The Study of Graduate Education at Stanford: Report of the Task Force on Women* (Stanford, CA: Stanford University Press, 1972).

30. *GRE: Guide to the Use of the GRE Scores in Graduate Admissions, 1970–1971* (Princeton, NJ: Educational Testing Service, 1971).

31. Astin, *The Woman Doctorate in America*, 103; Sharp, *Education and Employment*, 130; and Feldman, Saul D., "Impediment or Stimulant? Marital Status and Graduate Education," *American Journal of Sociology* LXXVIII, no. 4 (January 1973), 992. According to Feldman, the national survey of American graduate students in 1968–69, sponsored by the American Council on Education and the Carnegie Corporation, showed that women were slightly more apt to have fellowships and men to have teaching or research assistantships.

32. Allan Tucker, David Gottlieb, and John Pease, *Attrition of Graduate Students at the Ph.D. Level in the Traditional Arts and Sciences* (East Lansing: Michigan State University, 1964), Office of Research and Development, publication no. 8.

33. Michelle Patterson and Lucy Sells, "Women Dropouts from Higher Education," in *Academic Women on the Move*, ed. A. S. Rossi and A. Calderwood (New York: Russell Sage Foundation, 1973), 85–86.

34. See Lucy W. Sells, "Sex Differences in Graduate School Survival" (paper presented at meeting of the American Sociological Association, 1973, 9) in which the author analyzes the Carnegie Commission Survey of graduate students.

35. Cynthia F. Epstein, "Bringing Women In: Rewards, Punishments, and the Structure of Achievements," *Annals of the New York Academy of Sciences* CCVIII (March 1973): 62–70.

36. Sells, "Sex Differences in Graduate School Survival," 5.

37. Sells, "Sex Differences in Graduate School Survival," 5.

38. Astin, *The Woman Doctorate in America*, 20.

39. Feldman, "Impediment or Stimulant?" 984–85.

40. Joseph D. Mooney, "Attrition Among Ph.D. Candidates: An Analysis of a Cohort of recent Woodrow Wilson Fellows" (Department of Economics, Princeton University),

unpublished manuscript cited in Folger, Astin, and Bayer *Human Resources and Higher Education*, 286.

41. Feldman, "Impediment or Stimulant?" 988.

42. Folger, Astin, and Bayer, *Human Resources and Higher Education*, 285, report these figures for recipients of doctoral degrees in 1961 from departments ranked "distinguished" or "strong" by the American Council of Education.

43. The class of university where the doctorate was obtained may have different implications for the careers of women: see Bernard Berelson and Gary A. Steiner, *Human Behavior: An Inventory of Scientific Findings* (New York: Harcourt, 1964), 109ff; Lowell Hargens and Warren Hagstrom, "Sponsored and Contest Mobility of American Academic Scientists," *Sociology of Education* XL, no. 1 (Winter 1967): 24–38; and Crane, Diana, "The Academic Marketplace Revisited: A Study of Faculty Mobility Using the Cartter Ratings," *American Journal of Sociology* LXXV, no. 6 (May 1970): 953–64, for the familiar generalization that where one is trained determines where one ends up. This may not be true in all instances; women's preferences for employment may differ systematically from those of men and thereby attenuate the correlation between the rank of the university which awarded the degree and that in which the person is later employed. To the extent that discrimination persists, it will reduce the relationship even further. Current intensive efforts to recruit women to university teaching staffs may further modify the relationship by placing larger proportions of women in universities ranked higher than would be predicted from their doctoral origins, just as in the past rules regulating nepotism also modified that relationship.

44. Astin, *The Woman Doctorate in America*, 57; and Folger, Astin, and Bayer, *Human Resources and Higher Education*, 289.

45. Astin, *The Woman Doctorate in America*, p. 58. What is true for women with PhDs does not hold for the rest. David, "Career Patterns and Values,": in tables 6–2 and 6–3, finds in the analysis of the Post-Censal Survey of scientific and technical manpower that 7 percent of women with PhDs were not employed at the time of the survey, and 12 percent of those with master's degrees and 19 percent of those without any graduate education were not employed. In contrast with this, David finds only 2 percent, of the men, regardless of educational level, were unemployed. She also finds that female engineers are no more apt to leave the labor force than are other women with similar education.

46. Astin, *The Woman Doctorate in America*, 58.

47. Bayer, A. E., "Teaching Faculty in Academe: 1972–1973" (Washington, DC: American Council on Education, August 1973), *Research Reports* 8: 16.

48. Astin, *The Woman Doctorate in America*, 56; and David, "Career Patterns and Values," table 6–3.

49. Astin, *The Woman Doctorate in America*, 6–20; David, "Career Patterns and Values," table 6–18.

50. A. Brues, "Letter," *Science* CLXXIV, no. 3983 (May 1971): 515.

51. L. R. Harmon, *High School Ability Patterns: A Backward Look from the Doctorate.* Scientific Manpower Report no. 6 (Washington, DC: National Academy of Sciences/ National Research Council, 1965); Rossi, "Barriers to the Career Choice of Engineering, Medicine, or Science Among American Women," 67ff.; Astin, *The Woman Doctorate in America*, 71; and Richard R. Bolt, "The Present Situation of Women Scientists and Engineers in Industry and Government," in *Women and the Scientific Professions*, ed. Mattfield and Van Aken, 139–62.

52. Cole, *Woman's Place in the Scientific Community.*

53. David, "Career Patterns and Values," table 6–12; Harmon, *Profiles of Ph.D.s. in the Sciences*, 58–59.
54. Bayer, "Teaching Faculty in Academe," 23, 28.
55. Some persons allege that the contributions of women to collaborative efforts are less often recognized by co-authorship than those of men and that this contributes to their lower rates of publication. This may be so. One current fact of scientific life is that the contributions which count for most, for both men and women, are embodied in the scientific literature, and that the standing of scientists among their colleagues rests on these contributions as registered in the authorship of research publications. See Bernard, *Academic Women*, 152ff.; Lowell Hargens, *The Social Contexts of Scientific Research* (Madison: University of Wisconsin, 1971), unpublished PhD thesis, 288–289; and Barbara F. Reskin, *Sex Differences in the Professional Life Chances of Chemists* (Seattle: University of Washington, 1973), unpublished PhD thesis, 303ff. It is also alleged that women publish fewer papers than men because referees for scientific journals systematically discriminate against them in decisions on publication. The correlation between sex and published productivity does not however differ in fields in which "blind refereeing" is the custom and those in which referees are given the names of the authors whose papers they evaluate.
56. Cole, J. R., *op. cit.*, p. 27.
57. Rita James Simon, Shirley Merritt Clark, and Kathleen Galway, "The Woman Ph.D.: A Recent Profile," *Social Problems* XV, no. 2 (Fall 1967): 221–36, also report that female scientists regardless of marital status are less productive than men, but their samples seem to be incomparable.
58. Cole, *Woman's Place in the Scientific Community.*
59. These findings on female PhDs parallel Eleanor Maccoby, *The Development of Sex Differences*, where it is stated that the relationship between early school performance and later achievement is weaker for girls than for boys. See also Eleanor Maccoby, "Feminine Intellect and the Demands of Science," *Impact of Science on Society* XX, no. 1 (January–March 1970): 13–28.
60. Alan E. Bayer and John Folger, "Some Correlates of a Citation Measure of Productivity in Science, "*Sociology of Education* 39, no. 4 (Autumn, 1966), 381–390.
61. J. R. Cole and Stephen Cole, "Measuring the Quality of Sociological Research: Problems in the Use of the Science Citation Index," *American Sociologist* VI, no. 2 (February 1971): 25–29.
62. Although female scientists are statistically under-represented among those honored for truly significant contributions to science, it is not yet known whether this is the result of discriminatory practices or differences in scientific performance.
63. Harriet Zuckerman, "Nobel Laureates in Science: Patterns of Productivity, Collaboration, and Authorship," *American Sociological Review* XXXII, no. 3 (June 1967): 391–403; and Harriet Zuckerman, *Scientific Elite: Nobel Laureates in the United States* (Chicago: The University of Chicago Press, forthcoming).
64. See Bernard, *Academic Women*, 158; and Bayer, A. E., *op. cit.*, 1970, p. 18.
65. Derek J. de S. Price and Donald Beaver, "Collaboration in an Invisible College," *American Psychologist* XXI, no. 11 (November 1966): 1011–18.
66. John B. Parrish, "Women in Top Level Teaching and Research," *American Association of University Women Journal* LV, no. 2 (1962): 99–107; Bernard, *Academic Women*, 189–91; and Simon, Clark, and Galway,, "The Woman Ph.D.," 221–36. A. E. Bayer and H. S. Aston ("Sex Differences in Academic Rank and Salary Among Science Doctorates in Teaching," *Journal of Human Resources* April 1968), however, report equality for academic men and women in science but not in social science.

268   FREEDOM AND UNFREEDOM

67. Carolyn Cummings Perrucci, "Minority Status and the Pursuit of Professional Careers: Women in Science and Engineering," *Social Forces* XLIX, no. 2 (December 1970): 250; and Laura Morlock, "Discipline Variation in the Status of Academic Women," in Rossi and Calderwood, *Academic Women on the Move*, 268.

68. The association between rank and gender is stronger eight years after the doctorate than it is four years later, when more scientists, men and women, have been promoted. Cole, *Woman's Place in the Scientific Community*, 1974.

69. Morlock, "Discipline Variation in the Status of Academic Women," 268.

70. Astin, *The Woman Doctorate in America*, 57, 63; Cole, *Woman's Place in the Scientific Community*.

71. Astin, *The Woman Doctorate in America*, 27; David, "Career Patterns and Values," table 5–3.

72. For the general pattern, see the following: Victor R. Fuchs, "Differences in Hourly Earnings between Men and Women," *Monthly Labor Review* XLIV, no. 5 (May 1971): 9–15; Janice Fanning Madden, *The Economics of Sex Discrimination* (Lexington, MA: Heath, 1973); and Larry E. Suter and Herman P. Miller, "Income Differences between Men and Career Women," *American Journal of Sociology* LXXVIII, no. 4 (January 1973): 962–74. For particular manifestations in academic life, see Bayer and Aston, "Sex Differences"; Patrick McCurdy, "A Man and a Woman," *Chemical Engineering News* XXVI, no. 43 (October 1970): 26–28; and Michael A. La Sorte, "Academic Women's Salaries: Equal Pay for Equal Work?" *Journal of Higher Education* XLII, no. 4 (April 1971): 265–78. Two case studies of university salary scales show, however, that sex differences in salary can be explained by length of service, rank, and discipline.

73. Morlock, "Discipline Variation in the Status of Academic Women," 288. The reasons for these differences are not examined.

74. Bayer and "Sex Differences," 349, 353–354, is the most thorough investigation of salary differences to date; it does not, however, treat the sciences separately but covers academics in all fields.

75. Since differences in remuneration between the sexes often reflect part-time or part-year work, it is necessary to compare the salaries of those who work full-time throughout the year.

76. Morris Rosenberg, *Occupations and Values* (Glencoe, IL: Free Press, 1957), 49; Ralph Turner, "Some Aspects of Women's Ambition," *American Journal of Sociology* LXX, no. 3 (November 1964): 281–85; Sharp, *Education and Employment*, 78; David, "Career Patterns and Values," table 8–1. Ralph Underhill, "Values and Post-College Career Change," *American Journal of Sociology* LXXII, no. 2 (September 1966): 163–72, reports that in studies of values and occupational choice in which individuals can choose, their values almost always affect their decisions rather than the other way round.

77. Astin, *The Woman Doctorate in America*, 179.

78. Astin and Bayer, "Sex Differences;" and J. R. Cole, *Social Stratification in Science. Chicago: University of Chicago Press*, 1973.

79. Cole, *Social Stratification in Science*.

80. L. S. Fidell, "Empirical Verification of Sex Discrimination in Hiring Practices in Psychology," *American Psychologist* XXV, no. 12 (December 1970): 1094–98.

81. Arie Y. Lewin and Linda Duchan, "Women in Academia," *Science* CLXXIII, no. 4000 (September 1971): 892–95. Since the statistical findings on which the authors based their conclusions are not presented in this paper, the results reported should be treated with caution.

82. Women are somewhat more numerous in the Royal Society of London, making up about 3 percent of the fellowship. They became eligible for membership in 1919 when

it became illegal to disqualify persons on the basis of sex from membership in bodies having a royal charter. Kathleen Lonsdale, a crystallographer, and Margery Stephenson, a biologist, were the first women to be proposed for membership just twenty-five years later and were actually elected the following year, according to Lonsdale's own report in "Women in Science: Reminiscences and Reflections," *Impact of Science on Society* XX, no. 1 (January–March 1970): 45–60. Florence Sabin, an anatomist, became the first woman to be elected to the United States National Academy of Sciences in 1925, where she sat alone until Margaret Floy Washburn, a psychologist, was elected to membership six years later.

83. Alan E. Bayer, *College and University Faculty: A Statistical Description* (Washington, DC: American Council on Education, June 5, 1970), *Research Reports* 5: 13.

84. Bayer, A. E., *op. cit.*, 1973, p. 24. There is no systematic evidence available on the distribution of applicants for research funds by sex. Carl D. Douglass and John C. James, in "Support of New Principal Investigators by N.I.H.: 1966–1972," *Science* CLXXXI, no. 4,093 (July 1973): 243, report that the proportion of women among new principal investigators on projects funded by the National Institutes of Health has risen slightly from 6.2 percent in 1966 to 8.1 percent in 1972. These data, however, tell us nothing about the proportion of women applying for funds, nor the proportion of women among all recipients of grants. Given the NIH's prohibition on applications from part-time research workers and the fact that some universities will not permit members without regular academic appointments to apply for research grants on their own, there is reason to suspect that proportionately fewer women than men are in a position to apply for support. This matter is, of course, altogether different from the question of whether all applications for funds are evaluated on their merits. Douglass and James cite an unpublished study by Kaufman which is said to show that "the approval rate[s] of research project applications from male and female investigators do not differ significantly."

85. Briefly introduced in Merton's 1942 paper on the ethos of science, the notions of "accumulation of advantage" and of disadvantage in systems of social stratification, which relate to the notions of the self-fulfilling prophecy and the Matthew effect, have been developed in a series of investigations: Robert K. Merton, *The Sociology of Science: Theoretical and Empirical Investigations* (Chicago: University of Chicago Press, 1973), 273, 416, 439–459; H. A. Zuckerman, "Stratification in American Science," *Sociological Inquiry* XL, no. 1 (Spring 1970): 235–57, esp. 245; H. A. Zuckerman and R. K. Merton, "Age, Aging and Age Structure in Science," in R. K. Merton, *Sociology of Science* (Chicago: University of Chicago Press, 1973), 497–559, esp. 532; J. R. Cole and Stephen Cole, *Social Stratification in Science* (Chicago: University of Chicago Press, 1973), 237–47; Paul D. Allison and John A. Stewart, "Productivity Differences Among Scientists: Evidence for Accumulative Advantage," *American Sociological Review* XXXIX, no. 4 (August 1974): 596–606; Zuckerman, *Scientific Elite*, chap. 3, passim.

86. Merton, Robert K., "Columbia University Lectures" (New York: Columbia University, 1972, unpublished). See also his "The Self-Fulfilling Prophecy," in R. K. Merton, *op. cit.*, 1968, pp. 475–490, and "The Matthew Effect in Science," in Merton, *Sociology of Science*, 439–459. Jonathan R. Cole develops the formulation of the "haunting presence" in *op. cit.*, 1975.

87. Epstein, *Woman's Place*. Since part-time workers, almost by definition, contribute less to science and are less committed to it than those who work full-time, it is more or less inevitable that rewards should be concentrated among those who work full-time. As a consequence, if part-time employment is largely confined to women, the chances for equality of rewards between the sexes are small. Moreover, it may be difficult to do

270   **FREEDOM AND UNFREEDOM**

scientific research part-time and especially on command during the hours at which a babysitter is available. If this is so, then part-time work will not enhance women's opportunities for doing outstanding scientific work. Finally, settling for part-time work as a solution to the problems of female professionals puts off the more difficult question of how domestic obligations might be reallocated so that women's opportunities for occupational achievement might be made equal to those of men.

88. Beverly F. Porter, Sylvia F. Botish, and Raymond W. Sears, "A First Look at the 1973 Register," *Physics Today* XXVII, no. 4 (April 1974): 23–33, esp. 27; Astin, H. S., "Career Profiles of Women Doctorates," in Rossi and Calderwood, *Academic Women on the Move*, 139–62.

89. R. W. Hodge, P. M. Siegel, and P. Rossi, "Occupational Prestige in the United States: 1925–1963," *American Journal of Sociology* LXX, 3 (November 1964): 286–302.

CHAPTER 13

WOMEN IN SCIENCE

(1981)

JONATHAN R. COLE

Despite many recent advances, women are still less likely than men to be promoted to high academic rank, and few have full citizenship in the informal scientific community.

Within the past decade, we have witnessed a notable increase in both the absolute number of women entering scientific professions and in their proportion among recent PhDs in science. While these proportions hardly approach those recently obtained in medicine, law, or business, women no longer have only token representation in most fields of science.

The increased presence of women in science makes an understanding of their treatment within the scientific community a subject of substantial import. If the structure of opportunities for women in science differs from that of men, and if their patterns of scientific productivity differ from those of men, the scientific community may pay a price in reduced growth of knowledge resulting

---

American Scientist 69, no. 4 (July–August 1981): 385–91. This article is based on *Fair Science: Women in the Scientific Community* (New York: Free Press, 1979), on his keynote address at a Harvard University symposium, "Choices for Science," and on the Franklin Institute's Nathan Hayward Lecture. Research grants from the NSF and a Guggenheim Fellowship supported the work. He has published numerous books and articles in the sociology of science and has just completed two volumes for the National Academy of Sciences on resource allocation at the NSF.

272 FREEDOM AND UNFREEDOM

from barriers placed in the career paths of female scientists. The question is: What are the impediments in science, if any, to the full development of the talents of female scientists?

I want to consider here the question of where women in science stand today, and what their futures look like, but also to take a glance backward to gain some perspective on where we have come from. Our observations about women in science result to a notable degree from the accumulation of two basic types of knowledge, what William James (1885) distinguished as "acquaintance with" and "knowledge about." Acquaintance with knowledge "involves direct familiarity with phenomena that is expressed in depictive representation," or what Max Weber referred to as *verstehen*. Knowledge about "involves more abstract formulations which do not at all 'resemble' what has been directly experienced." Plainly, both forms of knowledge contribute to a full understanding of social phenomena such as the place and treatment of women in science. My knowledge, which derives to a significant degree from quantitative, empirical inquiries, must surely be classified more as "knowledge about" than as "acquaintance with." With this in mind, let us consider several features of the attitudes toward women and their structural position in science around the turn of this century.

A Glance Backward at Marginal Women

Until the twentieth century, science was populated almost exclusively by men, and thus the phrase "men of science" was almost equivalent to the non-sex-linked tag "scientists." What is remarkable is that for the first seventy years of this century there was little change in the proportion of scientists that was female. If there have been few women in science, there have been even fewer among the scientific elite. The rolls of the National Academy of Sciences and the list of Nobel laureates register very few women. As of 1980, only 35 of the 1,361 members of the National Academy, or 2.6 percent, were women; only five of the 345 scientists who had received the Nobel Prize, or 1.4 percent, were women (H. Zuckerman, personal, communication). How can we account for these facts?

Although we have observed over the past fifteen years marked increases in the proportion of women entering medical, law, and business schools, so that they now represent roughly one-third of the population of these professional schools rather than one-tenth, we have not witnessed as dramatic a rise in the proportions of women in the natural sciences. When we examine the figures at the point of entry into the scientific community—the awarding of the PhD—we find that women still constitute a very distinct minority of members, although the proportions have actually doubled in most scientific disciplines over this period. The difference is that an increase of more than 100 percent in the proportions of women earning Ph.D.s in physics and astronomy between 1968 and 1978 brings the proportion of all PhDs in these fields to only 4.9 percent. Similar stories can be told for mathematics, chemistry, and the biological sciences.

These statistics give rise to two basic, interrelated questions. Why have so few women chosen science as a career? Why have so few of the women who have entered science achieved notable distinction and rewards? Or put more broadly, what are the cultural, social, psychological, and economic forces that influence the career paths of female scientists? Until recently women were considered unfit for scientific work—and certainly for creative scientific work. Consider several examples of such opinion. Alphonse de Candolle devotes two pages to the place of women in science in his otherwise extraordinary 1885 work *Histoire des sciences et des savants depuis deux siècles.* Candolle remarks that the female mind "takes pleasure in ideas that are readily seized by a kind of intuition; a mind to which the slow method of observation and calculation by which truth is surely arrived at are not pleasing."

Candolle's ideas were not exceptional in their time. Within the scientific community in general, a significant research effort was being devoted to establishing the intellectual and psychological inferiority of women. Extensive work was carried out on cranial volume and its relation to intellectual capacity, for example. The weight of brains also became a focus of considerable attention, with men of science, otherwise noted for their care in weighing evidence, trying with some difficulty to weigh human brains and to draw the inference that women's smaller capacity was directly related to the lower average weight of their brains. As might be expected, evidence contrary to the desired outcome continually confronted these investigators. These various attempts to prove a physiological basis for the lack of female accomplishments in the arts and sciences led, of course, only to dead ends. Yet when at the turn of the century the German scholar Paul Moebius published his *The Physiological Feeble-Mindedness of Women*, there were fewer attacks on the conclusion than on the causal argument.

Nor were the founding fathers of sociology immune from such sensibilities. Emile Durkheim observed in his book *Suicide* (1897) that woman's "sensibility is rudimentary rather than highly developed." Herbert Spencer (1896), whose work had an extraordinary impact around the turn of the century, believed that women represent "a somewhat earlier arrest of individual evolutions." And finally, Auguste Comte, the putative father of sociology, worked out an elaborate rationalization of the intellectual inferiority of women (Lenzer 1975).

A final example of attitudes toward women in science in the early part of the century is offered by Otto Hahn, the renowned physicist, in his reflections about his early collaborations with his longtime colleague, Lisa Meitner. To be sure, Hahn speaks of life in Germany, but there is ample evidence that conditions were much the same for women working in American science. He recalls the difficulties Meitner faced in pursuing her research goals:

> The beginning was difficult for her. Emile Fischer, the director of the Chemical Institute [at Berlin], did not then accept women, but he did make a concession in her favor. With the condition that she was not to enter the laboratories where male students were working, she was permitted to work with me in the wood shop. In 1907 this was a really large concession . . . and in time he also developed an attitude of fatherly friendship toward Lisa Meitner. But the rule that she had to stay in the wood shop [which was later extended to include another basement room] remained in force. (Hahn 1966)

In short, women who entered science in the early part of this century held distinctly marginal positions. They were at once rejected by those in the general culture who felt that such activities were inappropriate for women, and at the same time they were not accepted as full members of the scientific community.

Women have, then, faced traditionally a triple penalty: (1) science was culturally defined as an inappropriate career for women, few women were recruited into science, and few sought it out; (2) those who surmounted the first barrier continued to be hampered by the belief that women were less competent than men at science; whatever the validity of this belief, it contributed to women's ambivalence toward work and reduced motivation and commitment to scientific careers; and (3) women encountered significant amounts of discrimination against members of their sex within the scientific community.

Women who even managed to overcome the initial hurdle of cultural resistance to scientific careers, and who were able to make it to the starting line with PhDs in hand, were in the true sense of the term "survivors." And while I believe much has changed since the early part of this century to transform women from marginal figures closer to full partners with men in the scientific community, it is plain that women scientists, even today, are still "survivors."

Women in Science Today

If this description represents the historical frame, we may ask whether the same barriers to equality continue to exist. Consider the following queries: (1) How much evidence is there for gender-based discrimination in contemporary science? (2) Is gender discrimination, if it is found to exist, uniformly distributed throughout the scientific community? (3) Does gender discrimination obtain when we consider access to higher education, receipt of scholarships and fellowship support, initial job location and hiring after the PhD, promotion to higher ranks and tenure, receipt of formal honorific awards, recognition among peers, receipt of research funds, the attraction of outstanding students and collaborators, access to outlets for scientific publications, access to masterful teachers, and equal pay for equal work?

Questions about the level and intensity of gender-based discrimination in science must be viewed in the context of a substantial body of research over the last decade that has shown science to be a highly meritocratic social system. These findings, however, were based almost exclusively on the study of male scientists, particularly those in physics and chemistry, where larger-scaled empirical inquiries using random sampling techniques rarely turned up any women in the first place. Nevertheless, I have begun to investigate the extent to which women are denied access to positions and to rewards in the scientific community and the extent to which they are subjected to discrimination.

First, I must note the limited state of our knowledge about the measurement or estimation of discrimination. Although each of us probably feels that we "know" discrimination when we see it, *verstehen* knowledge is not the same as estimation. And when we examine the literature on discrimination, we not only see multiple definitions of discrimination but, more important, we find few if any direct measures of discrimination. Virtually all quantitative, empirical studies of discrimination, including my own, measure it not through direct observation or measurement but residually. This is an important point to bear in mind and indeed is an important limitation in studies such as my own—a limitation that requires us all to be organized skeptics about our results.

Discrimination is not synonymous with inequality. There can be a great deal of inequality in an institution, as there surely is in science, without that inequality resulting from discriminatory behavior. Zero-order correlations and one-dimensional views of the world can tell us next to nothing about the presence or absence of discrimination. Of course, most social scientists today employ what I have called elsewhere "sophisticated residualism." They start with an inequality, perhaps income differentials between men and women, and they then introduce multiple variables that they believe can explain the original difference. The leftover, or residual, difference is then taken as an estimate of "discrimination." Are there better techniques for measuring discrimination? Apparently not, or at least none that has found even limited application.

Let me turn to results of what I have found about the position of women in contemporary science (a more detailed discussion is available in Cole 1979). How meritocratic is the scientific community in regard to its female members? I studied more than two thousand male and female scientists, matched initially in terms of the university from which they received their degree, the year they received the degree, the initial field of activity, and the specialty in which they did doctoral work. I traced the careers of these scientists for twelve to twenty years, collecting data as well on aspects of their social backgrounds, including their IQs and marital and family statuses; on their career histories, such as job changes, dates of promotions, prestige of their affiliations, honors and awards they received, and promotion to different academic ranks; and on their publication histories and the patterns of citations or references to their work. I also collected data on the "reputational standings" among a group of peers for a sample of roughly six hundred male and female scientists. Standard multivariate techniques were used to analyze these data.

The results of this investigation show evidence from the PhD onward of significant gender-based discrimination in the promotion of female scientists to tenure and high academic rank. Even after I have taken into account many other factors, such as career interruptions and the quantity and assessed quality of research performance of men and women, I find that women are still less likely than men to be promoted to high academic rank. And when they are promoted, it is not apt to happen as quickly. This finding should not be minimized. It is important because tenure and high rank not only represent security in the world of academic science; they are also requirements for full participation in the inner circles of scientific activity. Furthermore, academic rank plainly can influence income differences. The evidence suggests that the pattern of promotion to high rank has persisted for the past fifty years at much the same level. While we shall see that there are signs of substantial progress in other areas, there has been little movement here.

In almost all other areas of the scientific career following the PhD, there is little evidence of substantial formal status inequality between men and women. Consider a series of findings from both my own work and that of other scholars of the subject. First, in terms of measured IQ, women in science are certainly every bit the equal of men. In fact, in comparisons of the IQ scores of male and female PhDs earning degrees at science departments grouped in terms of their assessed quality, women consistently had slightly higher IQs at every level of department quality (Cole 1979; Harmon 1965). Second, there seems to be no systematic discrimination against women in admission to graduate science departments: they are admitted roughly in the proportion in which they apply. Third, there is no evidence that women are systematically discriminated against in the granting of graduate and postdoctoral fellowships (Cole 1979; Reskin 1976).

Fourth, when we examine the point of entry into the scientific community there is no evidence that women are shortchanged in appointments to the "distinguished," "strong," and "very good" departments in their fields. Little statistical evidence suggests that appointments at the level of assistant professor are meted out on the basis of gender. To be sure, we do not find nearly as many women as men in disciplines such as chemistry, biology, or astronomy, and this is true at the level of assistant professor as well as at higher ranks. But that observation does not signal discrimination, since in any particular group of PhDs far fewer women than men are entering the academic labor market.

Fifth, despite the paucity of women who have gained the most prestigious of all scientific honors, such as the Nobel Prize, the evidence suggests that, on average, women are no less apt to receive honorific recognition than are their male counterparts. This finding highlights a point often disregarded in examining women's place in science. Most of us have been trained to look only at the elite of science. We pay little attention to the rank and file. There are good reasons to dwell on the elite. The overwhelming majority of important scientific discoveries are produced by them. But if you examine the conditions and opportunities confronting most scientists, you find that few can expect to earn *any* significant honors; few women or men will gain any formal honorific recognition.

Sixth, I have found significant differences in the average "reputational standing" of men and women in the same fields. I have measured reputation by peer appraisals of research contributions made by a stratified random sample of scientists who work in the same specialty. On the basis of simple comparisons between men and women scientists, there seems to be a "cost" of being female in the process of building scientific reputations. But if we take into account the research performance of men and women scientists, in terms of both quality and quantity of output, the differential disappears almost completely: highly prolific female scientists have reputational standings among their colleagues similar to those of male scientists who are approximately equally prolific.

Seventh, inequalities in salaries of male and female scientists continue to exist, but recent larger-scaled studies by Centra (1974) and by Bayer and Astin (1968, 1975), among others, suggest that the earnings ratio of females to males, after taking into account a variety of explanatory factors, including research productivity, exceeds 0.9–that is, they virtually disappear. For example, Centra found that male and female full professors who have been employed at universities for only five or six years have essentially identical incomes (women earn 98.5 percent as much as men); for those employed from thirteen to fourteen years, women's salaries are 95 percent of men's.

More sophisticated work by Bayer and Astin in 1975 strongly suggests only small salary differences between men and women when a multivariate model is used to explain differences. Bayer and Astin found a −0.04 partial correlation between gender and salaries after controlling for nine variables in a multiple regression equation. This could easily have been a spurious association had not the researchers teased out the influence of differences in academic rank. In fact, female-to-male earnings ratios in academic science are considerably higher than in any other institutional sphere in which earnings have been compared (Lloyd and Niemi 1979).

Eighth, the social processes and background characteristics by which successful scientific careers are made and by which reputations are built are almost identical for men and women.

This brings me to in many ways the most intriguing and puzzling result of my labors. For every cadre of male and female PhDs for whom I have collected data, there is a consistent and patterned difference in research performance. Female scientists tend not to publish as much as their male counterparts who are of equal professional age, who come from similar educational backgrounds, and who are in the same specialties. This result holds for every group of male and female scientists since the turn of the century that I have studied. The patterns hold both for absolute number of published papers and for rates of publication. These patterns of productivity differential have been observed by many other investigators as well (Zuckerman and Cole 1975).

Research productivity in science has received generous attention. It is highly skewed: about 10 to 15 percent of all scientists produce roughly 50 percent of all scientific literature. Most scientists produce only three or four papers in a scientific career. The skewed distribution of research productivity is similar for both men and women. More important, there is a moderate and consistent correlation between gender and research performance, in the neighborhood of 0.3, with men consistently publishing more than women. When we focus on the impact of published research, similar differences are found. Finally, when we examine these performance differentials over time, we find that differences in research productivity seem to increase over the course of the scientific career.

There have been several hypotheses, and many speculations, to account for these patterned differences in research performance but few satisfying answers. Consider just several proposed explanations. It often has been asserted that the multiple role obligations of female scientists, particularly those of wife and mother, detract from the time needed to be more productive scientists. Surely the effects of marriage and families differ for men and women, even in today's world, and indeed this is a plausible hypothesis. Yet the data do not seem to support it. Not only do marriage and family life fail to impede scientific productivity of women, but they seem to be related to small increases in research performance. Female scientists who are married turn out to be significantly more prolific than those who are not; and women who are married with either one or two children are slightly more scientifically productive than unmarried women and only slightly less so than those who are married without children. However, there is a limit to all of this. Once a woman has had three or more children there is a decline in research output but not to a point significantly lower than that found among the unmarried women. Such results fly in the face of conventional wisdom, and we should therefore approach them with caution.

The relationship between gender, marital and family life, and scientific research performance typifies many other recent findings: they are somewhat counterintuitive and puzzling, and since they are based upon limited data, they call for further inquiries. For example: Does this pattern hold for women in different scientific fields? In all age groups? When we compare married women with families to those who do not have families, are we comparing the "same" women? Perhaps the women who continue to be active scientists and simultaneously manage their families are quintessential cases of superwomen. Or perhaps they have more help in carrying out their multiple tasks.

Some observers of science believe that the gap in research productivity is itself the consequence of subtle and not-so-subtle discriminatory practices within science, some of which may be consequences of women's lower academic ranks: limited access to the resources needed to carry out scientific research, to publication outlets, to government and other forms of support, to the best "human capital" within the graduate student population, and to willing colleagues with whom to collaborate, as well as excessive nonresearch demands made on women faculty members, such as higher teaching loads and more extensive committee assignments. There are almost no data available to test these conjectures.

Other hypotheses include claims about the debilitating effects on women's productivity of teaching-oriented college settings and of differences in motivation of male and female scientists. Again, there is limited information to test these conjectures. We do not really understand the social processes at work that influence the emergence and maintenance of these patterned differences in research productivity.

Full Citizenship?

From this brief discussion of the highlights of my empirical studies, we might conclude that status inequalities between male and female scientists have been reduced sharply over the past fifty to seventy-five years, and that women have moved from marginal positions toward the inner core of the community. In fact, there have been substantial reductions in most status inequalities, with the critical exception of promotion to high rank. And I believe the quality of experiences has improved substantially.

It is in the domain of informal activities in science that the biggest gaps between men and women remain. It is in the more intangible set of experiences associated with doing science from day to day that women rightly feel most excluded. To say that women of science have now entered the central scientific community, and that they have achieved formal equality with men in terms of many measurable aspects of the reward system, does not say that the opportunity structure for men and women interested in science as teenagers is equally open. Nor does it say that the women who have chosen science have an equal chance of ending up in the inner circles of science or that they will be equal participants in the "invisible colleges" of the scientific establishment. Resistance to full participation, to full citizenship of women in the scientific community, continues to exist. Some of the types and sources of that resistance are known, but many of the sources are little understood, if even correctly identified.

I suggested several years ago that some of the incredulity about these findings results from the haunting presence of functionally irrelevant statuses. I hypothesized that the salience of gender in contemporary society leads both men and women to construct post factum causal explanations of career decisions and formal inequalities in terms of gender. Gender is perceived as the principal reason for differences in achievement, and other alternative explanations are discounted. I now believe that the reasons for the incredulity go beyond this and have to do, at least in part, with the type of knowledge that can be obtained through the use of current quantitative, social science methods.

I believe that typical quantitative methods can estimate rather well features of the formal aspects of citizenship in science. They allow us to estimate status inequalities in affiliations, awards and honors, academic rank, and so on. But these techniques, at this stage of their development, do not allow us to measure adequately other, informal aspects of citizenship.

Until the early part of this century, the inequalities in positions held by male and female scientists were so dramatic that they resulted in mutually exclusive sets of statuses. Men controlled all of the means of scientific production; they dominated almost all positions of authority and power; they completely controlled the structure of formal opportunities. Women were totally marginal figures. Many of those fundamental cleavages began to break down early in the century, but sharp differences in social status remained. Although women had moved *into* the community of science, they were not *of* that community. If there has been movement toward greater application of meritocratic principles over the past twenty-five to thirty years, it has taken the form of reducing the level of formal status inequality between male and female scientists.

What has not been achieved to any significant degree is full citizenship for women in science. Many women continue to be excluded from the very activities that allow for full participation and growth, or productivity and change. These are, by and large, the informal activities of science—the heated discussion and debates in the laboratory, inclusion in the inner core or the invisible colleges, full participation in the social networks where scientists air ideas and generate new ones. These relationships involve what Mark Granovetter (1973) has called "the strength of weak ties." These are also the close collaborative relationships that grow up over time, which help to shape scientific taste and sharpen the eye for a good research problem. Women in science perceive that they remain excluded from those activities that define full membership in a community. I believe that there is much merit to those perceptions, although we have almost nothing more specific than anecdotal, autobiographical evidence to support them.

In fact, I believe that the differences in status inequality and social citizenship explain some of the apparent clash between the knowledge generated by *verstehen* and that found by means of larger surveys. *Verstehen* knowledge captures aspects of citizenship that quantitative data in their current form are simply not able to measure. The value of knowledge acquired through direct experience is that it allows us to focus on particular, detailed forms of inclusion and exclusion within the scientific community. Its shortcoming is that it cannot accurately represent the general level of status inequality that obtains in the larger social system.

Looking Forward

Several little-understood processes must be addressed if full citizenship, and indeed real meritocracy, is to be achieved in science. First of all are the processes of self-selection and social selection. Most of our intellectual energy in examining the position of women in science has been focused on the social processes within science that act as impediments to career development. For the most part, we have looked at processes of social selection, by which institutions and organizations choose among competing candidates for positions, awards, promotions, and salaries. We have neglected a host of extremely difficult questions about barriers faced by young women who are interested in science but who turn away from it as youngsters or as college students. We have concentrated on the survivors. In a sense, we start looking for problems at the finish line—after the race is already run. We need not rejoice over science approaching its meritocratic ideal when only 4 or 5 percent of the total population of physicists are women in the first place.

Perhaps the single most important question we can ask is: What forces persist in the general culture, in the general value system of American society, in the American family, in our schools, in the presentation of professional careers by the mass media, that turn young women away from the starting line? Young women tend to turn away from science at early ages. We know remarkably little about the factors that lead to this decision. We know that science has not been viewed as an appropriate career for women and that women were thought not to have the capacity to do outstanding scientific work. But we do not know much about how these values are transmitted to young women.

Understanding the decision-making of teenagers may be even more difficult than understanding other decision-making processes. Occupation choice has had a long history of being a particularly thorny area of research. Nonetheless, the problems are so significant, and the consequences so great, that a renewed effort must be mounted to understand better how our culture leads young women to view science as an inhospitable and unattractive environment. In the final analysis, opening up science to women may depend more on our getting a handle on these problems than on any other.

The second process we should examine is the accumulation of advantage and disadvantage. There is a tendency to view the position of women in science statically, rather than in terms of ongoing social processes. In fact, there is substantial evidence that advantages and disadvantages in science cumulate over time. Suppose the social system of science is less apt to support the education of women than men in terms of apprenticeships. If female scientists receive less support than their male counterparts, it should not surprise us to find them less scientifically productive. Their futures in fact become predictable. When they come up for promotion, their publication records are carefully reviewed and found inferior to those of men with the "same" types of background, and they lose out in the academic marketplace. The self-fulfilling prophecy, which is based on the assumption that women are less motivated, less productive, and less reliable scientific risks than men, is now strikingly supported by data. Plainly, if such a self-fulfilling prophecy operates at an early point in the careers of women, a "fair" judgment of their work later on will reinforce the expectation of low achievement for women.

Correlatively, if a group of youngsters are labeled early on as potential stars and are consequently given resources and disproportionate opportunities, it should not surprise us if later on in their careers they have more formidable research track records. If the initial assessment of who is apt to be a star is based on functionally irrelevant criteria such as gender, then the process of accumulating advantage can begin to enhance the career possibilities of men, while diminishing the chances of women. There is relatively little information about the processes of accumulating disadvantage and how it influences the careers of women in science.

Moreover, we do not know how aspects of women's lives outside science produce disadvantages that are felt during the course of a scientific career. Consider only one concrete example. Apparently, patterns of geographic mobility differ for male and female scientists, especially for married women. For whatever reasons, women feel more constrained than men in shifting job locations if they are part of a dual-career couple. Geographic mobility tends to be associated with the building of reputations in science, as well as with tangible rewards such as increases in salary. If women are less willing than men to entertain moves, and perhaps equally important, if they are also defined by others as less likely to be movable because of a spouse, then a process has been set in motion that may reinforce the accumulation of disadvantage. We must begin to study the careers of scientists as a set of dynamic and interacting processes that can quickly transform equality into inequality.

Differences in productivity are a third process that needs to be investigated. Are the differentials a thing of the past? What accounts for sex differences in published productivity? Do such differences in fact increase with age? Are recent female PhDs more like their age-mates among men in research patterns and performance than are women who received their degrees a generation ago? And what accounts for differences in the extent to which the works of men and women are cited in the research literature? How, if at all, do the work habits of men and women differ? What factors determine whether a woman will be a prolific or a silent scientist? How do these factors differ among the several scientific disciplines?

A fourth area to be studied is that of informal social networks. My work has not been fine-tuned enough to identify and describe precisely the detailed patterns of social interaction and sponsorship that are an essential part of successful careers. Being part of the proper social network, being linked to the "right" people, plainly goes some distance in determining the paths taken in a scientific career and contributes to making science fun. None of this, of course, is a substitute for talent. But given talent, it plays a significant role in launching and sustaining careers.

Look, for example, at the question of whether women have the same opportunities as men to establish apprentice relationships with older, eminent scientists. Whether we examine Auguste Comte's relationship with Saint-Simon, Fermi's with Corbino, Segré's with Fermi, Otto Hahn's with Rutherford, Lisa Meitner's with Otto Hahn, Mary Whiton Calkins's with William James, or the thousands of other master-apprentice relationships, we are dealing with an important mechanism of transmitting a scientific tradition from one generation to another. Even a cursory scan of the autobiographical histories of scientists reveals poignantly how sponsorships involve jealousies, ambivalence, conflict, admiration, and love—and how they are influential in producing in young scientists a sense for a good question or a key problem, a style of doing research or theorizing, a critical stance, and a way of teaching their own future intellectual progeny.

Are women less apt than men to spend extended amounts of time in informal scientific discourse with their teachers—and if so, why? Do senior professors engage in scientific collaboration with junior female colleagues? Do they ask them to join their laboratories or get them invited to conferences at which they describe their work to groups of scientists who have already gained some prominence and are influential in determining the next generation of esteemed and prominent scientists? What role, if any, does sexuality play in the relationship between male and female scientific collaborators or between mentors and students?

There are no multiple regressions that can describe the impact of these social linkages. There is work to be done here if we are to discover how the actual experiences of men and women scientists are both similar and different and, indeed, whether gender is of any consequence in the development of scientific careers. It remains unclear whether or not women are less apt than men to benefit from the operation of informal networks in science. I suspect that women are relatively deprived.

A fifth area needing work is the measurement of discrimination. Our attempts are still relatively primitive, and substantially more work must be done to discover new and improved methods of measurement.

Gender discrimination in promotion is a final process requiring investigation. And in the end I want to return to the beginning: the evidence that significant gender discrimination exists in promotion. Affirmative action efforts should concentrate on this critical pressure point; diffuse efforts tend to deplete the energy that is needed to attack the major problem of discrimination in academic science. Emphasis should be given to elite institutions of science, which have always set standards followed by others.

We have made a beginning toward true meritocracy for women in science. Historical materials and contemporary studies suggest strongly that we have moved a great distance toward equality of status for men and women. Important pockets of status inequality remain. In particular, colleges and universities have been remarkably slow to reward their female scientists with promotion to tenure and high rank. But there are two important areas where I believe change has been even slower. Within science, the failure of women to be accepted as full citizens with equal participatory rights in the informal activities of science results in their remaining in many ways second-class citizens. Such forms of exclusion and discrimination are difficult to overcome; they are not easily attacked through litigation and other formal sanctions. But they must be overcome before we can say that science approximates a meritocratic community, in which all citizens enjoy equal rights and opportunities.

And beyond these matters of attaining equality for those already in science, the larger question for tomorrow is how to alter American culture in such a way that science will become an attractive career for bright, young, energetic women. If we are looking toward the horizon, the most pressing problem is to increase sharply the numbers of young women who enter the community in the first place. We are abysmally ignorant of the social and psychological processes that influence these decisions. To remain so condemns us to being concerned with the welfare of 10 or 15 percent rather than what might be 50 percent of the scientific community.

References

Bayer, A. E., and H. S. Astin. 1968. "Sex Differences in Academic Rank and Salary Among Science Doctorates in Teaching." *Journal of Human Resources* 3: 191–201.

——. 1975. "Sex Differentials in the Academic Reward System." *Science* 188: 796–802.

De Candolle, M. A. 1885. *Histoire des sciences et des savants depuis deux siècles.* Quoted in H. J. Mozans, *Women in Science*, 392. Cambridge, MA : MIT Press, 1974 (originally published in 1913).

Centra, J. 1974. *Women, Men, and the Doctorate.* Princeton, NJ: Educational Testing Service.

Cole, J. 1979. *Fair Science: Women in the Scientific Community.* New York: Free Press.

Durkheim, E. 1897. *Suicide.* New York: Free Press, 1951.

Granovetter, M. 1973. "The Strength of Weak Ties." *American Sociological Review* 78: 1360–80.

Hahn, O. 1966. *Otto Hahn: A Scientific Autobiography.* New York: Scribner.

Harmon, L. 1965. *High School Ability Patterns: A Backward Look from the Doctorate.* Scientific Manpower Report no. 6. Washington, DC: National Academy of Sciences/National Research Council.

James, W. 1885. *The Meaning of Truth.* Reprint. London: Longmans, 1932.

Lenzer, G., ed. 1975. *August Comte and Positivism: The Essential Writings.* New York: Harper and Row.

Lloyd, C. B., and B. T. Niemi. 1979. *The Economics of Sex Differentials.* New York: Columbia University Press.

Reskin, B. 1976. "Sex Differences in Status Attainment in Science: The Case of the Postdoctoral Fellowship." *American Sociological Review* 41: 597–612.

Spencer, H. 1896. *Principles of Sociology*, 3rd ed., vol. 1. New York: Appleton.

Zuckerman, H., and J. R. Cole. 1975. "Women in American Science." *Minerva* 13: 82–102.

CHAPTER 14

THE PRODUCTIVITY PUZZLE

Persistence and Change in Patterns of Publication of Men and Women Scientists (1984)

JONATHAN R. COLE AND HARRIET ZUCKERMAN

Abstract

More than fifty studies of scientists in various fields show that women publish less than men. Moreover, correlations between gender and productivity have been roughly constant since the 1920s. The existence and stability of gender differences in productivity continue to be puzzling.

Drawing upon data on publications by and citations to 263 matched pairs of male and female scientists ($N = 526$) who received doctorates in 1969–70, we examine productivity and impact over the first twelve years of scientists' careers to determine whether disparities observed in prior studies persist and, if so, to what degree and why.

Aggregate gender differences in productivity in the cohort of 1970 are much like those in earlier ones. Women published slightly more than half (57 percent) as many papers as men, with that proportion decreasing somewhat as time passed. However, women now account for 26 percent of the most prolific scientists in the cohort (those who published at least 1.6 papers annually) compared to just 8 percent in the cohort of 1957–58 in comparable years. Since highly productive scientists contribute disproportionately to the literature and presumably to the development of scientific knowledge, the increased representation of women in this group is significant.

Motivation and Achievement 2: 217–58

Here, as earlier, women's papers had less impact than men's (59 percent as much), when impact is gauged by frequency of citation. Prior studies suggesting that women are cited less even when productivity is held constant are not borne out. Paper for paper, women are cited as often or slightly more often than men. However, since women publish less, their work has less impact in the aggregate.

Three explanations for gender differences in output and citation are examined. There is no support for the view that women publish less because they collaborate less often than men, when collaboration is measured by extent of multiple authorship. Nor are overall gender differences in citation attributable to gender differences in first authorship, Women are first authors as often as men. Last, there is support for a hypothesis attributing differentials in productivity to differential reinforcement. Women are not only less often reinforced than men, reinforcement being crudely indicated by citations, but they respond to it differently. Women were less apt to maintain or increase their output and more apt to reduce it than comparably cited men at the same levels of prior productivity. Reinforcement has less effect on later output of women than of men.

Although gender differences in output and impact are statistically significant, their substantive importance is not self-evident. It is not clear whether observed gender differences signal real disparities in contribution to science. Nor is it clear how productivity should be measured over the course of scientific careers. Questions are also raised about the nature of age, period, and cohort effects at the institutional and societal levels.

Introduction

"Work, Finish, Publish." It is said that a sign bearing these words hung in the laboratory of the great British physicist and chemist Michael Faraday. Now, as then, this directive calls attention to three fundamental requirements of the scientific role. In this chapter, we focus on publication. The reason for doing so is clear. Science is public, not private, knowledge; publication is a necessary step in the process by which new knowledge comes to be certified by scientific peers (Merton, 1938; Ziman, 1968). It should not be surprising then that sociologists and other observers of the scientific enterprise have focused considerable attention on publication practices and particularly on variations in such practices among those of differing age, field, degree of scientific achievement, social and educational origin, and sex.

More than fifty studies covering various time periods and fields of science report sex differences in published productivity, more specifically, that men publish more than women, even when age and other important social attributes are taken into account.[1] Moreover, gender differences in publication rates appear to have persisted for decades. So far, efforts to account for these differences have not been successful; their existence continues to be a puzzle.

Two classes of explanations have been examined empirically: one dealing with gender discrimination of various kinds and the other with women's greater obligations to marriage and family. The former, for example, has emphasized gender differences in access to the means of scientific production, including resources associated with high academic rank, university affiliation, grants, graduate students, and time for research, as well as differential treatment by referees and editors of scientific publications. The latter, by contrast, attributes the observed disparities to the time women devote to household and childcare at the expense of their research. Neither class of explanation has found much empirical support. Married women are as productive as single women and women with two or fewer children just as productive as those without children (Centra, 1974; Cole, 1979; Reskin, 1978a, p. 1241). Moreover, observed disparities in productivity between the sexes have not been eliminated by taking into account variables such as rank and institutional affiliation, although such disparities are reduced when this has been done.

Research on sex differences in role performance in science is, of course, politically charged and there are some who think it cannot yet be done. For example, the Committee on the Education and Employment of Women in Science and Engineering of the National Research Council (1979, pp. xiv, 87–88) observed, research productivity cannot be used yet as an overall comparative measure of male and female academic scientists' performance. In most fields in research universities, there are not yet enough women faculty who have held professional positions with the necessary perquisites long enough to make such comparisons meaningful.

Such statements confuse the existence of a phenomenon with its explanation. First, it is not the case that there are too few female scientists for quantitative comparative study. Individuals can be aggregated across universities to provide samples of sufficient size for statistical analysis. Second, even if the Committee's conclusion proves to be correct, it would be appropriate to see whether differences in rates of publication actually exist and, if so, to try to account for them in terms of the "perquisites" identified as significant. We do not think that it is premature to examine sex differences in published productivity; rather, their extent, sources, and consequences merit serious attention since studies to date have yet to provide satisfactory explanations for the observed disparities.

We therefore take up the productivity puzzle once again, this time focusing on a cohort of young scientists (those who received doctoral degrees a decade or so ago) to determine whether the sex differences reported again and again for older cohorts persist or whether signs of change toward greater sex equality can be detected. The question, of course, is whether younger female scientists

The Productivity Puzzle   289

who were professionally socialized in the early years of the women's movement and Affirmative Action have fared any differently from older women who began their careers under quite different circumstances. We also take up the related question of the comparative impact of publications by male and female scientists on the development of their fields. Earlier studies of impact, measured crudely by the number of citations in the scientific literature to male and female authors, seemed to show that papers by women had less impact than those by men even when disparities in productivity were taken into account (Cole, 1979; Reskin, 1978a).

More specifically, we examine the publication histories of matched samples of male and female scientists who earned PhDs in 1969 and 1970 in order to answer these questions:

- Are there gender differences in rates of publication and, if so, of what magnitude?
- If such differences are observed, are they more or less constant over the period under review (1968–79) or do they grow or diminish as time passes?
- How do "profiles" of published productivity, the distribution of high and low producers, among men and women compare?
- How do patterns of productivity in this cohort compare to those observed in earlier ones?
- Are female scientists given to collaborative publication to the same extent as men? Do such proclivities change over time?
- Is there evidence here for differences in the impact of publications by male and female authors and, if so, are they attributable to productivity differences alone?
- To what extent can differences in the impact of papers by men and women, if such differences exist, be accounted for by differences in the extent of primary authorship? (Since citations are available only for primary or solo authors, sex differences in authorship practices may produce artifactual differences in citation counts.)
- And finally, is early recognition of scientists' work related to their subsequent rates of publication?

Before we examine evidence bearing directly on these questions, a number of historical observations are in order.

Historical Contexts

For centuries women were considered unfit for intellectually demanding careers, science being no exception. Few women became scientists, and those

290 FREEDOM AND UNFREEDOM

who did encountered ideological and structural barriers to productive work. As late as the first decades of the twentieth century, scientists as well as the general public believed that half of the species was intellectually feeble and emotionally and physically frail.

By way of example, G. Stanley Hall, one of America's foremost psychologists, suggested (as quoted in Gould [1981, p. 1181; originally in Hall [1904, p. 194]) that the higher suicide rates of women were

> one expression of a profound psychic difference between the sexes. Woman's body and soul is phyletically older and more primitive, while man is more modem, variable, and less conservative. Women are always inclined to preserve old customs and ways of thinking. Women prefer passive methods; to give themselves up to the power of elemental forces, as gravity, when they throw themselves from heights or take poison, in which methods of suicide they surpass man.

Indeed, as late as 1924, Felix Frankfurter wrote in support of "protective legislation" for female workers: "Nature made men and women different . . . [and] the law must accommodate itself to the immutable differences of Nature."[2]

That was a heavy burden for the few women who embarked on scientific careers. Given such public sentiment about women, it is surprising that female scientists in the early decades of this century contributed as much as the evidence indicates they did.

For a small sample of fifty-three American men and women who received PhDs between 1911 and 1913, Cole (1979, p. 242) found that men were more apt to publish scientific papers in the first five years of their careers than women ($r = -.46$ where the coding scheme produces a negative correlation if men are more productive and a positive correlation if this is so for women). This difference between the sexes moderates somewhat as time passes; it is reduced to less than half its size at the end of ten years ($r = -.20$) and then rises slightly after fifteen years ($r = -.26$). Given women's exceedingly limited access to research facilities generally and their virtual exclusion from major professorial and research posts specifically, these differentials might well have been larger.

It is also surprising that the relationship between gender and published productivity remains so stable among cohorts entering science in the United States in successive decades, in spite of marked changes having taken place in the larger society, higher education, and the social organization of science. Again, using a coding scheme in which negative correlations signal greater productivity among men, Cole (1979, pp. 64, 242)[3] reports the following correlations between gender and productivity after the first fifteen years of scientists' careers:

Year of PhD	Zero-Order Correlation Between Gender and Published Productivity
1922	−.24
1932	−.30
1942	−.25
1952	−.27
1957–58	−.30

It appears, then, that men have published more copiously than women in each decade for which data are available. Moreover, these differences seem to hold for English and Canadian scientists as well as for Americans (Blackstone & Fulton, 1974, 1975; Endler, Rushton, & Roediger, 1978). So much then for the historical context.

There is now accumulating evidence that sex role attitudes of American men and women became more egalitarian in the 1960s and the 1970s, though not to the same degree or at the same velocity in all domains of social life (Mason, Czajka, & Arber, 1976; Duncan & Duncan, 1978). Such changes should mean that the youngest cohort of female scientists are less hobbled than their predecessors by sex role stereotypes, both their own and those held by men. To the extent that role performance is affected by such attitudes, gender differences in productivity should be reduced. Moreover, changes in sex role attitudes should also be followed by structural modifications of the sort that make it easier for young women to be productive scientists. To the extent that such structural modifications have been made, they too should reduce gender differences in productivity. What do the data on productivity show for the 1970 cohort? Is the productivity puzzle in the throes of solving itself?

The Data

Data were collected on productivity patterns of 263 pairs or a total of 526 male and female scientists in six scientific fields: astronomy, biochemistry, chemistry, earth sciences, mathematics, and physics.

The sample, drawn from *American Doctoral Dissertations* (ADD), is composed of all women listed in ADD who earned degrees in 1969–70 in five out of six fields.[4] In chemistry, female PhDs were sufficiently numerous to permit selection of every other woman. Male "matches" were selected from among those who received degrees from the same departments in the same years. Thus, the samples of men and women have the same educational origins and have received degrees in the same years in the same fields. In accord with the vast differences in the numbers of PhDs in these fields, the cases are unevenly distributed, with only 4 astronomers having been included along with 24 earth scientists, 134 chemists, 168 biochemists, 130 mathematicians, and 66 physicists.

Although this cohort is young, chronologically and professionally, it may not be young enough to show the effects of the social and cultural changes we seek. It might have been preferable to focus instead on the youngest cohort of scientists (those who earned PhDs in the 1980s) since they presumably would show these effects more strongly than older scientists. But the youngest cohort has the disadvantage for our purposes (unlike the cohort of 1970) of not being far enough along in their careers to provide sufficient data for analysis. Choosing the cohort of 1970 has an added benefit. We know from studies of scientific careers (Berelson, 1960; Cole & Cole, 1973; Crane, 1969; Gaston, 1978; Zuckerman, 1977) that scientists' standing in the stratification system is fairly well set by the end of their first decade of work. Thus, what we learn about members of this cohort will not only tell us about their past but will also provide a good basis for predicting their futures—although more in terms of their career attainments than extent of individuals' publications.

Drawing on the *Source Index* of the *Science Citation Index* (SCI) we traced the publication histories of men and women for a twelve-year span beginning in 1968 (one year prior to receiving their doctorates) up to 1979. These publication data are close to complete since the *Source Index* covers all the major and the great bulk of the minor scientific journals. In 1979 it recorded papers published in 2,993 scientific journals.[5] However, the *Source Index* does not list publication in book form. Although most sciences are, as Derek Price (1971) has observed, "papyrocentric" rather than "bibliocentric," given to paper rather than book publication, we cannot estimate the effects of this omission on productivity counts. Nonetheless, it is not immediately evident why one sex should publish books more often than the other. If the data are biased, then the bias should not systematically penalize or benefit men or women.

Data were collected on the total number of papers each scientist published in each of those twelve years. This total was subdivided into papers published by the scientist alone and those published in collaboration with others. We also noted the number of times the scientist in our sample was the prime author in the author set.[6] All citations, year by year, to the author's cumulative publications were enumerated as they were listed in the SCI, although self-citations were excluded from the counts.[7] Productivity and citation counts can, of course, be aggregated over any number of years within the twelve-year interval, enabling us to compare earlier and later patterns of scientific publication and citation. Data on the field and prestige rank of the doctoral departments (Roose & Anderson, 1970) from which these scientists received their degrees were also coded.

This set of data is limited. Information is not included on the social origins of the 526 scientists, their jobs, promotions, and honors. This phase of data collection is not yet complete. In fact, we do not know yet which of these scientists remained in academic life after receiving their PhDs. If unequal proportions of men and women have remained in academia, the results of this study could be biased, since academics tend to publish more than government and industrial scientists. Studies of the distribution of male and female PhDs among sectors of employment suggest, however, that sex differences in this regard exist but are not large (Astin, 1969; Harmon, 1965). By way of example, the National Research Council reports that 67 percent of the women who earned PhDs in 1977 in science and engineering were employed in educational institutions of all kinds, with 61 percent being located at four-year colleges and universities. Correlatively, 55 percent of the men were also employed in educational institutions and 53 percent at four-year colleges and universities. Not surprisingly, a smaller proportion of women held jobs in industry than men. Similar percentages of both sexes were employed in government, although men were more apt than women to work for the federal government. (Committee on the Education and Employment of Women in Science and Engineering, 1979, table 4.1, p. 58.) If women in this sample tend more often than men to be employed in academic jobs and if academic scientists publish more than others, then in the aggregate, differences between the sexes in sector of employment should, if they have any effect at all, work in favor of rather than against women.

Findings

Published Productivity of Male and Female Scientists

As the data in table 14.1 show, men were more prolific than women, on average, publishing 11.2 papers compared with 6.4 for women. Thus, the ratio of productivity of women to men was .57. (Comparisons of medians yield similar results: 7.6 papers for men and 3.2 for women, for a ratio of .42.) Putting aside the question of substantive import of such differences, these findings are, in a purely statistical sense, highly significant.

Gender differences are maintained when we examine rates of publication for different segments of the twelve years under review. During the first seven years of scientists' research careers, the "tenure-relevant" years, we find that the women publish about two-thirds (.63) as many papers as men on the average but only half (.51) as many in the next five years. Turning from means to medians, the female to male ratios are reduced: in the first seven years (1968–1974) to .51 and to .30 in the next five years (1975–1979). There are then increasing

294 FREEDOM AND UNFREEDOM

TABLE 14.1 Published Productivity of 526 Male and Female Scientists: Summary Statistics

	Men	Women	Ratio
	(*N* = 263)	(*N* = 263)	(Women/Men)
Mean Total Number of Papers (1968–1979)	11.2	6.4	.57
(S.D.)	(12.5)	(9.5)	
Median Number of Papers (1968–1979)	7.6	3.2	.42
Mean Number Early Papers (1968–1974)	5.7	3.6	.63
(S.D.)	(5.5)	(4.9)	
Mean Number Later Papers (1975–1979)	5.5	2.8	.51
(S.D.)	(8.5)	(5.7)	
Median Number Earlier Papers	4.4	2.2	.51
Median Number Later Papers	2.3	0.7	.30

differences in publication rates between men and women as they progress beyond the first seven years of their careers.

It also turns out that men publish more than women, in five fields out of six, although there is some variability in the ratios. Astronomy, with but four cases, is the exception. Although the female to male ratio of mean publications in mathematics is 0.46; in physics, 0.44; in chemistry, 0.48; in biochemistry, 0.68; and in earth science, 0.25, it is 5.7 for the four astronomers. In all fields but astronomy, which simply has too few cases for meaningful interpretation, the direction of the inequality in publication is consistent. There is sufficient variation in the ratios among the several fields to raise questions about whether the conditions for research productivity for women in biochemistry, for example, differ from those of women in the earth sciences, physics, or mathematics. This question cannot be answered with these data but we note that women are far more numerous in biochemistry and the biomedical sciences than in the physical sciences and mathematics, and it has been suggested that women encounter less discrimination when their relative numbers in social groups increase beyond a given point (Kantor, 1977).

Despite the variability among fields, we have pooled the productivity data for the analyses that follow. When field differences depart substantially from the aggregate data, such differences are noted.

Another way of describing gender differences in research publication is the zero-order correlation of −.21 for the entire sample of 526 men and women. That is, productivity is negatively associated with feminine authorship. The strength of this association is essentially unaffected by transforming the productivity counts into logarithms ($r = -.24$), or by eliminating from consideration those scientists who had failed to publish a single scientific paper in the twelve years under examination ($r = -.19$).[8]

These gender differences in research productivity in the cohort of 1970 are consistent then with data on earlier cohorts. Significant differences between men and women persist, and their extent is similar in magnitude to those found for groups of men and women who received PhDs as far back as the 1920s.

Measures of central tendency may, of course, mask important gender differences in the distribution of scientific productivity. As table 14.2 shows, the separate distributions of productivity for men and women exhibit considerable skewness. It also shows that twice as many female scientists failed to publish a single paper during the twelve years under review: 22 percent against 11 percent. The disproportionate representation of women among "silent" scientists is again consistent with other studies of earlier cohorts of men and women (Cole & Cole, 1973). Correlatively, women are underrepresented among "prolific" scientists, those who published approximately 1.5 to 2.0 papers annually or a total of at least sixteen publications for the twelve-year period. Although more than a quarter of the men published this many papers, about one-tenth

TABLE 14.2 Distribution of Published Productivity (1968–1979) for 526 Male and Female Scientists

Total Number of Papers: 1968–1979	Men %	Cum. %	Women %	Cum. %
0	11.4	11.4	21.7	21.7
1	9.1	20.5	10.6	32.3
2–3	12.2	32.7	19.8	52.1
4–6	13.3	46.0	17.9	70.0
7–10	17.5	63.5	11.7	81.7
11–15	9.9	73.4	9.5	91.2
16–20	8.0	81.4	2.4	93.6
21–24	7.9	89.3	1.1	94.7
25 or more	11.0	100.3	5.6	100.3
	100.3		100.3	
(N) =	(263)		(263)	

of the women did so. At the same time, the 70 "prolific" men (27 percent) published more papers overall than the remaining 193 men in the sample while the small proportion of "prolific" women scientists (7 percent) account for an even higher proportion of the total output by all women.[9]

The distributions in figure 14.1 show that the most productive 15 percent among men account for almost half (49 percent) of all the papers published by the men in the sample, whereas the most productive 15 percent among women account for an even larger proportion, 57 percent, of all papers published by women. These "profiles" of productivity show slightly greater skewness for women than men; "prolific" women scientists generate a larger share of all papers published by women than their prolific counterparts among men.

When the data for men and women are combined, what proportion of the top stratum of producers (taking 15 percent as an arbitrary figure) is composed of women? To be included in this group, scientists must have published as

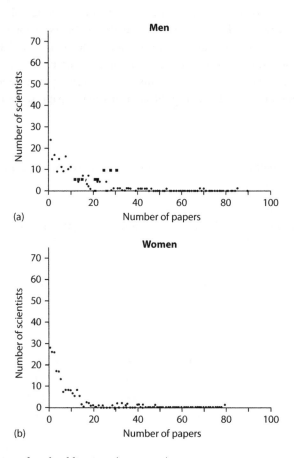

14.1 Distribution of total publications (1968–1979).

many as nineteen papers in the period being reviewed. Of this group of the prolific scientists, only twenty are women, or about one-fourth (24 percent).[10]

Paired Comparison of Scientists: Productivity

Thus far, we have focused mainly on aggregated productivity patterns of male and female scientists. But since we selected these scientists using a matched sampling procedure designed to hold constant field of PhD and place of graduate education, we are also able to compare the productivity of pairs of male and female scientists. Three aspects of the paired comparisons are of interest: first, the proportions of pairs in which scientists of one sex published more than those of the other or of those in which there were no such differences; second, the mean differences in output, pair by pair; and finally, the distribution of the differences between the pairs. Table 14.3, which presents paired comparisons, shows that in 60 percent of the 263 pairs, men published more than their female "matches," and in 35 percent of the pairs, women published more than their male matches. In 5 percent of the cases there were no differences.[11]

Table 14.4 shows the distribution of differences for pairs in which women outproduced men and men outproduced women. In more than half of the

TABLE 14.3 A Comparison of the Total Number of Papers Published by 263 Pairs of Male and Female Scientists, According to Differences within Pairs

Male-Female Comparison	Percentage	N Pairs
Men Published More than Women	60.5	(159)
Women Published More than Men	34.6	(91)
No Difference in Number of Papers Published	4.9	(13)
	100.0%	(263)

Total Cases: 526 Scientists

TABLE 14.4 Distribution of the Difference in Number of Papers Published by 263 Pairs of Male and Female Scientists*

	Differences in Numbers of Papers Published by Each Pair				
	1–5	6–10	11–20	21+	Total
Men published more than women	33	23	22	22	100% (159)
Women published more than men	56	21	12	11	100% (91)

*No difference for 13 pairs
Mean difference = 4.80 (men higher)
Median difference = 2.46 (men higher)

ninety-one cases in which women published more than men, the margin of difference was small (between one and five papers), but in about one-tenth of the cases, that difference was as much as twenty-one papers or more. When men published more than women, they were apt to do so by a greater margin. In one-third of the 159 pairs the difference was relatively slight (one to five papers), but in as many as one-fifth of the pairs, they did so by twenty-one papers or more. Men were substantially more apt to publish more than their female matches, and they were also more apt than women to do so by a considerable margin. Put another way, among pairs of men and women who received doctoral degrees from the same departments, men outpublished women 64 percent of the time and women outpublished men 36 percent of the time (excluding the thirteen ties). The data that we have in hand do not allow us to account for these differences.

Change in Productivity Distributions

Although these data largely conform to those recorded for earlier cohorts of scientists, one important change can be observed: the proportion of women turning up in the most prolific stratum of scientists has greatly increased. In an earlier study of men and women who received degrees in 1957–58, Cole (1979) found that approximately 13 percent of the men had published more than twenty papers during the first twelve years of their careers as against just 1.2 percent of the women.[12] In the recent cohort, however, prolific men and women are considerably more numerous, with 19 percent of the men and 7 percent of the women having published that many papers. Although data for just two time periods do not constitute a trend, this change may be significant since prolific scientists contribute far more than their proportionate share to the development of scientific knowledge.

The same findings can be reported in a different and, in this instance, preferable manner.[13] Comparing the proportions of women among "prolific" scientists in the two periods, we find that fully 26 percent of those who published as many as twenty papers in the cohort of 1970 are women as against just 8 percent in the 1957–58 cohort. When the ante is raised even further to those who published twenty-five or more papers, the feminine presence is even more marked: one-third of these in the cohort of 1970 are women.

Although important changes may be underway, differences in output between male and female scientists persist and are not well understood. As we noted, one important class of hypotheses for gender differences focuses on discrimination against women and their limited access to the means of scientific production. With the data now in hand, we can begin to see how well one version of the discrimination hypothesis actually squares with the evidence.

Gender differences in productivity have been attributed in part to differential access to scientific collaboration and thus to shared publication. Given the considerable role of collaborative research in contemporary scientific inquiry (many problems simply cannot be worked on by a single investigator), such differences in access, if they exist, would be important not only for women's rates of publication but also for their chance to participate in much of mainstream science. Reasons given for differential access vary. Some think that women have difficulty establishing collegial relationships with men (Kaufman, 1978; Reskin, 1978b). Others think that women have less access to appropriate collaborators because they are affiliated more often than men with small colleges and universities whose faculties are of limited size and scientific diversity. Whatever the reason, if women participate less often than men in collaborative research, then a larger share of their publications should be published under their names alone, regardless of field. What do the data actually show on the extent of solo authorship by men and women?

First, we find no differences in the average proportions of solo-authored papers published by men and women: 24 percent of papers by men have one author, as do 23 percent by women (after eliminating scientists who have not published at all from the computation of averages). In keeping with their greater overall output, men publish a larger absolute number of solo-authored papers than women: 3.09 papers as against 1.89 ($p < .01$). Nor do we find that women show a greater proclivity to solo authorship as time passes and they make the transition from student to mature investigator. The rates of solo authorship remain roughly the same in the later period (23 percent for men and 22 percent for women). These data are consistent with those reported by Over and Moore (1980) for psychologists in Australian universities. Although further inquiry may show that women collaborate less often than men, these data on authorship patterns provide no evidence for it.

Thus far, the data on gender differences in productivity raise a number of important questions, not all of which can be answered immediately. One set of questions deals with the character of scientific output and proper units of analysis and another with multiple consequences of observed sex differences in productivity.

Little work has been done on identifying the appropriate unit of analysis for studies of scientific productivity. Is the paper really the right item to analyze? Many scientists have told us that they publish "clusters" of papers on one or more related experiments and "strings" of papers on successive experiments. Do men and women have different propensities to produce clusters or strings of papers? Are their clusters or strings apt to be of different sizes? And if so, do gender differences in these propensities account in part for women's lower absolute number of publications? In turn, why should women exhibit different patterns of production?

300 FREEDOM AND UNFREEDOM

All apart from matters of measurement, to what extent do observed gender differences affect individual careers and the growth of scientific knowledge? Prior studies show that differences in output are correlated with differences in rewards allocated to male and female scientists (Cole, 1979; Reskin, 1978a). How important should it be that, in this sample, men publish 0.9 papers annually as compared to 0.5 for women (eleven papers in twelve years compared to six)? To be sure, the normative framework of science requires that rewards be in rough accord with individual contributions to science (Merton, 1942/1973) but it is not clear how differences in output at this level of magnitude (five papers on average in twelve years) should translate into differences in scientific contribution. Although these gender differences in productivity are statistically significant, it is not self-evident whether such differences should or should not be significant in allocating symbolic or material scientific rewards. Shifting our focus from individuals to the population of male and female scientists, are such differences consequential for the growth of scientific knowledge? Perhaps so, perhaps not. Many scientists complain that too many papers are published, that many are written to satisfy officers of granting agencies or are "requirements" for promotion and salary raises, regardless of their real contributions to the advancement of knowledge. It may be that men contribute more to cluttering the literature than women. And it may be that women publish a higher density of papers that are useful to their scientist colleagues. And it may be that there is a threshold of productivity that is required before an individual's work can become known and thus be useful. The fact is that we know little about matters of this kind.

We now turn to an analysis of citations to the work of the 526 scientists. This should give us some sense of the comparative impact of the publications of men and women on their fields.

Patterns of Citation to Scientists in the 1970 Cohort

Citation counts register the "impact" or "influence" of scientific publications, as we noted earlier. This is so because recognizing the cognitive contributions of others in footnotes or references is a well-established practice in science. Thus, the number of times a particular paper has been cited is a rough indicator of the number of different occasions on which other authors have taken note of it (Cole & Cole, 1971; Garfield, 1979, especially chap. 10). Citation counts for individual authors are, as we have said, significantly related to a variety of forms of scientific recognition such as prizes and awards, as well as to independent peer assessments of the significance of scientists' contributions. In fact, citation counts are a better predictor of influence or impact of contributions by individual scientists as they are measured by awards than are publication counts.

Output (measured by publication counts) and impact (measured by citation counts) have been shown to be highly correlated, by and large, from .5 to .75 depending on the sample being studied (Allison & Stewart, 1974; Cole & Cole, 1973; Gaston, 1978; Hagstrom, 1971; Long, 1978). The more scientists have published, the more apt they are to be cited by others. This is as much the case for this sample of young scientists as for earlier ones where the correlation between publication and citation counts runs to .62 ($p < .001$).

The earlier studies have also shown that impact and output are not the same, that impact is not merely a function of output. Scientists who publish a great deal also tend to publish particular papers which are influential in and of themselves (Garfield, 1981). The same pattern turns up in the cohort of 1970. If we look only at each scientist's most-cited paper—the one receiving the most attention from other scientists (here, in 1979)—the more prolific the scientists are, the more citations their most-cited paper receives ($r = .59$). If we aggregate the citations to scientists' two or three most-cited papers, the outcome is much the same ($r = .62$ and .63, respectively). Being highly cited therefore is not just the outcome of copious output but is also associated with having particular papers in print which have received a great deal of attention.

Since we know that women have published less than men, we would expect that they are also less often cited. Earlier studies report, however, that the association between gender and citations is modest and ranges from .1 to .3 in various fields and for various samples even though men are consistently shown to be more often cited than women. For example, Cole (1979, table 3–3, p. 64) found a zero-order correlation of −.19 between total citations and gender for the group of 1957–58 PhDs he studied, although the figure varied somewhat among disciplines ($r = -.33$ for 61 chemists; −.21 for 297 biologists; −.22 for 159 psychologists but was .21 for 44 sociologists). Cole also reports a correlation of −.14 between gender and citations for a sample of 248 biologists of various ages. His data are consistent with those reported by Reskin (1978a, p. 1240), who also finds differences in the extent of citation to the work of men and women chemists.

Do these small but significant gender differences in impact also appear in the cohort of 1970? They do but not to the same degree. Again, using a coding convention in which negative correlations indicate that men have been cited more often than women, we find that the correlation between gender and total citations for the twelve-year period is −.11 ($p < .01$), somewhat smaller than the correlations reported in earlier studies.[14]

These correlations cannot, of course, convey anything about orders of magnitude differences in citations between the sexes. As table 14.5 shows, women average 32.3 citations over the twelve-year period and men, 55.1, or a ratio (w/m) of 0.59. But women in the cohort of 1970 do relatively less well than men as time passes; in the first seven years, they received 0.66 as many citations as men but

302 FREEDOM AND UNFREEDOM

TABLE 14.5 Citations to 526 Male and Female Scientists: Summary Data

	Men	Women	Ratio
	(N = 263)	(N = 263)	(Female/Male)
Mean Total Number of Citations (1968–1979)	55.14	32.27	.59
(s.d.)	(135.4)	(69.8)	
Median Number of Citations (1968–1979)	13.67	5.40	.40
Mean Number: Early Citations (1968–1974)	16.51	10.86	.66
(s.d.)	(50.9)	(26.6)	
Mean Number: Later Citations (1975–1979)	38.63	21.40	.55
(s.d.)	(90.9)	(48.3)	
Median Number: Early Citations	3.61	1.34	.37
Median Number: Later Citations	7.42	2.41	.33

only 0.55 as many in the next five years.[15] We shall have more to say later about increasing disparities in citations to men and women and what consequences this pattern might have.

As we found in the case of output, there is considerable inequality in citations within each sex. Some women are very heavily cited, as are some men, and these scientists contribute disproportionately to the cumulative numbers of citations earned by those of the same sex. Table 14.6 and figure 14.2 show that the distributions for men and women are sharply skewed (Gini coefficient = .76 for women and .70 for men). Thus, women are no more like one another

TABLE 14.6 Distribution of Total Citations 1968–1979 to 526 Male and Female Scientists

Total Number of Citations	Men %	Cum. %	Women %	Cum. %
0	13.3	13.3	27.4	27.4
1–5	22.0	35.2	22.8	50.2
6–10	11.4	46.7	11.4	61.6
11–20	11.0	57.7	9.5	71.1
21–50	15.2	72.9	11.0	82.1
51–100	11.8	84.7	8.4	90.5
101 +	15.2	99.9	9.5	100.0
	100%		100%	
(N) =	(263)		(263)	

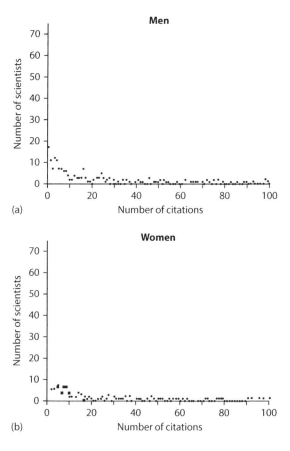

Figure 14.2

than are men when it comes to the distribution of citations; a few in each group receive the lion's share.

We noted that differences in cumulative rates of citation for men and women are statistically significant. But are they "real" or an artifact of sex differences in authorship practices and of procedures used in the SCI for recording citations? Since it was first published, citations have been enumerated only for solo-authored papers and those on which a given author was first in the author set. This convention means that citation counts are underestimated for the subset of authors who collaborate more than usual and whose names tend not to be first in the author sets.

Before we proceed to any further analysis of gender differences in citations, we need to know whether they simply reflect differences between male and female scientists in authorship practices. We have already seen that women are no more given to solo authorship than men. But some have claimed, for example, that women are less often first authors on collaborative papers because their contributions are taken less seriously than those by men (Epstein, 1970). It has also been suggested that women are less often heads of laboratories and therefore less often in the position to insist on primary authorship (Reskin, 1978b). What do the data show on such authorship practices of men and women in the cohort of 1970?

When it comes to rates of primary authorship, we find no differences at all between men and women. On the average, men and women take first position in author sets on 43 percent of their multi-authored papers. Nor are there any detectable differences in first authorship in the early and later periods, although both men and women show a decreasing tendency to be first authors as they mature, a finding consistent with earlier studies of authorship practices (Long, McGinnis, & Allison, 1980; Zuckerman, 1968).[16] And finally, when rates of first authorship and solo authorship are combined, we also find no gender differences which might account for differences in citation rates, since men occupy one or the other of these positions on 66 percent of their papers and women on 59 percent. For the cohort of 1970, then, gender differences in citation rates do not appear to be an artifact of authorship practices of men and women.

So far, the data on citations indicate that publications by men have more impact than those by women. To what extent are these differences a function of differences in output that we described earlier? Are publications by men cited more often simply because there are more of them to cite? The answer to this question is more complicated than it might at first appear.

One way to assess the influence of output on impact for men and women is to compare their average numbers of citations. Although men are cited more often overall, it turns out that there are no gender differences in average citations per paper. Women's papers averaged 5.02 citations and men's, 4.92 (p = n.s.). Paper for paper, then, women's publications are just as influential as those by men.[17] Moreover, this seems to be true in both the early and later periods. In the first seven years of their careers, women earned an average of 3.02 citations for each paper and men, 2.89. In the next five years, citations per paper increased for both groups to 7.6 for women and 7.05 for men. In short, it appears that men are cited more often largely because they publish more.[18]

Even so, it is still possible that we have not detected differences in impact between men and women who publish at different rates. The most productive women may be cited far more often than comparably productive men while less productive men are cited more than equally unproductive women. As Figure 14.3 shows, average citations per author are much the same for men and women at each level of productivity, with the women having a slight edge in each group. (But it should be noted that there are more women in the less productive group and only a third as many women in the highly productive group.) Again, it

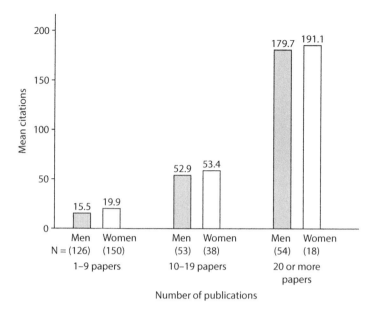

14.3 Mean number of citations to male and female authors, according to their productivity (1968–1979).

seems appropriate to conclude that gender differences in impact are largely a function of gender differences in output.

The picture becomes somewhat more complicated, however, when we look at those papers by men and women which have had the most impact, that is, have been most often cited. Taking citations in 1979 as an indicator of citations in other years (year-by-year correlations of citations are high for this cohort as for others), we find that most-cited papers by men have been cited more on the average than those by women, with papers by men averaging 3.7 citations and those by women, 2.7 citations, although the medians show somewhat smaller gender differences (1.2 vs. 0.7 citations). Combining citations to each scientist's two or three most-cited papers does not change the findings to any significant degree. Why should it be that women's papers on the average are cited as often as those by men but their most-cited papers are cited less often?

The answer seems to lie in the distribution of output and citations for men and women, as figure 14.4 shows. Scientists who are highly productive are highly cited, not only because they have published a great deal but also because some of their papers are themselves highly cited. This is as much the case for women as for men. The most-cited papers of highly productive female scientists are cited at about the same rate as such papers by equally productive

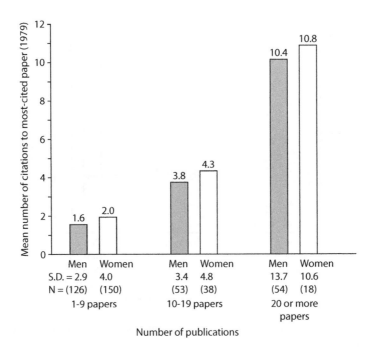

14.4 Mean number of citations to the most-cited papers of male and female authors (1979), according to their productivity (1968–1979).

men, 10.8 as against 10.4 times. At the same time, there are three times as many men (54) as women (18) in the highly productive group of scientists and it is this which produces the higher overall average citations for most-cited papers by men.

On the average, then, papers by women are cited as often as those by men. There is no indication of gender differences in impact when we examine average publications. Another conclusion to be drawn that is just as warranted is that men publish more and as a consequence are cited more in the aggregate. Later on, we shall have more to say about how gender differences and gender similarities may be interpreted.

Paired Comparison of Scientists: Citations

We turn now to patterns of citations to pairs of men and women. As we observed earlier, comparison of pairs allows us to see how scientists trained in the same departments fare when it comes to output and citations to their work. Table 14.7 presents data on the proportions of the 263 pairs appearing in each

TABLE 14.7 Comparisons of Citations (1968–1979) to Pairs of Male and Female Scientists Who Received Doctoral Degrees from the Same Departments

Result of Paired Comparison	Total Citations 1968–1979		Citations per Paper		Single Most-Cited Paper	
	%	(*N*)	%	(*N*)	%	(*N*)
Men More Cited than Women	60.8	(160)	52.5	(138)	44.5	(117)
Women More Cited than Men	31.6	(83)	40.7	(107)	31.6	(83)
No Gender Difference	7.6	(20)	6.8	(18)	24.0	(63)
	100.0	(263)	100.0	(263)	100.0	(263)
Mean Difference:	− 22.87		.10		1.00	
Median Difference:	− 8.27		20		27	

of three categories: those in which the male partner in the pair was cited more often, those in which the female partner was cited more often, and those in which there were no gender differences observed in citation.

The first column in table 14.7 shows that men were almost twice as apt as the women to have been more often cited (61 percent vs. 32 percent). However, as column 2 shows, when the average number of citations per paper is considered for each pair, the differences between men and women are substantially reduced. In 52 percent of the pairs, men had the higher average per paper as against 41 percent of the pairs in which the women's average was higher.

Turning from averages to the distribution of differences in citations per paper in table 14.8, we see that men and women look much the same when the *margin* of difference is considered. That is, when men are more often cited, the extent to which their citations exceed those of women is no greater than the margin of difference observed when women lead in citations. If anything, the margin of difference is slightly greater in those cases where women have been cited more often.

TABLE 14.8 Comparison of Citations (1968–1979) per Published Paper for Pairs of Male and Female Scientists According to the Extent of Differences between Pairs

Result of Paired Comparison	Extent of Difference in Per-Paper Citations				
	0.1–1.50	1.51–5.0	5.01–10.0	10.1 +	Total
Men More Cited than Women	38.4	34.1	15.9	11.6	100% (138)
Women More Cited	31.4	29.9	17.8	15.0	100% (107)

No Gender Difference = 18 Pairs

308 FREEDOM AND UNFREEDOM

One more comparison of citations to pairs of men and women merits comment. We wanted to see whether pairs of men and women differed when we compared citations to their most cited papers in 1979. Column 3 of table 14.7 shows that in 44 percent of the pairs the men were more often cited as against 32 percent of the pairs in which women were more often cited. This suggests that the gender differences we have observed for overall citation are somewhat reduced when we consider only the most heavily cited papers but that men also do better on this criterion.[19]

Citation Counts and Their Significance

Before considering changes in output for men and women over the course of time, some comments are in order on citation counts and the meaning of differences in citations among groups of scientists.

The data we have reported so far, which detail differences between men and women in the extent that their work is cited, do not lend themselves to unambiguous interpretation. For one thing, there is no agreement among sociologists and historians of science on the substantive significance of unequal rates of citation. We can say that citation counts are strongly correlated with scientists' reputation and the prestige of their honorific awards (Cole & Cole, 1973; Gaston, 1973; Zuckerman, 1977). However, we cannot say whether the gender differences in citations that we have described indicate that research by men really has greater impact than research by women. It is not at all clear that scientists whose work is cited an average of fifty-five times over twelve years (the figure for men in the cohort of 1970) have had substantially greater impact than those whose work is cited an average of thirty-two times (the figure for women). To make the point more generally, not enough is known about the connections between citation counts and contributions to scientific knowledge or about the comparative impact of work by scientists who are cited n times as against those who are cited a fraction of n times. We can say how both groups compare to the average cited scientist in the SCI but more information is needed than numbers of citations to assess comparative contributions to the development of scientific knowledge.[20] In short, we report gender differences in citation counts but underscore our conviction that the connections between such counts and the extent of contributions to knowledge have not been satisfactorily established.

Another set of questions about the interpretation of citation counts deals with their distribution among the set of papers published by each scientist. Have two scientists who have each been cited fifty times in a given year had the same impact when one has published twenty-five papers, each of which has been cited twice, and the other has also published 25 papers but has two papers which received twenty-five citations each, with the rest having received none at all? How are these scientists' patterns to be compared to others who also have received fifty citations but who have published only five or ten papers? Is the impact of the work published by all these scientists the same? We do not know. Preliminary evidence for physicists, collected almost fifteen years ago, suggests that those who were highly cited but who had a relatively modest output (a group dubbed "perfectionists") tended to have received greater formal recognition than scientists who had approximately the same number of citations but who had published more papers (these dubbed "prolific" scientists) (Cole & Cole, 1973). But the detailed relationships between the distribution of citations to scientists' publications, patterns of the relation between number of citations and quantity of published work, and appraisals of impact or influence of work remains substantially uncharted territory. This is still another reason why it is difficult to assess the precise meaning of the gender differences in citations that we have described.

And finally, we note that the attribution of citations to male and female authors contains its own ambiguity. Earlier, we reported apropos solo- and multi-authorship that 76 percent of the papers published by men in the cohort of 1970 and 77 percent of the papers published by women on the average were collaboratively authored. But in the absence of detailed information on the sex composition of sets of authors, we cannot say precisely how often papers "by men" and "by women" are cited. On the basis of their incidence in the population of scientists, men are probably more apt to work with men than they are with women. There are simply too few female scientists to make incidence of all-female and cross-sex collaborations as frequent as all-male collaborations. Numbers aside, it is still not clear how best to count citations when the sex composition of collaborative groups varies. At the minimum, classification of papers by the sex composition of the author set is needed to permit comparison of citation counts to subsets of authors.

Reinforcement of Published Output: Its Effect on Male and Female Scientists

Last, we take up the evidence bearing on a hypothesis which may account in part for gender differences in rates of publication. If women receive less response to their research contributions than men (are less often reinforced), they may become discouraged and publish less as time passes. If the hypothesis of differential reinforcement is correct, then we would expect to find declining rates of publication more common among women than men. This is precisely what we find. A much larger proportion of women published fewer papers in the second six years of their careers than did men. In fact, 50 percent of the women who had published almost a paper a year (five papers or more) in the first six years dropped below that number in the second six years in contrast to just 31 percent of the men. Correlatively, a much smaller proportion of women than men increased the pace of their output: 12 percent of the women did so as against 24 percent of the men.[21] Moreover, when productivity of men and women is compared over time, we found, as noted earlier, that disparities between the men and women increased. In the early period women published 63 percent as many papers as men and, in the later period, 51 percent as many. So far, then, the data are consistent with the hypothesis of differential reinforcement.

Lacking better indicators of reinforcement, such as appointments in prestigious departments, early promotion, research grants, honorific awards, and other symbols of the good opinion of colleagues, we use citations to scientists' research as an approximation. As we noted earlier, citations register the impact of a scientist's work or how much attention it receives. This is the case because poor or uninteresting research tends to sink without a trace; other scientists do not cite it. By contrast, research which provokes others sufficiently to try to disprove it does have impact and is cited. In this sense, other scientists pay attention to the work even if they are not sure it is correct.

Plainly, we do not know for sure whether citations provide incentives for further work, whether scientists are aware of the extent to which they are cited, and if so, how they experience citations by different classes of scientists. One citation from an important reference individual may be more sustaining than numerous citations from the rank and file. In the absence of fine-grained data on linkages between types of citations and reinforcement of scientific work, we use citation counts with caution.

We have already seen that women are less often cited in the aggregate than men and also that the extent to which this is so increases over time. As we noted, women received 66 percent as many citations as men in the early period but only 55 percent as many later on. These data then are also consistent with the hypothesis of differential reinforcement.

Prior studies have shown that early recognition, as indicated by use through citations, is positively related to later productivity. Scientists whose early publications evoke no response (are uncited) are less apt to continue to do the difficult work required for publication than those whose papers are recognized by citations (Cole & Cole, 1973, pp. 110–14; Reskin, 1978a, p. 1240). As it turns out, the data for the cohort of 1970 provides further support for this hypothesis.

Table 14.9 presents data on early publication and recognition and their relation to subsequent publication. First, note that rates of publication for the majority of scientists are quite stable during the two periods. The last row of table 14.9 shows that 62 percent of those who had published as many as five papers in the early years of their careers continued to publish that many later on. Similarly, it can be inferred from the same row that 83 percent (100 percent – 17 percent) of scientists who had published fewer than five

TABLE 14.9 The Effects of Early Recognition on Later Productivity Among Scientists According to Levels of Early Productivity

| Early Recognition: Citations: 1968–1973 | Percent Having Published 5 or More Papers in Later Period (1974–1979) | | |
| | Early Productivity: 1968–1973 | | |
	0–4 Papers	5 or More Papers	(N)
0	9	46	
	(185)	(13)	(198)
1–5	25	58	
	(114)	(55)	(169)
6 or more	29	66	
	(55)	(104)	(159)
Total (%)	17	62	
(N)	(354)	(172)	(526)

papers also continued at the same slower pace. Among those scientists whose publication rate changes, we find that reduction, not increase, is the more common pattern.[22]

The columns of Table 14.9 show the effect of differing degrees of early citation on scientists who had been differentially productive in the early period. It is evident from data in both columns that the more cited scientists have been, the more likely they are to move into or to remain in the prolific category. As many as 29 percent of those who had published four or fewer papers in the early period and who were cited comparatively often increased their published output, but just 9 percent of those who were uncited and unrecognized published more than they had earlier. Similarly, as many as two-thirds (66 percent) of those who had both been prolific and were cited comparatively often in the early period continued to be prolific in contrast to fewer than half (46 percent) of those who were equally as productive but who had not been cited at all.

Does recognition affect male and female scientists in the same way? The data in table 14.10 suggest that women in the 1970 cohort are *slightly more* responsive than men to the lack of reinforcement and *considerably less* responsive than men to positive reinforcement. Put in a different way, the women seem more readily discouraged and less readily encouraged by varying degrees of citation to their work. The first row of table 14.10 shows the effect of silence or the absence of citation. Among those men and women who published a

312 FREEDOM AND UNFREEDOM

TABLE 14.10 The Effects of Early Recognition on the Later Productivity of Men and Women According to Levels of Early Productivity

Early Recognition: Citations: 1968–1973	Percent Having Published 5 or More Papers in Later Period (1974–1979)				(N)
	Early Productivity: 1968–1973				
	0–4 papers		5 or more papers		
	Men	Women	Men	Women	
	(1)	(2)	(3)	(4)	
0	11	7	50	43	
	(75)	(110)	(6)	(7)	(198)
1–5	36	16	64	44	
	(53)	(61)	(39)	(16)	(169)
6 or more	39	22	73	54	
	(23)	(32)	(67)	(37)	(159)
(N)	(151)	(203)	(112)	(60)	(526)

small number of papers in the first six years, women were slightly less apt than men to increase their rate of productivity in subsequent years, and among those who start out at a reasonably fast pace (columns 3 and 4), women are slightly more apt than men to reduce their rate of publication.

A modest reinforcement (1–5 citations) has a more pronounced effect on subsequent publication by men than by women. This level of encouragement is more often followed by greater research activity among men than women. Indeed, 36 percent of the slow-starting men compared with only 16 percent of comparable women later increase their rate of publication after receiving modest reinforcement in the first six years of their careers.

Such modest reinforcement also appears to have a slightly more stabilizing effect on men than women. Among men and women who began by publishing five or more papers and whose work was modestly reinforced, 64 percent of the men compared with 44 percent of the women maintained that pace of publication. Among the comparatively prolific scientists who were also significantly reinforced in the early period (with six or more citations), 73 percent of the men compared with 54 percent of the women continued to publish at their earlier rate. Turning the data around, we see that almost half of the women who start off being comparatively productive fall back to a slower pace, despite the reinforcement received through citations, in comparison to slightly more than a quarter of comparable men.

These findings on differential responsiveness of men and women to varying degrees of reinforcement suggest that women may need more encouragement than men to maintain the level of publication they set for themselves earlier in their careers. These aggregated data are inconsistent with Reskin's (1978a, p. 1242) findings for an earlier cohort of chemists. She reports that women are *more* responsive than men to formal recognition in the form of citations although the models she uses are somewhat different from those presented here. When we examine this reinforcement pattern for chemists alone, we find results that are more consistent with Reskin's. This suggests that there may be significant differences in the effects of reinforcement in the several scientific disciplines. In attempting to predict later productivity, we may find substantial interaction between reinforcement through citations, scientific field, and gender. It is not clear at this juncture why this should be so nor is it clear whether the same pattern of response to recognition would be observed if we were in a position to examine the effects, for example, of recognition in the form of appointments, research support, or honorific awards. We do not yet have these data in hand. But we do know from extended interviews with scientists that many are aware of whether, when, and how others are making use of their work even if they do not know how many times they have been cited.

What then can we say about the hypothesis of differential reinforcement and its possible contribution to growing disparities in publication between men and women? First, the data are consistent with the hypothesis. Women are reinforced less than men, to the extent that citations are indeed a reasonable measure of reinforcement. Moreover, women appear to respond differently to the reinforcement they receive. At similar levels of reinforcement, women are less apt than men to maintain or increase their rate of publication.

The hypothesis of differential reinforcement only goes part of the way, of course, in accounting for increasing gender differences in publication. This social psychological explanation is consistent with the sociological explanation which emphasizes women's unequal access to the means of scientific production. There is ample evidence that women achieve high academic rank (and its perquisites) later than men (Zuckerman & Cole, 1975) and also that they are less apt to be promoted than comparably qualified men (Astin & Bayer, 1975; Bayer & Astin, 1975; Cole, 1979, p. 246). Since a variety of resources for research are associated with high academic rank, the difficulties and delays women encounter in achieving it may well contribute to their falling farther behind men in publication in the post-tenure years. When data are in hand on the career histories of the 1970 cohort, we should be able to say more about the relative contributions of structural impediments and differential reinforcements to disparities in publication between male and female scientists.

Summary and Discussion

By now it should be clear that social change has not eliminated gender differences in productivity and thereby solved the productivity puzzle. Disparities in publication between male and female scientists in the 1970 cohort are about as large as those observed in earlier cohorts. The data for the 526 men and women indicate that men publish almost twice the number of papers as women in the first 12 years of their careers. The correlations between gender and productivity (about −.2 to −.3) have remained fairly stable since the first decades of this century. Moreover, the findings for the cohort of 1970 are consistent with those reported in scores of other studies.

At the same time, these data on recent PhDs suggest that some important changes may be underway. We have found an increased proportion of women among the most prolific scientists (those who publish about two papers annually) and who contribute a hefty share of all publications in science. Thus an increasingly large subset of women scientists now resemble, for good or ill, those men who start publishing early and do so copiously year after year.

Citation differences between men and women seem to be largely a function of their different rates of publications. When average citations per paper are compared for men and women, we find no difference. This means, of course, that the papers published by women on average have as much impact as those by men, but overall, owing to the weight of numbers of men's publications, women's work has less impact.

In some sense, these findings raise as many new problems about the character of scientific productivity as they resolve. How important for individual reputations and for the development of scientific knowledge are strings and clusters of papers? What is the threshold of output required for a scientific contribution to become visible? Do thresholds differ among fields and historical periods as well as with the contributor's age and prior achievements? At what point do scientist's publications become overabundant, mere clutter in the literature?[23] And finally, since the per-unit effect of papers by men and women does not differ, sheer output may take on added significance in assessing their relative contributions.

Before we can be sure that differences in cumulative citations to men and women are really a function of the size of their respective bibliographies, additional data are needed. We need to know, for example, whether men and women are represented among authors of the most-cited papers in each field in proportion to their representation in those fields. Although a number of the most-cited papers are technical or instrumental rather than substantive contributions, most scientists would agree that they are important papers which have been useful and have had considerable impact. It would be helpful to have such data because the measure we use here, the average count per paper, provides no clues as to the distributions of citations to specific papers or subsets of them.[24] If women are represented among the authors of most-cited papers in proportion to their approximate numbers in each field or specialty, then there will be good reason to conclude that differences in citation levels are mainly a function of differences in output. If, however, women are underrepresented among authors of most-cited papers, we will have to be more tentative about this conclusion.

The data presented here allow us to discard for the moment several explanations for the productivity differentials. Women are just as apt as men to publish alone and, correlatively, they are just as apt as men to publish collaboratively. Differences in output are not a simple function of lack of access of women to collaboration. However, these data do not show whether men and women participate in collaborative groups of the same size and duration nor do they say anything about the relative ranks of male and female co-workers. We do not know whether men and women are equally apt to collaborate with status peers and with status unequals and what effect such relationships between co-workers have on published output.

We do know that women are as apt as men to be first authors on their publications, and this dispels the notion some have that women have fewer citations than men because they are less often first in author sets. But we still do not know whether women are less apt than men to be accorded first authorship on papers judged by the authors to be important contributions and, conversely, more often given "token" first authorship on papers judged to be routine.[25]

The data presented here also suggest that women are less likely than men to translate high early productivity into high later productivity when they have experienced early positive reinforcement in the form of frequent citation. The regression analyses presented in tables 14.11A and 14.11B summarize these results. In table 14.11A, where men and women are aggregated in one group, we find that the three independent variables—early productivity, gender, and early citations—each have significant independent effects on later productivity. Early productivity is, of course, the strongest predictor of later productivity, but both gender and reinforcement are "independent" predictors of later productivity. At the same time, gender has less effect than early reinforcement in the OLS (ordinary least squares) regression equation.

Table 14.11B presents the effects of early productivity and early citation on later productivity for men and women separately. Here we find a substantial reinforcement effect for men but no reinforcement effect for women. The regression coefficient in standard form for early citations is beta = .19 for men and beta = .02 for women. These regression results are, as we would expect, consistent with the results reported earlier in the paper.

What we do not know yet is why the early positive feedback has different effects for the men and women. Reskin (1978a) has suggested that early reinforcement is more effective for women than for men. Yet these results in the

316 FREEDOM AND UNFREEDOM

TABLE 14.11A OLS Regression of Later Published Productivity (1975–1979) on Early Productivity, Early Citations ("Reinforcement") and Gender ($N = 526$)

Independent Variables:	B	B	F-ratio*
Gender	−1.65	−.10	1.32
Total productivity: T_1	.86	.46	134.10
Total citations: T_1	.04	.15	14.23
(Constant)	3.63		
$R = .57$			
$R' = .32$			

*$f > 3.78 = p < .01$

TABLE 14.11B OLS Regression of Later Productivity on Early Productivity, Early Citations ("Reinforcement") for Men and Women as Separate Groups

	Men ($N = 263$)			Women ($N = 263$)		
Independent Variables:	B	β	F-Ratio*	B	β	F-Ratio*
Total productivity: T_1	.96	.46	12.32	.86	.56	55.79
Total citations: T_1	.05	.19	12.68	.01	.02	.07
(Constant)	1.44			0.62		
$R = .54$				$R = .58$		
$R_2 = .29$				$R' = .33$		

aggregate indicate the contrary; while reinforcement may be more important for women, it is apparently not translated into high continued output in some scientific fields.

Finally we must emphasize that while the difference between the sexes in research output is both significant and puzzling, gender per se does not explain much of the variance in published productivity or citations. Variability between the sexes in productivity is not nearly as great as variability within each sex.

Although we have shown that differences in research productivity persist among young male and female scientists, we have not explained them. The data necessary for an adequate explanation of the phenomenon will come from improved quantitative data, which expands the types of "explanatory" variables employed, and from increased attention to qualitative data including detailed accounts by scientists of the standards they subscribe to, the decisions they have made, and the constraints they have encountered which affect publication. We are now collecting these types of data.

It is premature now to speculate on how the productivity puzzle will be solved, but several points should be made to help redefine the pieces of the puzzle in such a way as to make its solution more likely.

First, structural determinants of differing rates of output have not received sufficient attention. As many working scientists know, the social structure and composition of research laboratories (the relative numbers of senior researchers, postdoctoral fellows, and graduate students, as well as the status differences within these groups) affect group and individual productivity.[26] Scientists who must work alone or with undergraduates are not only limited with respect to the sorts of problems they can tackle but also in the numbers of papers they can produce.

Second, it is often assumed that high rates of productivity are associated with long hours spent at the bench. It appears from the interviews we are now conducting with scientists that this is misleading and that roles in the laboratory change markedly with upward mobility and over the course of the scientific career. Although senior scientists may have begun by spending many hours at the bench and in the lab, their responsibilities for and involvement in the finished product usually change. Authorship for this group is almost always collaborative and no longer derives from their having "done the research" but instead from having set the problem for those who actually carry out the experiments, from raising the funds to keep the laboratory going and to support those at the bench, from discussion of results, and from helping to draft the papers submitted for publication. These patterns are apt to be conditioned by the character of research, by discipline or specialty, and by organizational context, but whatever form they take, scientists' involvement in research and participation in publication changes in ways that affect rates of publication but are not visible in straight counts of papers.

Third, publication counts alone provide no clues as to varying norms regarding publication in different fields and specialties and, more importantly, for different statuses and organizational settings. In some settings there are strong normative obligations to publish (and incentives provided for doing so), and in others there are almost as strong prohibitions (and probably some disincentives). Correlatively, certain statuses carry with them the obligation to publish and to help others to publish, principally students and postdoctoral fellows. Little attention has been paid to the way in which normative prescriptions and proscriptions contribute to scientists' output; these need to be understood better than they are now to account for differentials in output between men and women.

Clearly, considerable empirical work is needed on the fine structure of scientific productivity before persistence of gender differences in output in successive cohorts can be understood fully. Equally clearly, the theoretical contexts and implications of the productivity puzzle need further development. The questions we have investigated here are connected to the growing body of work on age stratification and the interplay of age, period, and cohort affects in social life (Riley, Johnson, & Foner, 1972; Riley, 1980; Zuckerman & Merton, 1972).

We sought to identify changes in the role performance of women in this cohort of young scientists because we assumed that they, more than women in earlier cohorts, would be affected by widespread cultural change in attitudes toward women. We suspected that being professionally socialized during the emergence of the women's movement might have led to their having more egalitarian attitudes, aspirations, and role definitions than women in earlier cohorts. We also thought that the same cultural changes would affect men in all age cohorts in ways that would facilitate women's role performance.

At the same time, the impact of these changes might not be as great in science as in the society at large, not because female scientists are more conservative than others but rather because the opposite may be so. Female scientists who received their PhDs two or three decades ago may have held more egalitarian attitudes than their age peers among women as compared with female scientists in the 1970 cohort relative to their age peers. Thus, the differences in attitudes between different cohorts of female scientists may be smaller than they are between cohorts of women of the same ages generally. Social and self-selection may produce cohort differences within institutions which are less (or more) marked than in the society as a whole.

Important countervailing forces have also been at work, which affected the cohort of 1970. They have been plagued by greatly reduced job opportunities in colleges and universities and by reductions in research funding brought about by changes in government policy and high rates of inflation. These changes in the opportunity structure for doing science have affected all cohorts but perhaps the youngest ones most strongly. If female scientists have benefited from Affirmative Action (and there is some question about whether this has been so in posts other than those at entry level), both men and women in the cohort of 1970 did not when they started out and do not now face futures as rosy as did their age peers who benefited from the expansion of higher education in the 1950s and 1960s. How these changes in opportunity structures have affected the role performance of successive cohorts of scientists is not known.

Finally, our studies of age, period, and cohort effects in science suggest the need for an extension in the analysis of these effects. Until now, attention has focused primarily on these effects as they operate at the societal level. A new perspective is required which differentiates such effects at the societal level from those which operate in particular institutions and, most important, treats the reciprocal relations between these effects at different levels of analysis. Cohorts of scientists are a strategic research site for the study of these reciprocal relations.

This is so because multiple layers of age, period, and cohort effects that interact with one another in science can be readily identified. By way of example, the sciences seem to have their own period effects separate from those which have impact on the whole society. There are times of much cognitive development (or little) and times when resources for research are ample (or meager). There also appear to be marked cohort effects in the sciences. Becoming a scientist in the 1960s when there were many posts being created was surely quite different from becoming a scientist in the 1930s when the Depression made itself felt in science and academic life as well as the economy at large. There are also cohort effects in the sciences which are associated with cognitive change. Thus, those scientists who came of age after relativity and the quantum revolution or after the discovery of the helical structure of DNA share perspectives and a research agenda different from those of their predecessors. And, finally, there may be institutionally specific age effects that are different from age effects that operate in other institutions or in the society at large. By way of example, the association between age and authority in science may be much weaker in science than is generally the case. Scientists who are thirty-five or forty years of age are considered fully qualified for elite status if their contributions merit it, whereas those of the same age may be considered still rather young in law or business or medicine regardless of their achievements.

These observations are meant to suggest that the productivity puzzle is connected to a series of theoretical questions about the relations between social and cultural change, aging, and cohort flow, questions we have only begun to articulate.

Faraday's edict, "Work, Finish, Publish," still holds—at least for most scientists—and now probably as much for women as for men. But since gender differences in published productivity persist, the productivity puzzle has yet to be solved.

APPENDIX A

The following empirical studies, mostly published since 1973, examine aspects of published productivity of male and female scientists. They vary in kind and quality. Some directly address the question of gender differences and similarities in the productivity patterns; others simply examine the extent to which female scientists are represented in scientific publications in rough proportion to their numbers in a field; still others focus on the reward system of science and use productivity of scientists, male and female, as one explanatory variable. The scientific fields represented differ; the quality and size, as well as the type, of samples vary. The adequacy of the methods and presentations also vary. This list (compiled by Jan Sedofsky) is confined to studies that indicate simple bivariate differences in research output by gender; whether or not the studies purport to "reduce" the differential through multivariate analysis is not considered here. For a list of pre-1975 papers on this subject, see Cole (1979).

American Astronomical Society, Report of the Committee on the Status of
   Women (1979)
Astin (1978)
Astin and Bayer (1975)
Astin and Bayer (1979)
Bayer and Astin (1975)
Blackbum, Behymer, and Hall (1978)
Blackstone and Fulton (1974)

Blackstone and Fulton (1975)
Bryson, Bryson, and Johnson (1978)
Bryson, Bryson, Licht, and Licht (1976)
Centra (1974)
Chubin (1974)
Clemente (1972)
Clemente (1973)
Clemente and Sturgis (1974)
Cole (1979)
Cole and Cole (1973)
Converse and Converse (1971)
Fitzgerald, Pasewark, Thomton, and Sawyer (1975)
Freeman (1977)
Guyer and Fidell (1973)
Hamovitch and Morgenstem (1977)
Hansen, Weisbrod, and Strauss (1978)
Hargens, McCann, and Reskin (1978)
Heckman, Bryson, and Bryson (1977)
Heins, Smock, and Martindale (1978)
Helmreich, Spence, Beane, Lucker, and Matthews (1980)
Katz (1973)
Ladd and Lipset (1976)
Loeb and Ferber (1973)
Over (1980)
Over (1982)
Over and Moore (1980)
Pasewark, Fitzgerald, Thomton, and Sawyer (1973)
Pasewark, Fitzgerald, and Sawyer (1975)
Persell (1978)
Reskin (1978)
Simon, Clark, and Galway (1967)
Teghtsoonian (1974)
Widom and Burke (1978)

Acknowledgments

This research was supported by a grant from the National Science Foundation (SES-80-08609). We thank the staff of the Center for the Social Sciences, Columbia University, and in particular Jan Sedofsky, Margaret Carey, and Madeleine Simonson, for their assistance. Jan Sedofsky compiled the Appendix.

322 FREEDOM AND UNFREEDOM

Notes

1. The number of studies reporting data on published productivity of male and female scientists has grown rapidly. More than 40 studies are listed in Appendix A. Just one of these is purported to show that women publish more than men (Simon, Clark, & Galway, 1967), and several others show no zero-order difference. In fact, Simon et al. do not show that women publish more than men, although they are often cited as demonstrating that this is so. As the authors say, "To summarize, of the four measures of productivity, the two most direct ones, numbers of articles and books published, married women publish as much or more than men, and unmarried women publish slightly less than men. The differences on the whole are not great" (231). Their paper presents an interesting case in the sociology of knowledge. It has often been misread as evidence for one social fact (that women publish more than men) when it does not actually provide evidence for that position. The misreading is consistent with a particular ideological stance. It turns out that the paper, which really shows one thing, acquires a quite different symbolic meaning. This new meaning is reinforced through the subsequent citations to it as a demonstration of a "falsely" attributed fact. Several questions arise from this single instance of the wider phenomenon of "mis-citation" and reification. First, what produces the original misclassification and citation? Is it a simple misreading of the text? Is it "intentional" and motivated distortion? Is it a socially conditioned misreading of the paper, influenced by strongly felt personal values and beliefs? In sum, how is the original error made? Second, how does the original error come to be perpetuated; how does it take on an independent life of its own? Plainly, this case is one among several classes of errors in references and citations, each of which would make interesting cases in the sociology of knowledge. For discussion and examples of other types of errors in use of evidence and in scientific scholarly research, see, among many others, Merton (1965; 1973, pp. 402–12); Altick (1963); Gould (1981). The list in Appendix A is not exhaustive. A significant number of these studies make use of the same data sets such as the data on the faculty collected by the American Council on Education and thus are not, in fact, fully independent inquiries.

2. Such attitudes were widespread in both the United States and Europe during the later part of the nineteenth and early twentieth centuries. One only has to look at the work of sociologists such as Le Bon, Durkheim, Spencer, Comte, and at the works of Freud and his contemporaries to get ample doses of such expressed beliefs. For examples, see Cole (1979) and Gould (1981).

3. The number of cases on which these correlations are based increases as we move toward the current period. As noted, for the 1911–1913 group, the number totals only 53; by 1952, the number is 383, and it is 561 for the 1957–58 group of scientists.

4. *American Doctoral Dissertations* (ADD) and *Dissertations Abstracts* (DA) list degree recipients annually and there is a great deal of overlap between their entries. However, we found delay in entry dates in DA with some 1970 PhDs not listed in DA until 1971. We therefore used ADD as our prime source for the sample of male and female scientists.

5. In 1968, the first year of our data collection, the *Source Index* covered 1,968 source journals. Coverage of journals has expanded yearly.

6. "Authorship" includes the following types of publications by a given scientist: first, papers that were written and published by the author alone and, second, all collaborative papers on which the scientist's name appears. If two scientists collaborated to produce one paper, that paper would count in the totals of each author. We did not

differentiate between collaborative papers involving different numbers of authors. A paper on which there were two authors or six were counted as one paper and contributed equally to the total publication counts for each scientist. Third, we made *no* distinction between collaborative papers on which there were only male authors, only female authors, or a combination of male and female authors. We know of no study that has distinguished between collaborative papers involving *only* women, *only* men, or the *combination* of men and women. This last distinction could be important. Studies of productivity and citations attribute authorship to men or to women but do not acknowledge the existence of a third category of papers authored by men and women jointly. Publication and citation counts for men and women, to be precise, should take into account this class of papers.

7. Vigorous discussion continues between those who advocate for the use of citation counts as measures of "influence" or "impact" and those who object to their use because scientists' citation practices are not well understood and because the procedures used to count citations introduce artifacts and errors that are hard to estimate (Cole & Cole, 1971; Garfield, 1979; Goudsmit, 1968). We have tried to minimize the more common errors and artifacts in citation counting (those introduced by homonyms, for example) by requiring detailed checking of the fields represented by citing journals listed under the name of a given scientist (for example, citations in astrophysics journals to a scientist we know is a biochemist were deleted). This may have produced underestimates of citations, but such errors are much smaller than those introduced by homonyms when such deletions were not made. We also carefully reviewed cases where copious citation was associated with sparse publication to be sure that references really were to scientists in the sample.

   It would have been preferable to have had complete citation counts for the full bibliographies of the scientists in the sample instead of being confined to "straight" counts for papers published as solo or first authors. (The SCI does not list citations to papers where a scientist was a secondary author.) But the disparity between "straight" and "complete" counts is small; the two are correlated on the order of .80 to .96 in various fields (Cole & Cole, 1973, p. 73; Long, McGinnis, & Allison, 1980, p. 134). This correlation is greater for younger scientists than older ones since the proclivity to first authorship declines with age. (Long, McGinnis, & Allison 1980, p. 139). Others, however, claim that first-authored and solo-authored papers are not a representative sample of authors' publications and would restrict the use of straight counts (Lindsey, 1980). The Institute for Scientific Information now has an algorithm for counting all citations to a given author, irrespective of his location in an author set. When this is made available to researchers, the controversy on straight counts should be moot.

8. This correlation is based on 439 cases (p < .001).

9. These results conform closely to patterns described by Lotka (1926) and later by Price (1963).

10. The total number of papers produced by the top 15 percent of male scientists exceeded the total number of papers by all of the female scientists.

11. When no differences were found, men and women may have published no papers at all or they may have published precisely the same number of papers.

12. Cole (1979, p. 63fn). These data are for matched samples of male and female PhDs in chemistry, biology, psychology, and sociology—different fields from those treated here. It should be noted that the increased presence of women among high producers may

324 **FREEDOM AND UNFREEDOM**

have resulted in part from the inclusion of biochemists in the 1970 sample. Biochemists tend to publish more papers than other scientists.

13. Since the fields represented in the two time periods differ, it is more appropriate to examine the sex distribution among prolific scientists than to examine the proportions of men and the proportions of women publishing as many as "n" papers.

14. The association between gender and number of citations is weaker for the first seven years ($r = -.07$) than for the next five ($r = -.12$), which is consistent with the increased difference between men and women in output that we noted earlier.

15. Comparisons of medians show somewhat greater inequalities, with the twelve-year ratio being 0.40; for the early years, it is 0.37, and for the later ones, 0.33.

16. Moreover, there were no significant gender differences in first authorship within four of the six fields; in one, the earth sciences, men are more often first, but in astronomy the opposite is true. The number of cases is so small in both fields that departures from the overall pattern should not be given much weight.

17. The fact that the median citations to all papers by women is 5.4, and not much larger than the average citations per paper for women, suggests of course that women do not publish many papers in the aggregate.

18. Average citations per paper for men and women are much the same in each of four fields: biochemistry, chemistry, mathematics, and physics. There are too few cases in astronomy and earth sciences to draw reliable conclusions.

19. In previous studies of citation practices, the use of most-cited papers represented an attempt to control for the effects of productivity on citation totals. Since most scientists are infrequently cited, the correlation between total citations to scientists' work in any given year, such as 1979, is highly correlated with the total number of citations to their most-cited paper. For these scientists that correlation was $r = .69$. To extend this a bit, the correlation between total citations and citations to the two most-cited papers is .73 and $r = .75$ between the total and the three most-cited papers. For men as a group, the zero-order correlation between total citations in 1979 and the citations to the top paper was .69 and $r = .71$ for women. When we extend the number of most-cited papers to two, the correlation for men is .73 and is .74 for women. And finally, the correlations between the three most-cited papers and the total are $r = .76$ for men and .75 for women.

20. The number of citations to the average cited author in the SCI has remained fairly stable for the last decade. In 1969, each author was cited an average of 8.12 times; in 1979, 8.05 times; and in 1980, 8.28 times. This number of citations is much higher than one would find if citations for all authors, cited or not, were computed since most authors are uncited in any given year.

21. At the same time, the majority of scientists, men and women, maintained about the same rate of productivity over this period. The data reported in this section have been divided into two equal periods of six years rather than being broken into the first seven and the later five years as was the case earlier.

22. Using five or more published papers as a cutting point between "lower" and "higher" productivity, we find that only 17 percent of the low producers in the initial six years became high producers in the next six, but 38 percent of those who began as high producers in the first six years move into the group of low producers in the subsequent six years. The correlation between early and later productivity is .60. The correlation between early and later productivity is actually higher for women than for men, that is, there is greater stability over time in the level of output for women than for men. This is shown in the following table:

Transition Probabilities of Male and Female Scientists for Stability and Change in Levels of Productivity between Earlier (1968–1973) and Later (1974–1979) in the Career

	T2 Productivity			
	Women		Men	
	Lower	Higher	Lower	Higher
	(0–4 papers)	(5 +)	(0–4)	(5 +)
TI Productivity:				
Lower				
(0–4)	.88	.12	.76	.24
		(203)		(151)
Higher				
(5+)	.50	.50	.31	.69
		(60)		(112)
		(263)		(263)

The data indicate that the primary reason for the greater association for women than for men is the clustering of 178 women among lower producers in both time periods.

23. These findings also suggest why many studies report that sheer output of papers has little independent effect on peer recognition when total citations are taken into account. This is so because citations are a function of total output to a significant degree and thus total output independent of citations has little influence.

24. Paired comparison of citation counts to the most-cited papers of the 1970 cohort showed that papers by men were cited more often in 44 percent of the pairs; and those by women in 32 percent. This provides a limited example of distributions of citations to particular papers by men and women.

25. Such calculations are made. Zuckerman (1968) found that Nobel laureates-to-be often exercised noblesse oblige in authorship by giving first authorship to younger colleagues—except on those papers they knew to be important and signal contributions.

26. Absolute group size alone, as Cohen (1980) has shown, is not associated with output.

References

Allison, P. D., and J. A. Stewart. "Productivity Differences Among Scientists: Evidence for Accumulative Advantage." *American Sociological Review* 39 (1974): 596–606.

Altick, R. D. *The Art of Literary Research*. New York: Norton, 1963.

American Astronomical Society. *Report of the Committee on the Status of Women*. Center for Astrophysics, Cambridge, October 1979.

Astin, H. S. *The Woman Doctorate in America*. New York: Russell Sage Foundation, 1969.

Astin, H.S. "Factors Affecting Women's Scholarly Productivity." In *The Higher Education of Women: Essays in Honor of Rosemary Park*, ed. H. S. Astin and W. Z. Hirsch. New York: Praeger, 1978.

Astin, H.S., and A. E. Bayer. "Sex Discrimination in Academe." In *Women and Achievement*, ed. M. T. S. Mednick, S. S. Tangri, and L. W. Hoffman. Sydney, Australia: Halsted Press, 1977.

326 FREEDOM AND UNFREEDOM

Astin, H.S., and A. E. Bayer. "Pervasive Sex Differences in the Academic Reward System: Scholarship, Marriage, and What Else?" In *Academic Rewards in Higher Education*, ed. D. R. Lewis and W. E. Becker. Cambridge: Ballinger, 1979.

Bayer, A. E., and H. S. Astin. "Sex Differentials in the Academic Reward System." *Science* 188 (1975): 796–802.

Berelson, B. *Graduate Education in the United States*. New York: McGraw-Hill, 1960.

Blackbum, R. T., C. E. Behymer, and D. E. Hall. "Research Note: Correlates of Faculty Publications." *Sociology of Education* 51 (1978): 132–41.

Blackstone, T., and O. Fulton. "Men and Women Academics: An Anglo-American Comparison of Subject Choices and Research Activity." *Higher Education* 3 (1974): 119–40.

Blackstone, T., and Fulton, O. "Sex Discrimination Among University Teachers: A British-American Comparison." *The British Journal of Sociology* 26(3) (1975): 261–75.

Bryson, R., J. B. Bryson, and M. F. Johnson. "Family Size, Satisfaction, and Productivity in Dual Career Couples." *Psychology of Women Quarterly* 3 (1978): 67–77.

Bryson, R. B., J. B. Bryson, M. H. Licht, and B. G. Licht. "The Professional Pair: Husband and Wife Psychologists." *American Psychologist* 31 (1976): 10–16.

Centra, J. A. (With the assistance of N. Kuykendall). *Women, Men, and the Doctorate*. Princeton, NJ: Educational Testing Service, 1974.

Chubin, D. "Sociological Manpower and Womenpower: Sex Differences in Career Patterns of Two Cohorts of American Doctorate Sociologists." *American Sociologist* 9 (1974): 83–92.

Clemente, F. *A Note on Sex Differences in Research Productivity*. Unpublished manuscript, 1972.

Clemente, F. "Early Career Determinants of Research Productivity." *American Journal of Sociology* 79(2) (1973): 409–419.

Clemente, F., and R. B. Sturgis. "Quality of Department of Doctoral Training and Research Productivity." *Sociology of Education* 47 (1974): 287–299.

Cohen, J. E. "Publication Rate as a Function of Laboratory Size in a Biomedical Research Institution." *Scientometrics* 2 (1980): 35–52.

Cole, J. R. *Fair Science: Women in the Scientific Community*. New York: Free Press, 1979.

Cole, J. R., and S. Cole. "Measuring the Quality of Sociological Research: Problems in the Use of the Science Citation Index." *American Sociologist* 6 (1971): 23–29.

Cole, J. R., and S. Cole. *Social Stratifiction in Science*. Chicago: University of Chicago Press, 1973.

Committee on the Education and Employment of Women in Science and Engineering. *Climbing the Academic Ladder: Doctoral Women Scientists in Academe*. Washington, DC: Commission on Human Resources, National Research Council, National Academy of Sciences.

Converse, P. E., and J. M. Converse. "The Status of Women as Students and Professionals in Political Science." *Political Studies Association* 4(3) (1971): 328–48.

Crane, D. "Social Class Origin and Academic Success: The Influence of Two Stratification Systems on Academic Careers." *Sociology of Education* 42 (1969): 1–17.

Duncan, B., and O. D. Duncan (with the collaboration of J. A. McRae). *Sex Typing and Social Roles: A Research Report*. New York: Academic Press, 1978.

Endler, N. S., J. P. Rushton, and H. L. Roediger. "Productivity and Scholarly Impact (Citations) of British, Canadian, and U.S. Departments of Psychology (1975). *American Psychologist* 33 (1978): 1064–82.

Epstein, C. F. *Woman's Place: Options and Limits in Professional Careers*. Berkeley: University of California Press, 1970.

Fitzgerald, B. J., R. A. Pasewark, L. Thomton, and R. N. Sawyer. "Research Activities of Experimental Psychologists." *Professional Psychology* 6(2) (1975): 114–15.

Freeman, B.C. Faculty women in The American University: Up the down staircase. *Higher Education* 6 (1977): 165–188.

Garfield, E. "The 1,000 Contemporary Scientists Most Cited 1965–78. Part I. The Basic List and Introductions." *Current Contents* 13 (1981): 5–14.

Gaston, J. *Originality and Competition in Science.* Chicago: University of Chicago Press, 1973.

Gaston, J. *The Reward System in British and American Science.* New York: Wiley, 1978.

Goudsmit, S. A. "Citation Analysis." *Science* 183 (1968): 28.

Gould, S. J. *The Mismeasure of Man.* New York: Norton, 1981.

Guyer, L., and L. Fidell. "Publication of Men and Women Psychologists: Do Women Publish Less?" *American Psychologist* 28 (1973): 157–60.

Hagstrom, W. "Inputs, Outputs and the Prestige of American University Science Departments." *Sociology of Education* 44 (1971): 375–79.

Hall, G. S. *Adolescence: Its Psychology and Its Relations to Physiology, Anthropology, Sociology, Sex, Crime, Religion, and Education* (2 vols.). New York: Appleton, 1904.

Hamovitch, W., and Morgenstern, R.D. "Children and the Productivity of Academic Women." *Journal of Higher Education* 48(6) (1977): 633–45.

Hansen, W. L., B. A. Weisbrod, and R. P. Strauss. "Modeling the Earnings and Research Productivity of Academic Economists." *Journal of Political Economy* 86(4) (1978): 729–41.

Hargens, L. L., J. C. McCann, and B. F. Reskin. "Productivity and Reproductivity: Fertility and Professional Achievement Among Research Scientists." *Social Forces* 52(3) (1978): 129–46.

Harmon, L. R. "Profiles of Ph.D.'s in the Sciences." In *Career Patterns Report No. 1.* Washington, DC: National Academy of Sciences, National Research Council.

Heckman, N., R. Bryson, and J. B. Bryson. "Problems of Professional Couples. A Content Analysis." *Journal of Marriage and The Family* 39 (1977): 323–30.

Heins, M., S. Smock, and L. Martindale, L. "Current Status of Women Physicians." *International Journal of Women's Studies* 1(3) (1978): 297–305.

Helmreich, R. L., J. T. Spence, W. E. Beane, G. W. Lucker, and K. A. Matthews. "Making It in Academic Psychology: Demographic and Personality Correlates of Attainment." *Journal of Personality and Social Psychology* 39(5) (1980): 896–908.

Honig, M., and G. Hanoch. *Age, Cohort and Period Effects in the Labor Market Behavior of Older Persons* (preprint no. 83). New York: Columbia University, Center for the Social Sciences.

Kantor, R. M. "Some Effects of Proportions on Group Life: Skewed Sex Ratios and Responses to Token Women." *American Journal of Sociology* 82 (1977): 965–90.

Katz, D. A. "Faculty Salaries, Promotions, and Productivity at a Large University." *American Economic Review* 63(3) (1973): 469–77.

Kaufman, D. R. "Associational Ties in Academe: Some Male and Female Differences." *Sex Roles* 4(1) (1978): 9–21.

Ladd, E. C., Jr., and S. M. Lipset. "Sex Differences in Academe." *Chronicle of Higher Education* 12(11) (1976): 18.

Lindsey, D. "Production and Citation Measures in the Sociology of Science: The Problem of Multiple Authorship." *Social Studies of Science* 10 (1980): 145–62.

Loeb, J. W., and M. A. Ferber. "Representation, Performance and Status of Women on the Faculty of the Urbana-Champaign Campus of the University of Illinois." In *Academic Women on the Move,* ed. A. S. Rossi and A. Calderwood. New York: Russell Sage Foundation, 1973.

Long, J. S. "Productivity and Aacademic Position in the Scientific Career." *American Sociological Review* 43 (1978): 889–908.

Long, J. S., R. McGinnis, and P. D. Allison. "The Problem of Joint-Authorship Papers in Constructing Citation Counts." *Social Studies of Science* 10 (1980): 127–43.

Lotka, A. J. "The Frequency Distribution of Scientific Productivity." *Journal of the Washington Academy of Sciences* 16 (1926): 317.

Mason, K. O., J. L. Czajka, and S. Arber. "Change in U.S. Women's Sex-Role: Attitudes, 1964–1974." *American Sociological Review* 41(4) (1976): 573–96.

Merton, R. K. *Science, Technology and Society in Seventeenth Century England.* New York: Howard Fertig, 1938.

Merton, R. K. *On the Shoulders of Giants: A Shandean Postscript.* New York: Free Press, 1965.

Merton, R. K. "The Normative Structure of Science." In *The Sociology of Science: Theoretical and Empirical investigations*, ed. N. W. Storer. Chicago: University of Chicago Press, 1973 (originally published in 1942).

Over, R. "Research Productivity and Impact on Men and Women in Departments of Psychology in the United Kingdom." *Bulletin of the British Psychological Society* 33 (1980): 385–86.

Over, R. "Research Productivity and Impact of Male and Female Psychologists." *American Psychologist* 37(1) (1982): 24–31.

Over, R., and D. Moore. "Research Productivity and Impact of Men and Women in Psychology Departments of Australian Universities, 1975–1977. *Australian Psychologist* 15(3) (1980): 413–18.

Pasewark, R. A., B. J. Fitzgerald, L. Thornton, and R. N. Sawyer. "Icons in the Attic: Research Activities of Clinical Psychologists." *Professional Psychology* 4(3) (1973): 341–46.

Pasewark, R. A., B. J. Fitzgerald, and R. N. Sawyer. "Psychology of the Scientist: XXXII. God at the Synapse: Research Activities of Clinical, Experimental, and Psychological Psychologists." *Psychological Reports* 36 (1975): 671–74.

Persell, C. *Explaining Productivity Among Male and Female Applied Researchers.* Mimeo, 1978.

Price, D. J. de S. *Little Science, Big Science.* New York: Columbia University Press, 1963.

Price, D. J. de S. "Citation Measures of Hard Sciences and Soft Science, Technology and Non-science." In *Communication Among Scientists and Engineers*, ed. C. E. Nelson and D. K. Pollack. Lexington, MA: Heath, 1971.

Reskin, B. F. "Scientific Productivity, Sex, and Location in the Institution of Science." *American Journal of Sociology* 83(5) (1978a): 1235–43.

Reskin, B. F. "Sex Differentiation and the Social Organization of Science." In *The Sociology of Science*, ed. J. Gaston. San Francisco: Jossey-Bass, 1978b.

Riley, M. W. "Age and Aging: From Theory Generation to Theory Testing." In *Social Theory and Research: A Critical Appraisal*, ed. H. Blalock Jr. New York: Free Press, 1980.

Riley, M. W., M. Johnson, and A. Foner (eds.). *Aging and Society: A Sociology of Age Stratification*, vol. 3. New York: Russell Sage Foundation, 1972.

Roose, K. D., and C. J. Andersen. *A Rating of Graduate Programs.* Washington, DC: American Council of Education, 1970.

Simon, R. J., S. M. Clark, and K. Galway. "The Woman Ph.D.: A Recent Profile." *Social Problems* 15(2) (1967): 221–36.

Teghtsoonian, M. "Distribution by Sex of Authors and Editors of Psychological Journals, 1970–1972: Are There Enough Women Editors?" *American Psychologist* 29 (1974): 262–69.

Widom, C.S., and B. W. Burke. "Performance, Attitudes, and Professional Socialization of Women in Academia." *Sex Roles* 4(4) (1978): 549–62.

Ziman, J. *Public Knowledge.* Cambridge: Cambridge University Press, 1968.

Zuckerman, H. "Patterns of Name-Ordering Among Authors of Scientific Papers: A Study of Social Symbolism and Its Ambiguity." *American Journal of Sociology* 74 (1968): 276–91.

Zuckerman, H. *Scientific Elite: Nobel Laureates in the United States.* New York: Free Press, 1977.

Zuckerman, H., and J. R. Cole. "Women in American science." *Minerva* 13(1) (1975): 82–102.

Zuckerman, H., and R. K. Merton. "Age, Aging, and Age Structure in Science." In *Aging and Society: A Sociology of Age Stratification*, vol. 3, ed. M. W. Riley, M. Johnson, and A. Foner. New York: Russell Sage Foundation, 1972.

CHAPTER 15

MARRIAGE, MOTHERHOOD, AND RESEARCH PERFORMANCE IN SCIENCE

(1987)

JONATHAN R. COLE AND HARRIET ZUCKERMAN

Women publish less than men, but marriage and family obligations do not generally account for the gender difference. Married women with children publish as much as their single female colleagues do.

S tudies of scientists' research performance, as gauged by their published productivity, find that women generally publish fewer papers throughout their careers than men matched for age, doctoral institution, and field. Various explanations have been proposed to account for this disparity in scientific publication, ranging from systematic gender discrimination to biological differences, as yet undemonstrated, in scientific aptitude.

One frequent explanation holds that women, far more than men, bear the burdens of marriage and childcare, and that this fact of social life best accounts for gender differences in scientific publication. Whether or not this is true, the belief that it is so affects women's career opportunities, their decisions, and the way they are treated.

We decided (as part of a larger investigation of the careers of male and female scientists in the United States) to test the counterclaim, made in earlier studies, that marriage and motherhood have no effect on women's research

---

Scientific American 256, no. 2 (February 1987): 119–25.

performance. We did so by assessing the dynamic relation of family life and women's research throughout their careers—an approach not taken by earlier investigators, who just correlated the number of papers published with current marital and parental status. Our study draws on interviews with 120 scientists: 73 women and 47 men. We wanted to know whether scientists (both male and female) believe marriage and parenthood is incompatible with a scientific career in general, whether this had been the case for them in particular, and what quantifiable effects (measured in number of publications) marriage and motherhood have actually had on the research performance of female scientists. Since men traditionally have not had primary responsibility for childcare, we focused almost entirely on women, comparing publication rates for those who are married and those who are single, those who are mothers and those who are childless.

Publication counts are, to be sure, an imperfect indicator of scientists' contributions. Yet such counts are highly correlated with better measures, such as peer evaluation; moreover, the extent to which scientists publish is of major consequence to their careers. We therefore took the extent of publication as a rough but serviceable gauge of research performance. We recognized that female scientists are, in some sense, "survivors." By definition they have passed through the rigors of graduate training, have earned a doctoral degree, and are employed in science. We did not seek to examine the impact of cultural expectations on women's chances of running this gauntlet; that would have called for investigating the processes by which women are winnowed out of scientific careers.

We chose subjects for the study by a stratified selection process that took into account gender, professional age, field of expertise, and scientific standing. To compare the effects of marriage and motherhood on women who earned degrees in different historical periods, we divided the women into three age groups: those who received their doctorates between 1920 and 1959, before the advent of the women's movement, and between 1960 and 1969 and 1970 and 1979, when the movement was getting underway and then becoming widespread. Eighty percent of the women were drawn from mathematics and the physical and biological sciences, and the remainder were from economics and psychology; the same 4:1 ratio was applied in each age group.

Scientists were further divided according to their peer recognition in relation to others of roughly the same professional age. The top tier of scientists (those we designated as "eminent") in the oldest group were members of the National Academy of Sciences or the American Academy of Arts and Sciences or were full professors in departments ranked in the top 10 in each field by national surveys of the quality of doctoral programs. In the intermediate age group, Guggenheim fellows or tenured professors in a top-10 department were classified as eminent. Younger scientists were considered eminent if they had held Guggenheim fellowships or were assistant or associate professors in a top-10 department.

Scientists not meeting those exacting criteria were designated as rank-and-file. They were randomly selected from lists of the faculty and research staff at accredited four-year colleges and universities in the same geographic regions as the eminent scientists. They were also matched to the eminent scientists for professional age and scientific field. Although these scientists were selected systematically, the criteria we applied, the small numbers of subjects, and our exclusion of some important groups (such as scientists working in industry) mean this is decidedly not a true random sample of all U.S. scientists.

Our subjects were asked about their research and publication histories and for comments on graphs we prepared that showed the number of papers they had published each year along with important events in their career and personal life.

Both male and female scientists reported having come up against the belief that marriage and motherhood cannot be meshed with a demanding scientific career. Not surprisingly, the oldest group of scientists encountered this belief most often. Before (and even shortly after) World War II, the proper priorities for women were widely held to be marriage and motherhood first and science second, and good science was believed to be all-consuming. The notion that women could simultaneously be traditional wives, traditional mothers, and productive scientists seemed to be patently absurd.

Such beliefs were articulated by a female zoologist who recalled that "if one had children and a working husband, it was not part of the psychology to suppose that one's job was anything more than a secondary consideration." Many scientists, of both genders, at the time shared these views. They believed most women could no longer be serious scientists once they were married. A distinguished female biologist who is now in her seventies said her female laboratory chief had been appalled that her protégée would marry: "She threw me out of the lab the minute she heard I was going to get married because that was treason against women." For one male chemist, marriage then meant that female scientists were "finished"; a male physicist said, "As soon as women got into domestic life, that was the end of it for all of them."

This climate of opinion meant that women determined to have serious research careers often did not marry. In the words of one biologist now in her seventies, "marrying was not considered the thing to do [for female scientists], In science, you're dedicated. You go into a shroud, you don't wear normal clothes. . . . You shouldn't get married; you shouldn't have children."

Not all women in science accepted those views, of course, and some did marry, have children, and continue to work. Yet until the end of World War II at the earliest, female scientists were few in number, and fewer than half of them were married. Married women with children were nearly invisible in American science. Such women were viewed as violating the prevailing family norms.

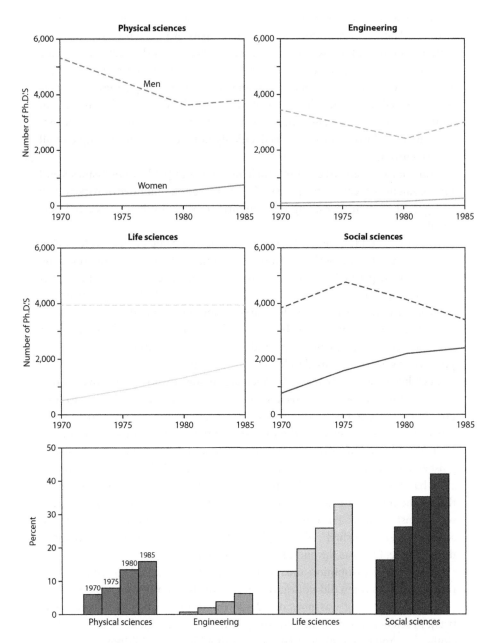

15.1 Number of PhDs awarded to women has increased since 1970 in spite of a general decline in the number awarded to men. The curves (*top*) show the annual number of U.S. doctoral degrees from 1970 through 1985. The percent awarded to women has risen sharply (*bottom*), but it remains low in the physical sciences and engineering.

The data are from National Research Council studies.

Although social attitudes concerning the roles of wives and mothers have changed significantly, even the youngest female scientists report that many people still consider marriage and motherhood to be incompatible with a scientific career. One young chemist says, "When I got pregnant, I was written off as a serious scientist . . . by lots of people." When those occupying positions of power and authority act in accordance with these beliefs, they severely limit the opportunities and careers available to married women.

To assess the actual impact of marriage and motherhood on women in science we needed answers to four questions: Are married women as a group less productive researchers than single women? Among married women, do those who have children publish fewer papers than those who are childless? Is there a drop in women's published research performance after childbirth? Does the number of children a female scientist has affect her research performance?

The publication and career histories of eminent scientists provided the first clues that marriage and children do not generally affect scientific productivity. On the average these eminent married women (and eminent women are just as likely to marry and have children as their rank-and-file counterparts) publish slightly more over their careers—not less—than eminent single women: an average of 3.0 papers per year compared with 2.2. Among the eminent married

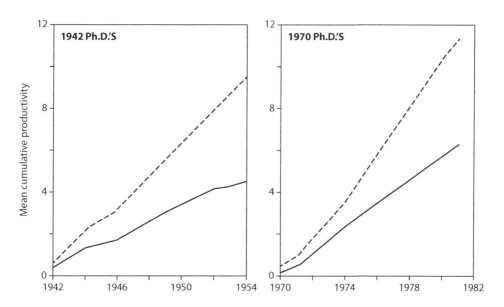

15.2 Male scientists publish more papers (*dotted line*) than women scientists do (*solid line*), and the disparity increases with time.

Data from large PhD samples reported in earlier studies by the authors suggest that the disparity has lessened somewhat since the 1940s.

women, those with children publish 2.9 papers annually and childless women publish 3.3. Moreover, during the three-year periods preceding and following the birth of first children the annual published productivity of these women does not fall but actually rises from 1.5 to 2.7 papers. Finally, the rate of publication of these scientists is unrelated to the number of children they have.

These statistical findings are plainly counterintuitive, and yet they are consistent with earlier cross-sectional studies. How well do they correspond with the subjective reports of women in the interviews? Do women think marrying and having children really is unrelated to the amount of research they publish and, if they do, how is one to account for that belief?

The publication histories of two older eminent women scientists, one with four children and the other with three, illustrate one general pattern [see figure 15.3]. Such scientists in general published less when they were younger and had young children; there is a marked upward trend in the number of papers published after the first decade of their careers. There are also year-to-year fluctuations—peaks and valleys—within the general upward trend. All these women do, to be sure, acknowledge that children take up a great deal of time. They are "a definite time commitment. That means you are doing less with other things"—but not less scientific research. The research continues.

How can it continue? These eminent women emphasize, first, that thinking about science goes on at home as well as at work. It does not end when they close the doors to their laboratories. "When the kids were small . . . I had ideas when I was washing the dishes and nursing the babies. Scientifically speaking, I did my best work during the period when the kids were coming." Second, if they have a scientist husband (and that is the typical situation), they talk about their research during "so-called off-time." Third, professional obligations other than research are far more limited for younger scientists than they are for older ones. "I spent more time doing science then than I do now. . . . The calls on my time [then] were my job and my kids. Now it's so many other things." Fourth, lower rates of publication in the early years are not necessarily attributable to the demands of motherhood but rather are characteristic of the beginning phase of a developing research program. As one physical scientist observed, "In the first years . . . we were building those enormous instruments. Developing the theory and experiment [took up] a lot of time, so there weren't that many papers." Another, commenting on the downturns in her own publication graph, said, "You are very busy in the valley."

According to these eminent scientists, marriage and motherhood did not reduce their published productivity. Should one believe their retrospective accounts? Perhaps their perceptions were not correct. After all, inspection of the graphs of older eminent married women with children shows that they indeed were less productive scientifically when they were young and had young

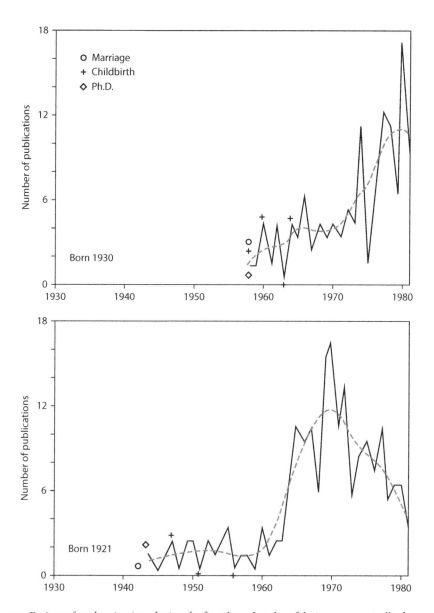

15.3 Eminent female scientists, during the first three decades of their career, typically show a general upward trend in the number of papers published annually. The publication histories of two married women, one with four children (*top*) and the other with three children (*bottom*), conform to this pattern; they show no lasting negative effects of marriage or children. The jagged curve indicates the number of papers published in each year; the smooth curve is calculated to indicate the general trend.

336 FREEDOM AND UNFREEDOM

children. Did having young children in truth affect their rate of publication, at least in the short run?

We thought the publication patterns of two groups of scientists who should not be affected by marriage or parenthood—eminent single women and eminent men whose wives took responsibility for looking after children—might further illuminate these first counterintuitive results. Do the presumably unencumbered scientists publish at a more rapid rate in the early years than women who had young children? The answer appears to be no [*see figure 15.2*]. Single women and married men are just as likely to show a low level of publication in the first decade of their careers. They are also as likely to show oscillations and an overall rising slope of publication with time. The fact that the early publication patterns of these two groups do not differ much from those of married women who have children lends credibility to the married women's observations.

Another question comes up, however. The eminent older women say that marriage and childbearing did not reduce their scientific productivity. If that is so, why do their publication rates increase as they pass beyond the child-care age—particularly in view of the additional distractions and responsibilities they say come with professional maturity and a higher degree of recognition? Part of the answer is that the opportunities for collaborative research increase as one's career progresses. Beginning scientists do most or all of the benchwork themselves; more established ones often assume major administrative roles and oversee the work that goes on in their laboratories. Their publication records reflect the resulting upsurge in collaborative research.

We should mention that the publication record of some eminent women scientists does not exhibit the typical rising slope with time [*see figure 15.5*]. These women too say they think marriage and motherhood had little bearing on their scientific productivity.

Our data seem to indicate, then, that older eminent women with children generally publish as much early in their careers as their unmarried counterparts do. Could we have made a critical error, however, in compiling and interpreting the data? Could it be that women who have children and yet remain scientifically productive are "self-selected," that is, are simply more talented scientists than those who choose to remain childless?

Although strict comparisons of scientific ability cannot be made, we can compare the publication rates of older eminent scientists who did and did not have children, focusing on the years before motherhood. To this end we matched the two groups of women roughly by their birth dates. The publication rate during the three years before women with children had their first child was compared with "equivalent years" in the life span of women without children. We found similar early histories: approximately 1.3 papers annually for women who subsequently had children and 1.6 for those who did not. In other words, older eminent women who eventually had children published inconsequentially fewer papers initially than women who never had children.

More important, might we be mistaken in concentrating on the histories of eminent female scientists instead of on those who might be more likely to experience the debilitating effects of marriage and motherhood on publication rates? The eminent women, after all, are very successful scientists; if marriage and motherhood had taken a toll in their case, such women would presumably not have been able to achieve the recognition they actually did achieve. Do the publication histories of other women scientists, those we have designated as rank-and-file, testify to the negative impact of marriage and childcare?

Rank-and-file male and female scientists do, of course, in general publish less than their eminent colleagues. Within the rank and file in our sample, married women did publish slightly fewer papers than single ones (an average of 1.1 a year compared with 1.7). But married women with children publish no fewer papers than married women without children; both groups average about one paper per year. As in the case of eminent women, their publication rates did not decline after children were born. Rank-and-file women averaged well under one paper per year (.2) in the three-year period prior to the birth of their first child and they averaged just under one paper per year (.8) in the three years following the birth.

Much the same impression is conveyed by these scientists' own testimony: having children did not significantly affect their records of research and publication. As one relatively unproductive behavioral scientist observed, "It didn't occur to me to stop working when I had [a child]. . . . In fact, right after she was born I wrote one paper and started to work on the next one, . . . so if anything . . . I seemed to work better (which is to say more efficiently albeit under greater pressure)."

Her account is consistent with accounts of other women, such as a biochemist who asserted that her publication rate was not affected by family obligations. "It's just fortuitous. . . . The kinetics of me as a parent and me as a researcher don't have a direct relationship. . . . One has not interfered with the other." Contrary to expectation, then, such women are no more likely than older eminent ones to say that marriage and family responsibilities account for their rate of publication, and the statistical data we have in hand bear out what they say.

Would we find a similar pattern in data for younger women? Because marriage and childbearing usually come early in a woman's career, the records for younger scientists should show their effects on published productivity, at least in the short term.

A behavioral scientist who is now a full professor in a high-ranking department, and who had just had a child, suggested that motherhood was unrelated to her pace of publication. "Having a child is draining in many ways, but not in terms of having affected my work, especially when I look at how much I've . . .

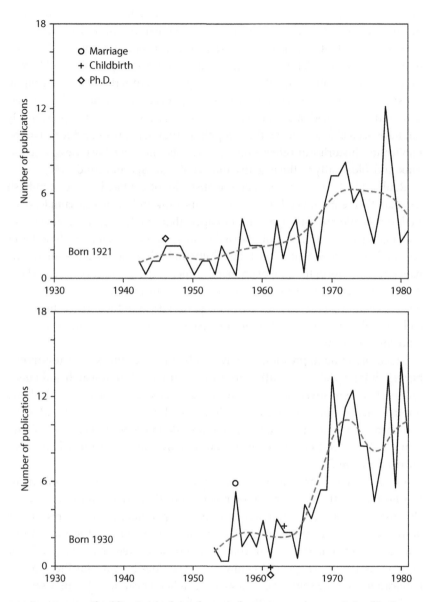

15.4 Lower rates of publication in the early part of a career are characteristic of both married men and single women. The publication profile of a distinguished woman biologist (*top*) who never married shows the same pattern of oscillations and an overall increase as the graphs of women who married and had children. The same pattern can be seen in the profile of an eminent male chemist (*bottom*). He published at a much slower pace when his children were young, although his domestic responsibilities were minimal.

done this year. No, it's really movies, social life and things like that. . . . I feel chronically slow and behind and this year I'm blaming it on the baby, but I realize that has nothing to do with it." In the past it was "too many graduate students, . . . [a] grant reviewing committee, [an] editorship [and now a] baby. So I have always had some kind of baby to blame it on."

Perhaps the limiting case is a woman who has an endowed chair in a major department. She has been married four times and divorced three times, and she has had four children by three different husbands. If marriage and mother-hood should bring a scientist's career to a halt, they should have done so here. Yet her published output has risen throughout her complex history. Ironically, the largest dip in her pattern came in 1979, one of the few years in which she did not get married, have a child, or get divorced [*see figure 15.4*]. Asked about her publication pattern, she replied, "Suddenly you're ready to report on three different projects and therefore [the papers] roll off the presses. . . . The ups and downs [have] nothing to do with the rest of my life."

Even so, pregnancy and its aftermath did interfere temporarily with research in the case of three of the thirty-seven women in our sample who had children. A female biologist told us, "I was one of those women who said I'm so well organized, I'll just drop a child and that will be it—but I didn't realize that hor-mones could do such a job on a person." Asked to explain why her publication rate declined only temporarily and not by much, she said, "I was lucky because by then I had people in my lab. They were working [and] productive. . . . But I found that for a whole year my mind wasn't functioning."

These longitudinal data indicate that marriage and children are not inimical to the published productivity of women in the aggregate. Although few people would question the fact that marriage and motherhood impose formidable responsibilities, apparently many female scientists can manage a career and family obligations simultaneously. How do they do it? And how, when conflicts arise between home and career, do they deal with them?

The answer can be found in part in how they manage their "status set," that is, the array of social positions each of them occupies (such as professor, laboratory director, wife, mother, and citizen). We focused on three inter-connected aspects of status sets: size (the number of positions held simulta-neously), congruence (the extent to which various obligations are consistent rather than in conflict), and the timing of the addition and deletion of status obligations.

At the extreme, several women, convinced that marriage is incompatible with scientific work, opted not to marry, in effect limiting their status sets. Yet three-fourths of the women in our sample did marry—a proportion that appears to be typical of female scientists in general now. For the majority of them the fundamental question about marriage was one of timing. As a young economist said, it "would also be a big career disadvantage for me [to

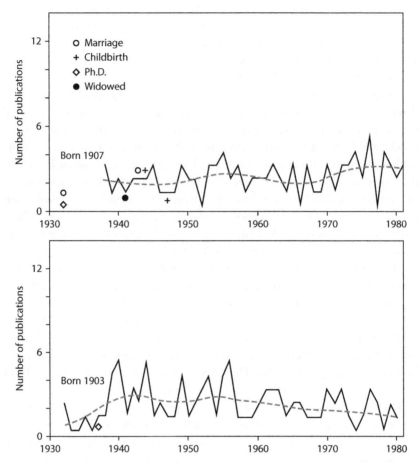

15.5 Some scientists publish at a rather constant rate throughout their career. These are profiles for two eminent women: one who married twice and had two children *(top)* and another who never married *(bottom)*. Although the annual number of publications fluctuates, the mean number in five-year intervals remains much the same for these scientists. This pattern appears as often among married women as among unmarried ones.

get married]. Once I have tenure and am more settled down in a university, it would be a little bit easier."

Two-thirds of these married women had children. Timing their arrival, many women say, helps one to maintain a research career. A renowned physical scientist delayed having her first child for nine years after marrying in order to prove herself as a professional. Many younger women said they were delaying motherhood until they receive tenure. Doubting that she could have a child and maintain the level of performance needed for tenure, a young biochemist noted, "My ideal scenario is to get a tenured position and then have a child or two."

All told, eminent and rank-and-file women had about the same number of children: an average of two, with none exceeding four. Our data show that annual rates of publication are virtually the same for women who have one child and for those with two or more.

There are aspects of marriage and motherhood other than timing that can make for congruent status sets. Close to four-fifths of the married women we interviewed were married to scientists (again, a proportion typical of married female scientists in general). Such "assortative," or selective, mating apparently gives these women (and men too) a variety of benefits, including ready understanding of their professional obligations and way of life. A molecular biologist observed that her husband could hardly be upset when she came home late because "he knew that no matter how well I had planned something, experiments do get delayed. I think that's made it a lot easier." Female scientists married to scientists publish, on the average, 40 percent more than women who are married to men in other occupations. The difference in publication rate may result from self-selection, congruence of values, or the flexibility of academic schedules.

Female scientists can also achieve congruence of status obligations by compartmentalizing their lives, but they report that compartmentalizing is not always feasible. In fact, many find it harder to keep their mind off work when they are at home than it is to keep their mind off children when they are at work. Moreover, every woman with children emphasized that she relied on some form of childcare or household help—necessary arrangements but fragile at best. The illness of a spouse, child, or housekeeper can throw the entire scheme awry.

In view of these difficulties, is it possible that female and male scientists manage to continue their research only by neglecting their spouses and children? Our study was not designed to answer that question. This much we do know: The divorce rate for both women and men is unrelated to published productivity.

Married female scientists with children do pay a price to remain scientifically productive. They report having had to eliminate almost everything but work and family, particularly when their children were young. As an eminent psychologist observed, what goes first is "discretionary time. I think I can only work effectively . . . 50 hours a week. . . . If I didn't have children, I'd probably read more novels . . . or go to more movies."

Loss of discretionary time not only affects leisure pursuits but also sometimes has serious consequences for women's research and careers, even if it has no significant effect on their rate of publication. Women scientists who adhere to rigid family schedules say they have lost the flexibility to stay late in the

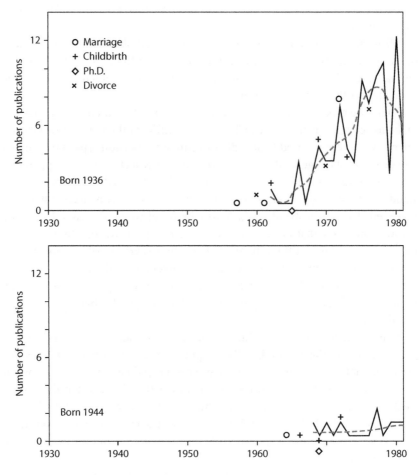

15.6 Published productivity is not related to family obligations, as is indicated by the publication histories of two women with strikingly different career patterns. One (*top*) is an eminent behavioral scientist whose productivity rose steadily in spite of the fact that she has been married four times (most recently in 1982) and divorced three times and has four children. The other (*bottom*) is an associate professor of chemistry and has three children. She attributes her low productivity to constraints other than domestic obligations. Her publication history is typical of many rank-and-file women scientists.

laboratory to work on an interesting problem; they report not feeling part of "the club," not having time for informal discussions with colleagues.

Other investigators have shown that only some 12 percent of female scientists stop work after getting their PhD. Surely some of them do so because of the intense conflicts that arise between science and parenthood. One woman who had left a promising research career for an administrative job said in a supplementary interview, "I was only in the lab [for] the hours that the children were in school. . . . I was working with really bright people who ground out the publications at a rate I couldn't keep up with. . . . It was too frustrating." A small subset of women, then, find that science and motherhood do not mix and alter their careers to give more time to their families.

Our study shows, however, that for most of these women science and motherhood do mix. Female scientists who marry and have families publish as many papers per year, on the average, as single women. Managing the simultaneous demands of research careers, marriage, and motherhood is not easy; it requires organization and an elaborate set of personal adaptations.

The results of this study should not be interpreted as meaning that marriage and children have no effect on the careers of women scientists. They do, but they generally do not take their toll on women's research performance. How then can the persistent disparity in rate of publication between male and female scientists be explained? Why do men publish substantially more papers over the course of their careers than women with comparable backgrounds? The difference is evidently not explained by marriage and motherhood. It remains a puzzle requiring further comparative inquiry into the research careers of all scientists.

CHAPTER 16

A THEORY OF LIMITED DIFFERENCES

Explaining the Productivity Puzzle in Science (1991)

JONATHAN R. COLE AND BURTON SINGER

This essay uses a new general theory of limited differences to propose an explanation for a long established, but poorly understood, pattern of scientific productivity.[1] The theory attempts to explain the empirical fact that male scientists, on average, publish about twice as many scientific papers as their female counterparts, and this disparity increases over the course of their careers.[2] Our aim here is to illustrate how a fine-grained explanatory theory of limited differences can account for this. We have chosen the productivity puzzle in science as a strategic research site, but the general theory of limited differences should apply to many other societal patterns of inequality and social stratification—from racial differences in income and occupational prestige over careers to differences in occupational choice among racial and gender groups.[3]

The first section describes the phenomenon to be explained. The second section presents the elements of the theory of limited differences and indicates how the structure of a kick-reaction system can explain the publication process in science. In the third section we review prior attempts to explain the gender-differentiated productivity patterns in science. The fourth section

---

J. R. Cole was supported by grants from the National Science Foundation (SES-84-11152), the Josiah Macy Jr. Foundation, and the Russell Sage Foundation. B. Singer was supported by grant ROI-HD19226 from the National Institute of Child Health and Human Development. Useful comments on earlier drafts were made by Salome Waelsch, Jeremiah Ostriker, Gerald Holton, Stephen Cole, John Bruer, Harriet Zuckerman, Joel Cohen, Margaret Marini, and participants at a Macy conference on women in science.

formalizes scientific development and the productivity of men and women in terms of the theory of limited differences; and in the fifth section we illustrate that process with detailed career constructions from a micro-simulation implementation of the theory. Section six explains how competition among scientists is the driving force affecting the dynamic features of the theory. We conclude with a discussion of testable features of the limited differences theory and outline a research agenda, focusing on measurement problems that must be resolved if the proposed theory is to be further validated or refuted.

The Skewed Distribution of Scientific Productivity and the Productivity Puzzle

Science is a highly stratified institution. A small proportion of scientists hold the lion's share of powerful and prestigious positions as well as honorific awards, and this inequality in rewards is paralleled by equally skewed rates of scientific productivity, that is, the numeric count of published scientific articles and books.[4] Most scientists publish a very limited number of papers; a small percentage publish a great number. Between 10 and 15 percent of all scientists publish about half of all the science produced.[5] This pattern is as true for female scientists as it is for male scientists. The theory of limited differences pertains to all PhD scientists, but since most do not produce more than three or four papers in a career, we intend to concentrate on the elite group of scientists who are the major producers. Therefore, it is this population of primary producers of science, that is, the upper tier of the stratification system, and the factors that influence their rate and amount of production that is the principal focus of this paper.

A second pattern of scientific productivity remains poorly understood. Male scientists publish more than females. This sex-related pattern has been demonstrated in more than fifty studies (see Cole and Zuckerman 1984). It is as true today as it was in the 1920s, 1930s, and through the 1960s.[6] To cite but one example, a recent summary of scientific productivity patterns for 526 "matched" men and women who received their PhDs in 1969–70 showed that for the twelve years following their degrees the female mean to male mean productivity ratio was .57 for published papers, .42 for median number of papers.[7] For each type of comparison, the gender difference increases over time (Cole and Zuckerman 1984).[8] Using mean numbers of papers, the ratio of publications of women to men in the first seven years of the career (i.e., the tenure-relevant years) was .63; for years 8 through 12 it was .51. The ratios of the medians change from .51 for the earlier years to .30 for the later years. This "fanning out" process of sex differences in publications for the 526 scientists and for four cohorts of matched men and women PhDs dating from 1932 to 1957 is illustrated in figure 16.1.[9] The picture of the 1970 cohort shows that almost all of the fanning action occurs among the top 25 percent of producers. The theory of limited differences attempts to explain these sex differences in scientific productivity.

We will also try to explain similarities that have emerged from studies of male and female scientists over the past two decades.[10] For example, there is virtually no association between sex status and (a) admission to graduate schools of varying prestige or assessed quality; (b) receipt of post-doctoral fellowships; (c) acceptance or rejection of manuscripts submitted for publication; (d) success rates for grant applications; and (e) number of early career honorific awards received. A priori, one might expect that there would be important disparities by sex in early career experiences that would reveal productivity

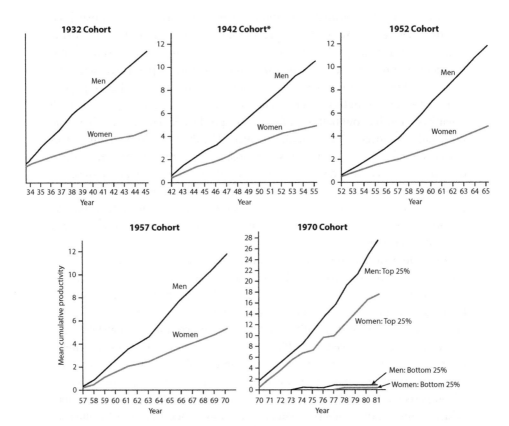

16.1 Publication histories for five cohorts of male and female scientists.

Source: J. R. Cole and H. Zuckerman, "The Productivity Puzzle: Persistence and Change in Patterns of Publication on Men and Women Scientists," in *Advances in Motivation and Achievement*, ed. P. Maehr and M. W. Steinkamp (Greenwich, CT: JAI, 1984), 217–56.

differences within a few years of PhD completion. That this does not occur is what brings forth the notion of a productivity puzzle.

Disparities only emerge gradually. They reveal themselves in *cumulative* numbers of publications, *total* citations to published work, promotion to tenured positions at the most prestigious science departments, and receipt of top honorific recognition, such as Nobel Prizes, Fields Medals, and Lasker Awards. Observations of small fragments in time of the careers of male and female scientists whose initial conditions at the start of graduate school are roughly the same reveal virtually no distinctions in productivity by sex. It is the cumulative, *long-term* nature of the development of productivity and, in turn, reward differentials that represents the challenge for an explanatory theory.

An obvious possible explanation for the gross disparities in productivity exhibited in figure 16.1 is simply sex discrimination. Although this has undoubtedly played some role—particularly in earlier cohorts—recent focused interviews of scientists revealed that most women indicated that they had not personally experienced discrimination.[11] Nevertheless, most had heard of "other cases" of sex discrimination in science. The theory proposed herein views sex discrimination as only one of many causes of the cumulative productivity differential between male and female scientists. There is partial empirical support currently available for a "limited differences" explanation. Full validation and further refinement of this theory is an important challenge for the future.

A Theory of Limited Differences: General Outline

We focus on the dynamics of individuals, each of whom is embedded in a network or networks of relationships constituting the social system of scientific specialties. System dynamics and details of the network structure of science are not part of the present formulation.

Individuals are exposed to a sequence of events of many different types, some or all of which may occur more than once, depending upon the substantive context. There may also be a priori order restrictions in time on some of the events; "acceptance into PhD institution," for example, must occur before "offer of post-doctoral position." Associated with each individual is an outcome variable—e.g., manuscript completions and manuscript publications in a science career setting, annual wages or annual income in studies of black-white earnings differentials, or scores on age-graded mathematics tests as in studies of U.S. vs. Japanese schools.[12] A "kick-reaction" pair corresponds to each event.[13] Examples, in the context of science careers, kicks (which may be positive, neutral, or negative) are acceptance or rejection by a top PhD institution; positive or negative funding decisions on grant applications; positive or negative publication decisions based on manuscript submissions; or marriage

348 FREEDOM AND UNFREEDOM

to a spouse who either hinders or facilitates the scientist's career. Associated with each kick is a positive, neutral, or negative reaction by the person who experiences it. This reaction acts—almost immediately, or with some delay— with other kicks and reactions to influence the outcome variable(s). Kicks and reactions thus have "memories."

The evolution of events, their associated kick-reaction pairs, and changes in levels of outcome variables are characterized as a vector stochastic process where the conditional probabilities of current state occupancy or changes in state are based on an individual's prior history, subject to the following general constraints:

i) with high probability, the kick-reaction sequences with memory, delays, and establishment of competencies for future kicks and reactions determine the outcome variable histories;

ii) with few exceptions, the influence over a short interval of any single kick-reaction pair on an outcome variable will be small (or "limited"). Two individuals with similar, even identical histories up to a given time, but who experience opposite kinds of kick-reaction pairs to any given event—e.g., (negative kick, negative reaction) vs. (positive kick, positive reaction)—will *not* exhibit dramatically different outcome variable dynamics over a short time interval in the immediate future;

iii) all or nearly all kick-reaction pairs influence durations until future events by small amounts. Recurrent events have gently changing duration distributions regulating inter-event intervals. Major changes in waiting-time distributions between events or changes in levels of outcome variables will occur over a long time or between pairs of events separated by "many" intervening events;

iv) there are a *few* special events for which, in distinguished subpopulations (call them A and B), the probability of a negative reaction to a negative kick for a member of group A exceeds the corresponding probability for a member of group B. Correlatively, the probability of an "improvement" in the outcome variable for an individual with a negative reaction to a negative kick on a special event is less than the corresponding probability for an individual with a positive reaction to a negative kick on the same special event, all other features of the past histories being the same;

v) conditional probabilities of specific changes in outcome variables at a given time are insensitive to the order of occurrence of a subset of the possible events that may have occurred in the past. The specific form of the kick-reaction pairs associated with these events will influence current outcome variable changes; however, their order of occurrence will not matter.[14]

When applied to science careers, this general framework implies that cumulative productivity differentials between male and female scientists—identified as group A and B, respectively in condition (iv)—are the result of small—or limited—differences in their reactions to a limited set of kicks. It is the cumulative effect of these small differences that produce, analogous to a "multiplier effect," major productivity differentials between men and women over a career.[15] Small within-sex differences in a few kick or reaction intensities also lead to large career productivity differentials in the population of female scientists alone and among male scientists alone. It is the specific substantive content of the kick-reaction pairs which influences productivity in science and which, more generally, gives specific content to the theory as it is applied to different phenomena.

Consider an example of a career event which is hypothesized to increase slightly the probability that female scientists will be less productive than men. A set of graduate students in physics must select PhD sponsors. All of the faculty in the physics department are men who are willing to sponsor the work of male graduate students; but 10–15 percent of them are unwilling to sponsor female graduate students. This is not an overwhelming disadvantage, but it is a disadvantage nonetheless. It will not affect most female scientists, who would not have wanted to work with that 10–15 percent of reluctant professors in the first place. Most women will not have any experience of a negative kick in the choice of sponsors. However, the entire set of women has a slightly smaller pool of eligible sponsors from which to choose. Some women would have selected these men but for their refusal to sponsor women. If among these sponsors there are some excellent and powerful scientists, then a small proportion of women students will experience a slight negative kick resulting from not studying with them. The reactions to this kick will vary among the women who experience it. Some will fight even harder against such discrimination; but a subset may have their aspirations dampened and motivation reduced slightly by the experience. Such a difference is small but has clear implications for future events in the careers of the affected scientists and to a limited degree is sex dependent.

Take another illustration. A slightly higher proportion of women than men of roughly equal quality have grant applications rejected for research support of the same size. Consequently, they have fewer resources for their research, less travel money for giving talks, papers, or attending conferences—in short, for becoming "visible" to their peers. The sex difference may be limited indeed, but the bias, where it occurs, may decrease slightly these women's productivity potential. In conjunction with prior disadvantages, the somewhat poorer probability of funding reduces the probability of quickly completing manuscripts and publishing papers.

These small negative kicks adversely influence productivity potential in the early phase of the career. They then influence a more significant event, the tenure decision. If tenure is denied, then further negative kicks and reactions can follow, exacerbating still further the difference in productivity potential.

In the case of scientific productivity, the major driving forces behind the manuscript production and publication processes are two interrelated goals:

350 FREEDOM AND UNFREEDOM

priority for scientific discovery and accompanying peer recognition (Merton 1957); and success in the competition for resources to pursue research at a high level. These primary goals are mediated by the action (kick)-reaction pairs experienced by individual scientists. Positive kicks and reactions are associated with increased incentive and hence with increased manuscript completion and publication. Negative kicks and reactions such as grant rejections and lack of peer recognition—for example, very low use of work by peers—serve us a major disincentive, often leading scientists simply to abandon the race for major scientific discoveries.

Prior Explanations for the Productivity Patterns

Of course, there have been a number of theoretical and empirical efforts to explain the skewed productivity among scientists and the puzzling sex differential. In addition, there are other classes of explanations that have not been applied to this problem but that are plainly theoretically relevant. Most efforts within the sociology of science have focused on three classes of explanations: theories of initial conditions, theories of evolving social processes, and structural constraints.[16] We briefly review these explanations and indicate their relationship to the theory of limited differences.

Theories of Initial Conditions: The Sacred Spark

Productive scientists, this orientation holds, are those with "a sacred spark," those with the aptitude to tackle and solve difficult problems. While few would question that variations in ability play a formidable role in distinguishing between creative and uncreative science, this position ignores the role of culture, social structure, and personality in contributing to the productivity process. It is problematic primarily because there is simply no evidence to support the claim that gender differences are related either to the initial physiological or to biological conditions that are claimed to result in the productivity differences we are attempting to explain.[17]

Theories of Initial Conditions: Psychological Traits and Socialization Patterns

Motivational intensity required for high levels of productivity is assumed a priori in this orientation to be dampened in women because of the formidable early cultural and structural barriers that women face and must hurdle before reaching the starting line for a predominantly "masculine occupation" (Berryman 1983; Kahle and Matyas 1985; Marini and Brinton 1984; Bielby 1991). Socialization processes lead young women to be less confident about their scientific ability, less assertive in advancing their ideas and opinions, and less apt to pursue their goals aggressively while simultaneously being more ambivalent than men about their work and family roles. In due course, women and men come to the starting line for scientific careers carrying baggage of substantially different weights. Differences in scientific production follow naturally from these differences in background and current attitudes and traits. But no mechanism is proposed to explain how these "initial conditions" facilitate or impede subsequent events which unfold during a career—in short, how they are linked to actual productivity. The theory of limited differences as specialized herein to scientific careers does deal with this latter process. It is incorporated in the differential probabilities of positive reactions to negative kicks on special events such as NIH grant decisions.

Theories of Evolving Social Processes: Reinforcement and Social Learning

Theories of "reinforcement" and "social learning" are based upon observations of the full process of scientific production over a span of years. Reinforcement theory assumes that high levels of initial productivity receive positive reinforcement (through conference invitations, citations, job offers, awards, etc.), which increases the probability of subsequent high scientific productivity (Cole and Cole 1973). Conversely, poor early performance, going unrecognized, is negatively reinforced and leads to lower future production. Social learning theorists hold that individuals' reactions to events, or stimuli, will be influenced both by their past experience with the stimuli, with cognitive processes that influence the perception and retention of the event, and with anticipated future effects of a particular response (Bandura 1986). As Bandura suggests,

> In the social learning theory view, people are neither driven by inner forces nor buffeted by environmental stimuli. Rather, psychological functioning is explained in terms of a continuous reciprocal interaction of personal and environmental determinants. Within this approach, symbolic, vicarious, and self-regulating processes assume a prominent role. (1977: 12–13)

Reinforcement processes and social learning undoubtedly operate to influence manuscript production, but as they have been formulated, they do not specify the emergent structural and cultural properties in social systems that operate to influence manuscript production, specifically sex differentials and the fanning out process. Furthermore, they emphasize the internalized psychological components of action rather than the dynamic structural bases for actions and reactions.

Theories of Evolving Social Process: Cumulation of Advantage

Processes of "cumulative advantage," first articulated by Robert K. Merton (1968), attempt to explain time-bounded patterns of skewed productivity (and recognition) in terms of increased opportunities for scientific resources, both capital and human, that accrue to those who are productive early on—and especially to the productive located at prestigious scientific institutions. Changing distributions of resources become the basis for creating even greater productivity distance between the "haves" and the "have nots."[18] The cumulative advantage literature has focused on the increasing inequality of scientific publications and citations but has failed to establish the crucial step-by-step linkage between the changing distribution of resources and productivity inequality. The growing inequality has been *assumed* to be the result of cumulative advantages (see Allison, Long, and Kraus 1982).

The theory of limited differences, while incorporating the idea of cumulative advantage, is much more fine-grained and specific about the mechanisms which generate productivity differentials. The kick-reaction sequences are the primitive ingredients in the theory of limited differences; there is no comparable *explicitly formulated* mechanism for the evolution of individual careers in the extant literature on cumulative advantage. In addition, the fact that reactions to particular events are allowed to depend upon prior events, past kick-reaction pairs, and initial conditions is what allows us to integrate socialization processes, psychological theory, and cultural value systems explicitly into the evolutionary dynamics of the limited differences formulation.

Furthermore, in the case of the scientific productivity differential between men and women, the theory of limited differences, through analysis of the sequencing of kick-reaction pairs, suggests a method for assessing whether or not one group rather than another will ultimately monopolize resources, rewards, and the productivity process. The theory allows us to determine the extent to which one group or another "dominates" the distribution of specific positive or negative kicks, and analysis of reaction systems enables us to identify differentiated responses by men and women to similar positive and negative career events.

Structural Constraints

Sex discrimination has been used to explain the sex differences in publications, and undoubtedly it has been a source of structural constraint for female scientists. Differential opportunities based on sex, or on other individual attributes that are unrelated to performance, can be translated into competitive advantages in the acquisition of resources and facilities necessary for high productivity.[19] In fact, discrimination, whether based on sex, religion, national origin, age, or race, enters the theory as a significant substantive element in determining limited differences.

But discrimination is viewed here as one among many sources of structural constraints affecting publication probabilities. For example, women's domestic responsibilities associated with child-bearing and child-raising could account for the lower rate of productivity of female scientists. This hypothesis has been studied in some detail. It turns out that women with children are as scientifically prolific, on average, as those without them (Cole 1979; Cole and Zuckerman 1987). But a small subset of women, distinctly limited in number but greater than the number of men, are adversely affected by building families, and this represents a limited difference that will influence productivity for that small subgroup. Thus, marital and fertility histories are a central feature of the application of limited differences dynamics to science careers.

Since scientific production is almost invariably carried out within social organizations, these organizational contexts can influence the form and substance of productivity. Some environments may be conducive to research; others hostile to it—and this holds for all scientists. But there may also be organizational structures that limit the productivity of women more than men. These range from barriers to participation to subtle exclusions from informal interaction within laboratory settings (Bielby and Baron 1984; Fox 1981a; Long 1978; Pfeffer 1982; Reskin 1978a,b). Only a rough beginning at empirical research aimed at measuring the actual effects of organizational structures on scientific productivity has been carried out (see Long and McGinnis 1981). These studies have been unable as yet to specify adequately how dynamic interactive processes between the "environment" and the individual influence productivity and, more specifically, what features of organizational environments adversely affect women's productivity relative to similarly situated men.

There are, then, a set of existing social and social psychological theories that purport to explain gender differences in scientific productivity and the fanning out process. Each of these theories has useful elements, but individually they take us only a limited way toward explaining the productivity patterns in question.[20] Aspects of each are incorporated into the theory of limited differences.

354 FREEDOM AND UNFREEDOM

We turn now to a description of basic elements in the theory and to its application to solving the productivity puzzle.[21]

Formalization of Science Career Development

We specialize and interpret the general outline of the theory given above in the context of science careers.

A Primary Event List and Delineation of Initial Conditions

A set of career events, hypothesized to be the basis of an explanation of the productivity puzzle in science, is listed in table 16.1. This is by no means an exhaustive list; however, it does contain the items which both empirical studies to date (see, among many others, Astin 1969; Astin and Bayer 1972; Bayer and Astin 1972; Centra 1974; Clemente, 1973; Cohen 1980; Crane 1969; Cole and Cole 1973; Gaston 1973; Hagstrom 1971; Hargens, McCann and Reskin 1978; Helmreich et al. 1980; Spence, Helmreich, and Stapp 1975; Spence and Helmreich 1978, 1979; Zuckerman 1977, 1989; Cole 1979; Allison and Stewart 1974; Long and McGinnis

---

TABLE 16.1 Events Influencing Scientists' Productivity Histories

E_1 = Decision on PhD institution

E_2 = Decision on PhD sponsorship

E_3 = First post-doctoral job or post-doctoral fellowship

E_4 = Publication decision: acceptance or rejection of paper

E_5 = Marriage or cohabitation

E_6 = Birth of child

E_7 = Perceived quality of research: critical reception of publications

E_8 = Funding decision

E_9 = Marital disruption or cessation of cohabitation

E_{10} = Tenure decision

E_{11} = Moderate honorific recognition (e.g., Guggenheim, Sloan fellowships)

E_{12} = Major honorific award (e.g., Lasker Award, NAS membership, Fields Medal, Nobel Prize)

E_{13} = Laboratory directorship

E_{14} = Job offer from outstanding department

E_{15} = Critical reception of paper prior to publication

Note: For each event E_i, there will be nine logical combinations of kick-reaction pairs.

1981; Cole, Rubin, and Cole 1978; Cole, Cole and COSPUP 1981; Reskin, 1977, 1978a, 1978b; Zuckerman and Merton 1971a, 1971b; Zuckerman and Cole 1975; Over 1982; and Over and Moore 1980) and prior theoretical proposals suggest should be the most important events.

There are some a priori order restrictions to be imposed on these events which indicate that some of them must occur in time prior to others. Introducing the relation < to mean "before," we require:

(i) $E_1 < E_2 < E_3 < E_{10}$

(i.e., acceptance into Ph.D. institution must occur *before* acceptance of Ph.D. sponsorship, which, in turn, must occur before a tenure decision.)

(ii) $E_5 < E_9$
(iii) $E_5 < E_6^{(1)} < \ldots \ldots < E_6^{(m)}$ ($E_i^{(j)}$ means j^{th} occurrence of ith event)
(iv) $E_2 < E_{11}^{(1)} < \ldots < E_{11}^{(v)}$
(v) $E_{11}^{(1)} < E_{12}^{(1)} < E_{12}^{(2)} \ldots$
(vi) $E_{15} < E_4 < E_7$

With the exception of these constraints, any ordering of events is possible in principle. Science careers will be assumed to start when an individual applies to a graduate program in some scientific field or specialty. The details of the process of self-selection which leads some individuals to this choice, as opposed to other career options, is an important topic which lies outside of the scope of the present formulation. Thus, gender differences in early socialization and a variety of attitudes and expectations about what is or is not achievable in a scientific career will be assumed to be the primary source of variation across individuals when the career process is initiated. Persons clearly differ in basic ability and motivation even when initially self-selecting to begin a science career; however, there are, as yet, no good measures of early ability and motivation which distinguish men from women at this stage. There are also no effective *early* screening measures which will indicate who among persons in the same discipline, prestige level of graduate school, and with comparable undergraduate record are likely to be the major producers of science in their cohort.

Outcome Processes

Two interrelated outcome variables will be central to the present specification of science careers: manuscript completions and publications. These variables are related in a publication process as delineated in figure 16.2.

When drafts of manuscripts are generated, they are frequently circulated among close peers for comment and criticism—this gives rise to the event, E_{15}.

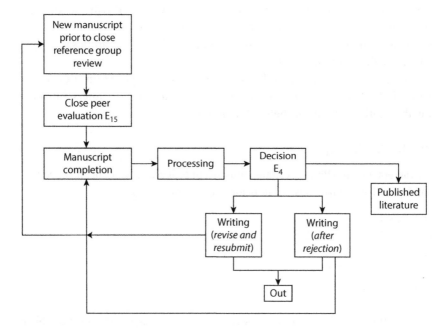

16.2 The process of scientific publication.

Following this event, manuscripts are completed and, with high probability, are submitted to journals for review, thus leading to a publication decision—the event E_4. A favorable decision is followed by a manuscript publication. However, an unfavorable decision puts manuscripts into a feedback loop which may lead to a revised manuscript completion and a subsequent publication or may lead to the scientist simply giving up on the paper. The process exhibited in figure 16.2 has separate compartments for "writing following an outright rejection for publication" and "writing following an editor's request for some revisions" because of the very different attitudes that scientists will have while in each of these regimes. This distinction then leads to different probabilities of manuscript completions and resubmissions, an important feature of a science career.

The probabilities associated with transitions along various paths in this set of events vary dramatically by field, specialty, and even research area. For example, the probability of publication given an initial submission in virology is approximately .9 or roughly .75–.9 in physics, while the same conditional probability is roughly .2–.3 in some subspecialties of sociology and economics.[22] Such differentials are apt to lead to different expectations by scientists in varying fields and hence to different intensities of reactions to rejected manuscripts.[23]

Thus the (submission → publication) link is primarily subject-matter determined while the (peer evaluation → completion), (completion → submission), (revision → submission), and (rejection → revision → submission) paths are much more heavily influenced by individual drive, motivation, and career aspirations.

Kick-Reaction Pairs

With each event from the list in table 16.1, as it occurs in an individual's evolving career, there is associated a kick-reaction pair. Kicks and reactions can each be of three kinds: positive, neutral, and negative.[24] We will denote these alternatives by the system of symbols:

Kicks	**Reactions**
k^+ = positive kick	r^+ = positive reaction
$k\cdot$ = neutral kick	$r\cdot$ = neutral reaction
k^- = negative kick	r^- = negative reaction

Thus, for each event there are nine possible kick-reaction pairs. The probabilities of occurrence of pairs such as (k^+, r^+) and (k^-, r^-) will be substantially larger than the probabilities associated with, for example, (k^+, r^-), (k^-, r^+), and $(k\cdot, r^-)$. A given kick-reaction pair will influence the outcome variables through conditional probability distributions whose structure is described in the next section. In addition, the probability distribution on kick-reaction pairs associated with a given event will depend on an individual's kick-reaction history and outcome history prior to the event in question.

There is a strong social psychological component to the reactions associated with particular kicks in the form of increased or decreased motivation. Motivational differences result not only from psychological sources but also from processes of socialization and social structure. Differences in socialization between men and women can lead to some differences in individual traits such as aggressiveness, competitiveness, self-confidence, degree of confidence, and comfort in an environment in which the individual represents a minority (Duncan and Duncan 1978; Maccoby and Jacklin 1975; Marini 1987). These may produce small sex differences in the distribution of expectations, aspirations, and motivation and somewhat different tolerance and resistance to negative events. In short, differences in reaction systems of male and female scientists may result from socialization processes. These processes are hypothesized to produce empirically identifiable differences in their reaction systems. Most male and female scientists may "look alike" in terms of their reaction systems, if for no other reason than that self- and social selection processes led to these similarities. However, scientists, male or female, with different types of reaction systems will respond in varying ways to the same kick. The actual impact of a kick depends upon the reaction to it.

Reaction systems, of course, also affect behavioral outcomes in an anticipatory way: fear of rejection forestalls action and produces avoidance behavior. Reaction systems affect and are affected by the cognitive styles of scientists. Some scientists will be risk adverse, fearing negative kicks. Others opt for tackling risky problems and take chances in their efforts to be published in the top journals or to be optimally funded for their research.

There are also structural constraints on flexible reactions to kicks that have little to do with psychological traits. Clearly, scientists in different social structural locations have differential opportunities to react positively or negatively to kicks (Fox 1983). Some are in situations where they can "do something about" a negative kick; others are not. Institutional structures not only affect reactions to kicks but influence the sequencing of future kicks and the duration of time between manuscript completions.

Social and cultural customs and mores, such as marital patterns, also can constrain types of reactions to kicks, as is the case when the geographic mobility of a woman is restricted by her spouse's job (Marwell, Rosenberg, and Spilerman 1979). Ceterus paribus, female scientists are more apt than men to be structurally constrained in their choices. For both men and women the sequence of reactions to the same or similar events will change with successive kicks.[25] In particular, the resiliency of positive or neutral reactions will diminish with a succession of negative kicks.

Probability Specifications and Memory Effects

We represent science careers in terms of (a) a sequence of early events—in particular, those which occur up to the first post-PhD position—where there is an accumulation of kicks and reactions which strongly influence subsequent mid-career development; (b) the period from first job beyond the PhD to first major award (this is where the basic publication record is established); and (c) the post-initial major award period, where substantial publication is reinforced, accelerated because of growing resources, or dampened because of increased obligations outside of the research role. Many scientists, even among prolific producers, will never move to phase c, but large proportions of those traveling in this fast lane will receive substantial honorific recognition.[26]

The Early Events Module

We consider the events E_1, E_2, and E_3, which are, of course, constrained by the order relation $E_1 < E_2 < E_3$. In addition, the events E_4, E_5, E_6, E_8, and E_9 may be interdigitated with $E_1 - E_3$ subject also to the order restriction listed earlier. Early event histories will consist of sequences of three or more events from the above list, and E_1, E_2, and E_3 must occur in each sequence. We denote by $|E_1, |E_2, \ldots$ the possible *sequences* made up of at most the above eight distinct events subject to order restrictions and allowing some events, such as birth of a child (E_6), to occur more than once prior to E_3.

For example, we may set

$$|E_1 = \{E_1, E_2, E_3\}$$
$$|E_2 = \{E_5, E_1, E_6^{(1)}, E_2, E_3\}$$
$$|E_3 = \{E_1, E_2, E_5, E_3\}$$

.

.

.

.

Then, within sequence $|E_i$ we denote the kick-reaction pairs by $(K_{j_1}^{(|E_i|)}, R_{j_1}^{(|E_i|)}), (K_{j_2}^{(|E_i|)}, R_{j_2}^{(|E_i|)}), \ldots$ where j_1 is identified with the subscript of the first event in $|E_1, j_2$ is identified with the subscript of the second event in $|E_1, \ldots$ etc. For example, in sequence $|E_2, j_1 = 5, j_2 = 1, j_3 = 6, j_4 = 2,$ and $j_5 = 3$.

With this notation at hand we represent the joint distribution of events and kick-reaction pairs as the product of conditional probabilities

$$\text{Prob}\left(|E_i; \left(K_{j_1}^{(|E_i|)}, R_{j_1}^{(|E_i|)}\right), \ldots, \left(K_{j_\mu}^{(|E_i|)}, R_{j_\mu}^{(|E_i|)}\right)\right)$$
$$= \prod_{l=0}^{\mu-1} \text{Prob}\left((K_{j_{\mu-l}}^{(|E_i|)}, R_{j_{\mu-l}}^{(|E_i|)}) \; \middle| \; \begin{array}{l} \text{kick-reaction pairs} \\ \text{prior to event } j_{\mu-l} \end{array}; \; |E_i\right) \text{Prob}\,(|E_i) \quad (1)$$
$$\text{where } \mu = \text{ number of events in } |E_i$$

Each kick-reaction pair can assume any one of the nine possible values (k^+, r^+), $(k\cdot, r^+)$, (k^-, r^+), $(k^+, r\cdot)$, $(k\cdot, r\cdot)$, $(k^-, r\cdot)$, (k^+, r^-), $(k\cdot, r^-)$ and (k^-, r^-). Numerical specification of $\text{Prob}(|E_i)$ is guided by empirical frequencies in existing surveys of scientists. The general conditional probabilities in equation (1) must be further restricted to conform to particular proposals about the influence of memory on current perceptions of kicks and associated reactions. Two specifications which are relevant for science careers are:

(A) For a given sequence, $|E_i$, past kicks and reactions prior to the lth event influence the probability of the lth kick-reaction pair only through the sums

$$\sum_{m-1}^{l-1} W_{j_m}^{(|E_i)} \mathrm{sgn}\,(K_{j_m}^{(|E_i)})$$

and

$$\sum_{m-1}^{l-1} V_{j_m}^{(|E_i)} \mathrm{sgn}\,(R_{j_m}^{(|E_i)})$$

where

$$\mathrm{sgn}\,(K_{j_m}^{(|E_i)}) = \begin{cases} +1 & \text{if} \quad K_{j_m}^{(|E_i)} = k^+ \\ 0 & \text{if} \quad K_{j_m}^{(|E_i)} = k \cdot \\ -1 & \text{if} \quad K_{j_m}^{(|E_i)} = k^- \end{cases} \tag{2}$$

and

$$\mathrm{sgn}\,(R_{j_m}^{(|E_i)}) = \begin{cases} +1 & \text{if} \quad R_{j_m}^{(|E_i)} = r^+ \\ 0 & \text{if} \quad R_{j_m}^{(|E_i)} = r \cdot \\ -1 & \text{if} \quad R_{j_m}^{(|E_i)} = r^- \end{cases}$$

The weights $\{W_{j_m}^{(|E_i)}\}$ and $\{V_{j_m}^{(|E_i)}\}$ indicate the relative influences of past events on the probability of a current kick-reaction pair. The weight sequences are associated with specific orderings of events—namely, $|E_i$—and need not be invariant under permutations of them. The parameterization (2) implies that the full past influences current probabilities—if all weights are non-zero—and that the longer the sequence the less influence any single kick-reaction pair in the past will have.

(B) If an early event and its kick-reaction pair only influence a specific future event, this effect is captured in the specification

$$\begin{aligned} &\mathrm{Prob}\,((K_l^{(|E_i)}, R_l^{(|E_i)}) \mid \text{kick-reaction history prior to } l^{th} \text{ event}) \\ &= \mathrm{Prob}\,((K_l^{(|E_i)}, R_l^{(|E_i)}) \mid (K_j^{(|E_i)}, R_j^{(|E_i)})) \end{aligned} \tag{3}$$

for a distinguished event—the jth event—occurring at an earlier time, $j < 1$. An example of this is where the jth event is "marriage by a woman scientist in undergraduate school" but to a man whose career imposes rigid geographical immobility for the couple. In terms of the event sequence formalism, this is a history for which $E_5 < E_1$. Now we define the lth event to be $E_3 = $ (offer of first post-doctoral position) and assume that the position is at an outstanding institution—thereby giving rise to a positive kick—but that it is located outside the geographical range which would preserve the husband's job. Thus, a negative reaction is associated with the positive kick—i.e., $(K_l^{(|E_i)}, R_l^{(|E_i)}) = (k^+, r^-)$. The marriage itself, at the time of its occurrence, is associated with

$(K_j^{(|E_i)}, R_j^{(|E_i)}) = (k^+, r^+)$. The dependency restriction (3) implies that all events other than the marriage have no influence on the current kick-reaction pair. The idiosyncratic detail of geographic immobility of a spouse is not formally incorporated in the probability specification; however, non-zero conditional probabilities for the sequence of kick-reaction pairs

$$(K_j^{(|E_i)}, R_j^{(|E_i)}) = (k^+, r^+) \rightarrow (K_l^{(|E_i)}, R_l^{(|E_i)}) = (k^+, r^-) \tag{4}$$

are interpreted to mean that some major obstacle associated with the marriage gave rise to the negative reaction on the lth event.

Mid-Career Dynamics

Development of manuscripts for publication usually begins prior to PhD completion in the sciences and, in some fields, even in undergraduate colleges. We assume that once the manuscript completion process begins, new manuscripts are produced at independent but not identically distributed intervals until the start of a first post-doctoral position. Kicks and reactions in the early events module are not assumed to influence manuscript completions prior to receipt of the PhD degree. However, kicks and reactions in the feedback loop of the publication process—figure 16.2—will slightly increase the manuscript completion rate when positive reactions occur and slightly decrease it when negative reactions occur.

Once the first post-PhD position is attained, then the waiting times between successive manuscript completions have means and variances which are functions of the cumulating kick-reaction experience to the full range of events listed in table 16.1. These means and variances decrease slightly with each positive reaction and increase with negative reactions. Thus, the intermanuscript completion intervals are decomposed into episodes separated by occurrences of events outside the publication module, and, condition on the kick-reaction pairs associated with these events, the conditional mean and variance of the waiting time distribution for manuscript completions is adaptively altered.

The cumulative number of manuscript completions and publications as well as their rate of occurrence in particular time intervals influences the probability of kick-reaction pairs on special events such as grant decisions and major and minor awards. Indeed, in the post-PhD regime, events occur in a continuously evolving stream where the inter-event time intervals and the character of the associated kick-reaction pairs is governed by the prior kick-reaction history and the productivity record. Qualitatively, r^+ reactions and increasing manuscript completions increase the probability of (k^+, r^+) on future events and the probability of r^+ when k^- occurs. Thus, past success generates resilience to future negative kicks, such as grant rejections. Waiting times until occurrence

362 **FREEDOM AND UNFREEDOM**

of both minor and major awards also depend on productivity and citation ranking of the individual scientist among peers in his (or her) subspecialty. For minor awards, the higher the ranking on at least one of these variables, the shorter the expected waiting time until reception of awards and the shorter the expected duration between successive awards.[27]

Major awards in most scientific fields are dominated by the most prolific and visible scientists—perhaps the top 10 percent. Major awards, such as Nobel Prizes, have a ratchet effect. Upon receipt of one, the influence of past history on the durations between manuscript completions is reset to a "post-award" level and no longer depends significantly on the earlier kick-reaction history.

Beyond the First Major Award

There is considerable variation in reactions by scientists to the receipt of major awards. Some continue research at an increased pace; others leave the laboratory altogether; still others have temporary reductions in scientific productivity followed by reestablishment of a prolific rate of publication.[28] Those who shift into administrative roles have dramatically reduced manuscript completion rates; their publication probabilities are assumed to be unrelated to past kick-reaction histories. For those continuing research as their primary activity, the previous reaction history no longer really influences manuscript completion rates. After receiving major awards, the primary influences in manuscript completion rates are assumed, a priori, to be kicks associated with grant rejections. Eminent scientists are not immune to negative peer reviews, lower than expected priority scores, and rejections of grant applications. While they tend to submit more proposals than their less distinguished colleagues, they generally have larger laboratories to sustain. Even the occasional rejection of a large budget proposal can represent a significant negative kick for the productivity of their labs. Indeed, the investment in large blocks of time to "keep the lab going" leads some of these eminent scientists to modify their future research aspirations and overall career goals. Finally, after receiving a major award, some scientists change specialties or fields of inquiry.[29] When this happens, we view the manuscript completion rate for these transfer scientists as roughly equivalent to a new PhD and with the same influence of negative reactions—if they occur—on their productivity.[30]

Limited Differences: Sources of Disparity between Groups

The formulation of the evolutionary dynamics of science careers in the previous section makes no distinction, in principle, between different

A Theory of Limited Differences 363

subpopulations—e.g., male vs. female scientists. Indeed, within each of these groups, the full range of qualitative principles listed as generic for the generation of productivity and kick-reaction histories is operative. Disparities between men and women are introduced as small (or limited) differences in probabilities associated with kick-reaction pairs for a small subset of the events in table 16.1. In particular we assume that

(i) For funding decision—event E_8—

$$\text{Prob}_{[women]} ((k^-, r^+) \text{ on } E_8 | \text{past history})$$
$$> \text{Prob}_{[men]} ((k^-, r^+) \text{ on } E_8 | \text{past history}) \tag{5}$$

Thus, given identical histories, women tend to have negative reactions to grant rejections more often than men.[31] Correlatively

$$\text{Prob}_{[women]} ((k^-, r^+) \text{ on } E_8 | \text{past history})$$
$$< \text{Prob}_{[men]} ((k^-, r^+) \text{ on } E_8 | \text{past history}) \tag{6}$$

(ii) $$\text{Prob}_{[women]} (k^- \text{ on } E_2 | \text{past history in early events module})$$
$$> \text{Prob}_{[men]} (k^- \text{ on } E_2 | \text{past history in early events module}) \tag{7}$$

This inequality is motivated by the fact that a small proportion of the outstanding scientists refuse to accept women as their students, as noted above, thereby limiting—by a small amount—advantageous post-doctoral positions and subsequent support groups recommending them for both minor and major awards.

(iii) $$\text{Prob}_{[women]} ((k^-, r^-) \text{ on } E_6 = \text{birth of a child} | \text{past history})$$
$$> \text{Prob}_{[men]} ((k^-, r^-) \text{ on } E_6 | \text{past history}) \tag{8}$$

(vi) $$\text{Prob}_{[women]} ((k^-, r^-) \text{ on } E_{10} = \text{tenure decision} | \text{past history})$$
$$> \text{Prob}_{[men]} ((k^-, r^-) \text{ on } E_{10} | \text{past history}) \tag{9}$$

(v) $$\text{Prob}_{[women]} ((k^-, r^-) \text{ on } E_{15} = \text{critical reception of paper prior}$$
$$\text{to publication} | \text{past history}) \tag{10}$$
$$> \text{Prob}_{[men]} ((k^-, r^-) \text{ on } E_{15} | \text{past history})$$

Correctively

$$\text{Prob}_{[\text{women}]}\,((k^-,r^+) \text{ on } E_{15}|\text{past history})$$
$$< \text{Prob}_{[\text{men}]}\,((k^-,r^+) \text{ on } E_{15}|\text{past history}) \tag{11}$$

Inequalities (i) and (v) imply that women tend to get more discouraged by negative decisions on grant applications and critical commentary about their work than men.[32] Although this is not universally the case, the consequence of the negative reactions is to slow down the manuscript completion rate by a small amount. Over a period of seven to ten years this can result in major disparities in productivity between otherwise indistinguishable men and women scientists. Thus, the full set of inequalities, (i)–(v), coupled with the conditional probability specifications, constitute the basic formalism of the theory of limited differences, as applied to science careers. Quantitative implementation of this formalism with a range of functional forms for the conditional probabilities based on past histories requires a *family* of microsimulation models, which will be reported on in detail in a later publication. The point of embedding this general evolutionary theory of science careers in a *family* of models is that the manuscript completion and publication histories are relatively insensitive to a diversity of perturbations in kick-reaction histories. This is a form of structural stability of science careers; that is, most small variations in the details of the kick-reaction histories do not lead to qualitatively different career paths.

Examples of Individual Histories and Their Interpretation

In order to clarify the character of microsimulation implementations of the theory of limited differences, we construct three hypothetical examples of science careers: one for a prolific and eminent male scientist; a second for a woman who is less prolific but eminent; and a third for a less productive and noneminent female scientist who might have been more prolific but for her action-reaction experiences. These hypotheticals represent only three of a myriad of possible careers and are intended to clarify the three interrelated sequences of kicks and reactions, completed manuscripts, and publications which develop over a career. They are portrayed schematically as shown in figure 16.3. The cumulative effect of the early kick-reaction pairs heavily influences the early and mid-career manuscript completion rate, based on the events E_1 (acceptance into PhD institution), E_2 (acceptance of PhD sponsorship), E_3 (first post-PhD job), and, if they occur prior to E_3, decision on first manuscript submitted for publication (E_4), E_5 (entry into first marriage or cohabitation), or E_6 (birth of a child).

The case history for the eminent male scientist begins by noting that his personal background and academic record prior to the PhD produced a sense

A Theory of Limited Differences   365

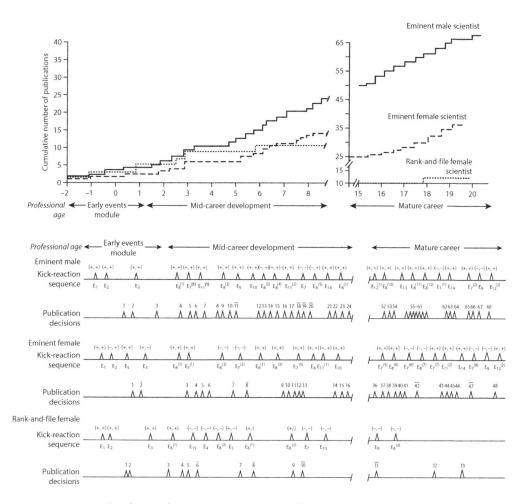

Figure 16.3 Simulation of manuscript completion and kick-reaction histories.

of great self-confidence in his scientific ability. His reaction system was geared toward success; he had high expectations for achievement. And indeed, his first three events are all positive, experiencing (k^+, r^+) pairs in terms of admission to the top PhD department of his choice, acceptance by a first-class sponsor, and receipt of a distinguished job upon completion of his degree. These kick-reaction pairs serve as a major cumulative influence on his rate of manuscript completion. This produces a strong incentive to succeed in competition with other scientists for important discoveries. The cumulated positive reaction intensities in the "early events module" determine the initial manuscript completion rate immediately following the first post-PhD job. This rate can, of course, be modified by later events. These early positive reactions then interact with the positive outcome and reaction of the scientist to having his first grant application ($E_8^{(1)}$) funded. This further increases the probability of high rates of manuscript completion and submission for publication.

The early events experienced by the eminent female scientist are similar to her male counterpart's (see case 2, figure 16.3). Her early educational achievements produced high personal expectations and lofty aspirations. She is confident about her aptitude for science but experiences some cross-pressure because she wants to mix a marriage and family life with a scientific career and has been led to believe that this may be risky for a woman with lofty scientific aspirations. Nonetheless, her confidence abounds and her reaction system leads her to be strongly motivated to succeed in science. She works in the same field as the eminent male scientist that we have just discussed. The female scientist experiences a (k^+, r^+) pair for acceptance into a top PhD institution. Immediately following graduate school, she marries a highly eminent scientist in her own field. In this case marriage represents an initial positive kick. She has increased opportunities to enter the network of leading researchers in her field—far better opportunities than those open to most other men and women of her professional age. These positive kick-reaction pairs represent strong incentives for her to begin publishing, which she does successfully. But she is married to a man whose career is firmly rooted in a very restricted geographical location. His job is not in close proximity to the outstanding academic or government research laboratories that have positions available that would best facilitate her career. The effect of what at the time appears to be a positive kick—i.e., entry into first marriage—is delayed until completion of the PhD, when it interacts with the woman being offered a position at a distinguished institution and her reluctantly declining the offer as a result of the geographic immobility of her husband. This discouragement, which is associated with the pair (k^+, r^-) for E_3, can serve to lower initially the aspirations of someone who might otherwise have been very highly motivated and skilled and with full capability of being one of the very best in her field. This negative reaction is interpreted as setting a lower initial manuscript completion rate than exists for the male scientist in case 1, who did not experience a dramatic negative reaction in the early events phase of his career.

Observe also that the female scientist experiences a negative kick from being turned down by her first-choice PhD sponsor [$E_2(k^-, r^+)$], who refuses to sponsor women, believing that they are poor risks who are apt to drop out of science to get married and raise children. For this particular woman the discrimination engenders a further fight to show the first-choice potential sponsor the error of his ways. For many women, however, this kind of negative kick can lead either to lower aspirations or, subsequently, to lower probability of finding a first job in a top post-doctoral position due to poorer training or lack of national influence of her sponsor. It also leads to a slight advantage for male scientists in the early career stage and later in regulating the probabilities of receipt of both major and minor awards.[33]

Returning to figure 16.3, observe that following E_3 there is very little difference in the manuscript completion and publication histories for male and female scientists except that the completion rate for the male (driven by the stronger initial cumulated positive reactions) dominates that of the female scientist. In addition, negative kicks from a few grant rejections do not lead to negative reactions by the male scientist, whereas they do lead to such reactions, with a slight accompanying reduction in the manuscript completion rate, for the female scientist. Recall that we are treating (k^-, r^-) pairs on grant decisions as proportionately slightly more frequent among women than men, assuming a priori that a slightly higher proportion of women are somewhat more vulnerable than men to intense negative reactions from grant rejections, leading thereby to a slight reduction in their relative productivity.[34]

For the female scientist, the early large negative reaction of not accepting an optimal first job (E_3) and mild discouragement from the grant rejection (E_8) leads to a slower rate of accumulation of publications relative to the male scientist working in the same field. The birth of a child (E_6) before the tenure decision does not produce a negative kick for the woman's productivity, but it does take her away from her department and colleagues and contributes to a delay in the decision on her tenure.[35] The female scientist receives tenure (E_{10}) but somewhat later than her male counterpart—after roughly fifteen completed manuscripts, compared with eleven for the male scientist. Thus, the intensity and set of consequences of her positive reaction to the tenure decision is less than it was for him. In terms of publication histories, this also reduces for some time her ability to attract the best graduate and post-doctoral students and to build the size of her laboratory. By the time the male scientist receives his first major award $(E_{12}^{(1)})$ some fifteen years after the PhD, the ratio of female to male publications is 30:52—or roughly .58. The accumulating publication disparity exhibited in these caricatures is a common but not universal feature of male-female differences among very eminent scientists. Each negative reaction for the woman on a few events contributes a small amount toward slowing down the manuscript completion rate—as it would for a man as well. Over a major portion of a career, say twenty years or so, this gives rise to a substantial disparity in lifetime productivity as measured by publication counts.[36]

For analytic purposes, the third phase of the career development process is assumed to begin after a scientist receives at least one major award. At this point a branching takes place. For the vast majority who continue to do scientific research, many begin to manage larger, quality laboratories. They obtain greater resources, their production of manuscripts increases dramatically, although their own relations to production often change. Thus, becoming the

368 FREEDOM AND UNFREEDOM

director of an excellent lab at a distinguished institution relatively early in a career often leads to substantial increases in output. Note that in our hypothetical examples, the eminent male scientist is made director of such a laboratory almost immediately after being honored with a major award; the woman does not receive this positive kick at all—although she might well have expected it. We assume that men are more apt than women to assume such directorships—and comparatively early on in their career.[37] This is viewed, in the present formulation, as a consequence of the slightly higher probability of outstanding male scientists having PhD or post-doctoral sponsors who are particularly influential and who facilitate the visibility of their intellectual progeny and sponsor them for minor and major awards at early ages. Thus, a virtually undetectable "limited difference" in the early events module can have major consequences in the later career stage.

The second woman (see case 3, figure 16.3) was also labeled as an exceptionally able student, although she retained a sense that her success was more a result of hard work than ability. Working in astrophysics, she attends a distinguished graduate school, holds a major pre-doctoral fellowship, and receives excellent training from her first-choice sponsor (E_2). She publishes papers with her sponsor before receiving the PhD, accepts a first job offer at the most distinguished department in the country (E_3), and finds herself among the brightest and most dedicated young scientists she has ever encountered. Competition is fierce in the fast lane in which she is traveling.

As an assistant professor, she is the sole author of two papers in prestigious journals and receives two years of support from the National Science Foundation (NSF) for her research. Her sense of competence increases, but a paper she thought offered a particularly novel solution to a long-established problem is received poorly by some of her distinguished colleagues who are important reference individuals for her (E_{15}) and is rejected by a major journal (E_4). She begins to question her originality compared to the other bright and seemingly indefatigable assistant professors in her department. This slight loss of self-confidence is exacerbated when a research proposal of hers is rejected ($E_8^{(2)}$). Her motivation to complete several manuscripts and submit manuscripts and grants for peer review is dampened, leading to delays in her submissions.

Her marriage to a nonscientist (E_5) and the subsequent birth of two children ($E_6^{(1,2)}$) does not result in loss of time in the laboratory, but it does mark the end of all her "discretionary" time. However, the termination of her grant and the rejection of a second manuscript reduces her motivation and her career aspirations and leads her to question whether she can maintain the pace of research required by her department. The final blow to her aspirations and motivations is her denial of tenure (E_{10}) by her distinguished department. Not satisfied with being simply run-of-the-mill, she cuts back significantly on the pace of her research—reducing still further the probabilities that she will continue to be a prolific scientist.

This scientist was headed for membership in the productive elite but experienced a set of slight negative kicks, which accumulated over time and interacted with her self-doubt about her ability to compete with the best young minds in her field. The several paper and grant rejections sting her; the denial of tenure is an intense kick. Together these events and the concomitant reactions lead to a longer time between completion and submission of manuscripts. She slowly moves out of the fast lane and never receives a major award. It is important to emphasize that precisely the same event history could be constructed for male scientists. The inequalities detailed above imply that the probability of the sequence of negative kicks and reactions are more apt to be part of the career histories of female than of male scientists.

Competition

> I was competitive beyond the run of younger mathematicians, and I knew equally that this was not a very pretty attitude. However, it was not an attitude which I was free to assume or reject. I was quite aware that I was an out among ins and I would get no shred of recognition that I did not force. (Wiener 1956: 87)

The formulations in this chapter may be viewed in many respects as a theory about the response of the community of scientists to an unstated driving force: competition. The social system of science is driven by competition in at least two forms. There is competition for ideas and hence priority in discovery and competition for the funds which are, in many instances, essential for the pursuit of particular lines of inquiry.[38] In the era of "little science," competition for ideas was the dominant form of this phenomenon. However, with the very large economic costs of resources for doing such things as high-energy physics via particle accelerators, astrophysics via satellite observations, or climatology via deep-sea sediment cores, competition for ideas is now augmented by and thoroughly intertwined with competition for funds.[39] This two-sided competition is particularly acute among the stratum of prolific, highly productive research scientists located at the major scientific institutions—those few who account for such a large proportion of all scientific discoveries.

Biographical reports and sociological studies ranging from large-scale surveys to studies of individual laboratories testify to the centrality of competition in the lives of scientists.[40] There are strong interactions between the action (kick)-reaction system as delineated in this chapter and competition processes, particularly due to the scarcity of resources for pursuing many types of scientific inquiry. Scientists' perceptions of the peer review systems of the National Institute of Health (NIH) and the NSF bring to life the interrelationship between the action (kick)-reaction system, competition, and scarcity.[41] Scientists' productivity is linked directly to keeping the laboratory operating at a high pitch, and it is becoming increasingly difficult and time consuming to obtain the necessary support.

Competition for ideas and priority in discovery exists for some problems, especially in the upper echelons of any scientific discipline, with the competing parties having nearly complete knowledge of what their competitors are doing. The quintessential example of this is, of course, the race for determination of the structure of DNA by the Watson-Crick and Pauling labs.[42] The intense transfer of information via frequent conferences, private laboratory visits, and even telephone conversations between competitors or their close collaborators plays a major role in structuring research agendas and in regulating the duration of time between experiment completion, manuscript submission, and publication.

The increasing awareness of the centrality of competition processes as a driving force in science has, unfortunately, not been accompanied by the extensive empirical research which is required to document the fine-grained relationships between competition and the events presented in table 16.1. Empirical research to date lacks specificity on the focus of competition, its types and intensities; it also lacks detail on the role played by reference groups and social networks in producing and maintaining competition. Furthermore, the fragmentary evidence currently at hand indicates that there is substantial heterogeneity across subspecialties in forms of competition. Because of the sketchy nature of the available evidence about competition in science and because of the complexity of the phenomenon itself, we have not attempted to formalize competition processes or their precise interrelationship to the action (kick)-reaction system already described. We view clarification of the details of competition processes in science as a topic of major importance for future elaboration of the theory of limited differences. In the present discussion, competition remains implicit in the action (kick)-reaction formulation.

Conclusions and Discussion

The theory of limited differences proposes an explanation for social patterns of group differences. It offers a theoretical explanation for dynamic patterns of increased differentiation, increased attenuation, or social stability in the relative standing of the groups over time. At a micro level of analyzing individual histories, it examines dynamic interactions in which small, limited differences in reactions lead to large changes in individual career histories over extended periods of time.

The theory avoids reliance on "causal" models that emphasize the action of one or two variables as determining agents or on a battery of correlates where no interrelationships have been either theoretically described or empirically demonstrated. The theory allows us to specify precisely the interrelationships between concrete events in the histories of individuals, a set of reactions to these experiences, and the short- and longer-term consequences on processes of differentiation in scientific productivity.

To test the theory, a program of research focusing on conceptual and methodological problems is required. Included in the portfolio of problems are determining the relationship between actions and reactions (and the adequate fine-grained measurement of the sequence of events); determining the relative intensities of a variety of actions and reactions that influence outcome variables; understanding the influence of the "anticipation" of events on the selectivity process; examining empirically the nature of time dependencies between events and their consequences; examining how action (kick)-reaction pairs are influenced by organizational and network structures; determining the precise relationship between micro-level outcomes and macro-level system outcomes; and determining precisely the relationships between structural analysis, social psychology, and culture.

The simple aggregate trends, exhibited in figure 16.1, indicating the increasing disparity ("fanning out") between cumulative publication counts of male and female scientists, can be modeled by exceedingly simple mathematical representations. Polynomial growth curves (Foulkes and Davis 1981; Ware and Wu 1981) with gender-specific parameters and Polya urn schemes (Feller 1968) with gender-specific selection probabilities are two of the most obvious possibilities. Unfortunately, simplistic models of this kind do not incorporate the fine-grained behavioral assumptions necessary to provide an *explanation* of the patterns in figure 16.1 in terms of more primitive psychological and sociological constructs. The theory of limited differences is one proposed explanation. It is highly *non*-parsimonious in terms of models which can account for these patterns, but, on the other hand, it is rooted in fundamental behavioral processes. Finally, it suggests that criteria in addition to an ability to reproduce those patterns should form the basis for assessments of whether or not empirical data can support the theory.

A minimum restriction is that we should require data on kick-reaction pairs associated with E_2, E_6, E_8, E_{10}, and E_{15} to support the inequalities (5)–(11) characterizing the sources of disparity between male and female scientists. It is important to emphasize that while gender differences in publication counts are not detectable over short time intervals, gender differences in the frequency of occurrence of kick-reaction pairs—i.e., for events E_2, E_6, E_8, E_{10}, and E_{15}—conditional on full or partial past histories should be ascertainable.[43] In addition, there should be *no* discernable gender differences in the frequency of occurrence of kick-reaction pairs for events other than those indicated above.

Having imposed this set of requirements on empirical evidence needed to support the limited differences theory, it is essential to address some basic—and as yet unanswered—questions about measurement processes. If we try to recover scientists' career histories from longitudinal surveys, then we need a defensible basis for structuring questions that will yield trustworthy responses for the nine types of kick-reaction pairs delineated herein. For events such as those in table 16.1, we must know how far back in time retrospective questions can be posed in a formal survey so that kick-reaction pairs can be defensibly recovered.

More basic than the above questions is the issue of just what one means by an accurate report of a reaction—i.e., whether r^+, $r\cdot$, r^-. There is no independent way to assess, for a given person, the accuracy of a reaction report *and* a statement of its impact on motivation to complete manuscripts. While we can observe the consequences for manuscript completions of the kicks, k^-, k^+, which are often readily ascertained regardless of the elicited reaction, defensible and relatively objective assessments of reactions are probably not achievable by standard survey instruments. The closest that one is likely to get to a "gold standard" for reactions is participant observation studies in which sociologists are members of a laboratory—as at Rockefeller University, a Hughes Institute, Fermi Lab, or the Stanford Linear Accelerator Center—where it is possible to observe (unobtrusively) in detail, and continuously over time, the behavior of scientists following receipt of kicks. The observed behaviors would then lead to characterizations and designations of r^+, $r\cdot$, r^- by the observer; and this would represent the standard for comparison against scientist-elicited responses. There are already some participant observation studies of this kind (see, among others, Latour and Woolgar 1979; Knorr-Cetina 1981; Gilbert and Mulkay 1984) not conducted with an eye toward kick-reaction measurement but certainly allowing for classification of behaviors and assignment of reaction types.[44]

Many more participant observation studies must be carried out if there is to be deep understanding of the psychological and social processes that are the basis for science careers. Furthermore, there is no substitute for this kind of study if there is to be a clear understanding of the competition processes which drive science careers. An unobtrusive observer, witnessing laboratory discussions of what competitors are doing and listening to the debate and rationale for problem choices, is a central feature of the measurement processes which can either support or refute the limited differences theory. An additional strategy for ascertaining reactions would be to have temporally specific interviews with both the scientist whose reactions are being measured and the fellow scientists who are themselves "witnesses" of the reactions. Through intensive questioning of role partners, it may be possible to increase the reliability of the participant observer's judgment of reactions to specific kicks.

Fine-grained nuances must be ascertained if kick-reaction designations are to be trusted; and it may be that standardized questionnaire surveys will be of limited value relative to within-laboratory participant observation studies. In particular, we expect that in the course of developing tests of the limited differences explanation of science careers, it will be necessary to develop further and elaborate on the structuring and analysis of vignettes (Rossi 1979). We envision the vignettes being prepared by the on-site observers in laboratories.

Shifting from measurement issues back to limited differences theory per se, there is another aspect of choice behavior by scientists that is not reflected in the theoretical formulation presented in this chapter but that deserves precise formalization as part of a research agenda for the future. The missing ingredient is the notion of a scientist's anticipation of future kicks of either positive or negative type and the influence of such perceptions on current motivation, hence on his (or her) manuscript completion rate. Evidence from focused interviews (Cole and Zuckerman 1987) suggests that the perceptions about future events which influence productivity are unions of events and their associated kicks rather than the precisely timed single events and kick-reaction pairs which govern the career history constructions described in this chapter. Whether a scientist perceives future positive or negative kicks on an event such as a grant decision, or candidacy for awards such as Guggenheim fellowships, or election to a professional society, depends on *both* past personal kick-reaction history *and* a consideration of what the competition is doing scientifically and receiving in the way of rewards. It will also be governed by his (or her) perceptions of the composition of the judges who will act on the proposal or application (Cole 1987). Assessments of anticipation of future events and the influence of these perceptions on a scientist's productivity will almost certainly require participant observation studies of the kind mentioned above. A full delineation of anticipatory processes and their interaction with the kick-reaction paradigm and limited differences explanation for productivity differentials between men and women is, in our opinion, a major task for future theoretical and empirical development. The present essay should be viewed as a first step in an extensive program aimed at a much deeper understanding of the characteristics of scientific careers.

Notes

1. A formal presentation of the general theory is currently in preparation.
2. The male–female difference exists within every productivity stratum, for example, when career publication totals are divided into quartiles or total publications in the twelve to fifteen years following the PhD or longer periods of time (see figure 13.1).
3. On strategic research sites, see Merton 1987.
4. Throughout this paper "scientific productivity" will refer to the number of scientific articles that are published within specific units of time. Whether we discuss total counts

374 FREEDOM AND UNFREEDOM

or papers per year, we refer to the number of papers published. There is a large literature on problems in measuring scientific productivity and its relationship to both the quality of scientific work and its impact. See, among many others, Cole and Cole (1973), Cole (1979), Gaston (1973, 1978), Allison and Stewart (1974), Long (1978), Long and McGinnis (1981), Reskin (1977, 1978a, 1979), Andrews (1979), Allison (1980), and for recent groups of PhDs, Cole and Zuckerman (1984). Suffice it to say, publication counts are strongly correlated with impact as measured by peer appraisals and by citations, as well as with the prestige of honorific awards.

5. See Price (1963). Subsequent studies demonstrated that this pattern obtains in every scientific discipline studied, and for the United States and all other nations whose scientific output have been examined. While we have charted these patterns well, there have been no successful attempts to explain them.

6. Cole (1979) reports data on the relationship between sex status and publications for matched samples of male and female PhDs who received their degrees in the same year and from the same science department in 1922, 1932, 1942, 1952, and 1957–58. The association is illustrated for each of these distinct cohorts in figure 16.1.

7. Scientists were drawn from six fields: astronomy, biochemistry, chemistry, earth sciences, mathematics, and physics. Pairs of scientists were matched in the sense that they were selected from the same departments in the same years. Analysis was performed both on the aggregates of 263 pairs and on individual pairs. The results were much the same regardless of the type of comparison.

8. While patterns of citations to published science by men and women look much the same as the productivity patterns, evidence suggests that female scientists publish articles that receive just as many citations per article as do their male counterparts. Thus, the differential in citations appears to result from the greater total output of the male scientists.

9. For each of the five pictures shown in figure 16.1, the random samples of scientists were matched by year of PhD, field, and department of PhD. Where possible, men and women were matched by specialty at the time of receiving their degree. Publication data were obtained from abstracts. For a complete description of these samples see J. R. Cole (1979).

10. For descriptions of the diverse set of samples that we have collected data on, see Cole and Cole (1973, 1976, 1985); Cole, Rubin, and Cole (1978); Cole, Cole, and Simon (1981); Cole, Cole, and the Committee on Science and Public Policy (1981); Cole (1975); Cole (1979); Cole, Cole, and Dietrich (1978); Zuckerman and Cole (1975); Zuckerman and Merton (1971a, 1971b); Zuckerman (1970, 1977).

11. This observation was made repeatedly in recent extended interviews with 123 men and women of science conducted by Harriet Zuckerman and J. R. Cole.

12. Plainly, there are many other outcome variables in science, such as appointments to various positions and peer recognition. In this chapter we focus exclusively on the research role and on those who are the major contributors (operationalized by producers of x or more papers within y years of receiving the PhD) to the development of science through publication. This is a small percentage of PhD recipients in science and is nearly invariant across PhD cohorts from 1920 to the present. There are, at present, no obvious early screening criteria to ascertain (at PhD completion) who will fall into this group.

13. The term "kick" is drawn from the physical science literature and is used here in a completely neutral way since it refers to a positive or negative event or perturbation. Kicks are experienced by men and women, and no invidious comparison is intended by the use of this term.

A Theory of Limited Differences 375

14. Condition (*v*) implies that every detail of past history will not influence future events in a unique and idiosyncratic fashion. The same outcome history can arise in a multiplicity of ways.

15. The theory of limited differences calls forth a series of interesting metaphors drawn from the biological and physical sciences. We need only look to Darwin's *On the Origin of Species* (1859) for a clear articulation of the effects of small cumulating differences, which may not even be distinguishable for substantial periods of time, becoming the basis for highly notable variations in species. In his chapter on natural selection, or the survival of the fittest, Darwin has many references to the influence of small differences over time. Consider only one:

> During the modification of the descendants of any one species, and during the incessant struggle of all species to increase in numbers, the more diversified the descendants become, the better will be their chance of success in the battle for life. Thus the small differences distinguishing varieties of the same species, steadily tend to increase, till they equal the greater differences between species of the same genus, or even of distinct genera.

E. O. Wilson's concept of "multiplier effects" also recognizes how small differences can interact with the environment to produce larger effects:

> A small evolutionary change in the behavior pattern of individuals can be amplified into a major social effect by the expanding upward distribution of the effect into multiple facets of social life. . . . Multiplier effects can speed social evolution still more when an individual's behavior is strongly influenced by the particularities of its social experience. (Wilson 1975: 11–13)

Finally, we find parallels to the concept of an action-reaction system in the experimental embryologist C. H. Waddington's concept of "competence" developed almost fifty years ago. In his discussion of competence, Waddington (1940) notes,

> In the first place, it is a state of instability, since it involves a readiness either to react to an organizer and follow a certain developmental path, or not to react and to develop in some other way. . . . One can compare a piece of developing tissue to a ball running down a system of valleys which branches downwards, like a delta. . . . The tissue, like the ball . . . must move downhill, but at some points there are two downhill paths open to it. At such branching points, it may sometimes require a definite external stimulus, such as evocator substance, to push the tissue in to one of the developmental paths; in such a case, competences which occur later along this path will only be developed if the evocator has acted. In other cases, a certain path may be followed merely because an evocator has failed to be present, and then the subsequent competences may appear to develop autonomously. (45)

The theory of limited differences calls forth a series of additional metaphors drawn from other scientific disciplines. The image of the controlled chain reaction is one such metaphor. There an initial action can of course balloon quickly into a large difference if many "kicks" for one group line up positively and all of the kicks for the other group are negative. In such a limiting case, enormous differences between men and women would occur as their careers unfold. But the qualitative data suggest that men *and* women experience both positive *and* negative kicks that we hypothesize affect scientific productivity and career advancements. In fact, the chain reaction which might lead from small initial differences to enormous disparities is modulated by a set of competing and conflicting positive and negative forces. These are metaphorically the barium rods which slow down or even halt the initial chain reaction. Consider another metaphor. We place a big stone on top of a hill; we let it begin to roll down. Depending on tiny impulses it gets, it moves one way or another and will end up in a very different place at the bottom, depending on the smallest chance variations. Each small perturbation changes its trajectory for the future. And the longer the run down the hill, the larger the possibility of spreading apart from the initial path. Still another metaphor is drawn from the kinetic theory of gases or fluids model. Here each scientist is viewed as a molecule. External events move the molecule in one direction or another. The path varies according to the number and types of pushes.

While these concepts help convey the image of actions and reactions as well as the concept of the long-term larger effects of initially small differences, in important respects each fails to capture a critical feature of the theory of limited differences. The fundamental distinction lies of course in "consciousness," that is, the ability of scientists to react to events in nonmechanistic ways that are not akin to reactions by either particles, molecules, or the biological systems described by Waddington or Wilson. Thus, these concepts drawn from biology are at best weak analogies to the distinctly socially structured action-reaction system developed here. Salome Waelsch brought Waddington's work to our attention and helped make us aware of the centrality of the reaction system for the theory of limited differences.

16. Here a truncated description is required. An in-depth critical appraisal of these earlier orientations will be published elsewhere.

17. There have been no agreed upon measures that predict scientific talent, imagination, or aptitude. IQ scores, at best a weak measure of scientific ability, have been found, first, to be uncorrelated with sex, as well as with publication counts and citations. Bayer and Folger (1966) found a correlation of .05 between IQ scores and citations to scientists' work (see also Harmon 1963, 1965); Cole (1979) found a correlation of −.03 between publication counts and IQ for the first thirteen years of the careers of male and femals scientists receiving their PhDs in 1957–58. Although men tend to have higher scores than women on the mathematical portion of the SAT and GRE, the explanation for this difference remains unclear. There is no evidence that after the groups are socially and self-selected into PhD programs that this difference is reflected in subsequent performance.

18. For elaborations upon Merton's work, see, among others, Cole and Cole 1973: 237–47; Allison and Stewart 1974; Allison, Long, and Kraus 1982; Zuckerman 1977; Cole 1979; Mittermeir and Knorr 1979; and Zuckerman 1989.

19. In the first half of this century, the application of nepotism rules, of quotas on having members of certain religious groups, and of an unwillingness to have women working in certain laboratories represented discrimination that significantly influenced the scientific productivity and career histories of women and men who were adversely affected by these discriminatory practices.

20. Although these theoretical concepts have been used to explain sex differences in productivity, they are applied, almost invariably, as a fortiori or post factum interpretations of observed patterns. There has rarely been an attempt to test precisely these theoretical interpretations. Either data do not exist for direct tests of the theory, or the tests have been carried out with imprecise and often questionable "proxies" for key variables.

21. It is important to assert at the outset that the stochastic process formulation with strong memory effects and dependence among multiple variables developed herein represents

A Theory of Limited Differences   377

a mathematical formalization of a very specific theory. Much of the modeling activity in the contemporary sociology literature is *not* of this character and is of an exploratory data analytic type where the goal is to assess which combination(s) of an a priori list of variables are the important influences on a given outcome variable(s). With this particular goal, standard regression modes with interaction terms including dynamic autoregressive models are the most prominent tools. This class of models, however, does not include a formalization of the limited differences theory. Indeed, the standard strategies for incorporating interactions among variables—i.e., as multiplicative terms—in regression models are too crude to represent the more subtle nonlinearities in the limited differences theory.

22. Zuckerman and Merton, *Minerva*, 1971b. The probability of a manuscript being published is largely a function of the effort by the scientist to see the paper through to publication. In fields and specialties with high specific journal rejection rates, the decision to resubmit an article either to the same journal or to a different one almost invariably leads to some form of publication. This may not always be in the journal of first choice, but it will result in publication. Zuckerman and Merton show in their study of *Physical Review* that eminent scientists not only published more than run-of-the-mill scientists but submitted about twice as many manuscripts for publication over a nine year period: 4.1 for those of the highest rank; 3.5 for the intermediaries; and 2.0 for physicists of the third rank. And the most prolific physicists submitted papers to *Physical Review* at a rate twelve times that of the rank-and-file (in Merton 1973: 479). A full empirical refinement of the publication process outlined here is an important agenda item for future research.

23. A more fine-grained classification involving intensities of kicks and reactions is both possible and meaningful; however, the coarse categories—positive, neutral, and negative—will be utilized to simplify the theoretical formulation herein and focus on the principal concepts.

24. The types of actions and reactions and their sequencing will vary from one historical period to another. Figure 16.1 illustrates a historical pattern of sex differences in scientific publications dating back to the 1930s. Although the aggregate level pattern persists, this does not mean, of course, that the cultural, social structural, or psychological factors that produce the patterns have remained constant. On the contrary, historical evidence suggests that the structure of action (kick)-reaction pairs, and in particular, their intensities, have changed in the past fifty years. The historical changes will he captured in the transformation of the kick-reaction pairs and, indeed, the replacement of some by others.

25. The analytic division of science careers into three phases is appropriate because there is evidence in the sociology of science literature that early events are critical in shaping subsequent probabilities for publication and rewards. The analytic phasing discussed here may not be appropriate when considering other dynamic processes of differentiation and subsequent fanning out or attenuation of the differences.

26. In a study of physics awards, only one-third of scientists report any awards; the most prestigious awards were monopolized by a small subset of scientists (see Cole 1969; Cole and Cole 1973; Zuckerman 1977).

27. Zuckerman 1977, chap. 6.

28. The possible examples here are numerous. To cite only one, Donald A. Glaser, the inventor of the bubble chamber, shifted from physics to biology after receiving the Nobel Prize.

29. This assumption needs testing, since undoubtedly the prestige of the Nobel Prize and other major awards cuts across fields and may in fact increase the initial probabilities

378 FREEDOM AND UNFREEDOM

that manuscripts produced by major award winners in the new field will meet with a more positive reception than those produced by recent PhDs in the field.

30. In all inequalities in this section, the past histories for the men and women are identical in the conditioning events. Thus, comparable histories still yield gender differentiated responses.

31. Empirical evidence supporting these limited differences comes from extensive focused interviews with eminent and rank-and-file male and female scientists conducted by Jonathan R. Cole and Harriet Zuckerman. It should be emphasized that many men and women react in precisely the same way to negative kicks. Indeed, the distributions probably show more similarity than difference; the differences in probabilities are not large.

32. This example suggests that discrimination is one of the fundamental sources for negative kicks in the sequence of action-reaction pairs. Plainly, discrimination need not be rampant for it to have a notable cumulative effect on the publication probability of a subset of women or men who suffer from the initial negative kick and from the negative reactions in terms of motivation and future aspirations.

33. There is actually some empirical support for this assumption. In terms of subsequent publication rate, men scientists are affected more positively by peer recognition in the form of citations than are women. Conversely, women are more adversely affected than men by a lack of peer recognition. In other terms, positive reinforcement has less of a positive effect on productivity for women than for men; negative reinforcement has a greater negative effect for women than men. Cf. Cole and Zuckerman (1985).

34. There is a growing literature that indicates that marriage and family obligations affect women's careers but not in terms of published productivity (see Cole 1979; Cole and Zuckerman 1987).

35. The empirical fact is that some eminent women receive more positive and fewer negative kicks than some eminent men. For this subset, the summary of kick-reactions would indicate that these women tend to produce more manuscripts than the men. It is by no means invariably the case that the careers of women show more negative kicks and negative reactions than men. In general, however, this has tended to be the case, and we hypothesize that the cumulation of these micro-level limited differences explain the macro-level disparity in publications.

36. There is, of course, a subset of eminent scientists who accept offers as administrators and public servants after achieving lofty recognition. Their publication rate is usually drastically reduced and, in some instances, virtually eliminated. Thus, it is necessary in our formal specifications of career development to allow for the termination of a publication history following the receipt of one or more major awards.

37. For an extended discussion of priority disputes in science, see Merton 1957.

38. Cole and Cole 1981, 1985; Cole, Cole, and Dietrich 1978; Cole, Cole, and Simon 1981.

39. Merton 1957; Hagstrom 1965; Latour and Woolgar 1979; Knorr-Cetina 1981.

40. The contemporary situation is placed in bold relief by the eminent biochemist Arthur B. Pardee:

> At the heart of current problems [in maintaining high scientific quality and productivity] are the difficulties and uncertainties every scientist faces in obtaining research funds. . . . A scientist perceives now that he has a small probability of getting a grant funded. He cannot afford to be without funds for a year or more if his application fails, because continuity is essential for progress and to retain highly trained, key personnel. So he writes [multiple] proposals in the hope that one of them will be lucky. . . . Fund raising rather than research becomes his major preoccupation. . . . Talents of fine scientists are a rare commodity; wasting

them is a very costly proposition. . . . A less evident but also highly important consequence is the diminution of scientists' self-confidence and morale. Rejections by the funding system of one's best ideas are extremely discouraging. We will see scientists in increasing numbers decide that they are in a rat race; they will slow down or get out. Some of the best unfortunately will be among them. (As quoted in Cole and Cole 1985, 28).

Similar opinions were voiced by many of the productive male and female scientists interviewed over the past four years.

41. Many other less well-known examples could be cited; for example, the competition between the labs of Andrew Schally and Roger Guillemin to identify the structure of TRF(H), thyrotropin releasing factor (hormone). See Latour and Woolgar 1979.

42. Verifying or refuting this claim is a major research task for the future. The supporting evidence to date is primarily from focused interviews which were not designed a priori to assess these points (see Zuckerman and Cole 1987). A content analysis of the interviews, and twenty years of study of the scientific community, often with various forms of quantitative data, suggested inequalities (5)–(11) and provided informal support for the limited differences explanation of figure 16.1.

43. Of course, even participant observation studies have the weakness of imputation, that is, the observer imputing reactions that are translated through his or her own set of constructs.

CHAPTER 17

FREEDOM GAINED AND FREEDOM LOST IN AMERICAN SCIENCE

(2002)

JONATHAN R. COLE AND DARIA FRANKLIN*

Her wings are cut and she is blamed for not knowing how to fly.

Simone de Beauvoir, *The Second Sex*

The American research university has dominated the world of higher learning for the past century. The twentieth century can be called the age of American universities. Our universities have been perhaps the greatest engine of social and economic change that we have witnessed since at least the seventeenth-century English scientific revolution. Among the literally thousands of important discoveries, the discovery of the insulin gene, the nicotine patch, antibiotics, the Richter scale, the origins of computers, the algorithm for Google, the basic RNA science that led to some of the COVID-19 vaccines, the fetal monitor, the pap smear, the cure for childhood leukemia, the methods of scientific agriculture, the electric toothbrush, even the Heimlich maneuver and Gatorade all came from our research universities. We, as a nation, have produced scholars and scientists whose work,

---

We want to thank particularly Harriet A. Zuckerman for her work on women in science and who (with Jonathan R. Cole) conducted many of the extensive interviews quoted in this essay.

* This collaboration resulted from equal contributions by each author.

discoveries, and ideas have transformed the world.[1] Science in the academy is often identified as the institution in our nation that most closely approximates the ideal of universalism.[2] The central question in this essay is: Have we idealized the scientific community? And if so, what are some of the ways that it has harbored and reinforced significant unfreedom within its borders? Freedom matters. Unfreedom matters as well.

The ambiguous word "unfreedom" refers to those mechanisms by which some members of society—the members of the larger society or members of, say the scientific community, impoverish, limit, or eliminate the options of other individuals or groups: *non-coercive threats to freedom*. It is also produced by deep-seated values embedded in cultures—leading individuals and groups to define freedom in ways that entail, as a matter of course, that certain behavior is accepted and becomes normative. Sometimes the opposite of freedom is not coercion but a subtler form of impoverishing of options. This happens often in political economy—such as when freedom is threatened by an authoritarian government—and in the concerns of the environment, where our option of breathing fresh air or drinking clean water, for instance, is increasingly under threat, rendering us less free therefore but where no one is coercing anyone.

Decision-making takes place thousands of times each day within the world of science. Not only are critical choices made about the pursuit of scientific inquiry itself and decisions about what to study and what to or not to publish in scientific papers, what lines should be drawn between fact and fiction, or when to end an experiment. But decisions about the fate of scientists themselves—who should be hired, who should be promoted, who should receive honorific recognition, who should receive major increases in salaries, who should receive grants for their research, and who should receive telescope time at national laboratories—are as continual in science as in any other walks of life. Observers of science often idealize it, taking the institution as a model of academic freedom, personal freedom, and a place with few fetters on lines of scientific inquiry. Science is the place to find "truth" without all of the human biases and presuppositions that might affect its pursuit. It is the American institution which, in theory, rewards people on the basis of their talents and achievements.

Like so much in our history, we tend to hide the ways our nation has fallen short of its ideals of freedom. This is true for science as well. The human side of science is often omitted from its examination. We shall discuss these shortcomings through a lens that looks at the place of women in the scientific community in the first half of the twentieth century, prior to World War II.[3] It has become popular sport to enumerate all the ways that research universities fall short of its ideals. But as we discuss the shortcomings in this one area, we do so while recognizing that the American research university has undergone many changes for the better over the past seventy-five years and remains the benchmark for excellence in higher education throughout the world.

The essay is divided into four parts. First, we limit our discussion here to the first half of the twentieth century, the time period that the historian of science Derek de Solla Price termed the Age of Little Science.[4] Our focus here is on women and science. Second, we try to capture ways "unfreedom" impoverished the options of qualified women and shaped their lives in science. In this part we turn to the stories of lives and careers in science as they were told by prominent female scientists. Third, we discuss the influence of the cultural and political landscape of the time on the implementation of the norm of universalism. Fourth, and finally, we will discuss how prevailing values and norms created an institutional environment that has made it difficult for some groups, particularly women and people of color, to succeed in doing science. They were "outsiders" rather than "insiders"—they could be in science, but they could never be of science. The mechanisms that kept them on the margins, as we shall see, were largely in the hands of academic institutions and less so in the hands of academic scientists.

More specifically, freedom and universalism in science applied only to men until well into the twentieth century. It was an age of white, male, privileged freedom. Women were not part of the measurement of universalism or meritocracy because they were located outside the definition of scientist.[5] Much the same, only more striking, has been true for the apparent contrast between the treatment of African Americans with "white freedom" that simply defined people of color as outside the definition of political liberty in a democratic society—a dilemma for philosophers like Locke and Rousseau and politicians for whom issues of political and social rights and issues of property were salient, while racial unfreedom became a striking anomaly in their work on liberty and freedom.[6] The practices that effectively banned women from regular careers in science after having completed their professional training on an equitable basis did not present a need for institutions and members of the scientific community to reconcile their discriminatory practices with claims of impartiality because of the presumption that the norms of science, of which universalism is salient, govern the relationship only between scientists themselves, and women were *not* scientists. But during this period, as we shall see, the number of anomalies—women with great talent who managed to find ways to practice science professionally—presented an increasing paradox of how the powers within the institution could claim its universal character when it clearly relegated most women aspiring to be scientists to inferior positions within the community.

As historian Eric Foner has said, "No idea is more fundamental to Americans' sense of themselves as individuals and as a nation than freedom" (Foner, 1999). Since freedom, itself, is a complex set of values, norms and beliefs, it defies easy definition. But Americans cling to it as a defining quality that sets us apart from other nations despite the plethora of historical examples of the way freedom in our country has been abridged from well before the American Revolution. But science has been viewed, stereotypically, as one of our institutions that comes closest in practice to our mythical ideal. Let's see how this stereotype, like all, is based partially on fact and also on fiction.

I. The Age and the Problem

In the age of little science, great, defining discoveries could be made with pencil and paper or, as in biology, in a small laboratory "closet," such as the famous Columbia "fly room."[7] There Thomas Hunt Morgan and his students carried out their now renowned genetics experiments on *Drosophila* flies from 1911 to 1928. In 1931, the botanist Barbara McClintock, while working in the cornfields, used brown paper bags attached to her belt to capture and scribble down ideas and observations that would subsequently show that chromosomes exchange parts; in the late 1930s, Max Delbrück, the physicist turned Nobel Prize–winning biologist, started the phage group as a small informal seminar; working as a small group at Rockefeller University, in 1944, Oswald Avery, Colin MacLeod, and Maclyn McCarty set the stage for the discovery of the structure of the DNA molecule; and in the 1940s George Beadle and Edward Tatum, working with bread mold, discovered the way genes regulate biochemical events in the cells— they formulate the theory of "one gene, one enzyme."

During the decades leading up to these discoveries, women were fairly well represented in secondary and post-secondary education. From 1900 to 1950, the proportion of female high school graduates always outpaced that of men.[8] At the level of the baccalaureate, the proportion of women steadily rose: from 19 percent in 1900 to 41 percent in 1940.

The greater number of women receiving education was one of the new forms of social arrangements and social structures that came with urbanization, industrialization, and bureaucratization of the late nineteenth and early twentieth centuries that led America into modernity. A modern American woman could live and work in the city, attend high school and college, pursue a career in nursing or teaching, or join a voluntary association or a social movement.[9] The new woman felt free to choose her lifestyles and fashion, her manners and morals. She could wear shorter hair and expose her knees, become a flapper or a vamp, listen to jazz, choose professional life over a family, and attend public spaces, such as entertainment centers and department stores. The historian Marilyn Ogilvie explains, "Educational reforms fell in lockstep with political ones. By the early twentieth century, women were active in most fields of science, although certain areas such as the biological and the human sciences were better represented than the physical sciences and mathematics." Among PhD recipients, women fared better in 1920 than they did in 1960.[10]

384 FREEDOM AND UNFREEDOM

The changing social landscape for the definitions of American womanhood did not change the way the state viewed women. The government remained conservative in its assumptions about women's nature and maintained that men and women were different and therefore should be treated differently. In 1905, the Supreme Court in *Lochner v. New York* (198 U.S. 45, 1905) addressed the number of hours bakers could work. The result, based on the Court's interpretation of the 14th Amendment, reinforced the right of employers to hire workers for up to ten hours a day and sixty hours a week. Three years later, as part of the Progressive Movement, in the landmark Supreme Court case *Muller v. Oregon* (208 U.S. 412, 1908), the lawyers and social scientists for the plaintiff argued for protective legislation for women. They produced statistical evidence—ultimately included in what we now refer to as the Brandeis brief—illustrating that long hours and strenuous work conditions can harm the procreative function of women. Therefore, protecting women was a matter of state interest:

> That woman's physical structure and the performance of maternal functions place her at a disadvantage in the struggle for subsistence is obvious. This is especially true when the burdens of motherhood are upon her. . . . The physical wellbeing of woman becomes an object of public interest and care in order to preserve the strength and vigor of the race. . . . Still again, history discloses the fact that woman has always been dependent upon man. . . . It is still true that, in the struggle for subsistence, she is not an equal competitor with her brother.

The Court found an exception to Lochner, but it did not overturn it.

Even winning the right to vote in 1920 was no fait accompli. Gaining new political rights and stepping outside the traditional women's roles led to conservative reactions and inspired a new cult of domesticity.[11] Through the 1920s the White House and the courts were not hospitable to reforms and did not provide further legislative support for women. Women's choices were still bound by rules and customs that limited women in most areas of public life. Husbands still controlled family earnings; women were barred from entering the majority of professions; they were solely responsible for illegitimate children, yet they had no options for childcare.[12] Even expanded entry into the workforce for women stalled by 1930, and women were largely finding jobs rather than building professional careers. The Depression deteriorated the scant female employment opportunities as women became increasingly seen as competition to men on the tightening labor market. Married women were particularly under attack, as they were assumed to have a partner with income. The results of the ROPER poll for *Fortune* magazine in 1936 revealed that only 15 percent of the nation approved of married women having full-time jobs outside the home (Erskine, 1971).[13]

The women whose stories we examine in this essay began their lives in science in the late 1920s and 1930s. They were the products of the greatest social changes that allowed them to imagine lives outside of home, but they also witnessed how short that life span of opportunities was. In 1937, commenting on the women's experiences in the early twentieth century, literary scholar Marjorie Nicolson wrote,

> I find myself wondering whether our generation was not the only generation of women which ever found itself. We came late enough to escape the self-consciousness and belligerence of the pioneers, to take education and training for granted. We came early enough to take equally for granted professional positions in which we could make full use of our training. This was our double glory.[14]

Nicolson's views underlined the gains rather than the resistance to women in the academy during this time of expanded educational opportunities. Universities certainly were hardly receptive to women in almost any capacity, much less as full citizens in the scientific community. As the women we will encounter came of age, there was nothing that approximated shared governance at our top research universities. Most places of higher learning—including women's colleges—were ruled by white men: the benefactors who started the university, the boards of regents, the president, and the leading administrators. There was no protective legislation for members of the faculty. During and after the World War I and the first Red Scare, many faculty members whose political or academic views offended presidents, chancellors, and trustees of universities were unilaterally fired. The American Association of University Professors' first Statement of Principles, in 1915, did not weigh heavily in these dismissals. Few universities were deeply committed to the principles of academic freedom and free inquiry.

One would not expect many women to have received tenured positions or honorific recognition within this male bastion of authority and power—even when they clearly merited it. Consider the following facts. Roughly the same number of women won Nobel science prizes between 1994 and 2019 as had won the prize in the previous ninety years. Between 2000 and 2019, more than twice as many women were elected members of the American Philosophical Society as in the previous seventy years. The National Academy honored a majority of its female members after the beginning of the twenty-first century. Since 1936, Maryam Mirzakhani and Maryna Viazovska are the only female recipients of the Fields Medal, awarded every four years to the most outstanding mathematicians under the age of forty. Of the forty-three recipients of the John Bates Clark Medal, for the most outstanding work by an economist under the age of forty, only five are women, and all of them received the award in the last fifteen years.[15] Of the leading research universities in the first half of the twentieth

386 FREEDOM AND UNFREEDOM

century, Columbia's first tenured professor, the anthropologist Ruth Benedict, was elevated to this position in 1937; Harvard awarded tenure to a women for the first time in 1948, when the historian Helen Maud Cam joined the club of 180 arts and sciences male professors; Yale's first female tenured faculty member, Bessie Lee Gambrill, a specialist in elementary education, received tenure in 1952, after three decades of teaching and doing research at the institution; finally, it took Princeton until 1968 to grant its first tenured position to the sociologist Suzanne Keller. Women, in short, remained part of the scientific "outer circle" throughout most of the twentieth century. One can point to a few exceptions, but on closer inspection, even those extraordinary women of science faced many more obstacles than men.

Let's turn now to two stories and some common "patterns" that we garner from the women of this generation.[16] The stories below are meant to be suggestive about the world that female scientists faced and their adaptive behavior in the period before World War II. Despite the limitation on their options, the unfreedom that they faced, these women carved out exceptional, if unusual, careers.[17]

II. Short Stories of Unfreedom

Our stories focus both on the formative years and later career opportunities and choices of two geneticists: Barbara McClintock and Salome Gluecksohn-Waelsch.[18] Both women came of age and contemplated professional careers in the mid-1920s and 1930s and were finishing their formal education as the country and Western world were entering the Great Depression. They also represent the two most common "portraits" of women in science of the first half of the twentieth century: unmarried and childless and married to a scientist, with children.[19]

Barbara McClintock (June 16, 1902–September 2, 1992)

Barbara McClintock might well be the best-known and most celebrated female scientist in twentieth-century America. From the outset of her career as a graduate student at Cornell University, everyone knew they were interacting with a rare intellect, a bit of an odd person, but who had enormous capabilities. She had the right stuff. Her colleagues at Cornell, who also worked as cytologists examining the genetics of maize plants, knew it. "Hell, it was so damn obvious," Marcus Rhoades said of McClintock, "she was something special . . . quick, imaginative, and perceptive. Almost instantly she grasped the significance of a new observation or recently discovered fact" (Federoff & Botstein, 1992). Most others who worked in the field of genetics also knew that they were interacting with someone special—even if many of them could not quite understand what she was doing.

Like most great minds in science, McClintock questioned the received wisdom of the times.[20] There was a general belief among geneticists that the genes on chromosomes were static—they did not move around.[21] McClintock pioneered the cellular analysis of genetic phenomena in corn. Working alone, she conducted the painstaking work on multiple generations of maize plants. She knew her corn plants. She "had a feeling for the organism," as Evelyn Fox Keller put it. In the 1930s, as a young woman, McClintock, who was not more than five feet two or three, wore wire-rimmed glasses, and could be found in her army-like attire meticulously examining corn plants at Cornell University's labs and later, since 1941, in Cold Spring Harbor laboratory. With exceptional discipline and imagination, and free from the pressures of the dogma of her field, she discovered that certain pieces of chromosome 9 could break off and exchange parts among chromosome pairs. In the 1940s, as a mature scientist, she demonstrated that chromosomes have "switches" that can turn genes on and off and allow them to move to different places on the chromosome.[22] This process is known as "transposition" or "jumping genes." In other terms, she was interested in how genes on a chromosome could move during plant reproduction and how that was affected by mutations.

For this work, McClintock was the third woman elected to the National Academy of Science in 1944. So, her peers knew how exceptional her scientific originality was. According to McClintock, her election to the Academy at an early age was not a blessing:

> I didn't want the Academy. . . . I was not elated. I felt caught. I said to myself, "Caught! Caught! For the first time I'm caught." I got into doing interesting things, but I didn't have that same sense of freedom that I'd had. I just knew I couldn't let the women down. You can't walk out. I couldn't walk out the way I'd been walking out.[23]

She saw no hope for women in science at the time. In answering the question whether there was anything that could be done to change the status of women in science, she insisted,

> No, no. I felt that you couldn't change this. The time has to be ripe and it wasn't ripe. [The walls and barriers were too high.] They were just too high. Why get disturbed? You worked within the particular area you could work within but have your own standards within it to do what you wanted to do. This is where the freedom came in. The barriers meant that I had lots of freedom to walk around them and do something else.

But where did this woman, who turned unfreedoms to freedoms, come from? How did she "make" it? McClintock was born in Hartford, Connecticut, in 1902, but her family moved to Brooklyn when Barbara was six. She attended public schools and graduated in 1919 from one of New York's best public schools, Erasmus Hall High School. Unsurprisingly, McClintock was an excellent student. She especially enjoyed the physics class, which gave her a taste for science. She was quite uncertain in what direction she wanted to turn, but she did intend to go to college. Her mother, a piano teacher, was set against it; her father, an army doctor, who had served in Europe during World War I, got home in time to insist that McClintock go to college. Her family life, where the four children were very much part of the decision-making, proved extremely important in producing the trajectory of her education. It was important that the young women had a means for independence, but the form of work tended to be traditionally gendered work. As she put it,

> I had the support of my father, the complete support. My mother—if she could have done it without raising trouble—she'd have stopped me [from going to Cornell]. It was a real fear on her part that I'd be a professor. My mother hid a lot of ambivalent ideas and one of them was a fear of professional women.[24]

We begin to see the full force of the culture on individual histories when McClintock was at Cornell, one of the most progressive American universities at the time. After completing her bachelor's degree, she began working on her PhD in the Department of Agriculture. Here she begins to encounter "unfreedom." McClintock was able to obtain some highly prestigious postdoctoral fellowships to pursue her work in genetics, but, unlike some of her male colleagues, she was not allowed, for example, to enter the faculty dining room or to hold any significant position at Cornell. She majored in botany, although she had already made friends with a set of exceptionally able Cornell geneticists who were approximately her age. When asked about the vision of her own career, McClintock opined,

> I didn't have any projections. I just figured that a person could do a lot of different things. In those days there was no difficulty in getting jobs. I could do something. You could even wash dishes. But you wouldn't starve. There was absolutely no idea of a career. That was out. I was turned down for fellowships, not all, but some. They didn't think you were going to stay, and therefore it wasn't worthwhile giving it to you. And that was alright with me. I mean, I didn't get mad at that. That was part of the time.[25]

During her early postdoctoral years, McClintock had two exceptionally talented colleagues roughly her age: Marcus Rhoades and George Beadle. Since

Beadle was so similar to McClintock in age, specialty, and work, consider this one example of the trajectory of a male scientist with a female—Beadle compared with McClintock.[26] He was at Cornell studying maize plants at the same time as McClintock. He too would win a Nobel Prize for his science.

The careers of Beadle and McClintock are illustrative of certain aspects of unfreedom faced by female scientists in those early years. Born in 1903 into a family of farmers in Wahoo, Nebraska, Beadle did his undergraduate work at the University of Nebraska, and then he went to Cornell for graduate study in genetics. He received his PhD in 1931, four years after McClintock received hers. They became part of the maize group at Cornell. McClintock's friendship and work with Beadle and Rhoades lasted while they were at Cornell and for a long time after the group was separated. They were the core of the young faculty. They would construct their own seminars and invite young students, while denying access to older professors. As McClintock said, "The three of us worked together very, very well." Both Beadle and McClintock received prestigious postdoctoral fellowships. Beadle would spend the four years following his PhD at the California Institute of Technology (Caltech) and then become an assistant professor of genetics at Harvard. No such position was available for McClintock. The University of Missouri, which also had a superb maize group at the time, hired McClintock as an assistant professor that same year. Although her research was progressing well, she was not happy with the prospects at Missouri, even if she were to be granted tenure. If anything, she did not want it as it would come with greater teaching and administrative load, impeding her research. McClintock left Missouri on June 1, 1941. Having spent a few months with her colleagues at Columbia University, in December she received an invitation from Milislav Demerec, the newly appointed acting director of the Department of Genetics at Cold Spring Harbor, to be a salaried guest investigator, and by December 1943, the president of the Carnegie Institution, Vannevar Bush, confirmed McClintock's regular appointment.[27] In 1944, she became the third woman elected to the National Academy of Science. George Beadle became a member the same year.

Meanwhile, by 1946, having spent one year at Harvard and nine years working with Edward Tatum on the major discovery of "one gene, one enzyme" at Stanford University, Beadle was back at Caltech, as chair of its Division of Biology. He remained there until 1961, when he became the president of the University of Chicago. Before his presidency, in 1958, he received the Nobel Prize. He waited for it for thirteen years. McClintock waited for hers for thirty-nine years. She received it in 1983 when she was eighty-one years old.

The unfreedom that McClintock experienced took many forms—partly because of the structure of academic life as well as the cultural beliefs about women in science. She experienced significant resistance to her novel ideas. Whether this was a question of appropriate skepticism about what she had produced, ignorance on the part of other scientists, or a common case of a "premature discovery" is difficult to discern. Here is how McClintock expressed the obstacle course that she had to run and her response to it. She was emphatic:

> Oh yes, yes. The resistance was because you couldn't communicate. No words would do it. . . . After I got my doctor's degree, I stayed at Cornell to do a job there and it was extraordinary that it wasn't understood. From the genetic point of view, it was so obvious to me and to the people who worked in the Morgan group. . . . It was obvious to them.

McClintock felt the difference between her treatment and that toward her male colleagues:

> I had been ostracized at that time by everybody in the department. They called me crazy and just irrational. They didn't understand the very basic thing that was being done. And this fellow, who had been with the Morgan group, knew right away that it was basic. And then he explained it to other people and then I was re-associated with the group again. But it made no difference to me whether or not I was with the group. I was just astounded. That was my first rather strong, rather difficult experience. And I've had several others. One lasted nearly twenty years, or more.[28]

McClintock had turned, in her own mind, a set of actual restrictions on her options on the basis of gender into its opposite—a sense of freedom. She often used the word *freedom* in this context:

> Well, if you have to think—this is in those days, and now too—you have to think of a career, you select it because you're following what other men have followed. Very few women would have a guide in this respect. You would have to make it if you wanted it. Or you just did what you wanted to do, as I did. So that gave you the freedom. Now, another thing. If a man started a profession and went along part-way and then quit, that would be a black mark according to other people. But if a woman did that, no black mark at all, because they were doing it all the time. That is, they were starting something and then getting married and leaving. And nobody blamed them. Nobody called them defective in one way or another in their personality or strength.
>
> The absence of requirements to follow a certain line and continue with it. Also, the ability to move into any direction regardless was free for me. And it wouldn't have been free for some of the men. I don't think they would have enjoyed ridicule. I didn't care if I had it.

We might interpret McClintock's response, as an individual's reducing of cognitive dissonance and "justifying" what has been externally imposed on her as an unanticipated gift. It could be a form of rationalization of existing limited options. But in her case, given that she was a scientist who had ideas outside of the prevailing paradigmatic structure of the field (Duran, 1998), and who tended to work alone, she may well have, in fact, felt free. It is an aspect of unfreedom that we have to consider: freedom resulting from institutional exclusion.[29] As long as women made no claims on institutional resources and status, they could practice science freely, independent from the demands of the field and institutions. They could "afford" working in non-mainstream sub-fields and pursue riskier tasks and questions. At the conclusion of the interview with McClintock, conducted in 1981, she described what doing science meant to her: "On Mondays through Saturdays I talk to the maize plants; on Sundays, the maize plants talk to me." There was little doubt this was true.

McClintock's story illustrates that the structure of academic life did not consider women to be legitimate members of the scientific community, regardless of their demonstrated capabilities. Representatives of the administrative power structure, who were mostly men, could acknowledge the brilliance of a women scientist like McClintock, but that did not mean that they would "waste" time and scarce resources on appointments to regular academic lines. Where the scientific community represented faculty members, such as in professional associations and societies, they were more likely to honor women with awards and recognition, but this too often came later for women than for men. It was always men who set the terms of scientific discourse and debate, and it was male administrators who decided how scientific institutions should work and who could be their members. They defined the criteria of access to the institutions, which was incompatible with being female. Women knew they had no chance of access to resources or autonomy even if they were to be full members of the scientific community. They had no choice but to be "outsiders," seeking for the freedom to practice science beyond the boundaries of the scientific community drawn up by men in charge of institutions.

Salome Gluecksohn-Waelsch, an Émigré Scientist (October 6, 1907–November 7, 2007)

Salome Gluecksohn began her scientific career in Germany, which, before Adolf Hitler came to power in the 1930s, was the clear international leader in developing outstanding research universities. Born five years after Barbara McClintock, she would be a part of the great intellectual migration from Germany to England and the United States after the National Socialists first passed laws that fired Jewish scientists and scholars and denied them access to any positions at universities.[30] The intellectual migration of scientists and scholars, many of whom were Jewish, helped American science grow rapidly. Yet, for many it represented a "double negative," or what is now referred to as the experience of the effects of intersectionality.[31] Waelsch was one who managed to emigrate to the United States, and we include her sketch, in part, because she *was* an immigrant woman and was also Jewish. She also, unlike McClintock, who remained single her entire life, was married to a scientist and had children.

Salome Gluecksohn-Waelsch was one of the world's great geneticists and a leader for some fifty years in the field of developmental genetics. The geneticist Virginia E. Papaioannou, in her biographical essay about Gluecksohn-Waelsch, succinctly described her scientific contributions and the brilliance of her work:

> Her career was remarkable not only for its longevity—she continued experiments well into her 90s—but also for ushering in new ways of approaching developmental biology in mammals. In her studies of the *T*-complex in mice she made us aware of naturally occurring mutations as nature's own experiments that allowed the investigation of the normal role of genes in the events of morphogenesis. . . . Throughout the decades that saw a blossoming of the entire field of genetics, Salome Gluecksohn-Waelsch's work tackling some of the most perplexing problems in mammalian genetics firmly established the mouse as a model organism, not only for studying development, but also for the eventual application of molecular biology techniques to development. Her published work is a beautifully coherent and rigorous opus.[32]

But for the fact that Waelsch had a set of truly exciting and interesting science teachers at university, she might not have turned to science. She became interested in embryology when taking her first course at the University of Freiburg and turned to the most exciting experimental embryologist in Germany—Hans Spemann.

The Weimar years in Germany were considered by many intellectuals and scientists as an unusually creative and open time, with opportunities for Jews and others who were typically outsiders. But life for German women of science was only little better during the Weimar Republic than it was in the United States.[33] Hans Spemann, who would soon receive the Nobel Prize, was, according to Waelsch, "a strong male chauvinist." One way this was manifested was in the problems he targeted for his male and female PhD students. "You were told what to work on for your PhD thesis. He gave me a problem that was very boring. In retrospect it was an insult to have been given such a project for my dissertation. Whereas a young man who was my colleague was given a very exciting problem."[34] After finishing her PhD degree in 1932, Gluecksohn-Waelsch moved to the University of Berlin, where she was a research assistant in cell biology for a year, and then moved back to Freiburg to marry the promising biochemist Rudolph Schoenheimer. Nepotism rules prevented her from holding an appointment at the University of Freiburg, where her husband was the head of physiological chemistry. When Hitler took power in 1933, the couple moved to the United States, where Schoenheimer got a position at Columbia's College of Physicians and Surgeons. Widowed in 1941, she married the Columbia neurochemist Heinrich Waelsch, with whom she had two children, in 1943.

After moving to the United States, for the first three years, Salome had no job. Then, in 1936, she began working in L. C. Dunn's Columbia laboratory—a non-faculty position she held for seventeen years. Dunn, an outstanding developmental geneticist who received virtually every honor conceivable except the Nobel Prize, was not an exploiter of female scientists. Quite the opposite. But he was working within a social system that limited female scientists' options. There were hiring and promotion rules, such as anti-nepotism constraints, which prevented married couples from working in the same department or even university.

Dunn's path was clear. He followed it, and made notable discoveries, while earning prestigious positions and honors. Waelsch, his equal or better as a scientist, experienced various forms of unfreedom. If only as research associates, Dunn hired women and allowed them to work on important research questions and to publish their work under their own names. But he stuck by university rules, and categorically told Waelsch and other women that there was no chance for their career advancement. Few within the power structure thought this was unjust. When asked whether she expected to have job opportunities, Waelsch reflected on the gendered nature of academic structure:

No. I didn't have a job in Germany either, because universities had a rule against nepotism. The most I could do was to continue working with Spemann on a joint project, in which my husband would also have been involved. But it was without pay. So, I didn't have a job.

Asked whether it was acceptable to work and be a wife but not get paid for it, Waelsch had this to say:

Here, at Columbia, I worked. But don't forget what time that was. It was the Depression and there was no money. So, I was glad that there was someone who let me work in his lab. And when I came to Dunn I worked in his lab without salary for a year and then he asked me how would I like to have a salary. But there was no NIH, there was no Cancer Society, there were no outside agencies and money didn't exist. That I could perhaps have a better job wasn't anything that occupied my mind. I was perfectly happy working there in Schermerhorn

394 FREEDOM AND UNFREEDOM

Hall at Columbia, for $1,500 a year, and I really enjoyed the work thoroughly. It was only later that Heini Waelsch called my attention to the possibility that I could have a faculty appointment. It is not quite true that this hadn't occurred to me, because I did go to Dunn and tell him that there had been young men here during my time who had climbed the academic ladder and I asked him why I still remained a research associate. He told me there was no chance for advancement for me.

Waelsch reflected on where this position originated: "I would think that it was the University's. In Dunn's case I don't think he theoretically approved of that position, but he certainly wasn't willing to fight it."[35]

In 1955, when the Albert Einstein School of Medicine was founded, Waelsch was appointed there as an associate professor. Three years later she was promoted to full professor, and by 1963 she was the founder and chair of the Department of Genetics. She would go on to receive many forms of honorific recognition, including election in 1979, at seventy-one, to the National Academy of Sciences. In fact, in 1988, when she was eighty-one, Salome Waelsch received NIH funding for five more years. At the time of this award, she had one of the longest continuous grants from the NIH.[36] Yet, even she, the woman who made a mark in the field of genetics, had to deal with various forms of unfreedom in order to practice science. None of these unfreedoms affected men of equal or lesser talent.

Common Conditions and Shared Experiences

Barbara McClintock and Salome Gluecksohn-Waelsch's professional histories represent the common experiences of female scientists during the first decades of the twentieth century.[37] The majority of highly successful female scientists saw other women in their departments while pursuing their doctorate degrees. The biochemist Sarah Ratner, for example, remarked in the early 1980s, "[Now] everybody is talking about women and the biological professions as if this great interest was born yesterday. . . . When I was a graduate student which—I began in the early 30's—50 years ago, there was a good representation of women." Similarly, describing her graduate school experience at the University of California, the physicist Melba Phillips recalled, "There were at least seven women taking their PhDs. There were women, and nobody told you not to do [physics because you were a woman]."

Seeing women around, however, did not translate into aspirations for professional scientific careers. The majority of women claimed indifference toward professional recognition, and they, at best, often aspired to be teachers. Grace Hopper, a highly prolific early computer scientist who in 1944 worked on a team that developed the Mark I computer, said, "I fully expected to spend the rest of my life [teaching at Vassar]." She moved to private industry and became part of another team that developed UNIVAC I, and then she managed the development of the first COBOL compiler, which proved that programming language based on English was possible. Those women, who dared to imagine themselves as professional scientists, saw the true state of the field and took a rational approach. Commenting on her education, Berta Scharrer, who would go on to cofound the discipline of neuroendocrinology, explained her thinking: "I also took the precaution of getting those courses that would have made it possible for me to teach in the gymnasium later on, if no academic possibilities existed. Because the situation for an academic career for a woman was very, very poor at that time. I knew that."

These scientists almost never framed their experiences as careers hindered by gender discrimination. Instead, they attributed their uneven careers, dismal or no pay, and slow promotion to factors outside of the institutions. Melba Phillips, for example, understood her professional experiences in the following way: "I don't think that being a woman hindered [my career] particularly. I think that all sorts of circumstances and my own talents and so forth and so on hindered it." The freshwater ecologist, botanist, and limnologist Ruth Patrick, to whom recognition did not come until 1960, opined, "I don't think it was just being a woman, at all. I think it was both [being a woman and working in a marginal subfield]." Ratner, while pursuing her PhD at Columbia University, stated that she experienced no prejudice on the basis of gender. However, she added, "There probably was more anti-Semitism." Grace Hopper spent no time thinking about matters of gender: "At UNIVAC, I was always so busy to get more budget, more people, more computer time, more this and that, fighting for something, trying to put a new idea across. If prejudice was there, I never saw it. And it may have been because I never had time to look." When asked why there was a paucity of women in engineering and computing, Hopper responded, "Not in business data processing. In the academic world, yes. In the business world, no. It's much more a woman's world in the business world." Then she quipped, "Oh, the academics have all their usual prejudices."

Even when they acknowledged inequalities of treatment and opportunities, these women took unfreedom for granted. It was not a time for activism in the academic world, and they did not see the limitation of their options as reasons to resist the existing order of things. Ratner framed her position this way:

And none of the women talked about women's rights or the "professional woman." [Science] was what we wanted to do and so we did it. You know [of the disparate treatment and discrimination], but somehow or other you don't take the bull by the horns. I mean, you wouldn't march up to someone and protest and say, "why am I not included?" or "I need more money," something like that. Sometimes I have felt that I was also being exploited. I was very glad to have those positions. They were very good labs, I was learning new things, but I wasn't paid what a man would have been paid. It used to be said—and it still is—that women, particularly married women, that it isn't so important whether they have a salary equal to that of men because they don't need it. And sometimes women themselves would accept the situation.

Scharrer, who was a Jewish immigrant, expressed a similar view of being in the "outer circle":

> All I wanted to do was work and not make waves. It was not necessarily liked that I worked without pay, you see. I was given a room in Cleveland, I was given some space in Denver, and that was already considered a little bit generous on the part of the university.

The limitations at the career level, as these women saw them, had little to do with their relations with colleagues and peers. On the contrary, the women working in those days said they had positive experiences at the individual level. "It is my friendship with people in the scientific world that have been the greatest help to me. And they have been terrific. I mean, all of my colleagues have been great," Patrick asserted. When Hopper was asked what the environment was like at Yale, where there were only two women, she said, "Everybody but one professor was [receptive of females]." Phillips had a similar experience: "I didn't suffer any discrimination as far as the relationship with the fellows was concerned. I don't think it existed."

Some of the women had experiences similar to McClintock's and observed that institutional exclusions paradoxically resulted in increased freedom to work on their own terms and take greater risks. Waelsch described her time as a young scientist as a period of her greatest freedom: "I was free, I mean, I had absolutely no restraints on my activities and so I did enjoy it very much. But I was sort of like a private scholar, you know, who doesn't have to worry about the future." Scharrer commented, "But the advantage was that my time wasn't taken up by anything. You know, I was on no committees. . . . I felt completely free. I could do what I liked. And I did! And I wasn't pressured." Reflecting on the conditions for the generations of women that came after her, she, however, stressed that greater inclusion didn't lead to greater freedom for women. She explained, "But a young academic—especially a woman at that time—would have had to follow [career] principles, and get something safe and run-of-the-mill, rather than experimenting [in an] unpromising field."

These patterns, arising from the cultural values in the years of Little Science, require further research. An open question is: Were the violations of the norm of universalism a function primarily of faculty values and beliefs or of the institutional values at the university level—or both?

III. The Paradox of Individual Particularism and Institutional Universalism and Its Correlative, The Paradox of Individual Universalism and Institutional Particularism

In the past four decades a good deal has been written about the system of social stratification in science. In the 1970s, social scientists explored whether or not the system approximated a universalistic ideal. Some empirical work concluded that it did, finding that the disparities in recognition resulted from a process of cumulative advantage and disadvantage (e.g. Cole & Cole, 1973). Robert K. Merton and Harriet Zuckerman developed the concept of *The Matthew Effect*, taken from the Gospel of Matthew: "For to everyone who has will more be given, and he will have abundance; but from him who has not, even what he has will be taken away" (Matthew 25:29, RSV) (Merton, 1968). They applied this concept to scientific rewards. If those who gained by the process were more meritorious to begin with, then the norm of universalism would not necessarily be abridged. Other social scientists argued that it was futile to imagine that science can follow its own normative prescriptions, and by 1970 some studies argued that the actual behavior of scientists didn't reflect the Mertonian normative prescriptions (Barnes & Dolby, 1970). Another group of social scientists stressed that social environments, cultures, ideologies, politics, economic incentives, and sensibilities of scientists shape the way science is organized, managed, and conducted, and scientists are more likely to rely on flexible "ideological-rhetorical" repertoires and draw from different combinations to account for particular situations and actions (Mulkay, 1974). In the late 1980s, it was suggested that universalism could exist at the aggregate level and particularism at the level of individuals (J. R. Cole, 1989).

The earlier positions overlooked the combined influence of institutional sorting and of social networks on particular decisions. Although decisions may be based on criteria that are independent of extraneous factors, particularism could enter the system because of scarcity of resources. After the initial sort, or a series of sorts, is made on a population of applicants, aspirants, job seekers, grant applicants, and so on, there still remains a pool of qualified individuals that exceeds by a significant amount the number of positions, honorific awards or federal grants available for distribution. Although these individuals need not be alike along a host of important dimensions, they all have legitimate claims to be part of the pool of qualified candidates. Given the imbalance between the number of those qualified and the resources and positions available the open question becomes: on what bases do individuals make decisions as scientists and as representatives of institutions?

Could individual scientists, guided by the normative principles of universalism, be in favor of opening up positions for women in the academy, and, at the time, abide by the institutional power of trustees and presidents of the universities? These power relations, like all cultural norms, have been, of course, subject to evolution and change. But in dealing with particular forms of constrained options facing women in the first half of the past century, we should consider the distinction between *individual universalism* and *institutional particularism*.

Today, presidents rule by authority, largely by the will of the governed. But, from 1890 to World War II, there was far less shared governance of universities. In the past, presidents, trustees, and donors ruled by sheer power. Consider just two examples of attitudes and policies about women in the student body and on the faculty held by prominent presidents. Charles William Eliot, one of Harvard's great presidents, led the institution from 1869 to 1909. So distinguished a man as Eliot elicited a two-volume biography by Henry James III (Houghton).[38] As great as many of Eliot's policies were for the history of higher education, he let his views about women at Harvard be known immediately in his inaugural address:

> The attitude of the University in the prevailing discussions touching the education and fit employments of women demands brief explanation. America is the natural arena for these debates; for here the female sex has a better past and a better present than elsewhere. Americans, as a rule, hate disabilities of all sorts, whether religious, political, or social. Equality between the sexes, without privilege or oppression on either side, is the happy custom of American homes. While this great discussion is going on, it is the duty of the University to maintain a cautious and expectant policy. The Corporation will not receive women as students into the College proper, nor into any school whose discipline requires residence near the school. The difficulties involved in a common residence of hundreds of young men and women of immature character and marriageable age are very grave. The necessary police regulations are exceedingly burdensome. The Corporation are not influenced to this decision, however, by any crude notions about the innate capacities of women. The world knows next to nothing about the natural mental capacities of the female sex. Only after generations of civil freedom and social equality will it be possible to obtain the data necessary for an adequate discussion of woman's natural tendencies, tastes, and capabilities. Again, the Corporation do not find it necessary to entertain a confident opinion upon the fitness or unfitness of women for professional pursuits. It is not the business of the University to decide this mooted point.[39]

Thirty years later his views on women continued to be—in the words of Bryn Mawr College's legendary president, M. Carey Thomas—"the old argument of the rural deans." At the inauguration of Caroline Hazard in 1899, among other things, Eliot had this to say about women:[40]

> It would be a wonder, indeed, if the intellectual capacities of women were not at least as unlike those of men as unlike those of men as their bodily capacities are. It remains to discover and apply the best means and methods for making a college for women a perfect school of manners. Everybody knows that the influence of women depends more than that of men on bearing, carriage, address, delicate sympathy, and innocent reserve; that manners, in short, are much more important to the influence of women than they have been to the influence of men in the actual world—not that they ought to have been, but than they have been. A man relies on his strength; a woman on more delicate qualities.

One more example, among the many that we could produce, will suffice to illustrate the prevailing views of people of significant power and influence in the academy at the turn of the twentieth century—and to suggest what the cultural norms were for the women thinking of careers in science or the academy during the first decades of the twentieth century. Here are just some of the words uttered in 1906 by William De Witt Hyde, President of Bowdoin College:

> What, then, is the womanly as opposed to the manly ideal of scholarship? What is the beneficent ordering of intellectual consumption? . . . A very important, a very arduous, if less conspicuous and less popular part remains, and probably will remain almost exclusively in the hands of men—the part of productive scholarship. By productive scholarship is meant the power to grasp as a whole some great department of human knowledge; keep abreast of every advance that is made in it; from time to time add some contribution to it, and above all so vitally to incorporate it, so vigorously to react upon it, and so systematically to organize it, that the scholar puts his individual stamp upon it, and compels whoever would master the subject to reckon with the individual form which he has given to it. Productive scholarship of this high sort is very rare, whether in men or women. Its price is very high—in time and strength, in withdrawal from other interests and concentration upon one's chosen subject, in sacrifice of domestic and social claims.[41]

These were the men who held power at our great universities and colleges. They, of course, had their vocal critics in the likes of Thorstein Veblen and the muckraker Upton Sinclair. These two, among others, saw the powerful presidents as part of the larger American plutocracy of the Carnegies, Rockefellers, Morgans, and their crowd, as having too much power in setting university governance structures. Sinclair, for example, referred to Nicholas Murray Butler, the president of Columbia University from 1902 to 1945, as "Nicholas Miraculous"—"a man with a first-class brain, a driving executive worker, capable of anything he puts his mind to, but utterly overpowered by the presence of great wealth." These men of power set the norms and values of the major universities. They carried the cultural imperatives with them and stamped them on the universities. The faculty, if and when they thought otherwise about women in the academy, had little power to affect change—and they rarely tried. The consensus was that women could not be the legitimate members of the scientific community and their access to institutional structures had to remain marginal.

The typology presented in figure 17.1 allows us to locate a large portion of women and men of science during the Age of Little Science. The majority of men would be located in the "George Beadle Quadrant." Male scientists were more likely to experience universalism at both institutional and individual levels. They could reasonably expect that their claims and results would be judged on the basis of objective scientific criteria and that the rewards and resources they would receive from peers and institutions would match their scientific contributions. That, of course, would not have been true for male scientists who were of color or were Jewish, for example, but those who constituted the majority of white and privileged scientists had all the reasons to expect impartiality of judgments.

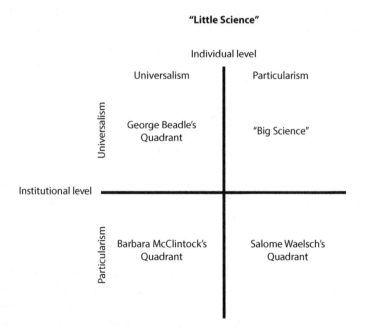

17.1 Typology of scientists during the Age of Little Science.

The "Barbara McClintock Quadrant" represents individuals—whether female or belonging to other minority groups—who experienced formidable institutional limitations on their options even when they were recognized for their ability and achievements by their peers. While this remains an open question, if further research reinforces this observation, it would suggest that the institutional barriers to freedom in science in the first decades of the twentieth century may have been greater than the limitation on options imposed by individuals. One indicator of this is that many women of this generation were sooner to become high-ranking members of professional scientific associations in their respective fields and receive professional awards than acquire permanent positions at their institutions. In scientific societies and associations, it was peers who made the decisions, not administrative authorities, who could exclude scientists from the institutional structures not on the basis of talents and achievements but on the basis of gender. In other words, at the institutional level, women were subjects to particularism, but their professional networks and individual-level connections were spaces where they were more likely to find recognition and legitimacy as scientists and where they experienced Mertonian universalism.

The "Salome Waelsch Quadrant" represents the vast majority of female scientists whose options were constrained by both other scientists within the community and by powerful leaders of universities. Waelsch fits here for the better part of her long career, which she spent in Germany and then at Columbia, without much recognition or opportunities for advancement. Married women and women with children were more likely to experience particularism at both levels. Waelsch received recognition for her scientific work only when she moved to Albert Einstein University, at the age of forty-eight.

The typology, although presented in static form, should be seen as part of a dynamic process of changing culture, values, and norms. As the institutions and society changed in terms of their values—and as science moved with the larger society—there was greater movement toward institutional universalism, but individual particularism has persisted. The "Big Science Quadrant" represents a space within the community that would increasingly be open to both female and male scientists during the second half of the twentieth century and where the struggles over access and treatment of minorities by and within the scientific community persist.

IV. The Role of Culture

The issue of gender in science began to gain serious scholarly attention in the second half of the 1970s. Since then, there have been many conceptual and empirical efforts to explain differences in career outcomes of male and female scientists. The majority of the research fell into four broad theoretical approaches. In earlier times, the prevailing research dealt with theories of initial conditions, such as early socialization patterns, psychological traits, and reinforcement (see discussions in J. Cole & Singer, 1992; Hanson, 1996; Seymour, 1995; Tidball, 1989; Vetter, 1996). Other researchers focused on institutional and structural constraints, such as inequalities in education, resources, and mobility, exclusion from professional networks, and motherhood penalties (Epstein, 1992; Fox, 1989; Reskin, 1977; *Women*, 1987). Within the same theoretical frame, we have encountered theories of cumulative advantage and disadvantage used to explain various and growing differences in productivity and career outcomes.[42] Related to the idea of cumulative disadvantages is a theory of limited differences—where in a stochastic process over time, the careers of men and women diverge because women experience more, even if small, negative feedback than men (J. Cole & Singer, 1992). Other scholars looked for causal mechanisms within discriminatory practices at the individual and institutional levels, such as stereotyping, biases, and harassment (Berg & Ferber, 1983; Connell, 1987; Hollenshead et al., 1996; MacKinnon, 1979). Finally, some of the strongest progress in studies of scientific representation has been made by the feminist scholars in the 1980s and 1990s.[43]

Here, we go beyond these more commonly used concepts and theories to elaborate on and underscore the influence of culture in comparing and understanding scientific careers—specifically the role that non-coercive forms of unfreedom play in shaping the lives of would-be or actual female scientists. As we have seen in our few truncated case histories, culture contributes strongly to the shaping of paths taken and not taken by young women considering their own life choices.[44] Unfreedom begins when girls are told that "girls don't do science" or when they can't imagine a scientist is a woman (Chambers, 1983).[45] Unfreedom exists when female scientists are offered inferior laboratory space and support; when they are denied appointments and promotions, especially to tenure; when they experience harassment in the workplace; when male colleagues don't take women's ideas as seriously as men's; when women have a more difficult time getting their ideas published; when it is harder for them to receive research grants or fellowships; when university resources are not provided to balance family life and careers; and when women are excluded from professional networks, be they formal or informal.

The influence of culture may include both exogenous and endogenous forces: those influences in the broader society and those within the smaller scientific community.[46] These forces produce an ethos for a nation. The ethos of America as a special or exceptional nation, or a city on the hill as it is taught in our schools and enters our textbooks, is part of our rhetoric and produces in our citizens many false beliefs. It is, nonetheless, a powerful part of our culture and once embedded it is difficult to dislodge. At its worst, these cultural beliefs are exported as hegemonic imperatives. It is to this difficulty to change that we briefly turn.

Who Owns the Null?

We are born into a culture that provides us with a set of historical beliefs and values that can be easily translated into hypotheses, since few are proven facts or theories. In quasi-scientific experiments we set up the opposite of the belief and then try to overturn or falsify it. This is referred to as the "null hypothesis." The null hypothesis states the hypothesis of zero difference or equality. This can be contrasted with the research hypothesis that involves a statement of expected differences. For example, suppose that we believe that the scientific community treats women unfairly in hiring, promotions, and salaries, as well as in terms of various forms of honorific recognition. To test this hypothesis, we proceed to set up the null hypothesis: that women are treated fairly in science. To overturn the null—to demonstrate that freedom and equality exists for women in science—we must collect sufficient evidence to demonstrate that the null hypothesis of equality must be rejected, that it is not true. The tests used for demonstrating that the null has been overturned is of no concern here. What is important is that finding sufficient empirical evidence to reject, or overturn, the null hypothesis, given the limited power of social science theory and our inability to identify adequate methods and techniques that can be applied to complex social situations, is very difficult to do. So, in a society or community that holds a set of beliefs that produces unfreedom, whoever controls the definition of the null hypothesis is apt to carry the day.

Whether or not the challenges to core beliefs and principles rest in the larger political and social arena or within institutions like science, they involve conflicting views of basic principles and what is required to prove that one or another organizing principle in our culture is right or wrong.[47] Today, as in the past, there is a political drama unfolding over who owns the null—who gets to define the "truth" that must be falsified.[48] The attacks come from both the Right and the Left. The questions may range from how much free expression should be tolerated on campuses to what can and cannot be published in scientific papers.[49] This, of course, has been a recurring historical theme at American universities and more generally. If this conflict exists today, the unfreedom of women in science during the days of little science was believed to be based on "facts." Those facts became part of the cultural norms and the belief system. Behavior to a significant degree conformed to these normative expectations. If you believe that it is a fact that women are less able to do science than men and because of those beliefs you establish a structure that reinforces your belief, you have the perfect conditions for a self-fulfilling prophecy—and the perpetuation of the prior belief system. Once the self-fulfilling prophecy is established, we have the sources of great inequalities that are justified as a result of inferior competency rather than prejudice or discrimination.

In the years between 1900 and 1950, women were assumed to be unable to do advanced science; it was a male profession. If they did fight against the null, they would find cultural and structural obstacles in every aspect of scientific life because they turned out to, in fact, work on "smaller" problems, were less scientifically productive, some got married and had children that hypothetically would end any meaningful career options, and so on. Cultural beliefs about African Americans, about Jews, about immigrants, and about women in other fields dominated the prevailing modes of thinking in the United States. They still do.

To reiterate: It is far simpler to create a fiction and perpetuate it than to overturn it with the facts. Perhaps this is part of what Daniel Kahneman and Amos Tversky referred to as "the status quo bias," but it is undoubtedly more than that. Consider a few examples of cultural beliefs that are taken as facts by many of our citizens that have been extremely difficult to overturn. Think of the small study that claimed that vaccinations were correlated with an increased probability of autism. Since its publication in 1998, it was retracted and debunked many times, its author's medical license was revoked, yet it is still cited in the debates against childhood vaccinations. Or take the belief that dietary cholesterol can significantly increase the risk of cardiovascular disease. Even though the 2015–2020 Dietary Guidelines for Americans removed the recommendation to restrict dietary cholesterol, the link persists in the minds of the general public. Similarly, the beliefs about women and science run deep—and are difficult to overturn. Much has been done in the past fifty years to overturn some of the beliefs about women's aptitude and ability to do science, but the residues from the nineteenth and twentieth centuries persist.[50] And American textbooks, for example, are still more likely to feature Madonna at length than tell students about Marie Curie and her discoveries.[51]

* * *

Each of the scientists pictured in this essay was able to "win" while holding a "losing" hand. They had talent and tenacity. Even in their triumphs they had to accept and acknowledge the way that science worked against the interests and desires of women in the larger society and within their own community. They were often left out of the "invisible colleges" where a lot of the important conversations about work went on and where the exchange of ideas took place. They tried to turn evidence of unfreedom into its opposite. But most women could not overcome the cultural divides that they faced—from early childhood to the building of careers. In the age of little science, women worked in backwater ponds chasing diatoms rather than in fly rooms breeding *Drosophila*. If they were very talented, they worked alone among the corn plants, not in seats of authority in growing research universities. If they chose to work, and were incredibly able, there might be a spot for them at a woman's college or as a research associate working in the basement of some makeshift laboratory. Often, they could not get proper jobs because of nepotism rules or because their superiors and the administrative authorities expected them to leave and take up the life of wife and mother. A few, some of whom we have met here, one way or another remained in science. They were lucky to secure funding, find themselves in supportive environments, have the option of working for meager or no pay, or have interests in non-mainstream subdisciplines. They often were treated as equals by many colleagues, but administrators, those who held power, did not see them as valued members of the scientific community. The ones that we've met and whose lives we have examined so briefly were rarities. Most, even if they got to the starting line, opted out. The professional world they lived in defined them as "outsiders." The culture denied them the belief that they could make it among the men of science. Most were right. Their options were very limited. They suffered from institutionalized unfreedom and particularism.

Notes

1. For a detailed discussion of these university-based discoveries, see part II in J. R. Cole, *The Great American University* (New York, PublicAffairs, 2012).
2. Universalism, one of the four basic norms of science, defined originally by Robert K. Merton, refers to the belief that the scientific community judges truth claims and the substance of discoveries on the content of the claims and the demonstration of their validity, not on an any particular characteristic of the scientist: race, gender, nationality, age, etc.
3. When the senior author began to study women in science during the mid-1970s, few other social scientists focused on their treatment within the scientific community. His interest in women in science, in particular, derived from a more theoretical interest in how the reward system of science fell short of its normative ideal of universalism. Today, much has changed, including the publication of biographical and autobiographical accounts of female scientists, of great success stories that led women to produce scientific breakthroughs, and accounts of the way women continue to draw the short end of the stick within the community. It is no longer a lonely enterprise. We are indebted to the historians of science Margaret W. Rossiter and Londa Schiebinger, physicist Evelyn Fox Keller, philosopher Sandra G. Harding, sociologist Mary Frank Fox, and many other female scholars for their groundbreaking work on women in science.
4. In his book *Little Science, Big Science* (1963) the historian of science Derek de Solla Price showed that 80 to 90 percent of all scientists who ever lived were alive in the 1960s. He defined this historic evolution of science as a movement from "Little Science" to "Big Science."

5. We have shied away from using the term "meritocracy." Its meaning implies the tension between the desire to quantify merit in absolute terms and its instrumental character, and the discussions of merit-based assessments and quantifiable output are beyond the scope of this essay.

The term "scientist" was first introduced in print by the English historian of science William Whewell, in 1834, when upon reviewing Mary Somerville's treatise *On the Connexion of the Physical Sciences* he proposed to replace the then common "men of science" to "scientist." At that time the term caused scholarly disagreements. The term did not become commonly accepted until 1882. For a more detailed discussion, see David Wootton, *The Invention of Science. A New History of the Scientific Revolution* (New York: HarperCollins, 2015), 27–28.

6. For a full discussion of racism and political and social freedoms see Tyler E. Stovall, *White Freedom: The Racial History of an Idea* (Princeton, NJ: Princeton University Press, 2021).

7. The fly room, described as a dingy, closet-like space at Columbia University, still exists today. Several Nobel Prize winners did their work in these cramped quarters; the place was filled with excitement and some resentment during its operation.

8. The proportion of people between the ages of five and twenty enrolled in schools between 1900 and 1920 rose from 50.5 to 64.3 percent. In 1920, 60.5 percent of all high school graduates were women (n = 311.000). In 1950, the proportion of women was smaller, at 52.4 percent, but women were still better represented then men (n = 2.889.000). Thomas D. Snyder, *120 years of American Education: A Statistical Portrait*. U.S. Department of Education, Office of Educational Research and Improvement, National Center for Education Statistics, 1993.

9. These new freedoms were accessible to mostly white and affluent women. Women of color and working-class women remained bound to low-paying menial jobs and housework.

10. In 1920, 15 percent of all doctorate degrees were conferred to women. In 1960, the proportion of women with PhDs was 10 percent (U.S. Department of Education, National Center for Education Statistics, Earned Degrees Conferred, 1869–70 through 1964–65).

11. For a detailed discussion see Megan McDonald Way, *Family Economics and Public Policy, 1800s–Present: How Laws, Incentives, and Social Programs Drive Family Decision-Making and the US Economy* (New York: Springer, 2018), 134.

12. For a detailed discussion see Sara Evans, *Born for Liberty* (New York: Simon and Schuster, 1997), chap. 8.

13. It has to be noted that despite the stagnant economy and limited opportunities married women still managed to make some progress in gaining employment through the Depression. See Winifred D. Wandersee Bolin. "The Economics of Middle-Income Family Life: Working Women During the Great Depression," *Journal of American History* (1978): 60–74.

14. Marjorie Nicolson, *A University Between Two Centuries* (Ann Arbor, MI: 1937), 414.

15. Susan C. Athey in 2007; Esther Duflo in 2010; Amy Finkelstein in 2012; Emi Nakamura in 2019; and Melissa Dell in 2020.

16. All women who appear in this essay were interviewed by Harriet Zuckerman and the senior author of this essay between 1978 and 1982. For a variety of reasons, these interviews, which remain in the form of recordings and their transcripts—as well as a good deal of biographical information—have been used in only a few essays. We will refer to those interviews and other sources in this section of the essay. Support for that interview project came from the National Science Foundation. The sample was by no means

a random or stratified random sample. Eminent women are oversampled. The male scientists interviewed were roughly matched to the women in terms of discipline and professional age, but they too were not randomly selected within groups. All of the scientists quoted in this essay are no longer alive. We have chosen to use their names, which we would not have done if they were alive today.

17. Because of the cultural norms and the values within the United States at the time, most women self-selected themselves out of science as an occupation. But that self-selection was based on perceptions of the proper role for women in society and the absence of values of inclusion in academic science.

18. We intentionally chose the stories of women from the same field. The differences in their backgrounds, character, circumstances, and choices highlight the broad commonalities of experiences of women during the first few decades of the twentieth century.

19. In the sample of women born between 1902 and 1921 ($n = 21$), 28.5 percent were never married, 71.5 percent were married to another scientist, and 47.6 percent had children.

20. The philosopher Jane Duran points out that Evelyn Fox Keller, the feminist scholar who studied the work of Barbara McClintock extensively, stressed that the first reaction to the chromosomal alteration that McClintock encountered was that it "couldn't" be happening. Duran maintains that it was precisely McClintock's ability to break away from her previous training in positivist tradition that allowed her not to treat what she was observing as an error in the data. See Jane Duran, *Philosophies of Science/Feminist Theories* (London: Routledge, 2018), 46.

21. The system that a scientist chooses to work on can be of essential importance to the results achieved. Those cytologists who chose the maize plant took on a very difficult system, primarily because you only had one or two "crops" each year to examine in experiments and analysis. Those early geneticists who chose *Drosophila* flies as a system, like the Morgan lab at Columbia University did, could breed the fruit flies many more times in a short period of time to examine hypotheses or theories of the gene. That gave them a definite advantage in making rapid scientific advances—if they had people of rare curiosity and intellect working on the system.

22. She published her findings over several years but gave her presentation at the Cold Spring Harbor Symposium in 1951 "The Origin and Behavior of Mutable Loci in Maize," *PNAS* 36, no. 6 (1950): 344–55.

23. Unless stated otherwise, all the quotes in this essay are from Jonathan Cole and Harriet Zuckerman's personal interview with McClintock conducted in 1981, before she received a Nobel Prize.

24. A supporting father combined with a somewhat reluctant mother was not an uncommon motif in the interviews that Cole and Zuckerman produced. It seems that in a disproportionate number of cases, fathers without sons treated their female children (or at least one of them) as if they were a son.

25. Even though the Rockefeller Foundation funded her later research, McClintock's application for a grant from the foundation right after her PhD contains a note: "The applicant is a woman and may leave the field anytime."

26. In this essay, we don't intend to compare the careers of the women and men. The sketches of career paths for men are meant to highlight the sharp differences in the opportunity structures and power within the scientific community.

27. Vannevar Bush was an exceptional "truffle-dog" for smelling out talent. He and others saw McClintock's brilliance, understood the resistance to her discoveries in parts of the scientific community when she discovered transposons, and offered her a full-time staff position at Cold Spring Harbor. He was willing to support this unique scientist

who began her quest to find a "controlling mechanism" that produced the mutable loci of genes in maize plants. Of course, this was the same man who would lead much of the nation's World War II scientific efforts, become a confidant of Franklin D. Roosevelt, and author the transformative science policy document *Science, the Endless Frontier.*

The detailed historical record of McClintock's transition from Missouri to Cold Spring Harbor between 1940 and 1943 can be found in Lee B. Kass, "Missouri Compromise: Tenure or Freedom? New Evidence Clarifies Why Barbara McClintock Left Academe." *Maize Genetics Cooperation Newsletter* 79 (2005): 52.

28. The long-lasting resistance to her idea she is referring to is "the one that was the controller gene action."

29. This is not true for only female scientists. Many male scientists have been known to experience this as well. For example, Stanley B. Prusiner, who made the extraordinarily original discovery of prions, was not afforded many options to work on his ideas while operating within the paradigmatic, dogmatic system surrounding concept of causes of diseases. After decades working in relative isolation at UCSF, with few resources from the major supporters of biological research, he discovered that proteins, as well as bacteria and viruses, cause certain form of neurological brain diseases. The community was highly skeptical of his results, and while its criticism may have led Prusiner to refine his research, he worked with very limited options.

30. There is a good deal of literature on the intellectual migration in the 1930s and 1940s from Germany and parts of Europe controlled by the National Socialist Party. Perhaps the single best source for autobiographical essays about the migration by those who went through it can be found in Donald Fleming and Bernard Bailyn's *Intellectual Migration* (Cambridge, MA: Belknap Press of Harvard University, 1969). The migration of over half of Germany's top physicists and other renowned scientists was only part of the story. Among the scientists who left were Leo Szilard, Max Delbrück, Enrico Fermi, and, of course, Albert Einstein. Among the other creative minds that left Germany, Austria, Poland, Russia, Italy, and France were the writers and playwrights Thomas Mann and Bertolt Brecht, the artists Wassily Kandinsky and Max Ernst, the architects Ludwig Mies van der Rohe and Walter Gropius; the art historians Erwin Panofsky and Walter Friedlaender; and the composers Paul Hindemith, Béla Bartók, Arnold Schoenberg, and Igor Stravinsky. Social scientists, such as Paul F. Lazarsfeld and Theodor Adorno, also saw the writing on the wall. Some did not and they suffered under Hitler's rule. There were also the children of the émigrés, who went on to distinguished careers. Gerhard Sonnert and Gerald Holton, *What Happened to the Children Who Fled Nazi Persecution* (New York: Palgrave Macmillan, 2006) and Laura Fermi, *Illustrious Immigrants: The Intellectual Migration from Europe 1930–41* (Chicago: University of Chicago Press, 1968) have written about the general phenomena, while Hannah Gray has recounted her experiences in an exceptionally fine autobiography: *An Academic Life* (Princeton, NJ: Princeton University Press, 2018).

31. The term was coined by the Columbia legal scholar Kimberlé Crenshaw in 1989.

32. Virginia E. Papaioannou, "Salome Gluecksohn-Waelsch. 6 October 1907—7 November 2007," *Biographical Memoirs of Fellows of the Royal Society* 19 (2019): 153–71.

33. Great female scientists, like the Austrian-Swedish Lisa Meitner, a physicist of the first rank, worked with the enormously talented chemist and Nobelist Otto Hahn. But she found herself working in a basement laboratory. Whatever options existed during the Weimar Republic narrowed or closed with the triumph of the Third Reich.

34. Quoted from Harriet Zuckerman, Jonathan R. Cole, and John T. Bruer, *The Outer Circle: Women in the Scientific Community* (New York: Norton, 1991), 72.

35. Zuckerman, Cole, and Bruer, *The Outer Circle*, 81–83.

36. Zuckerman, Cole, and Bruer, *The Outer Circle*, 71.

37. Harriet Zuckerman and Jonathan Cole interviewed seven women and two men who were born before 1910. It is difficult to conclude that the patterns that we have identified in this section would hold for a larger and more varied population of female scientists of this period. Here, we are discussing patterns based on a small number of high achieving female scientists. The scientists interviewed included Barbara McClintock (1902), genetics; Sara Ratner (1903), biochemistry; Berta Scharrer (1906), endocrinology (immigrant, Jewish); Grace Hopper (1906), computer science; Salome Glueck-sohn-Waelsch (1907), genetics (immigrant); and Melba Phillips (1907), physics. The two men were Chandra Subrahmanyan (1910), astrophysics (immigrant); and Tjalling Koopmans (1910), mathematics and economics (immigrant). For a more comprehensive study of female scientists see the work of Margaret Rossiter.

38. Henry James III (May 18, 1879–December 13, 1947), a Pulitzer Prize winner for Biography (1931), was the nephew of the novelist Henry James.

39. Charles Dudley Warner, *The World's Best Literature: Ancient and Modern*, 46 vols. (J. A. Hill & Co., 1902).

40. M. Carey Thomas was president of Bryn Mawr College from 1894 to 1922; Caroline Hazard was president of Wellesley College from 1899 to 1910.

41. William De Witt Hyde, *The College Man and the College Woman* (Boston: Houghton Mifflin, 1906), 206.

42. Although the concept was originally introduced by Robert Merton, many others have used it. See, for example, Thomas A. DiPrete and Gregory M. Eirich, "Cumulative Advantage as a Mechanism for Inequality: A Review of Theoretical and Empirical Developments," *Annual Review of Sociology* (2006): 271–97.

43. See in particular the works of Mary Frank Fox, Donna Haraway, Sandra Harding, and Evelyn Fox Keller.

44. Culture, as Raymond Williams noted in his celebrated work *Keywords* (1976), is "one of the two or three most complicated words in the English language." Without going into the details of the ongoing scholarly debates around the definition and meaning of culture, in this essay—considering its focus on the autobiographical narrative evidence—we adopt Stuart Hall's "socioanthropological" approach to culture. According to Hall, culture is "*both* the meanings and values which arise among distinctive social groups and classes . . . [*and*] the lived traditions and practices through which those 'understandings' are expressed and in which they are embodied."

45. See also an overview of the nearly 80 studies that reported the results of the "draw-a-scientist" experiment: Miller, David I., Kyle M. Nolla, Alice H. Eagly, and David H. Uttal, "The Development of Children's Gender-Science Stereotypes: A Meta-analysis of 5 Decades of US Draw-a-Scientist Studies." *Child Development* 89, no. 6 (2018): 1943–55.

46. Exogenous factors include broad social belief systems, societal values, norms of behavior, expectations within society that become embedded through socialization and other means into the minds of most of a society's people. Endogenous factors are here more localized to the values, norms, and culture within the scientific community. Sometimes these are consistent with exogenous factors; sometimes they are at variance with them.

47. For an extended discussion of this idea, see Jonathan R. Cole, "Dilemmas of Choice Facing Research Universities," in *Research Universities in a Time of Discontent*, ed. Jonathan R. Cole, Elinor G. Barber, and Stephen R. Graubard (Baltimore, MD: Johns Hopkins University Press, 1993), 11–23. This follows closely those earlier ideas.

410 FREEDOM AND UNFREEDOM

48. This theme of who owns the null has been implicit in many works of contemporary literature. The best depiction of this conflict can be found in David Mamet's play *Oleanna*. Philip Roth has made it a theme in *The Human Stain*, and it is a central theme in J. M. Coetzee's novel *Disgrace*.

49. Cole, "Dilemmas of Choice," 50, 13–14.

50. Among other works, see Sue Rosser (ed.), *Women, Science, and Myth: Gender Beliefs from Antiquity to Present* (Santa Barbara, CA: ABC-CLIO, 2008); Donna Haraway, *The Haraway Reader* (New York: Routledge, 2004); Londa Schiebinger, *The Mind Has No Sex? Women in the Origins of Modern Science* (Cambridge, MA: Harvard University Press, 1989); Maralee Mayberry, Banu Subramaniam, and Lisa H. Weasel (eds.), *Feminist Science Studies: A New Generation* (Hove, UK: Psychology Press, 2001). A lot of studies examine cultural and structural factors. For example, Ruth Watts, *Women in Science: A Social and Cultural History* (New York: Routledge, 2007); Sally Gregory Kohlstedt, "Sustaining Gains: Reflections on Women in Science and Technology in 20th-Century United States," *NWSA Journal* 16, no. 1 (2004); Catherine Cronin and Angela Roger, "Theorizing Progress: Women in Science, Engineering, and Technology in Higher Education," *Journal of Research in Science Teaching* 36, no. 6 (1999): 639–61.

51. See, for example, Jonathan R. Cole, "Two Cultures Revisited," *National Academy of Engineering* 26, no. 3–4 (September 1996). See also, on issues of race, the recent work done by the historian Donald Yacovone, who has been studying the way race has been portrayed historically in textbooks. When you examine the treatment of science in the most widely adopted secondary school textbooks, the first thing one notices is the almost total absence of a discussion of science. If you want to understand "anti-science," you might begin here. When you compare the content devoted to the pop culture (in this case references to the singer Madonna) to that of female scientists, there is simply no comparison. As might be expected, there is virtually no mention of a female scientist at all other than an occasional passing reference to Marie Curie. Not only does Madonna win hands down over all references to women but also over any discussion of Watson and Crick's discovery—or the myriad other revolutionary scientific discoveries in the twentieth century, except perhaps for the creation of the atomic bomb. When authors of these texts were asked why this was so, the authors almost invariably replied that the publisher told me that "science doesn't sell." Moreover, the authors of these texts knew very little about science in general and about women in science in particular. The reference to the sales factor also suggests what goes into history texts is often driven by their commodification and by another factor—the power of state boards that control the purchasing of texts for schools and public libraries in their state. Thus, Texas has a significant impact on the content of texts by suggesting that the state will not adopt a text unless certain content is modified or dropped. That includes the way women are portrayed.

Works Cited

Barnes, S. B., and R. G. Dolby. (1970). "The Scientific Ethos: A Deviant Viewpoint." *European Journal of Sociology* 11(1): 3–25.

Berg, H. M., and M. A. Ferber. (1983). "Men and Women Graduate Students: Who Succeeds and Why?" *Journal of Higher Education* 54(6): 629.

Chambers, D. W. (1983). "Stereotypic Images of the Scientist: The Draw-a-Scientist Test." *Science Education* 67(2): 255–65.

Cole, J. R. (1989). "The Paradox of Individual Particularism and Institutional Universalism." *Social Science Information* 28(1): 51–76.

Cole, J. R. (2012). *The Great American University: Its Rise to Preeminence, Its Indispensable National Role, Why It Must Be Protected.* New York: PublicAffairs.

Cole, J. R., and S. Cole. (1973). *Social Stratification in Science.* Chicago: University of Chicago Press.

Cole, J., and B. Singer. (1992). "A Theory of Limited Differences: Explaining the Productivity Puzzle in Science." In *The Outer Circle: Women in the Scientific Community*, ed. H. Zuckerman, J. R. Cole, J. T. Bruer, and Josiah Macy Jr. Foundation, 277–310. New Haven, CT: Yale University Press.

Connell, R. (1987). *Gender and Power: Society, the Person, and Sexual Politics.* Stanford, CA: Stanford University Press.

Duran, J. (1998). *Philosophies of Science/Feminist Theories.* London: Routledge.

Epstein, C. F. (1992). "Constraints on Excellence: Structural and Cultural Barriers to the Recognition and Demonstration of Achievement." In *The Outer Circle: Women in the Scientific Community*, ed. H. Zuckerman, J. R. Cole, J. T. Bruer, and Josiah Macy Jr. Foundation. New Haven, CT: Yale University Press.

Erskine, H. (1971). "The Polls: Women's Role." *Public Opinion Quarterly* 35(2): 275–90.

Federoff, N., & Botstein, D. (Eds.). (1992). *The Dynamic Genome: Barbara McClintock's Ideas in the Century of Genetics*, illustrated ed. Woodbury, NY: Cold Spring Harbor Laboratory Press.

Foner, E. (1999). *The Story of American Freedom.* New York: Norton.

Fox, M. F. (1989). "Women and Higher Education: Gender Differences in the Status of Students and Scholars." In *Women: A Feminist Perspective*, ed. J. Freeman, 217–35. Mountain View, CA: Mayfield.

Hanson, S. L. (1996). *Lost Talent: Women in the Sciences.* Philadelphia: Temple University Press.

Hollenshead, C. S., S. A. Wenzel, B. B. Lazarus, and I. Nair. (1996). "The Graduate Experience in the Sciences and Engineering: Rethinking a Gendered Institution." In *The Equity Equation: Fostering the Advancement of Women in the Sciences, Mathematics, and Engineering*, ed. C.-S. Davis. Hoboken, NJ: Jossey-Bass.

MacKinnon, C. A. (1979). *Sexual Harassment of Working Women: A Case of Sex Discrimination.* New Haven, CT: Yale University Press.

Merton, R. K. (1968). "The Matthew Effect in Science: The Reward and Communication Systems of Science Are Considered." *Science* 159(3810): 56–63.

Mulkay, M. (1974). *Science and the Sociology of Knowledge.* London: Routledge.

Reskin, B. F. (1977). "Scientific Productivity and the Reward Structure of Science." *American Sociological Review* 42: 491–504.

Seymour, E. (1995). "The Loss of Women from Science, Mathematics, and Engineering Undergraduate Majors: An Explanatory Account." *Science Education* 79(4): 437–73.

Tidball, M. E. (1989). "Women's Colleges: Exceptional Conditions, not Exceptional Talent, Produce High Achievers." In *Educating the Majority: Women Challenge Tradition in Higher Education*, ed. C. Pearson, D. L. Shavlik, and J. G. Touchton, 157–72). London: Macmillan.

Vetter, B. M. (1996). "Myths and Realities of Women's Progress in the Sciences, Mathematics, and Engineering." In *The Equity Equation: Fostering the Advancement of Women in the Sciences, Mathematics, and Engineering*, ed. C.-S. Davis. Hoboken, NJ: Jossey-Bass.

Women: Their Underrepresentation and Career Differentials in Science and Engineering. Conference proceedings, p. 18771. Washington, DC: National Academies Press, 1987.

PART III

CONSENSUS IN SCIENCE

JUDGMENT AND CHOICES

CHAPTER 18

PEER REVIEW AND THE SUPPORT OF SCIENCE

(1972)

JONATHAN R. COLE, STEPHEN COLE, AND LEONARD RUBIN

A statistical analysis of the evaluative procedures on which the National Science Foundation bases its funding decisions provides no evidence to substantiate recent public criticisms

For more than twenty-five years the National Science Foundation has played a major role in the expenditure of public money for the support of science in the United States. Currently, the NSF accounts for about 20 percent of the funds distributed by the federal government for basic scientific research and more than 30 percent of the federal funds allocated for such research at universities. The NSF awards its grants on the basis of a decision-making process commonly known as peer review. The term is derived from the fact that the government officials responsible for deciding which investigators receive grants rely on the evaluations of other investigators in the same discipline.

In recent years the peer-review system has been attacked for a variety of reasons by certain members of both the scientific community and Congress. Hearings on the alleged inequities of the peer-review system were held two years ago by a subcommittee of the House Committee on Science and Technology.

Scientific American 237, no. 4 (October 1977): 34–41.

416   CONSENSUS IN SCIENCE

In an effort to assess the validity of the public criticisms of the peer-review system raised in the Congressional hearings and elsewhere we have been engaged for more than a year in a sociological study of the operation of the peer-review system at the NSF. This study, which is being conducted for the National Academy of Sciences, is supported by grants from the NSF; we have nonetheless had complete autonomy from the NSF in conducting our research. Our results to date have yielded little evidence in support of the main criticisms that have been made of the peer-review system. On the contrary, we have tentatively concluded that the NSF peer-review system is in general an equitable arrangement that distributes the limited funds available for basic research primarily on the basis of the perceived quality of the applicant's proposal. In particular, we find that the NSF does not discriminate systematically against noneminent scientists in the ways that some critics have charged. This is not to say, of course, that there are not errors in individual cases.

How does the NSF peer-review system work? To begin with, a scientist who wants to obtain NSF funds prepares a written proposal describing his past research, his qualifications, and the new research he intends to do if he receives funds from the NSF. This proposal is usually submitted to the NSF through the scientist's institution, in most cases a university.

The staff of the NSF is divided into approximately eighty program areas corresponding to the various scientific disciplines and subdisciplines. (The chemistry section, for example, is divided into eight different programs.) When a research proposal comes to the NSF, it is assigned to the appropriate program and is thereafter handled by an employee of the NSF called the program director. On receiving a proposal the program director generally looks it over to determine its specific subject area. He then selects a number of reviewers who are sent the proposal by mail. The reviewers are asked to rate the proposal as being excellent, very good, good, fair, or poor and in support of their rating to present written comments evaluating the proposal. In some programs an independent evaluation of the proposal is also made by a panel of scientists who meet with the program director three times a year in Washington.

The NSF explicitly states to its reviewers the criteria that should be applied in evaluating the proposals. The main criteria are (1) the significance of the scientific investigation described in the proposal; (2) the ability of the applicant to carry out the proposed research; and (3) the capacity of the applicant's institution to support the type of research in question. Where all these factors are roughly equal, another set of criteria, including the geographic location of the applicant's institution, may be considered. Heavy emphasis is placed on the quality of the work described in the proposal and on the past research performance of the applicant.

The most fundamental criticism made of the NSF peer-review system is that it leads to inequitable decisions. Critics charge that scientists who are most capable of advancing science are sometimes denied grants and that scientists who are doing less significant work are given grants. Former Representative John B. Conlan of Arizona, for example, asserted at the Congressional hearings that peer review is essentially an elitist system run primarily for the benefit of a clique of eminent "old boys." He said, "I know from studying material provided to me by the NSF that this is an 'old boy's system' where program managers rely on trusted friends in the academic community to review their proposals. These friends recommend their friends as reviewers. . . . It is an incestuous 'buddy system' that frequently stifles new ideas and scientific breakthroughs, while carving up the multimillion-dollar Federal research-and-education pie in a monopoly game of grants-manship."

Critics in and out of Congress maintain that the main organizational condition that gives rise to this unfair distribution of support is the extraordinary power in the hands of the program directors to decide who should get funds. The program director is alleged to be at the center of the old-boy network in which reviewers favorably evaluate the proposals of their friends, eminent scientists favorably review the proposals of other eminent scientists, and funds are denied to scientists who are not part of the exclusive old-boy system.

Further abuse is said to be possible because the reviews received by the program director are only advisory, leaving him free to ignore them, and because the program director can predetermine the outcome by selecting a biased group of reviewers. The critics argue that knowledgeable program directors deliberately select reviewers who will be either hard or easy on a particular proposal. Even if the program director feels compelled by the reviews to support a proposal he dislikes, he can effectively stifle the research by reducing the size of the budget. The program director can supposedly do so because there are no effective checks on his power either inside or outside the NSF. In short, there is no appeals system to challenge the decisions made by the program director.

Critics assert further that the NSF cloaks its activities in secrecy in order to protect the old-boy system, refusing to allow Congressmen or others to see verbatim reviews or to learn the names of the reviewers of particular proposals. This protective shield of confidentiality enables the system to function unchecked and prevents effective oversight of the NSF by Congress. The ultimate consequence is that the peer-review system actually stifles innovative research, since the eminent scientists who serve as reviewers are likely to reject ideas that differ from their own.

In our study of the peer-review system we decided to limit ourselves at first to an examination of how peer review works in just those NSF programs responsible for the funding of basic research. We have not studied peer review in the NSF's applied-research programs or in its educational programs. Furthermore, we chose a sample of only ten basic-research programs for detailed study: algebra, anthropology, biochemistry, chemical dynamics, ecology, economics, fluid

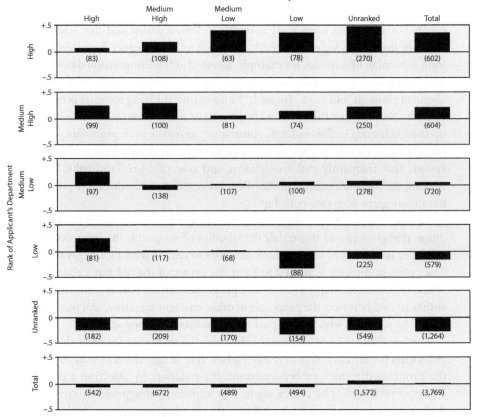

18.1 Statistical analysis of 3,769 peer-review ratings given by various mail reviewers to 1,200 applicants for basic-research grants from the National Science Foundation in the fiscal year 1975 was aimed at testing the "old boy" hypothesis, which holds that the proposals of eminent scientists are apt to be rated more favorably by eminent reviewers than by other reviewers. The ratings in the ten different program areas studied were first converted into standard scores in the following manner: Within each field the mean rating was set at zero, and the rating received by an applicant was then expressed in terms of the corresponding number of standard deviations above or below the mean rating. A high number means a comparatively favorable rating, and vice versa. Both the applicants and the reviewers were separately classified according to the prestige of their current academic department, as determined in an independent survey. Thus, the entry in the upper left-hand corner of figure 18.1 signifies that there were eighty-three reviews by reviewers in high-ranked departments of proposals submitted by applicants from high-ranked departments; on the average these reviews yielded ratings that were .05 of a standard deviation above the mean. Since it appears that proposals from applicants in high-ranked departments are actually rated lower by reviewers from high-ranked departments than by reviewers from lower-ranked departments, in this sample at least the data offer no support for the old-boy hypothesis. The analysis does show that applicants from high-ranked departments are slightly more likely to receive favorable ratings than are those from unranked departments, but there is no evidence that this outcome is the result of inequitable treatment.

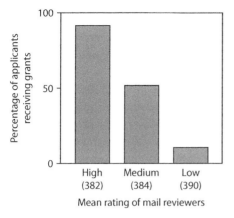 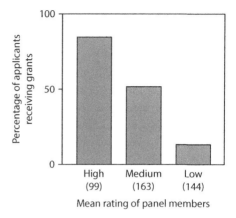

18.2 NSF program directors appear to rely heavily on the evaluations of the peer reviewers in deciding whether or not a research proposal is to be funded. As figure 18.2 shows, among the 382 applicants who received comparatively high ratings from the mail reviewers, 92 percent were awarded grants, whereas among the 390 receiving low mean ratings only 10 percent received grants. Similarly, as the figure shows, among those proposals that received comparatively high ratings from an independent panel of peer reviewers, 84 percent were funded, whereas among those that received low panel ratings only 12 percent were funded. Evidently peer-review ratings are the most important determinant of the program director's decision.

dynamics, geophysics, meteorology, and solid-state physics. Because our intensive analysis included only about an eighth of the NSF's basic-research programs our results may not be generalizable for the entire organization. We are currently conducting follow-up studies of other programs.

Our investigation has combined both qualitative and quantitative sociological techniques. We began by conducting seventy in-depth interviews with scientists involved at all levels of the peer-review system, including program directors, former program directors, mail reviewers, review-panel members and supervisory-level NSF officials. We also scrutinized more than 250 specific research proposals, read all of the peer-review comments on those proposals, and examined all of the correspondence between the applicant and the program director. In some cases in which our analysis of the applications raised specific questions about how the peer-review system worked in that particular situation, we went back and reinterviewed program directors with the files in hand.

In addition, we conducted a quantitative analysis of 1,200 applicants to the NSF in the fiscal year 1975. (Roughly half of the applicants were ultimately awarded grants.) The purpose of the quantitative study was to identify those characteristics that were correlated with the receipt of a grant from the NSF.

Were Representative Conlan and the other critics of peer review correct in their assertion that eminent scientists have a great advantage in the competition for funds and that less eminent scientists, particularly younger ones, are at a serious disadvantage? We shall try to answer this question by summarizing below some of the results obtained so far in our study.

One of the main charges of the critics is that the NSF program director can predetermine the outcome of the peer-review process by sending a proposal to scientists who he knows in advance are biased either in favor of the proposal or against it. We shall call this view the old-boy hypothesis. Presumably the proposals of eminent scientists who are members of the old-boy network are sent to other eminent scientists who give their eminent colleague a favorable evaluation. In return, of course, the reviewers expect reciprocity when their proposals are sent to other members of the old-boy club. Equally important, the proposals of less eminent scientists, who are not part of the network, are sent to scientists who will give them lower evaluations than they deserve. Although we have no direct evidence that the program directors either do or do not select reviewers with a certain outcome in mind, we can see if the outcomes are consistent with the old-boy hypothesis. Are the proposals of eminent scientists actually rated more favorably by eminent reviewers than by other reviewers?

To test this hypothesis we classified both the applicants and the reviewers according to the prestige of their current academic department, as determined by a survey conducted in 1969 by the American Council on Education. The ratings given to the applicants by the reviewers in the ten programs we studied were standardized separately before being combined into one large table (see figure 18.1). For example, there were a total of 83 cases in which an applicant from a high-ranked department had his proposal reviewed by someone who was also from a high-ranked department. The number associated with this particular applicant-reviewer pair (+.05) indicates the average rating (in standardized units) given by high-ranked reviewers to proposals from high-ranked applicants. The higher the number, the higher the rating.

In general we found that applicants from high-ranked departments received slightly better reviews of their proposals than applicants from medium-ranked and low-ranked departments. Furthermore, it appeared that high-ranked reviewers tend to be slightly more lenient with proposals than low-ranked reviewers are. These results, in and of themselves, cannot be interpreted as offering support for the old-boy hypothesis. For example, the fact that eminent scientists tend to get higher ratings could simply be a result of the higher quality of their proposals or of the belief on the part of the reviewers that the eminent scientists are in fact better able to carry out the proposed research.

In order to explore the matter more deeply we next conducted a statistical analysis of variance that compared the observed mean rating for each applicant-reviewer pair with the expected mean rating, assuming no bias. The results of this analysis indicated that in general reviewers from high-ranked departments were not disproportionately favoring proposals from applicants in similarly high-ranked departments. We conducted this analysis separately for each of the ten programs. In only one program were reviewers at high-ranked departments detectably more lenient toward the proposals of their colleagues at similarly high-ranked departments.

Another statistical analysis of variance tested the reviewers' bias in terms of geographic location and of the relative eminence of the reviewer and the applicant. It showed no significant tendency for scientists in one part of the country to favor proposals from colleagues in their own region or for eminent scientists to favor the proposals of eminent scientists over the proposals of less eminent scientists. Thus, even if it were true that the program directors at the NSF were attempting to manipulate the outcome of the peer-review process by their selection of reviewers (and our qualitative findings indicate that it is unlikely), the quantitative data suggest that they have not been successful.

One reason it is difficult to test the validity of the old-boy hypothesis is the absence of conceptual clarity in the charge. What is referred to by the old-boy label? There are at least three possibilities. The term could refer to investigators with a common view of their field who will only appraise favorably work that is done by people with similar views. It could refer to networks of friendships: scientists who know one another, who "grew up" together or attended the same schools, and who tend to fraternize and also to favor one another's proposals. It could refer to social position: scientists at a given level of eminence might tend to favor the proposals of others who are similarly situated in the hierarchy of science, even if they have no personal contact with them. Critics of the peer-review system never specify clearly which form of old-boyism is undermining the peer-review system. The data reported here allow us to examine the assertion that persons of similar rank, similar intellectual background, and similar repute favor one another's proposals, but we do not have in hand data for examining forms of old-boyism that may be connected with friendship patterns.

How do the characteristics of the applicants affect the peer-review ratings they receive? Critics of the peer-review system say that regardless of the quality of proposals eminent scientists enjoy an advantage over those who are not eminent. In the final analysis, these critics contend, the peer-review system results primarily in eminent scientists at high-ranked departments having an unfair advantage in grant approval over less eminent scientists at lower-ranked departments. To test this "rich get richer" hypothesis we combined the applicants from all ten programs into one large standardized sample. The 1,200 applicants in the sample were characterized by nine variables that established their status in the social system of science. Each of these characteristics was then tested separately to see if it provided evidence in support of the rich-get-richer hypothesis.

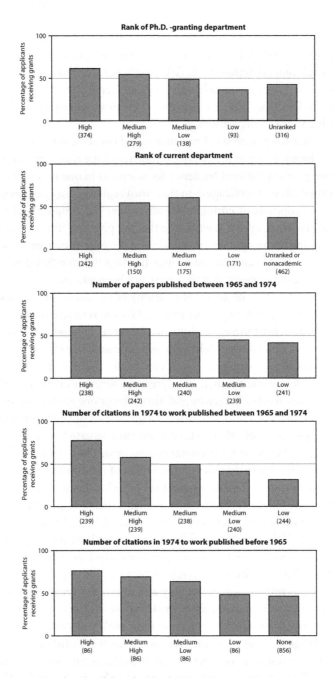

18.3 Characteristics of successful applicants for NSF grants in 1975 are summarized in figure 18.3. Among the characteristics represented here are rank of PhD-granting department (top), rank of current department (second from top), number of scientific papers published between 1965 and 1974 (middle), number of citations to work published between 1965 and 1974 (second from bottom), and to work published before 1965 (bottom).

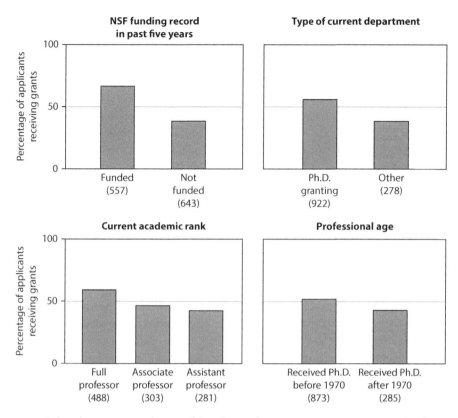

18.4 Other characteristics of successful applicants for grants in 1975 are represented in figure 18.4. Characteristics include past five years' funding record (top left), type of academic department (top right), academic rank (bottom left), and professional age (bottom right).

For example, we characterized the applicants according to the graduate departments from which they obtained their doctoral degree to see if scientists who come from prestigious PhD-granting departments tend to get higher ratings than those who come from less prestigious departments. The applicants were also classified according to their current academic departments in order to test the assertion that applicants in high-ranked departments have an undeserved advantage over applicants in low-ranked departments. We classified the applicants according to their current academic rank in order to see if assistant professors are any less likely to receive grants than associate professors or full professors. In addition we classified all the applicants according to their professional age, their published scientific works, the number of citations of their published works, and whether or not they had received NSF funds in the past.

424 CONSENSUS IN SCIENCE

The rich-get-richer hypothesis would suggest the existence of strong correlations between all of these variables and the ratings the applicants received on their proposals. There are, indeed, reasons other than old-boyism for this expectation. For one thing scientists who in the past had done research that other scientists had valued highly could reasonably be expected to write proposals that would be more likely to be rated highly. Moreover, since the NSF explicitly instructs reviewers to regard past performance as one of the major criteria in determining a rating, reviewers could be expected to give higher ratings to scientists with a superior "track record."

The data, however, provide little support for the rich-get-richer hypothesis. Our results show only weak or moderate correlations between each of the nine "social stratification" variables and the ratings received on proposals. The most highly correlated variable was the number of citations in the 1975 *Science Citation Index* of work published between 1965 and 1974. Even this rough measure of the significance of recently published work is not correlated very strongly with the ratings, explaining only 6 percent of the variance in the ratings. The correlations between the other variables and the ratings are all surprisingly low, explaining only an additional 5 percent of the variance in the ratings. In the end 89 percent of the observed variance in the ratings is left unexplained by the nine variables.

These results ran so counter to our expectations that at first we suspected they might have been caused by some methodological error. A thorough review of our correlation and regression procedures, however, left the results intact. In fact, the validity of our findings has been corroborated by a recent study conducted by members of the NSF's own chemistry section. Their independent analysis yielded results that were virtually identical with our own. It is difficult to avoid the conclusion that there is no substantial correlation between peer-review ratings received by applicants for NSF grants and statistical indicators of their professional status or past scientific performance. Scientists whose published work is frequently cited were only slightly more likely to receive favorable ratings than scientists with only a few citations or none.

It still appeared possible, however, that the weak correlations we observed could have resulted from a lack of agreement among the reviewers. For example, if an applicant with a large number of citations of his work received very favorable ratings from some reviewers and very unfavorable ones from others, that could account for a weak or nonexistent correlation between citations and ratings. How much agreement was there among the various reviewers of a given research proposal?

To answer this question we first determined the mean standard deviation of the reviewers' ratings, a quantity that can be taken as an approximation of the degree of agreement in a given field. This number varied from a low level of .31 in algebra to a high level of .69 in ecology and meteorology. (A low mean standard deviation corresponds to a high degree of consensus, and vice versa.) This approach could itself be flawed, however, if one were to fail to take into account the mean rating of the reviewers in each field. Clearly if there is a general tendency in a field to restrict the range of evaluations to either high or low scores, there would be less chance for variations in the ratings. We therefore relied on a statistic called the coefficient of variation, which is simply the mean peer-review rating divided by the mean standard deviation. In general we found that there was a good deal of agreement among the mail reviewers in all ten fields and little systematic variation among the fields. The coefficient of variation ranged from a low of .13 in economics to a high of .30 in ecology.

To test further the notion that the weak correlations we observed resulted from a lack of agreement among the reviewers, we examined the correlations between the mean rating received by a proposal and several characteristics of the applicant. If the weak correlations had resulted from a lack of agreement among the reviewers, the associations between mean ratings and individual characteristics would be substantially higher, since mean scores are almost invariably more strongly correlated with any given variable than are individual scores. When the mean rating was used as the dependent variable in a statistical regression analysis, we obtained results similar to those obtained in our original analysis. The highest correlation was found between citations of recent work and the mean rating, followed by the correlation between past funding history and the mean rating. Although this method of analysis had the effect of increasing the amount of variance explained by the characteristics of the applicants from 11 percent to 16 percent, the great bulk of the observed variance in the ratings remained unexplained. The new analysis supported the conclusion that the weak correlations observed were not a result of a lack of agreement among reviewers.

In short, these data suggest that the mail reviewers are not strongly influenced by the professional status of an applicant in evaluating a proposal. On the contrary, they appear to be much more likely to be influenced by their perception of the quality of the research proposed. One crucial question remained: How is the program director's funding decision related to the reviewers' ratings on the one hand and to the characteristics of the applicants on the other?

Critics of the peer-review system contend in effect that the decisions of the NSF program directors depend more on who you are than on what you propose to do. So far our data have tended to refute this version of the old-boy hypothesis. Before this refutation can be established conclusively, however, we must establish that the peer-review ratings are the single most important determinant of the program director's funding decision and that the characteristics of the applicants have little independent effect on the outcome.

The NSF states clearly that the reviews by either the mail reviewers or the panel members are advisory and the program director has the final responsibility for deciding whether or not a proposal is to be funded. Our data show that the program directors in fact rely very heavily on the evaluations of the peer reviewers. For example, among those applicants who received comparatively high mean ratings from the mail reviewers, 92 percent were awarded grants, whereas among those receiving low mean ratings only 10 percent got grants. Among the group who received mean ratings in the middle ranges about half were awarded grants. Similarly, among those applicants who received comparatively high panel ratings 84 percent were funded, and among those who received low panel ratings only 12 percent were funded (see figure 18.5).

What types of scientists were successful in receiving grants from the NSF in 1975? Of those applicants who obtained their degrees from the highest-ranked graduate departments 62 percent were awarded grants, compared to 38 percent of those who graduated from the lowest-ranked departments. Similarly, 74 percent of the applicants currently employed in the highest-ranked departments were funded, compared with 38 percent currently in either unranked departments or nonacademic institutions.

Recent NSF funding history and frequency of citations of recent work both had a moderate influence on the probability of receiving a grant. Among applicants receiving the most citations to recently published work roughly

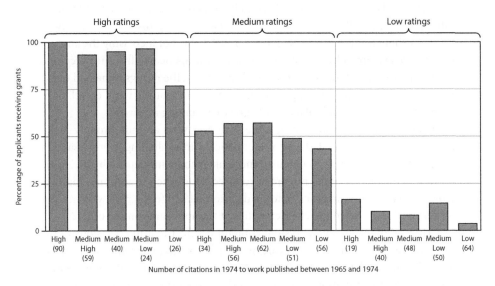

18.5 Independent effects of a scientist's past achievements on the probability of receiving an NSF grant are represented in figures 18.6 and 18.7. The applicants were divided into three groups: those who received comparatively high mean ratings from mail reviewers, those who received medium mean ratings, and those who received low mean ratings. Within each category the probability that particular scientists—in this case those with different numbers of citations of their recent work—would receive grants was then calculated. The results show that scientists whose work is frequently cited have a slight competitive advantage in the competition for funds.

three-quarters were awarded NSF grants; among those receiving the least citations of recent work less than a third received grants. The number of papers published and the number of citations of work published before 1965 were less strongly associated with the receipt of a grant. Other attributes of the applicants, such as their professional age or their academic rank, had a minor effect on the probability of receiving a grant.

The effect of professional age on the probability of receiving an NSF grant is particularly noteworthy. When we began our study many scientists indicated that they believed it was more difficult for younger scientists to obtain NSF funds. Our interviews with program directors, on the other hand, revealed that they perceived just the opposite. Because there is a commitment on the part of the NSF to help young, talented scientists get started, several program directors said that in the case of roughly equal peer reviews they would prefer to fund younger applicants. As it happens, the perceptions of both the applicants and the program directors are mistaken. The data we have gathered indicate that professional age has almost no effect on either the peer-review ratings or the final funding decision.

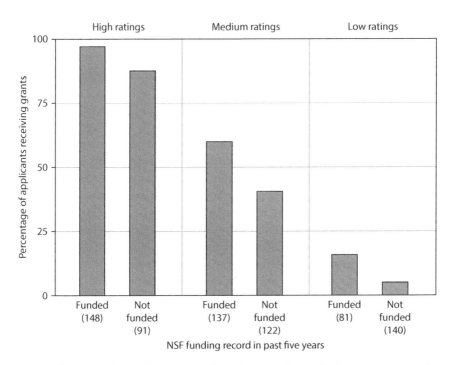

18.6 Similar accumulative advantage is indicated, among those scientists whose proposals received medium or low peer-review ratings, for applicants who had been funded by the NSF in the past five years. Again, a good record appears to produce a slight advantage.

The overall pattern of our data suggests that scientists with an established track record, many scientific publications, a high frequency of citations, a record of having received grants from the NSF, and ties to prestigious academic departments have a higher probability of receiving NSF grants than other applicants do. Nevertheless, the granting process is actually quite open, and there is nothing approximating a scientific caste system. Even among the most frequently cited scientists who apply for support an appreciable number do not receive grants, and among the group with the fewest citations to their work a significant number do receive grants. There is no evidence that scientists who have received grants in the past are guaranteed continued support or that those without a past funding record have no chance of obtaining current NSF funding. Indeed, given the heavy emphasis the NSF places on past performance as one of the two most important criteria in evaluating research proposals, it is somewhat surprising that measures of past scientific performance do not show a stronger influence on the probability of receiving a grant.

It should incidentally be noted that the data presented here allow us to answer two distinct questions. The first is: How well do the social characteristics of scientists and their previous record predict peer-review ratings and the probability of funding in general, that is, when we examine the entire sample of applicants? The second is: Are there substantially different probabilities of receiving high ratings or a favorable decision for the most eminent applicants compared with the least eminent applicants, that is, when we compare relatively small subsets of the sample? The answers can be different depending on which of these two questions we ask.

For the sample as a whole status differences are not good predictors of ratings. Consider a concrete example of what we mean by focusing again on the relation between the rank of an applicant's current department and the final funding decision. First recall that 55 percent of all 1,200 applicants received NSF grants; if one had to predict whether an individual applicant had received a grant, to predict in every case that he had received one would make one right on 55 percent of the applicants and wrong on 45 percent. The question is: How does knowledge of the rank of an applicant's department increase the ability to predict whether he received a grant? To estimate this we examine each of the five classifications of departmental rank. In the two bottom categories, where a majority did not obtain support, we would guess that all applicants did not receive grants: in the other three categories, where a majority received support, we would do better to guess that all received support. That would result in correct predictions in 63 percent of the cases. When we subtract from this total the proportion (55 percent) that we would have guessed correctly without any information about the individual's departmental affiliation, we get an estimate of the increase in predictability that results from knowledge of rank of department: in this case an increase of 8 percent, which is not an extraordinary increase in predictability.

Suppose, on the other hand, we want to know whether scientists in the highest-ranked departments have a better chance of receiving NSF support than those in unranked departments or in a nonacademic setting. If we compare the percentage difference between these extreme subgroups, we find a substantial 36-point difference. In other words, some percentage differences do appear large in the extremes, but that does not mean the characteristic is a good predictor of a decision for the entire sample. Of the variance that can be accounted for in funding decisions, the peer-review rating is by far the best predictor.

The well-documented social process referred to by sociologists of science as "accumulative advantage" would lead one to expect that eminent scientists have a better-than-average chance in the competition for NSF funds. Accumulative advantage in this context means that a scientist who has been rewarded at one stage in his career has an enhanced probability of being rewarded at a later stage, regardless of the quality of his scientific work in the interim. The concept explains in part the increasing inequality in rewards that is observed as an age cohort of scientists moves through time.

According to the concept of accumulative advantage, the initial social status of a scientist influences the probability of his obtaining a variety of forms of recognition, including the esteem of his colleagues, an association with centers of excellence in the academic world, and the resources and facilities necessary for productive scientific work. For example, young scientists who are trained in the best university science departments, and particularly those who have been apprenticed to leading scientists, have a better chance than less well-placed students of equal ability to secure first jobs at prestigious institutions. Once established in these positions they have a better chance than their peers to obtain support for their research. With greater support they have an enhanced opportunity for making significant scientific discoveries and publishing the results. And once the results are published they have still greater chances for future success. To the extent that this process works to the advantage of scientists who are initially well placed in the social system of science it also works to the disadvantage of their peers who are not so fortunate.

By taking the mean peer-review rating received by an NSF research proposal as a rough measure of the quality of the proposal we attempted to determine the independent effect of a scientist's past achievements on his receiving a grant. We first divided the applicants into three groups: those who received comparatively high mean ratings, those who received medium mean ratings and those who received low mean ratings. Within each category we calculated the probability that scientists who had had different numbers of citations of their recent work would receive grants. We then considered only the group of proposals that received the highest peer-review ratings. Of this group 100 percent of the quintile with the highest number of citations were awarded NSF grants. In the lowest quintile 77 percent received grants. This finding leads to two conclusions: (1) the mean peer-review rating is a far more important determinant of whether a scientist receives a grant than is the number of citations of his recent work, and (2) within each category of mean ratings the number of citations of recent work has only a slight influence on the probability of approval.

We next considered the cases of those scientists whose proposals received low ratings. A substantial majority of all the proposals in this category were declined, but the number of citations made little difference. Within the group of proposals that received low ratings 16 percent of the scientists with the most citations received grants, compared with 3 percent of those who received the fewest citations.

The foregoing data offer some limited support for the concept of accumulative advantage. Scientists whose recent work has been frequently cited have a measurable advantage in the competition for current funds; this advantage is, however, very slight. The process of accumulative advantage is somewhat more evident among those scientists whose research proposals received medium peer-review ratings but who had been funded frequently by the NSF in the past five years. Among scientists whose proposals received medium ratings, for example, 61 percent of those who had been funded within the past five years were awarded a current grant, whereas only 41 percent of those who had not received funds from the NSF in the past five years were awarded a current grant. Clearly a good funding record gives rise to a slight competitive advantage.

We also examined the independent effect of an applicant's current academic department on the probability of his being awarded an NSF grant. Here the story was somewhat different. The rank of a scientist's current department apparently has almost no effect on the probability of his receiving a grant independent of the peer-review ratings received by the applicant's proposal. Of the scientists in the highest-ranked departments whose proposals received comparatively low ratings 6 percent were awarded grants, a figure no different from that found among applicants in lower-ranked departments. In the competition for current funds, therefore, a scientist's past performance as measured by citations of his work and his recent NSF funding record does lead to a very slight accumulative advantage, but his academic affiliation does not appear to give him any advantage.

The results of our study of the operation of the peer-review system in the basic-research programs of the NSF are consistent thus far with other recent findings in the sociology of science, which suggest that the scientific enterprise is an exceedingly equitable, although highly stratified, social institution in which the individuals who produce the work that is most favorably evaluated by their colleagues receive the lion's share of the rewards. Further study of the equity of research-fund distribution will address two basic problems not yet

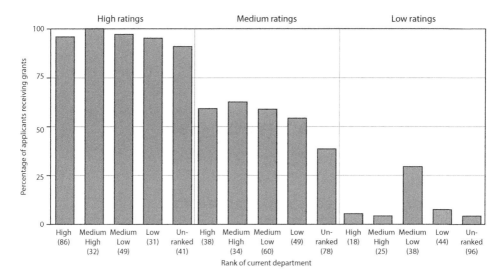

18.7 No independent effect was detectable in this similar statistical analysis, which measured the influence of an applicant's current academic department on the probability of being awarded an NSF basic-research grant. Apparently, current academic affiliation does not give an applicant any competitive advantage independent of the peer-review ratings that were received by his research proposal.

considered. In the first phase of our study we relied on the peer-review ratings elicited by the NSF program directors as an indicator of quality and found those ratings were strongly related to the actual funding decision; now we are submitting proposals to independent review panels in order to obtain independent appraisals of their quality. Finally, having learned that peer-review ratings are strong predictors of funding decisions, we are interested in whether or not they also are good predictors of future scientific performance, and so we are studying how the ratings and recent research performance compare as predictors of future research performance.

CHAPTER 19

THE "ORTEGA" HYPOTHESIS

(1972)

JONATHAN R. COLE AND STEPHEN COLE

Citation analysis suggests that only a few scientists contribute to scientific progress.

Most scientists are aware that science is a highly stratified institution. Power and resources are concentrated in the hands of a relatively small minority. For the past several years we have been studying the social stratification system of science.[1] Most of our research has concentrated on the social processes through which individual scientists are evaluated to discover why some scientists rise quickly to positions of eminence and others remain relatively obscure. Two conflicting theories explain social mobility in science. According to one theory the stratification system of science operates on strictly universalistic criteria: the scientists who publish the most significant work receive the ample recognition they deserve; those not publishing significant work are ignored. According to the other theory, a small elite at a handful of universities and government-supported laboratories control the social institutions of science in such a way as to perpetuate their own ideas and assure the social mobility of their intellectual children. The results of our research have for the most part supported the former theory. We have found that quality of published research explains more variance than any other variable on several types of recognition.

Science 178, no. 4059 (October 1972): 368–75.

Contributions from Scientific Strata to Progress in Science

Whereas most of our previous research has dealt with the processes through which individuals find their level in the stratification system, in this article we analyze another problem. We present data evaluating the comparative contributions of the various scientific strata to scientific progress, indicating whether progress is built on the labor of all "social classes" or is primarily dependent on the work of an "elite." In the past, historians and philosophers of science have attributed much of the growth of science to the work of the average scientist, who, it is suggested, has paved the way with his "small" discoveries for the men of genius—the great discoverers. This hypothesis is asserted in many sources but perhaps no more clearly than in the words of José Ortega y Gasset:

> For it is necessary to insist upon this extraordinary but undeniable fact: experimental science has progressed thanks in great part to the work of men astoundingly mediocre, and even less than mediocre. That is to say, modern science, the root and symbol of our actual civilization, finds a place for the intellectually commonplace man and allows him to work therein with success. In this way the majority of scientists help the general advance of science while shut up in the narrow cell of their laboratory, like the bee in the cell of its hive, or the turnspit of its wheel.[2]

Ortega seems to be suggesting that average scientists, working on relatively unambitious projects, make minor contributions but that, without these minor discoveries by a mass of scientists, the breakthroughs of the truly inspired scientist would not be possible. Thus, the work of the great scientist is built upon a pyramid of small discoveries made by average scientists. This view of science is widespread. Some even go so far as to maintain that scientific advance is more dependent upon the small discoveries of the many average scientists than upon the breakthroughs of the great scientists. Lord Florey, a recent president of the Royal Society, expressed this point of view:

> Science is rarely advanced by what is known in current jargon as a "breakthrough"; rather does our increasing knowledge depend on the activity of thousands of our colleagues throughout the world who add small points to what will eventually become a splendid picture much in the same way the Pointillistes built up their extremely beautiful canvasses.[3]

In the view of science, of course, a number of assumptions are made. Consider two: it is assumed (i) that the ideas of the average scientist are both visible and used by the outstanding scientist; and (ii) that the minor work is necessary for the production of major contributions. In short, it is proposed that the work of

434 CONSENSUS IN SCIENCE

the average scientist is indispensable if science is to advance. Little empirical evidence exists to substantiate these widely held beliefs. We shall examine data bearing upon the validity of this view of scientific progress. To make an empirical test of this conception manageable, we confine ourselves to one of its several aspects and to only one field of science. We examine the work of several groups (samples) of physicists and analyze what work these men used in making their discoveries.

We do not intend to suggest that great discoveries in science by an Einstein or a Lee and Yang are not preceded by numerous "smaller" discoveries or that great discoveries do not in turn stimulate a multitude of lesser ones.[4] We will suggest that even the scientists who make these "smaller" discoveries come principally from the top strata of the scientific community. In the proper perspective of the history of science, "normal science," as Kuhn defines it, is not done by the average scientist but by the elite scientists.[5] Indeed, in the longer perspective, the work of many of today's outstanding scientists, such as Nobel laureates and members of the National Academy of Sciences, may turn out to be minor footnotes in the history of science.

The question that we consider is how many scientists are contributing through their published research to the movement of science and how many are not. There are, of course, many ways to contribute to scientific progress other than through published research. The scientists who are primarily teachers, administrators, or technicians may play crucial roles in scientific development. We do not intend to downgrade the importance of these roles. Nevertheless, it is still valid to ask how many scientists contribute to scientific progress through their published work if we keep in mind that to the list of contributors of this type we must add the names of contributors of other types.

Price, following Lotkà, has estimated that the number of scientists producing n papers is approximately proportional to $1/n^2$.[6] This inverse square law of productivity estimates that, for every 100 authors producing one scientific paper, there are only 25 who produce two, 11 who produce three, and so on. Using Price's model we can estimate that roughly 50 percent of all scientific papers are produced by approximately 10 percent of the scientists. What remains problematic is the extent to which the 10 percent of the scientists who produce 50 percent of the research publications are dependent on the other 90 percent of research scientists and the 50 percent of the total research they produce. If the bulk of the scientific community produced work that is rarely used—that is, is infrequently cited in the work of outstanding scientists—the indication may be that their work does not materially advance the development of science. The basic question to ask is what the intellectual sources of influence on the production of scientific research of varying quality may be. If Ortega is correct, the work of scientific frontiersmen will to some extent be dependent upon the work of the vast majority of physicists.

Citation Practices of Academic Physicists

We collected data to illustrate the citation practices of academic physicists. One set of data consists of the citations made by 84 university physicists in their paper most heavily cited in the 1965 *Science Citation Index* (SCI).[7] We consider this to be the physicist's outstanding piece of work as gauged in 1965.[8] The 84 physicists are, in fact, a subsample of a sample of 120 university physicists chosen from a sampling frame in which the population of university physicists was stratified along four dimensions: age, prestige rank of their university department, productivity, and the number of honorific awards received. A second set of data consists of information on a one-third random sample of the scientists cited in the best paper of each of the 84 physicists. For the sample of 385 cited authors we collected data that enabled us to locate them in the stratification system.[9]

A basic assumption in this analysis is that the research cited by scientists in their own papers represents a roughly valid indicator of influence on their work. Of course, not all citations represent direct and specific influence. Everyone knows that scientists occasionally make ceremonial citations to friends, colleagues, or eminent people in the field. Sometimes a citation to an expert in the field serves the function of legitimating the new paper. Even when we cite work that has influenced us, it is difficult for the reader to know when it represents a significant, even necessary, antecedent to our work as opposed to a tangentially relevant piece of work, in which we are merely demonstrating our "knowledge of the literature." Furthermore, relevant and influential material is passed from one scientist to another through private communication, which, though often mentioned in today's age of Big Science, sometimes does not show up as a citation. However, a reasonable case can be made that citations generally represent an authentic indicator of influence.[10]

Let us consider the process through which we decide what to cite in our papers. Some part of our citations will be very clear-cut. We will, of course, cite papers that contributed directly to the current state of knowledge in our problem area. In this paper, for example, such a citation would be to the work of Derek de Solla Price. Another group of references, however, would be more questionable, namely, those to people who have done work in the area but have not had a direct influence on the paper. The reason for citing these rather than others may be that we tend to cite scientists having the highest visibility. Scientists gain visibility originally by publishing significant research. After such visibility is gained, they enjoy a halo effect as their research gains additional attention because of their visibility. Thus, if we consider the sum of a scientist's citations, some part will be due to the halo effect. But the size of the effect will probably be directly related to the significance of the scientist's research. The processes of objective evaluation of contributions and the subjective working of the halo effect combine to create substantial gaps between the number of citations received by members of the elite and the average scientist.

The halo effect would cause us to cite a scientist whose work was not directly influential. But we are primarily interested in situations in which work that is directly influential is not cited. The norms of science require scientists to cite the work that they have found useful in pursuing their own research, and for the most part they abide by these norms. Moreover, the audience of the work generally takes citations as an indicator of influence. We only have to think of the number of times we have taken a quick glance at the acknowledgments and references in books and papers with the intent of noting the influence on a piece of work to realize that, at the very least, citations do indicate intellectual connections.

Sometimes, however, a crucial intellectual forebear to a paper is not cited. The omission is rarely due to direct malice on the part of the author but more often to oversight or lack of awareness. It occurs most frequently when a scientist's work has had such a deep impact on the field that the ideas have become part of the accepted paradigm and explicit citation is not considered necessary. Only the work of a handful of scientists ever achieves this status, and they generally receive very heavy citation anyway. (The work of Einstein, for example, was cited 281 times in the 1970 edition of SCI.) We can assume that omitted citations to less influential work are random in nature; even if we may fail to cite the important work of a particular scientist, others will not likely make the same error. In general, the procedure of using citations as an indicator of influence probably errs on the side of overinclusion rather than exclusion of significant influences.

Characteristics of Cited Authors

The characteristics of the sample of 385 authors cited in the best papers of eighty-four university physicists are presented in table 19.1. We wish to compare the characteristics of these cited authors with those of the population of physicists. This is difficult because many of the population parameters are unknown. We have, therefore, chosen a sample of 1,308 university physicists as the comparison group; for a full description of this data set see (2). This sample, of course, is itself an elite group and far from representative of the more than twenty-five thousand American physicists. As the data indicate, physicists in the top strata are far more likely to be cited than those below the top. Whereas 73 percent of the sample of 1,308 university physicists had no awards listed after their names in AMS, only 33 percent of the cited authors had no awards.[11] The same results are found when we examine citations to the work of the cited authors and the university sample. On average, the cited authors

TABLE 19.1 Marginal Distributions of the Social and Individual Characteristics of the Authors Cited in 84 Papers and Comparative Figures for the Entire Field of Physics

Characteristics of cited authors		
Category	Percent	Comparative "population" statistics (%)
Current affiliation		
University	72	
College or nonacademic research laboratories	10	43
Industry	10	34
Government	8	11
N	(385)	(26,698)
Rank of department of those in academic departments		
Distinguished (top nine)	60	21
Strong and good	23	42
Lesser universities and colleges	17	37
N	(299)	(1,308)*
Number of honorific awards		
0	32	73
1	18	15
2 or 3	23	9
4 or more	27	3
N	(385)	(1,308)*
Quality of scientific output (number of citations: 1965 SCI)		
Under 15	25	67
15 to 29	33	25
60 or more	43	8
N	(385)	(1,308)*

* These figures are drawn from the sample of 1,308 university physicists.

Source: *American Science Manpower 1964: A Report of the National Register of Scientific and Technical Personnel*, NSF-66-29 (National Science Foundation, Washington, DC, 1966).

438 CONSENSUS IN SCIENCE

received 119 citations to their life's work in 1965; all the authors listed in the 1965 SCI received a mean of six citations. Further, although only 8 percent of the 1,308 university physicists averaged sixty or more citations, 43 percent of the cited authors exceeded this lofty number.[12]

The data in table 19.1 lead to the conclusion that most of the work used by university physicists in their best papers is produced by only a small proportion of those who are active in the field. It is equally important, however, to note that a significant *minority* of cited work is being produced by non-elite physicists. So far, we have not made any distinctions among the citing papers. We have merely considered the references in the "best" papers of a stratified random sample of university physicists. Many of these best papers may have been of relatively little significance. If the Ortega view of science is correct, we should find the top papers making just as much use of the work of little-known physicists as the less significant papers. We shall present three sets of data to test this hypothesis. Obviously, the number of citations to these best papers varied greatly. Some papers received only one or two citations; others received more than twenty or thirty. We shall now see the extent to which authors of papers of varying quality depend upon the work of elites and non-elites.

As the data of table 19.2 indicate, highly cited papers, more often than those receiving few citations, make use of high-quality work produced predominantly at the nine most distinguished departments.[13] We see that a mere 7 percent of the citations in the most highly cited discoveries go to scientists working in the lower-prestige university departments and colleges, whereas 60 percent are to scientists at distinguished departments. Even in the papers receiving less than ten citations, the work of scientists at top universities is cited much more frequently than that of those in departments of lower prestige. The best papers cite other significant papers predominantly. Forty-four percent of the citations in papers receiving twenty or more citations and 33 percent in those receiving less than ten citations go to the work of scientists who have received sixty or more citations. Finally, note the extent to which papers of high quality rely on the work of Nobel laureates and National Academy members. Forty-five percent of the citations in these papers go to the work of no more than two hundred scientists and their collaborators. Although the papers of lower quality do not cite these "elites" to the same extent, the work of the elites receives proportionately greater use in these papers as well.

Throughout this article we have defined important discoveries simply by the number of citations they have received. As a further test of the Ortega hypothesis, we asked a well-known physicist to list the five most important contributions to elementary particle physics in the last ten years. We admit that in many ways this procedure falls short of the rigorous study needed to test the hypothesis further. It would be useful, for example, to have a broad, stratified panel of judges evaluate the merits of various pieces of research and then look

TABLE 19.2 The Distribution of Citations in Individual Papers of Varying Quality According to the Characteristics of the Cited Scientists: High, 20 or More Citations; Medium, 10 to 19 Citations; Low, 0 to 9 Citations

	Quality of citer's "best" paper		
Characteristics	High (%)	Medium (%)	Low (%)
Rank of department			
Distinguished	60	50	36
Strong/good	14	19	19
Lesser	7	12	18
No academic affiliation	19	19	27
N	(95)	(139)	(151)
Quality of scientific output (number of citations)			
High (60 or more)	54	48	33
Medium (15 to 29)	28	30	36
Low (less than 15)	18	22	31
N	(95)	(139)	(151)
Prestige of highest award			
Nobel Prize, NAS member	45	32	25
Other honorific awards	15	8	12
Only fellowship,* no awards	40	60	63
N	(95)	(139)	(151)

* Fellowships such as the Guggenheim, Sloan, Rockefeller, and Fulbright were here considered as honorific awards as distinct from other postdoctoral fellowships.

at the citation patterns in papers judged to be of highest impact. It is noteworthy, however, that the five papers chosen by our informant received a mean of sixty-seven citations in the 1965 SCI.[14] We examined all the journals and private communication citations in these five papers and then located the cited authors in the stratification system of physics. The five papers cited a total of fifty-one articles (not counting self-citations) involving 126 authors, of whom nineteen were located at foreign universities and foreign research laboratories.

The data corroborate the earlier findings. Of the 107 American scientists cited in these five "pathfinding" papers, all but one were located at one of the top nine physics departments in the United States or at such distinguished laboratories as Brookhaven or the Lawrence Radiation Laboratory. All fifty-one articles were produced at one of these top nine departments or laboratories. The average number of citations to the cited authors is equally impressive. This group had a mean number of citations of sixty-nine in the 1965 SCI; 74 percent

440 CONSENSUS IN SCIENCE

of these authors had more than sixty citations to their work in 1965.[15] Among the 107 cited authors were a number of younger and not yet widely recognized scientists who were coauthors of more eminent colleagues. The mean number of citations to either a single author or the most highly cited author of collaborative papers is 134.

Additional Test of the Ortega Hypothesis

Since this type of subjective sampling procedure may indeed be methodologically suspect we decided to perform one final test of the Ortega hypothesis. We replicated the essential aspects of the study design using a set of independent data. We had a complete list of all papers cited three or more times in *Physical Review* in 1965; it contained more than three thousand scientific articles and substantive letters. A few of these papers were cited often; most received less than five citations. Since we are primarily concerned with the pattern of citation in influential papers, we initially examined the ten papers that were most often cited in *Physical Review*. After identifying these "super" papers, we listed the scientific articles that the authors of these influential papers cited. Finally, we counted the number of citations received in 1965 by authors of papers cited in the "super" articles.

This procedure can be clarified by reference to a specific example. Murray Gell-Mann produced the most heavily cited article on the list, receiving a total of forty-nine citations in *Physical Review* in 1965. We listed the references in Gell-Mann's paper. A total of thirty-three publications, or fifty-five scientists, were cited in the paper. We then noted the number of citations that the life's work of each of these scientists had received in the 1965 SCI. The same process was followed for all the scientists cited in the ten super papers. Thus, even though we were examining only the ten most highly cited papers, we studied a total of 299 authors.

The results obtained from this replication offer further evidence in support of the earlier findings. It turns out that authors cited in these ten papers were scientists who, on average, had produced truly outstanding scientific work. In 1965 the 299 cited scientists produced research that received an average of 135 citations. Since this average includes citations received by beginning scientists yet to make their mark, who are collaborating with their more eminent colleagues, the statistic is actually lower than it would be otherwise. In fact, if we take only single-authored papers and the most-cited author in each collaborative paper and compute the average number of citations to the author's life's work, the mean is increased to 175 citations. Clearly these data lend added weight to the counter hypothesis that work used by the producers of outstanding research is itself produced by a small minority of scientists. The work of the average researcher is rarely the work that is influential in the production of high-impact scientific research.

A question remains to be answered. What is the quality of the research cited in the work of the physicists whose papers receive fewer citations than those of the ten "super" papers? From the same *Physical Review* list we drew a small random sample of papers that had received from 23 to 3 citations in 1965. Papers that received 23 citations were of approximately the same impact as some of the top ten, since the citations to the "super" papers ranged from 49 to 24. This small sample consisted of 36 papers, within which references were made to 492 communications. We computed the number of citations in 1965 to the 837 physicists who authored these 492 papers. We examined citation rates to cited authors who produced single-authored papers and those in collaborative teams whose work was most often cited. The data presented in table 19.3 suggest that even authors of less than super quality papers were predominantly influenced by work of elites. Whereas the top ten papers made use of work produced by physicists who received an average of 175 citations to their life's work, the average quality of cited work found in papers receiving 5 to 9 citations is not appreciably lower. Only when we examine the citation patterns in papers receiving 3 or 4 citations is the average quality of work cited significantly lower. But, even in this group, the scientists cited are among the elite insofar as the quality of their work goes.

These high averages are not due to a handful of extreme individuals. Forty-one percent of the 837 cited physicists received more than 100 citations; another 13 percent, from 60 to 69. Thus, a total of 54 percent received more than 60 citations, a figure similar to those presented in table 19.2. Only 11 percent of the cited authors received less than 5 citations to their life's

TABLE 19.3 Citation Patterns Found in Papers Cited in the 1965 *Physical Review*

Citations to paper in the 1965 *Physical Review* (No.)	Citations in 1965 to the life's work of major authors* cited in the papers	
	Mean No.	No. of major authors
24–49	175	(174)
20–23	169	(88)
10–17[†]	158	(215)
5–9	149	(124)
3–4	85	(65)

* By major authors is meant all single authors and for collaborative papers the author whose life work has received the highest number of citations in the 1965 SCI.

[†] No papers were cited 18 to 19 times in *Physical Review* in 1965.

442 CONSENSUS IN SCIENCE

work, and 90 percent of the scientists comprising this 11 percent were coauthors on papers for which one of the other authors was more heavily cited. In short, there were virtually no cited authors whose work was not of above average quality.

Consider once again a set of comparative statistics: (i) about one-half of all the papers published in the more than 2,100 source journals abstracted in the SCI do not receive a single citation during the year after it is published; and (ii) the average cited author in the 1965 SCI received a mean of 6.08 citations to his life's work. These data offer further support for the hypothesis that even the producers of research of limited impact depend predominantly on the work produced by a relatively small elite.

Conclusions

Let us consider, then, some general conclusions that may be drawn from the findings reported in this study. The data allow us to question the view stated by Ortega, Florey, and others that large numbers of average scientists contribute substantially to the advance of science *through their research*. It seems, rather, that a relatively small number of physicists produce work that becomes the base for future discoveries in physics. We have found that even papers of relatively minor significance have used to a disproportionate degree the work of the eminent scientists. Although the conclusions of this paper may be reasonably clear, the implications of these data for the structure of scientific activity, at least in physics, need careful consideration.

Consider only one problem emerging out of the findings that needs a great deal of further research: the size of the research establishment of modern science. If future research on other fields of science corroborates our results, we may inquire what it implies about the relationship between the number of scientists and the rate of advance in science and whether it is possible that the number of scientists could be reduced without affecting the rate of advance. The data would seem to suggest that most research is rarely cited by the bulk of the physics community and even more sparingly cited by the most eminent scientists who produce the most significant discoveries. Most articles published in even the leading journals receive few citations. In a study of citations to articles published in *Physical Review*, we found that 80 percent of all the articles published in the *Review* in 1963 were cited four times or less; 47 percent, once or never in the 1966 SCI. Clearly most of the published work in even such an outstanding journal makes little impact on the development of science. Thus, the basic question emerges, whether the same rate of advance in physics could be maintained if the number of active research physicists were to be sharply reduced.

Several criticisms of our position are possible.

1) The data indicate that about 15 to 20 percent of the work cited in significant discoveries is produced by "average" scientists. We do not know whether the important discoveries could or could not have been made if only the work of eminent scientists had been considered. It might be maintained that the 20 percent of references produced by, let us say, 80 percent of researchers are just as crucial for scientific advance as the 80 percent of references produced by 20 percent of researchers. To suggest a reply to this first criticism we must make explicit an idea implicit in much work done in the sociology of science. Our entire analysis is dependent upon the assumption that no single scientist, elite or non-elite, is crucial for scientific advance. The study of independent multiple discoveries leads to the conclusion that, if a particular scientist had not made a particular discovery, another would have.[16] If the scientist who makes a discovery had not made it, it would be only a matter of time—probably a relatively short period—before the discovery would be made by another scientist.

The history of science is replete with examples of discoveries made independently by two or more scientists within a short period. Merton has suggested that multiple discoveries are the norm rather than a rare occurrence. Furthermore, many discoveries that are not multiples are forestalled multiples, for most scientists will stop working on a problem when they learn of the success of a competitor. As we learned from *The Double Helix*, if Watson and Crick had not made their historic breakthrough it probably would have been made in short order by Pauling. Most scientists working on important problems realize that many others are working on the same problems. Indeed, chance often plays its part in determining who makes a discovery first.

If the work done by any scientist, elite or non-elite, can be replaced by work done by other scientists, how then do we evaluate the extent to which a particular scientist is necessary for scientific advance? Merton (*19*) defines the scientific genius as a man who is involved in multiple multiples—the functional equivalent of many other scientists (*19*). Although no one citation or one man is crucial for any scientific discovery, the scientist who writes one paper that is cited once in an important discovery is less crucial than the scientist who writes many papers that are cited many times in many important discoveries.

Even though it might be maintained that all the work referred to in a paper is necessary for the production of that discovery, it does not therefore follow that all the individuals cited were essential for the discovery. Although all scientists are replaceable in the sense that other scientists would eventually duplicate their discoveries, some scientists have many more functional equivalents than others. For example, it would be relatively difficult to replace the work of a Murray Gell-Mann but not so difficult to replace that of a scientist who is cited once in one of Gell-Mann's papers. If the less distinguished scientists have many functional equivalents, so do the many laboratory technicians and staff workers who often perform vital tasks in the making of scientific discoveries. We are not saying that the tasks are unnecessary but that there are many people who could perform them.

We draw an analogy that may crystallize the point. Sanitation workers perform socially useful and necessary functions. Without them a complex industrial society would not function very smoothly. A prolonged strike of those workers would probably create more chaos than a strike of teachers, social workers, or even perhaps nurses and doctors. Yet the job that they do could be performed by the National Guard, whereas the jobs performed by professionals could not be handled by untrained people. The sanitation worker is given little prestige in the hierarchy of occupations not only because of the lower salary and poorer working conditions he has relative to a doctor, a lawyer, or a scientist but also because he has many more functional equivalents in the social system than a professional man.[17] It is far easier to find replacements for the individual sanitation worker than for the individual scientist or doctor. The same principle operates within a single occupation. Within science some men are more easily replaced than others. We suggest that it may not be necessary to have 80 percent of the scientific community occupied in producing 15 or 20 percent of the work that is used in significant scientific discoveries, if perhaps only half their number could produce the same work.

2) A second possible criticism of our analysis is that we have dealt with only one generation of influence. Untested in this paper is the possible "filtration" of ideas from the lower to higher levels of the stratification system. The filtering process may take a number of "generations" of papers before the low-impact papers have an influence on important discoveries. Further, in the process of filtration a minor contribution may be entirely absorbed by the next generation paper that makes use of it. Thus, only a single citation might be necessary for a piece of work to become part of the stockpile of knowledge. A minor contribution then might ultimately have an effect on the production of a great idea through a "great chain" of papers. The first links in the chain would be concealed from our vision because they were not cited by later generations. What is clearly called for is a study of the sociometrics of multiple generations of papers, in which we would examine the number of scientists added to the list of those who influence discoveries as we add new generations of papers.

We are now tracing patterns of influence. As we go back we add new names to the matrix. But, in line with the assumption of the "replaceability" of scientists, we would argue that the crucial question is not how many new scientists, but how many central names, are added to the matrix. We might define a scientist who appears three times or more as "central." We would guess that as new scientists are added to the matrix the proportion of central scientists will drop off sharply and soon hit zero. We hypothesize that we will not have to look at many generations of influence before we find that all new names added to the matrix are appearing only once. Kessler found the same pattern in his study of citations in *Physical Review:* 95 percent of the references were to articles published in the *Review* itself plus fifty-five other journals. He suggests,

> The same list of 55 journals . . . will account for the majority of references year after year. The remaining 5 percent of the references is to a large and ever-growing list of rarely used sources. . . . This list has no stability in time; each new volume examined is destined to carry 96 percent of the references in the subsequent 35 volumes. As we examine those volumes, 78–96, it is clear that, although the list of new titles never ends, their contribution to the total reference literature is comparatively small.[18]

3) An additional criticism of this article could be that we have considered only the research function of scientists. As we just pointed out, scientists can, of course, make important contributions to the advance of science through excellent performance in other roles, like teaching and administration. However, just as it would be incorrect to ignore these important roles, it would also be an error to assume that separation of these roles is necessary. Possibly the same scientists who produce the most significant research are also doing the most significant teaching and administration.

Let us look at the teaching function performed by scientists. If the assumption is correct that it is primarily elite scientists who contribute to scientific progress through their research, we should be primarily concerned with the teachers of future members of the elite. We know from qualitative sources and statistical studies of Nobel laureates, National Academy members, and other eminent scientists that the great majority of scientists who end up in the elite strata are trained by other members of the elite.[19] In fact, 69 percent of current members of the Academy and 80 percent of American Nobelists received their doctorates from only nine universities. It might be facetiously asserted that the best way to win a Nobel Prize is to study with a past laureate. Analysis of the graduate schools attended by physicists whose work is heavily cited indicates that a large majority of scientists who turn out to be productive get their doctorates at the top twenty graduate departments. We would not claim that unproductive scientists teaching at low-prestige institutions serve no function: they may, for instance, serve the truly important function of educating nonscientists to the objectives and methods of science. Nevertheless, little evidence exists that they contribute to the progress of scientific research through their teaching.

4) Another possible criticism could be that, even if all our hypotheses were supported by the necessary extensive future inquiries, we would still be left with a critical and difficult problem before any policy implications that may be found in these data could possibly be acted upon. We would still have to

identify correctly the scientists who would go on to produce important scientific discoveries. We would need a set of accurate predictive measures that could identify at an early age the students who would produce truly significant discoveries. Although the problem lies beyond our current capabilities, we believe that it would not be as difficult to solve as it seems at first.

We pointed out above that the majority of scientists who contribute to scientific progress are educated at a small number of graduate institutions. Probably most of the exceptions chose to attend institutions of lower prestige for idiosyncratic personal reasons rather than because they were not admitted to one of the leading departments. If each field had only about twenty graduate departments, all individuals showing any talent and interest in it would by necessity have to apply to one of these institutions. It is unlikely that, if twenty departments each admitted between twenty-five and fifty new graduate students each year, thus potentially reducing by a factor of two the number of doctorates granted, many students who in fact had the potential to make important scientific contributions would be denied access to graduate education. For example, if twenty graduate departments of physics admitted only fifty students a year, forty of whom were to receive their doctorates in due course, these twenty departments would produce eight hundred PhDs each year, or about half the total number of American physics doctorates awarded in 1970.[20] A reduction in the absolute number of training centers would not imply a reduction in the competition between these universities for talented researchers or students.

It is a well-known fact that new PhDs in science, especially in physics, are having a difficult time finding jobs. Most projections of supply and demand for scientists are not optimistic.[21] One way to handle this inequity in supply and demand is to cut back sharply the number of PhDs we are producing. The data we have reported led to the tentative conclusion that reducing the number of scientists might not slow down the rate of scientific progress. One crucial question remains to be answered: If the number of new PhD candidates is sharply reduced, will there be a reduction in the number of truly outstanding applicants or will the reduction in applicants come from those we would now consider borderline cases? This is not a matter of social selection, for we believe it possible for academic departments to distinguish applicants with high potential. It is a matter of self-selection. A reduction in the size of science might motivate some very bright future scientists to turn to other careers.

The ability of an occupation to attract high-level recruits depends to a great extent on the prestige of the occupation, working conditions, and perceived opportunities in the occupation. We, of course, do not intend to suggest the advisability of any policy that would either reduce the prestige of science or the resources available to scientists. What we are suggesting is that science would probably not suffer from a reduction in the number of new recruits and an increase in the resources available to the resulting smaller number of scientists. Perhaps the most serious problem that science faces today in recruiting is the perceived reality that there are few jobs available to new PhDs. Reducing the size of science so that supply would be in better balance with demand might ultimately increase the attractiveness of science as a career.

Notes

1. J. Cole and S. Cole, *Social Stratification in Science* (Chicago: University of Chicago Press, 1973); S. Cole and J. Cole, *American Sociology Review* 32, no. 377 (1967); S. Cole and J. Cole, *American Sociology Review* 33, no. 397 (1968); J. Cole and S. Cole, *American Sociology Review* 6, no. 23 (1971).
2. J. Ortega y Gasset, *The Revolt of the Masses* (New York: Norton, 1932), 84–85.
3. This quotation appears in J. G. Crowther, *Science and Modern Society* (New York: Shocken, 1968), 363.
4. For detailed and informative discussions of fluctuations in rates of discoveries in the history of science, see P. A. Sorokin and R. K. Merton, *Isis* 22, no. 516 (1935); P. A. Sorokin, *Social and Cultural Dynamics*, vol. 2 (Englewood Cliffs, NJ: Bedminster, 1962); and J. Ben-David, *American Sociology Review* 25, no. 828 (1960). For a qualitative treatment of the same idea, see G. Sarton, *History of Science and the New Humanism* (New York: Holt, 1931), especially 34–42.
5. T. S. Kuhn, *The Structure of Scientific Revolutions* (Chicago: University of Chicago Press, 1962). We use the term "elite" here and throughout in a statistical sense to refer to the small group of eminent scientists who publish the most, are most frequently cited, and occupy the most prestigious positions. In fact, Zuckerman notes that this statistical elite does form a fairly cohesive social group. See H. Zuckerman, "Stratification in American Science," in *Social Stratification: Research and Theory for the 1970s*, ed. E. Laumann (Indianapolis, IN: Bobbs-Merrill, 1970), 239.
6. D. Price, *Little Science, Big Science* (New York: Columbia University Press, 1963), 43ff; A. J. Lotka, *Journal of the Washington Academy of Science* 16, no. 317 (1926).
7. The *Science Citation Index* (SCI), published in Philadelphia by the Institute for Scientific Information under the direction of Eugene Garfield, lists all references made each year in more than two thousand journals. Much research has been done showing that the number of citations a scientist's work receives is a roughly valid indicator of its quality. Let us consider just one piece of validating evidence for this measure. Recipients of the Nobel Prize are generally regarded as having contributed greatly to advances in physical and biological sciences. Since the number of Nobel Prizes is limited, however, there may be other like-sized aggregates of eminent scientists who have contributed as much. Nevertheless, the laureates as a group can be safely assumed to have made outstanding contributions. The average number of citations in the 1961 SGI to the work of Nobel laureates (who won the prize in physics between 1955 and 1965) was 58, as compared with an average of 5.5 citations for other scientists. Only 1.08 percent of the quarter of a million scientists who appear in the 1961 SCI received 58 or more citations. We thought it possible that winning the prize might make a scientist more visible and lead to a greater number of post-prize citations than the quality of his work warranted. We therefore divided the laureates into two groups: those who won the prize five or fewer years before 1961 and those who won it after that year. The 1957–1961 laureates

448 CONSENSUS IN SCIENCE

were cited an average of 42 times in the 1961 SCI; the future prize winners (those winning the prize between 1961 and 1965), an average of 62 times. Since the prospective laureates were more often cited than the actual laureates, we concluded that the larger number of citations reflects the high quality of work rather than the visibility gained by winning the prize.

8. Since we were primarily interested in the influences upon the pure research of these scientists and were using citations to measure that influence, we decided to omit any "review article" from consideration, for we were not interested in a review of the literature and thought that citations to an enormous amount of literature of this type would distort our results. Therefore, if a review article was the most heavily cited, we chose the next most heavily cited paper to include in our sample. In looking up the cited authors within these individual papers we were faced with the problem of what to do with collaborating authors. We decided to treat a paper as a single unit and locate information on all collaborators in the research team. To some extent, junior collaborators, many of whom were students of the senior authors, dropped out of our sample because no information could be found for them in *American Men of Science* (New York: Cattell Press). But it can be seen that the average level of eminence of cited authors is weighted against our hypothesis to some extent, since our sample included junior men of distinctly less eminence than their senior collaborators.

9. One of the limiting features of this study has to do with the collection of information on the scientists who are being studied. Perhaps the best source for independent information short of a questionnaire sent to a sample of physicists is the *American Men of Science* (AMS). The sample includes only men and women listed in AMS. Only about one-half of all cited authors appeared in these volumes. We wanted to see whether there were systematic differences in the types of scientists who appear in AMS and those who do not. We found, of course, that a large proportion of the men who were not found in AMS were foreign scientists. The second largest group of individuals who could not be found turned out to be students at the institution where the research was being done. It is not infrequent that older, more eminent scientists collaborate with their doctoral students. It should be added that AMS tends to include more academic scientists than scientists in industrial concerns. We did find that the average number of citations to men in AMS was approximately 1.5 times that of those not found in it. We also found, as was to be expected, that the cited authors in the best work were more often found in AMS than those cited in work of lesser quality. To the extent that the AMS does not include the less eminent members of the scientific community our sample of cited authors overrepresents eminent scientists.

10. The extent to which unpublished work is being cited in leading journals is increasing rapidly, at least in physics. Second to articles published in *Physical Review* (American Institute of Physics, New York), private communications are the most-cited source of information in contemporary physics.

11. It would probably be safe to assume that more than 90 percent of the population of physicists have no awards.

12. Inclusion in the scientific elite could be a function of longevity if the bulk of citations went to older scientists. The data do not support this possibility. In fact, the pattern of citations by scientists in various age groups suggests that older scientists tend to cite work by older scientists; younger scientists tend to cite most often the work of other young scientists. More than 50 percent of the cited authors, however, were under fifty years old.

13. The so-called "Cartter" rankings of leading departments of physics were based on evaluations of eighty-six institutions "that reported the award of one or more doctorates in

physics from July 1952 through June 1962." The ratings were based on a scale ranging from 5 (highest) to 1 (lowest). All institutions with a mean ranking greater than 4.0 were called "distinguished." There were nine such physics departments. A. M. Cartter, *An Assessment of Quality in Graduate Education* (Washington, DC: American Council on Education, 1965).

14. These five articles included, for example, Lee and Yang's now famous paper on parity conservation. Three of the authors turned out to be Nobel Prize winners; the others, members of the National Academy of Sciences.

15. Nobel laureates in physics who received their prize between 1950 and 1964 averaged 130 citations to their life's work in the 1965 SCI.

16. R. K. Merton, "Singletons and Multiples in Scientific Discovery: A Chapter in the Sociology of Science," *Proceedings of the American Philosophical Society* 105, no. 470 (1961): 470–486.

17. See K. Davis and W. E. Moore, "Some Principles of Stratification," and, for a critique, M. M. Tumin, "Some Principles of Stratification: A Critical Analysis," both reprinted in R. Bendix and S. M. Lipset, eds., *Class, Status, and Power: Social Stratification in Comparative Perspective* (New York: Free Press, 1966).

18. M. M. Kessler, "Some Statistical Properties of Citations in the Literature of Physics," Report (Cambridge: Massachusetts Institute of Technology, 1962).

19. H. A. Zuckerman, "Nobel Laureates in Science: Patterns of Productivity, Collaboration, and Authorship," *American Sociology Review* 32, no. 391 (1967).

20. National Research Council, *Summary Report 1970, Doctorate Recipients from United States Universities* (Washington, DC: National Research Council, 1971).

21. D. Wolfle and C. V. Kidd, *Science* 173, no. 784 (1971).

CHAPTER 20

CHANCE AND CONSENSUS IN PEER REVIEW

(1981)

STEPHEN COLE, JONATHAN R. COLE, AND GARY A. SIMON

Summary. *An experiment in which 150 proposals submitted to the National Science Foundation were evaluated independently by a new set of reviewers indicates that getting a research grant depends to a significant extent on chance. The degree of disagreement within the population of eligible reviewers is such that whether or not a proposal is funded depends in a large proportion of cases upon which reviewers happen to be selected for it. No evidence of systematic bias in the selection of NSF reviewers was found.*

The National Science Foundation (NSF) employs one form of the peer review system in making research grants. For each application for a grant, an NSF program director selects a group of scientists, generally four or five, who are knowledgeable in the relevant subject matter, to act as referees. Each reviewer is sent a copy of the proposal and asked to evaluate it on the basis of its scientific merit and the ability of the principal investigator. Ability of the principal investigator is generally defined as the quality of his or her recent scientific performance. Each reviewer is asked to make substantive comments and to assign one of five ratings to the proposal: excellent, very good, good, fair, or poor. We ask whether the procedure employed by NSF is an equitable and a rational one.

For the past five years, as consultants to the National Academy of Sciences' Committee on Science and Public Policy (COSPUP), we have been conducting a study of NSF's peer review.[1] This work has been divided into two phases.

In this article we report on the second phase of that extended inquiry. Since they represent the point of departure for the experiment described here, we recapitulate briefly the principal results of the first phase, which were based on seventy-five extended interviews with NSF staff, on analysis of 1,200 proposals drawn from ten NSF programs, and on the substantive comments of reviewers of 250 of these proposals:

1) There is a high correlation between reviewer ratings and grants made. If one attaches numerical values to the ratings, say from 10 for poor to 50 for excellent, the mean scores predict with a high degree of accuracy which proposals will be funded and which will be denied. Whether or not NSF program directors actually compute statistical averages from the ratings and use them in decision-making, the statistical average of the ratings turned out to be highly correlated with the actual decision rules employed by the program directors.

2) For the 1,200 proposals there was not a high correlation between grants awarded and measures of the previous scientific performance of the applicants. This result was unexpected since one of the stated evaluation criteria is the ability of the applicants to conduct the research proposed.

3) Reviewers at major institutions did not treat proposals from applicants at major institutions more favorably than did reviewers from lesser institutions. In fact, there was a tendency in the opposite direction.

4) Professional age (length of career) had no strong effect on either ratings received or the probability of receiving a grant.

5) There were low or moderate correlations between reviewer ratings (and the funding decision) and the following characteristics of the applicants: prestige rank of current academic department, academic rank, geographic location, NSF funding history over the previous five years, and locus of PhD training.[2]

Because proposals from eminent scientists do not have substantially higher probabilities of receiving favorable ratings than proposals from scientists who are not eminent, we concluded that the peer review system employed by NSF was essentially free of systematic bias. We now want to take up the further question of whether the system as currently employed is, in addition to being equitable, a rational one. In particular, we are concerned with the role of chance in obtaining an NSF grant. A rational system would minimize random elements and maximize the influence of both the quality of the proposal and the ability of the principal investigator to perform the research.

The COSPUP Experiment

The second phase of the study was designed to tell us, among other things, whether or not the program directors were predetermining funding decisions

452 CONSENSUS IN SCIENCE

by their selection of reviewers—that is, whether independently selected panels of reviewers would reach similar conclusions.

In the spring of 1977, the NSF provided us with 150 proposals—fifty each from the programs in chemical dynamics, economics, and solid-state physics—upon which decisions had been made recently; half the proposals in each program had been funded and half had been declined. We then obtained other reviewers for those proposals. In order to select the new reviewers we utilized a panel of ten to eighteen experts in each of these fields, most of them members of the National Academy of Sciences.[3] Each proposal was sent to two members of this panel; each selected six or more reviewers for it. This gave us a list of approximately twelve reviewers for each proposal.

Some have argued that the highly specialized state of modern science would not permit more than a dozen or so scientists to be capable of reviewing any given proposal. The COSPUP experiment enabled us to test this hypothesis. If the number of eligible reviewers was, in fact, small, we would expect that a fairly high proportion of the original NSF reviewers would also have been selected by the experimental selectors. In each of the three programs, about 80 percent of the NSF reviewers were not selected by either of the two COSPUP selectors, about 15 percent were selected by one of them, and about 5 percent were selected by both. These data suggest that the pool of eligible reviewers for most proposals includes at least ten members and, given the low overlap rates we found, we would predict that if other equally qualified selectors were employed we would find it to be substantially larger than twenty.[4] Of course, in actual practice there is not a clear distinction between eligibles and noneligibles, and the numbers certainly vary according to subspecialties. Since the pool of eligible reviewers for most proposals is substantially larger than the actual number of reviewers used by NSF, and since there was little overlap between NSF and COSPUP reviewers, we want to consider the extent to which the two sets of reviewers agreed upon the merits of the proposals.

In general, the COSPUP reviewers tended to give slightly lower scores than did the NSF reviewers (see table 20.1). For example, for the fifty chemical dynamics proposals, the grand mean of the NSF reviewers' ratings was about 38 on the 10-to-50 scale, and that of the COSPUP reviewers was about 35. The experimental reviewers may have been slightly harsher in their evaluations because they knew that their ratings would have no effect on the careers of the applicants. The correlations between the mean NSF rating and the mean COSPUP rating for each proposal are moderately high (.60, .66, and .62). Proposals that are rated high by NSF reviewers tend also to be rated high by the independent sample of reviewers used by COSPUP. The match is, however, less than perfect.

TABLE 20.1 Correlation of Mean Ratings of NSF Reviewers and COSPUP Reviewers on Grant Applications (N = 50) in Each of Three NSF Programs Numerical values were assigned to the original ratings as follows: excellent = 50, very good = 40, good = 30, fair = 20, and poor = 10. Figures in parentheses are standard deviations.

	Mean ratings		
Field	NSF	COSPUP	Correlation coefficient
Chemical dynamics	37.7 (±5.85)	35.0 (±6.45)	.595
Economics	33.6 (±9.76)	31.5 (±9.28)	.659
Solid-state physics	38.2 (±6.15)	35.5 (±6.33)	.623

The findings presented thus far do not address one of the fundamental questions for evaluating the peer review system at NSF: How many funding decisions would be reversed if they were determined by the COSPUP ratings rather than by the procedures followed by NSF?

In figure 20.1 we show the rank order of the proposals in each program according to the NSF mean ratings and the mean ratings of the COSPUP reviewers. (Half-integer ranks are the result of ties.) Since the mean ratings generally determine which proposals are funded, and half were funded and half declined, decisions on proposals which were ranked in one-half of the range of scores by NSF reviewers and in the other half by COSPUP reviewers would have been reversed by the COSPUP ratings. There were differences of that degree in the ratings of approximately one-quarter of the proposals.

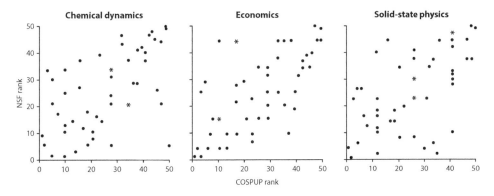

20.1 Rank order of proposals according to mean ratings by NSF and COSPUP reviewers. N = 50 in each program. (Asterisk indicates two proposals with identical ranks.)

Reversals

The NSF is faced, of course, with a zero-one decision rule: to fund or not to fund a proposal.[5] It follows that proposals with mean rankings that are fairly close, or virtually identical, may fall on opposite sides of the dividing line. Therefore, it was almost inevitable that we would find some reversals.

In determining what should be classified as a reversal we rank-ordered the proposals according to their mean COSPUP ratings and assumed that those with the top twenty-five scores would be funded and the bottom twenty-five would be declined. We then compared the COSPUP ratings with both the actual NSF decision and the decision NSF would have made if it had relied solely on mean ratings. The NSF funding decision was highly, though not perfectly, correlated with the mean ratings that NSF reviewers had given the proposals; hence, the two comparisons yield approximately the same results (see table 20.2).

If decisions on the fifty proposals were made by flipping a coin, we would expect to obtain a 50 percent reversal rate, on average. Correlatively, if the COSPUP reviewers were to rate the proposals in such a way that there was complete agreement with the NSF reviewers on which proposals were the top ones and which were the bottom, the reversal rate would be zero. Thus, we would expect to find a reversal rate somewhere between zero and 50 percent. In fact, the reversal rate turns out to be between 24 percent and 30 percent for each of the three programs computed in each of the two different ways. That is,

TABLE 20.2 Percentage of NSF Outcomes (Mean Rating of NSF Reviewers or Actual Funding Decision) Reversed in COSPUP Rank-Order Quintiles and Overall Reversals are shifts from the top 25 positions in the COSPUP rank order to the bottom 25 or vice versa.

NSF outcome	Quintile based on COSPUP rating					Overall ($N = 50$)
	1	2	3	4	5	
Chemical dynamics						
Mean rating	26	24	60	20	20	30
Decision	26	24	60	20	20	30
Economics						
Mean rating	20	45	30	45	0	28
Decision	5	45	28	42	0	24
Solid-state physics						
Mean rating	23	22	49	34	6	27
Decision	16	24	43	29	11	25

on twelve to fifteen of the fifty proposals in a program the COSPUP reviews led to a different decision from that of the NSF reviews.

We would expect to find some reversals around the cutting point—for example, to find that a proposal ranked 24 by NSF was ranked 26 or 27 by COSPUP. We want to examine the extent to which reversals were common not only at the midpoint but at a distance from it. This is shown in table 20.2 by the reversal rates within quintiles.[6] In chemical dynamics and solid-state physics we find, as expected, the highest reversal rate in the middle quintile. A 50 percent reversal rate for this quintile would not be surprising. In chemical dynamics it is 60 percent in both comparisons and in solid-state physics 49 percent and 43 percent. In economics, on the other hand, we find higher reversal rates in the second and fourth quintiles than in the third. In all three programs there are more than a few reversals in the first quintile. There are, in fact, proposals that were rated in the top quintile by NSF reviewers that would not have been funded had the decision depended on the appraisals of the COSPUP reviewers.

There are several possible explanations for the reversals. Differences between NSF procedures and COSPUP procedures will be considered first. If the two sets of reviewers used different criteria in appraising proposals, the outcome could have differed significantly, creating reversals—for example, if one group of reviewers based their ratings strictly on evaluations of the proposal and the other primarily on the past track record of the applicant. Since the two groups of reviewers were given identical instructions about the criteria, it is unlikely that there were systematic differences of that kind.

Another possible procedural cause of reversals might be obtained if NSF and COSPUP selected different types of reviewers. Reviewer differences rather than proposal differences could then result in reversals. Since a comparison of the characteristics of the two groups of reviewers showed few differences, it is likely that they were drawn from the same population.

Assuming that reversals did not result from the procedures employed in the experiment, we are left with two possible substantive explanations. Reversals could result from bias in the way in which the reviewers were selected by either the NSF program director or the COSPUP experiment. If, for example, the NSF program director purposely selected reviewers who would give unrepresentative negative or positive ratings to a proposal, this could create a reversal.

Second, reversals could have resulted from disagreements among fairly selected reviewers using the same criteria. If there is substantial dissensus in the population of eligible reviewers of a given proposal, then it would be possible for equally qualified and unbiased groups of reviewers using the same criteria to differ in the mean rating.

Consider a hypothetical proposal for which there is a population of approximately one hundred eligible reviewers. If all were totally agreed about its merits, then any sample of four or five selected at random would agree among

456 CONSENSUS IN SCIENCE

themselves, and two independently selected samples would not r each different conclusions. However, if the population of eligible reviewers had substantial disagreement about the proposal, two randomly selected samples could yield different mean ratings possibly leading to different outcomes for the proposal. Our data indicate that the reversals in this experiment were a result of such disagreement.

Consensus

In order to determine the extent to which the reversals could be explained by bias or disagreement we used analysis-of-variance techniques. Because we did not want to make the usual statistical assumptions (such as normality) which must be made in a standard two-way analysis of variance, we used a components-of-variance model that did not require some of these assumptions but would be useful in answering the same substantive question.

In order to assess the relative magnitude of contributions of the proposal evaluation method and the reviewer to the variation in ratings, we represent the rating y_{ijk}, given by the kth reviewer under method i to proposal j, by

$$y_{ijk} = a_i + b_j + c_{ij} + e_{ijk}$$

where a_i is the overall average rating by evaluation method i ($i = 1$ for NSF and $i = 2$ for COSPUP), b_j is the differential effect of proposal j, c_{ij} measures the extent to which the rating on proposal j depends on the evaluation method, and e_{ijk} is the effect caused by the kth reviewer of proposal j by evaluation method i.

We consider a_i to be a fixed quantity and the remaining terms to be random with means equal to zero. Then we can decompose the variance associated with proposals under evaluation method i into three terms:

$$\text{Var}\left(Y_{ijk}\right) = \sigma_p^2 + \sigma_I^2 + \sigma_{R,i}^2$$

where $\sigma_p^2 = \text{Var}(b_j)$ reflects the intrinsic variability of the proposals: $\sigma_I^2 = \text{Var}(c_{ij})$ is the variability associated with the interaction of proposals and evaluation method; and $\sigma_{R,i}^2 = \text{Var}(e_{ijk})$ is the reviewer variance for method i.

If σ_p^2 is large relative to σ_I^2, $\sigma_{R,1}^2$, and $\sigma_{R,2}^2$, we interpret this to mean that it is relatively easy to distinguish the proposals independent of the evaluation method. However, if σ_I^2 is of the same order of magnitude as σ_p^2, this would suggest that dependence between proposal and evaluation method is masking some of the intrinsic proposal variability. As a consequence, the proposals would be ranked differently under the two evaluation methods. If, as actually occurs in these data, $\sigma_{R,1}^2$ and $\sigma_{R,2}^2$ dominate σ_I^2 and are of the same

TABLE 20.3 Components of Variance of NSF and COSPUP Ratings

| Field | Proposal variance $\hat{\sigma}^2_p$ | Reviewer variance | | Interaction variance $\hat{\sigma}^2_I$ | Method difference $\hat{a}_1 - \hat{a}_2$ |
		NSF $\hat{\sigma}^2_{R,1}$	COSPUP $\hat{\sigma}^2_{R,2}$		
Chemical dynamics	23.67	55.91	56.67	1.18	2.73*
Economics	58.33	89.22	96.25	0.00[†]	2.14*
Solid-state physics	24.43	48.93	50.24	0.17	2.72*

*NSF is higher.
[†]Computed as −1.36.

magnitude as σ^2_p, then reviewer variability will be so pronounced that two different evaluations will give dissimilar rank orders.

The estimates of σ^2_p, σ^2_I, $\sigma^2_{R,1}$, and $\sigma^2_{R,2}$ are presented in table 20.3. The dependent variable for the analysis is the rating given the proposal by a reviewer. If we consider all the variance in an entire set of reviews (for example, all reviews done by both NSF and COSPUP reviewers for the fifty proposals), we want to know the sources of variance. There are four possible sources of variance, two of which turned out to be trivial in this study. Consider these four sources and the estimated effects for solid-state physics. The results for economics and chemical dynamics have parallel interpretations.

First, reviewers' responses to proposals differ because proposals differ in quality. That is easily dealt with statistically by taking as a rough indicator of the quality of a proposal the mean of all its ratings by both NSF and COSPUP reviewers. This leads to a measure of the variation in quality of proposals (σ^2_p) that can be compared with other sources of variation. The estimated proposal variance for the solid-state physics proposals was 24.43.

Second, the NSF review procedures and the COSPUP procedures were not identical. On average, there may be systematic differences between NSF reviewer responses to all proposals and COSPUP reviewer responses. In fact, this "method effect" can be observed in the differences in the mean ratings of proposals by NSF and COSPUP reviewers. As noted above, the COSPUP reviewers were on average slightly harsher than NSF reviewers. In the NSF-COSPUP comparison the estimated overall difference is 2.72 points, with NSF higher. Since funding decisions are based on rankings, this method effect is not important (but we did not ignore it in the mathematical analysis).

Even after compensating for the average methods effect, reviewers may disagree in their ratings of a proposal because they are members of two groups selected differently—NSF reviewers as opposed to COSPUP reviewers. This

458 **CONSENSUS IN SCIENCE**

"interaction" effect (σ_I^2) between proposals and evaluation method is important. It is the key component in estimating whether there appears to be any systematic bias among NSF program directors in the selection of reviewers. If there was bias in the selection of NSF reviewers, or if the two groups of reviewers had significant differences in the way in which they evaluated the proposals *due to any reason*, we would expect the interaction effect to be large. If it is large, then the NSF reviewer group and the COSPUP reviewer group evaluated proposals differently. If it is small, they did not and we would not be able to detect any bias in the selection of the NSF reviewers. It turns out that the estimated interaction σ_I^2 is trifling for each of the three fields, so there is no evidence of disagreement between the two selection methods aside from apparent disagreement resulting from the reviewer variability.

Finally, variation that remains is denoted by $\sigma_{R,i}^2$ and measures the reviewer variation within a given evaluation method i. The reviewer variances were estimated to be 48.93 and 50.24 for solid-state physics. These numbers are rather larger than the estimated proposal variance of 24.43. Thus, the reviewer brings to this process a higher variance than does the proposal. Of course, the average of several reviewers will have a lower variance; indeed, the average of four reviewers will have a variance of $48.93/4 = 12.23$ (NSF) or $50.24/4 = 12.56$ (COSPUP), but these are still not tiny compared to the proposal variance. This hard fact explains why the data exhibit so many reversals; they reflect substantial reviewer variance and not any fundamental disagreement between NSF and COSPUP reviewing methods or substantive evaluations. We may therefore conclude that there was no systematic bias in the way in which NSF reviewers were selected or in the way the two groups of reviewers made their evaluations.

To explain the reversals, then, we must look at two sources of variance: differences among the proposals and differences among the reviewers of a given proposal (see table 20.3). In the two physical sciences the variance among reviewers of the same proposal is approximately twice as large as the variance among the proposal means; in economics the reviewer variances are about 50 percent larger than the proposal variance. If the pooled proposal mean (the mean of both sets of ratings in each comparison) is taken as a rough indicator of the quality of the application, we can see that the variation in quality among the fifty proposals is small compared to the variation in ratings among reviewers of the same proposal. We have treated the reviewer variances as rough indicators of disagreement among reviewers. In all three fields there is a substantial amount of such disagreement. It is the combination of relatively small differences in proposal means and relatively large reviewer variation that creates the conditions for reversals.[7]

The substantial disagreement among reviewers of the same proposals can be shown by a simple one-way analysis of variance for each group of reviewers (see table 20.4). About half of all the variance in ratings is seen to result from

TABLE 20.4 Percentage of Total Variance in Reviewers' Ratings Accounted for by Differences among Reviewers of Individual Proposals. The number in parentheses is the total number of reviewers. For each field there were 50 proposals.

Field	Percent of total variance	
	NSF	COSPUP
Chemical dynamics	60 (242)	53 (213)
Economics	51 (192)	49 (190)
Solid-state physics	43 (163)	47 (182)

disagreement among reviewers of the same proposals. We replicated this one-way analysis of variance for the ten research programs studied in the first phase. In each of these programs we found that reviewer disagreement accounted for the largest share of the total variance in reviewer ratings. The within-proposal variance accounted for 35 to 63 percent of the total variance in the ten programs. Contrary to expectation, there was no less consensus in the social science fields of anthropology and economics than there was in the natural sciences.[8]

Another way of conceptualizing the relatively low consensus among reviewers of the same proposal is to think of placing all the NSF reviews of the fifty chemical dynamics proposals in a hat and drawing two at random. If we did this a large number of times we would find, on average, an expected absolute difference in the ratings of 9.78 points. Now, if we placed all the reviews of a single proposal in a hat and drew out two, we would find, on average, after multiple trials, an expected absolute difference of 8.45 points.[9]

Conclusions

We have shown that the reversals observed in the COSPUP experiment can be explained by the substantial disagreement among reviewers of the same proposal. If getting an NSF grant were an entirely random process, we would have found a reversal rate approximating 50 percent. If, instead of conducting an independent evaluation of the proposals, we had simply flipped a coin to determine which of the fifty proposals evaluated by the NSF would be funded, we would have obtained a 50 percent reversal rate. The difference between what we would expect from the coin flip and what we observe with the data can be viewed as a measure of what we "buy" from the peer review process. Since the reversal rate is about 25 percent, we may conclude that the fate of a particular grant application is roughly half determined by the characteristics of the proposal and the principal investigator and about half by apparently random elements which might be characterized as the "luck of the reviewer draw."

Although we conclude that the funding of a specific proposal submitted to the NSF is to a significant extent dependent on the applicant's luck in the program director's choice of reviewers, this should not be interpreted as meaning either that the entire process is random or that each individual reviewer is evaluating the proposal in a random way. In order to clarify the way in which the luck of the draw works, we must look at the sources of reviewer disagreement.

Some of the observed differences among scores given to the same proposal by different reviewers is undoubtedly an artifact of what anthropologists refer to as intersubjectivity. That is, two reviewers may translate their substantively identical opinions differently; reviewer A's opinion is expressed as an "excellent" and reviewer B's as a "very good."

The great bulk of reviewer disagreement observed is probably a result of real and legitimate differences of opinion among experts about what good science is or should be. This became evident from the qualitative comments reviewers made both on the proposals studied for the COSPUP experiment and on those studied in the first phase of the peer review study. Contrary to a widely held belief that science is characterized by wide agreement about what is good work, who is doing good work, and what are promising lines of inquiry, our research both in this and other studies in the sociology of science indicates that concerning work currently in process there is substantial disagreement in all scientific fields.[10]

As long as substantial reviewer disagreement, whatever its source, exists the fate of a particular proposal will depend heavily upon which reviewers happen to be selected. The element of chance would, of course, be reduced by increasing the number of reviewers; the larger the sample of reviewers the less likely it is that the sample mean will differ significantly from the population mean. It remains unclear what types of disagreement would be obtained if we examined other forms of peer review, such as the study section method used at the National Institutes of Health. If we found less reviewer disagreement in that context, would that indicate that study sections are a method for achieving intellectual consensus on the relative merits of research proposals, or would it simply reflect "artificial" consensus resulting from the influence of nonintellectual forces that are part of group dynamics? Our data cannot speak to this question since we did not examine peer review in the form used at NIH.

We must begin to question whether a system in which funding decisions depend to a significant degree on chance is the most rational one.[11] Here we will conclude with two observations. First, given the importance of chance in the current process, clearly the more proposals a researcher submits the higher the probability of being funded. In fact, eminent scientists may be more likely to be funded than less well-known ones not because their probability of success is greater for each submitted proposal but because they submit many proposals and are not deterred by an individual rejection. Second, the primary way in which the effect of chance might be reduced might be to give more weight to criteria for which there would be greater agreement than there is on the proposal. For example, it might be easier for scientists to agree upon the value of recently completed work than upon the value of a proposed piece of work.

Several important questions arise from the finding that there is a substantial random element in who gets an NSF grant. What degree of precision should we expect from the peer review system? Is it not healthy for science to have substantial disagreement among scientists who evaluate proposals rather than a single, agreed-upon dogma? At what point does disagreement become dysfunctional for the development of science? A distinction must be made between the effect of randomness in the peer review system on individual applicants and the effect on science itself. Plainly, the random element can be frustrating and debilitating for individual scientists trying to obtain financial support for their work, but it may have little effect on the rate of development of science as a whole. One clear disadvantage for science of the current peer review system is that it compels even our most talented scientists to spend substantial amounts of time and energy writing proposals, time and energy that might be more fruitfully spent doing research.

Notes

1. S. Cole, L. Rubin, and J. R. Cole, *Peer Review in the National Science Foundation: Phase 1 of a Study* (Washington, DC: National Academy of Sciences, 1978). Peer review in the form used at the National Institutes of Health (NIH) was not studied. The NIH study section form of peer review differs significantly from the form of peer review at NSF. The results reported in this article apply only to the basic science programs at NSF and may not be generalizable to any other form of peer review.

2. The report of the first phase of that study (1) includes detailed descriptions of the central role of the NSF program directors in the decision-making process, the role of peer review panels, the influence of budget size on decisions, the effects of "self-selection" processes on peer review ratings, an analysis of problematic and unproblematic cases, and the correlates of peer review ratings and funding decisions. See also S. Cole, L. Rubin, and J. R. Cole, "Peer Review and the Support of Science: *Scientific American*, 237 (4): 34–41, 1977; "Peer Review in the National Science Foundation: Phase One of A Study Committee on Science and Public Policy," *National Academy of Science* 237, no. 34), 1978.

3. The 150 proposals were reviewed under two different conditions in the COSPUP experiment, only one of which is reported on in this article. Under one condition (the one dealt with here), the COSPUP reviewers were sent the proposal exactly as it was received by NSF, with the name and institution of the principal investigator listed on

the title page, and were given exactly the same evaluation criteria to follow as were NSF reviewers. The participants in the experiment were told that their opinions would not influence the funding decision, which had already been made.

Under the second condition, the proposal was altered so as to try to hide the identity of the principal investigator. A report on this part of the experiment can be found in J. R. Cole and S. Cole, *Peer Review in the National Science Foundation: Phase 2 of a Study* (Washington, DC: National Academy of Sciences, 1981). The principal result of this part of the experiment was that it proved almost impossible to remove all evidence of the applicants' identity without destroying the integrity of the proposals; the reversal rates for anonymous proposals were somewhat higher than for identified proposals; and the reasons for reversals were the same. A report on results of this part of the experiment is in preparation.

The reviewer selectors received a copy of the proposal that did not include the name of the principal investigator and from which all references to the principal investigator's prior work had been deleted. It was therefore possible for the reviewer selector to name the principal investigator as a possible reviewer. In these cases, of course, the principal investigator's name was removed from the list of reviewers. We also removed from the list anyone who was at the institution of the principal investigator and all who had reviewed the proposals for NSF.

4. The size of the pool (N) can be estimated as follows. This approximation assumes that each COSPUP selector makes six equiprobable choices from the pool of N. Suppose that individual A had been selected by NSF. Then P (A will be selected by COSPUP selector) = $6/N$. Let $p = 6/N$. Then P (no overlap) = $(1 - p)^2$; P (one overlap) = $2p (1 - p)$; P (double overlap) = p^2. The proportions $(1 - p)^2$: $2p (1 - p)$: p^2 correspond closely to the proportions observed, when $p = 0.1$. This suggests that $N = 60$. The approximation is not perfect, of course. The field of economics surprisingly produces many double overlaps.

5. NSF may decide to fund a piece of the proposed scientific work, reducing the amount of the grant accordingly. Here we do not differentiate such a grant from a full grant.

6. We classified reversals within quintiles as follows: (i) The proposals were grouped into quintiles based on COSPUP rank, the first (best) quintile containing proposals with ranks 1, 2, . . ., 10; (ii) a proposal was counted as reversed if it was in the upper 25 by one set of ratings and in the lower 25 by the other set; and (iii) where there were ties in the mean ratings crossing quintile boundaries, proposals were apportioned among the categories involved. This rule results in noninteger numbers of reversals.

7. The relatively small differences in proposal means may result from processes of self-selection; that is, perhaps only relatively good scientists apply to NSF. This self-selection may result in attenuation of variance in reviewer ratings.

8. In all situations there were about the same number of proposals and reviewers per proposal, so the R^2 values are comparable. See J. R. Cole and S. Cole, "Which Researcher Will Get the Grant?" *Nature* 279, nos. 575–576 (1979).

9. If one takes two independent observations from a normal population with standard deviation σ, the expected absolute difference is $2\sigma / \sqrt{\pi}$. This statement is a reasonable approximation even when the population values do not follow a normal distribution. The numbers given in the text reflect a reviewer standard deviation of 7.49 and a proposal standard deviation of 4.36. Note that $(2 / \sqrt{\pi})$ 7.49 = 8.45 and $(2 / \sqrt{\pi})$ $(4.36^2 + 7.49^2)^{1/2} = 9.78$. The value $(4.36^2 + 7.49^2)^{1/2}$ reflects the fact that a randomly selected review incorporates both the proposal standard deviation and the reviewer standard deviation.

10. S. Cole, J. R. Cole, and L. Dietrich, "Measuring the Cognitive State of Scientific Disciplines," in *Toward a Metric of Science: The Advent of Science Indicators*, ed. Y. Elkana, J. Lederberg, R. K. Merton, A. Thackray, and H. Zuckerman, (New York: Wiley, 1978), 162.

11. A paper on how the system could be changed and how alternative systems might be more reasonable is in preparation.

12. The research was conducted under a contract between NSF and COSPUP. The statistical model was designed by J. Kiefer. We thank L. Cronbach for help in applying the model and for reviewing the results, the many members of COSPUP for their many useful suggestions in the design and analysis of the experimental material, and I. M. Singer for his support, without which the experiment could not have been carried out.

CHAPTER 21

NSF PEER REVIEW CONTINUED

(1982)

STEPHEN COLE, JONATHAN R. COLE, AND GARY A. SIMON

[These four comments are in response to the paper reproduced here as chapter 20.]

There is an important gap in the evaluation of the peer review process in the National Science Foundation (NSF) reported by Cole, Cole, and Simon (20 Nov., p. 881). Estimates of random error by means of correlations are a function of the "range of talent" in the population, but the authors neglected this feature of their data. For example, if an institution of higher education accepts almost every student who applies for admission, that is, the institution has almost as many openings as applicants, the mean level of talent will be low and the variance large. In such a case, the correlation between two different but equally valid methods of measuring the talent of the applicants will be high. If another institution is highly selective, having available few openings for the number of potential applicants, there will be a great deal of self-selection in the decision to apply. The mean level of talent in the applicant pool will be high and the variance low. The random error in the units of measurement of talent may be the same in the two institutions, but the correlation between two equally valid measures will be substantially lower in the selective institution.

It is probable that the populations of research applications to NSF from which the authors drew their samples are like the applicants to the selective

---

Science 215, no. 4531 (January 1982): 344, 346, 348.

institution of higher education just described. In fact, the authors recognize this in their reference 7, in small type at the end of the article, but they do not describe the implications of self-selection and the resulting homogeneity of proposal quality for the size of correlations among raters. Had there been full discussion of these issues, the data would have been better understood and erroneous national publicity might have been avoided.

It may not be possible to quantify precisely the effect of restriction of range of talent in the present instance, but data pertaining to it can be obtained. How large is the pool of potential applicants in each of the several disciplines? Is there evidence that the quality of the people who apply to NSF is higher than the quality of those who apply elsewhere or who do not apply at all? If the NSF research budget were to be doubled next year (and other sources of funding remained constant), one would expect that the accuracy of funding decisions as measured by the correlation between two independent and equally valid assessments would increase substantially. A drastic reduction in NSF funding (while other funding remained constant) would reduce the accuracy of the funding decision measured in the same way. By a similar line of reasoning, the accuracy of NSF funding decisions was probably higher in 1969 than it is today.

It is ironic, and may even present Congress with a catch-22 situation: reduced funding, though it may lead to greater care in review and to less measurement error, will lead to a lower correlation between ratings of proposals and thus to seemingly greater error.

LLOYD G. HUMPHREYS
Department of Psychology, University of Illinois at Urbana-Champaign, Champaign 61820

The conclusion that high variability in reviewers' evaluations of research goals gives rise to a significant element of chance (luck) in whether or not a proposal is funded may be valid for the system the COSPUP experiment tested. However, some programs at NSF use a much more intensive system. For example, all programs in the Division of Earth Sciences—where I was geochemistry program director from 1976 to 1978—use, in addition to mail peer review, a proposal review panel. The proposal review panel is composed of a group of scientific peers (five in the case of geochemistry), who provide an additional layer of judgment and selection on top of the mail reviews. This additional layer is, in my opinion, the most valuable part of the peer review system. Until a proposal is reviewed in a panel meeting the people who read and evaluate it are all acting as individuals: each mail reviewer reads the proposal in isolation, stares at the wall for a while, and records his or her judgment. Members of the review panel periodically receive copies of all proposals under consideration.

All members read all proposals, with each member having primary responsibility for proposals appropriate to his or her subfield. The panel then meets with the program director, and the proposals are discussed and rated (on the same scale system used for mail reviewers) one by one. The panel members have available to them the opinions of all the (highly variable) mail reviewers, as well as each other's opinions, which develop by discussion.

It seems to me not hard to see that adding the panel review process to the mail peer review system should result in a much more rational selection of proposals for funding. I firmly believe that it does so, having observed NSF programs that operate without as well as with such panels.

I suggest that COSPUP test the full NSF peer review system. If the three programs whose review system was examined (chemical dynamics, economics, and solid-state physics) use panel review, COSPUP could select a second set of panels to complete the process and then compare the results with those of NSF. If the programs they chose do not (or did not) use panel review, the study should go back to square one, and programs representative of all NSF review systems should be tested. I have no doubt that the fuller process would prove to have far more repeatable outcomes. I also believe that it results in the funding of better science.

JOHN HOWER
Department of Geology, University of Illinois at Urbana-Champaign,
Urbana 61801

The article by S. Cole et al. opens with the assertion that NSF "employs one form of the peer review system in making research grants." The process described is in fact not the only form of peer review used by NSF. In biochemistry and biophysics, for example, this system of ad hoc reviewers is combined with review by a panel that contains expertise across the whole field. Where the reviews written independently by expert panel members and ad hoc reviewers are in substantial agreement, as is the case more often than not, funding decisions are relatively routine. It is when there are disagreements in the responses of experts that this system displays its great strength. The reasons behind the differing responses are addressed and debated, and a well-considered decision is reached that goes far beyond the blind averaging of scores. I believe that this procedure results in a much smaller chance component in making funding decisions than does the use of either ad hoc reviewers alone or a panel alone and probably represents the best selection system available.

JOHN WESTLEY
Department of Biochemistry, University of Chicago,
Chicago, Illinois 60637

It is true, as the authors point out, that the present peer review system "compels" people to spend substantial amounts of time and energy composing proposals. Obviously, most of those hours are spent by authors of the middle three quintiles of proposals. Equally obviously, many hours are also so spent by "our most talented scientists." That even this is a "clear disadvantage . . . of the current peer review system" is not clear, however. As Dr. Johnson pointed out (in a rather different context), nothing so clears a man's mind as the knowledge that he is to be hanged on the morrow. It is easy to have an idea of what one wants to do without having a *clear* idea of what one wants to do; I suspect that wording proposals is often no mere exercise in composition but an important element in the process of giving one's ideas substance.

HENRY E. KYBURG JR.
Department of Philosophy, University of Rochester,
Rochester, New York 14627

With Atkinson (Letters, 18 Dec., p. 1292) and other members of the scientific community we regret that some members of the media have misread and incorrectly interpreted the results presented in our *Science* article. For instance, nowhere do we say that receipt of an NSF grant depends mostly on "luck." We did point out that getting an NSF grant was about half a result of the "luck of the reviewer draw" (in that we saw about 25 percent reversals rather than the 50 percent expected by chance alone), and we also pointed out that it was half a result of agreement among reviewers. This means that the current peer review system is decidedly superior to one based on random selection. Some of the critics of our article want to emphasize the 50 percent that is due to agreement rather than the 50 percent that is due to chance. But not one of them denies the validity of the main finding: that a new set of reviewers would produce a reversal rate of 25 to 30 percent and that reversals are largely a result of differences in the evaluation of these proposals by sets of apparently unbiased referees.

In discussing these findings with the press we emphasized that this study suggests the importance of more widespread funding of scientific research. If it is difficult to determine which project will lead to a major breakthrough, granting agencies should fund a wide range of research so as to reduce the probability that development of important ideas will be delayed because of lack of support. Lack of consensus is an inherent characteristic of science rather than of the NSF peer review system; thus, we have never said or implied that the NSF was not efficiently and equitably run or that some other type of peer review system could overcome the problems resulting from lack of consensus.

We are perplexed by Singer's statement (p. 1292) that our article "misrepresented" the results of the COSPUP report we coauthored with the committee (1).

The conclusions reached in that report are virtually identical with those reached in our article. In his letter Singer selects one part of the findings that emphasize consensus (comparing COSPUP ratings with NSF decisions) and ignores those that don't (comparing COSPUP ratings with NSF ratings). Singer suggests that the proposals in the top quintile have only a small probability of being reversed but in the report itself a concluding section written by the committee states, "It is clear, in short, that evaluative differences are not confined to a limited group of proposals of seemingly marginal value. Many projects given the highest ratings by some groups of expert reviewers would receive ratings from other, similarly constituted groups that would be too low to permit funding" (1, p. 58). It is important to recognize that a proposal rated in the top quintile by one group of NSF reviewers could have fallen into a lower quintile if rated by another group of reviewers. If the fact that the same proposal would either be funded or declined depending upon which group of reviewers was selected does not indicate that getting an NSF grant depends on the "luck of the reviewer draw," we would welcome another interpretation.

The standard deviations Cronbach (p. 1294) computes for the average of four reviews (3.8, 3.5, and 4.8) are far from small given the proposal standard deviations for the three fields (4.9, 4.9, and 7.6). Moreover, a substantial minority of proposals had three or fewer NSF reviewers. The scatterplot that Cronbach presents, showing mean ratings obtained by COSPUP reviewers against those obtained by NSF reviewers in the field of chemical dynamics (see fig. 21.1), illustrates a higher level of agreement between the two groups than figure 21.1 in our *Science* article, where proposal ranks are plotted. Chemical dynamics happens to be the field for which the plot of raw averages looks most pleasing. If Cronbach had chosen the identical scatterplot for the field of solid-state physics (see fig. 21.2), he might not have come to the same conclusion.

We agree with Cronbach that reviewer disagreement should not be disparaged as "random or nonrational." In fact, in the conclusion to our article we emphasize that our findings "should not be interpreted as meaning either that the entire process is random or that each individual reviewer is evaluating the proposal in a random way. . . . The great bulk of reviewer disagreement observed is probably a result of real and legitimate differences of opinion among experts about what good science is and should be." We thank Cronbach for pointing out the arithmetic error in our reference 9.

We will comment briefly on several other points made in the letters about our article:

1) Although Atkinson is correct in stating that the NSF decisions are not based upon numerical ratings in any technical sense, the Phase One report (2), based upon 1,200 proposals submitted to ten different programs, showed that an average of 67 percent of the variance in decision was explained by the mean

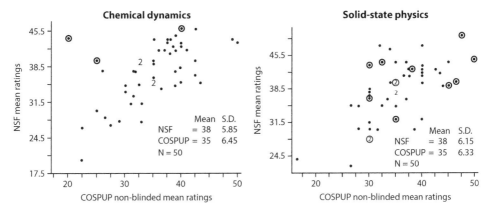

21.1 and 21.2 (left) Scattergram of mean NSF ratings with mean COSPUP ratings of individual proposals. Encircled dots indicate proposals that received two or fewer reviews from either COSPUP or NSF reviewers. Reproduced from J. R. Cole, S. Cole, with the Committee on Science and Public Policy, *Peer Review in the National Science Foundation: Phase Two of a Study* (Washington, DC: National Academy of Sciences, 1981), 28. (right) Scattergram of mean NSF ratings with mean COSPUP ratings of individual proposals. Encircled dots indicate proposals that received two or fewer reviews from either COSPUP or NSF reviewers. Reproduced from S. Cole, L. Rubin, and J. R. Cole, *Peer Review in the National Science Foundation: Phase One of a Study* (Washington, DC: National Academy of Sciences, 1978), 30.

rating. Of the 150 proposals used for Phase Two (1), on only twelve did the decision differ from the decision that would have been made by using mean ratings in a mechanical way.

2) Making verbatim reviews available and encouraging appeal of a decision the applicant believes to be unfair may be improvements in peer review provided that appeals receive equitable review. A study showing the number of appeals, the type of applicants who appeal, and the outcome of these appeals would be enlightening.

3) We strongly agree with Humphreys' point that the significance of our results must be interpreted in the light of self-selection of NSF applicants. (Space restrictions prevented us from expanding on our references 2 and 7, which dealt with self-selection.) It is possible that if we had asked a random sample of American scientists to write and submit proposals there would have been greater variance in the proposal means and a corresponding reduction in the ratio of reviewer variances to proposal variances. Therefore, the relative significance of chance might have been reduced. However, this is not the situation NSF actually faces.

4) We agree with Hower and Westley that the use of panels improves the peer review process. The way in which panels work is discussed in the Phase

One report (2). For one field included in the experiment, economics, NSF did use a panel. In that field the rate of reversals for COSPUP ratings compared with NSF mail ratings (28 percent) was very similar to that for COSPUP ratings compared with the NSF decisions (24 percent), decisions which were influenced by the panel. We note that the panel-augmented decision agrees strongly with the NSF mean ratings. The panel does reduce the reversal rate somewhat in the top quintile. There is no evidence yet that the reversal rate would have been lower if the COSPUP experiment had used either a substitute program director or a panel to make the decisions. A detailed examination of substantive comments made by reviewers for cases in which differences between the COSPUP and the NSF reviewers would have led to reversals suggests that the reversals were a result of legitimate intellectual differences rather than of "errors" by reviewers.

5) We agree with Kyburg that writing a proposal can be a very useful experience. However, it can also become an end in itself, resulting in a displacement of goals in which scientists spend almost as much time applying for funds as using them to produce new science.

We are pleased that Singer notes that the full report deals with questions other than funding reversals. COSPUP decided not to include in the Academy reports our analyses of additional topics which we believe shed light on peer review. These include a discussion of the effects of self-selection; data on peer appraisals of the reputations or "track records" of NSF applicants and a comparison of consensus on reputations with consensus on proposals; and an analysis of pooled data on the probability of a reversal as a function of the number of reviewers as well as the variance structure of ratings of the proposals.

Finally, it should be noted that, although our experiment was based upon only 150 cases, the conclusions on consensus replicated those from the Phase One data on 1,200 proposals (3). The variance structures of reviewer ratings in the ten fields studied were remarkably similar to the data produced by the experiment. Since reversals were found to be substantially explained by lack of reviewer agreement, we believe we would have found a similar reversal rate if the experiment had been replicated on the 1,200 Phase One cases.

S. COLE
Department of Sociology, State University of New York
J. R. COLE
Center for the Social Sciences, Columbia University
G. A. SIMON
Department of Applied Mathematics, State University of New York

References

1. J. R. Cole, S. Cole, with the Committee on Science and Public Policy, *Peer Review in the National Science Foundation: Phase Two of a Study* (Washington, DC: National Academy of Sciences, 1981).
2. S. Cole, L. Rubin, and J. R. Cole, *Peer Review in the National Science Foundation: Phase One of a Study* (Washington, DC: National Academy of Sciences, 1978).
3. J. R. Cole and S. Cole, "Which Researcher Will Get the Grant?" *Nature* 279, nos. 575–576 (1979).

CHAPTER 22

EXPERTS' CONSENSUS AND DECISION-MAKING AT THE NATIONAL SCIENCE FOUNDATION

(1985)

JONATHAN R. COLE AND STEPHEN COLE

"Work, Finish, Publish." A sign bearing this advice is said to have hung in the laboratory of the great British physicist and chemist Michael Faraday. Today, as in the nineteenth century, Faraday's motto, whether or not it was an explicit directive for his colleagues, focuses our attention on three essential components of scientific activity. And today, in the age of Big Science, perhaps more than yesterday, financial resources are essential for meeting Faraday's prescription.

Consequently, scientific funding agencies, such as the National Science Foundation (NSF), are in structurally central positions in the scientific community. They supply significant portions of the fuel needed to drive the scientific machinery through Faraday's three stages. Of course, some pathfinding science is conducted without large-scale funding. But this is increasingly rare.

We thank members of the National Academy of Sciences Committee on Science and Public Policy (COSPUP) for their help in the collection and exploration of the data reported here. Members of the Committee collaborated on the publication of the report published by the Academy.

We thank the staff of the Center for the Social Sciences, Columbia University, for assistance in preparing this chapter, which was presented to the Rockefeller Foundation Conference, "Developments in Scientific Information Systems," New York, NY, October 26, 1982.

Portions of this chapter were published in S. Cole, J. R. Cole, and G. A. Simon, "Chance and Consensus in Peer Review," *Science* 214 (November 1981): 881–86. Copyright 1981 by the American Association for the Advancement of Science (AAAS). Reprinted by permission of publisher.

The contemporary situation is placed in bold relief by the eminent biochemist Arthur B. Pardee:

At the heart of current problems [in maintaining high scientific quality and productivity] are the difficulties and uncertainties every scientist faces in obtaining research funds. . . . A scientist perceives now that he has a small probability of getting a grant funded. He cannot afford to be without funds for a year or more if his application fails, because continuity is essential for progress and to retain highly trained, key personnel. So he writes proposals in the hope that one of them will be lucky. . . . Fund raising rather than research becomes his major preoccupation. . . . Talents of fine scientists are a rare commodity; wasting them is a very costly proposition. . . . A less evident but also highly important consequence is the diminution of scientists' self-confidence and morale. Rejections by the funding system of one's best ideas are extremely discouraging. We will see scientists in increasing numbers decide that they are in a rat race; they will slow down or get out. Some of the best unfortunately will be among them.[1]

Because funding agencies have such great influence on the conduct of science and on individual careers, it is appropriate to examine thoroughly their methods for deciding to support one scientist or laboratory rather than another.[2]

This chapter focuses on the level of expert consensus in the NSF peer review system, the system used to decide on the allocation of roughly $1 billion annually for scientific research and training.[3] Here we attend to the following set of questions:

If peer review is based upon the independently arrived at judgments of experts as to the quality of research proposals, to what extent do these experts agree or disagree in their evaluations? How would two independently selected groups of reviewers evaluate the same proposals? If they would reach different decisions, how can we explain their disagreement? Would there be more or less agreement if peer reviewers were asked to simply evaluate the significance of an applicant's past contributions rather than a specific proposal?

Peer Review at the National Science Foundation

The National Science Foundation employs one form of the peer review system in making research grants.[4] For each application for a grant, an NSF program director selects a group of scientists, generally four or five, who are knowledgeable in the relevant subject matter to act as referees. Each reviewer is sent a

474 CONSENSUS IN SCIENCE

copy of the proposal and asked to evaluate it on the basis of its scientific merit and the ability of the principal investigator. Ability of the principal investigator is generally defined as the quality of his recent scientific performance. Each reviewer is asked to make substantive comments and to assign one of five ratings to the proposal: excellent, very good, good, fair, or poor.

It is critically important to emphasize at the outset that in the larger study of the NSF peer review system, we found, in general, that it was remarkably free of systematic bias. There was no support, for instance, that an "old boys network" operated to manipulate the distribution of research grants—as some scientists and members of Congress had apparently believed. Several specific results from our earlier work represent important contextual material for the present discussion. Therefore, we enumerate five findings obtained from analysis of 1,200 proposals drawn from ten NSF basic research programs and from extensive content analysis of substantive reviews associated with 250 of these proposals.

(1) There is a high correlation between reviewer ratings and grants made. If one attaches numerical values to the ratings, say from 10 for poor to 50 for excellent, the mean scores predict with a high degree of accuracy which proposals will be funded and which will be denied. Whether or not NSF program directors actually compute statistical averages from the ratings and use them in decision-making, the statistical average of the ratings turned out to be highly correlated with the actual decision rules employed by the program directors.

(2) For the 1,200 proposals there was not a high correlation between grants awarded and measures of the previous scientific performance of the applicants. This result was unexpected because one of the stated evaluation criteria is the ability of the applicants to conduct the research proposed.

(3) Reviewers at major institutions did not treat proposals from applicants at major institutions more favorably than did reviewers from lesser institutions. In fact, there was a tendency in the opposite direction.

(4) Professional age (length of career) had no strong effect either on ratings received or on the probability of receiving a grant.

(5) There were low or moderate correlations between reviewer ratings (and the funding decision) and the following characteristics of the applicants: prestige rank of current academic department, academic rank, geographic location, NSF funding history over the previous five years, and locus of PhD training.[5]

Because proposals from eminent scientists do not have substantially higher probabilities of receiving favorable ratings than proposals from scientists who are not eminent, we concluded that the peer review system employed by NSF was essentially free of systematic bias.

The COSPUP Experiment

In a second phase of the study, we were concerned with, among other things, whether program directors were predetermining funding decisions by their selection of reviewers, that is, whether independently selected panels of reviewers would reach similar conclusions. As a by-product of an experiment to test this concern, we became interested in the role of chance in obtaining an NSF grant.

In the spring of 1977, the NSF provided the Committee on Science and Public Policy (COSPUP) of the National Academy of Sciences with 150 proposals— fifty each from the programs in chemical dynamics, economics, and solid-state physics—upon which decisions had been made recently; half the proposals in each program had been funded, and half had been declined. To select new reviewers we chose from a panel of ten to eighteen experts in each of these fields, most of them members of the National Academy of Sciences.[6] Each proposal was sent to two members of this panel, who then selected six or more reviewers for it. This gave us a list of approximately twelve reviewers for each proposal.

Some have argued that the highly specialized state of modern science would not permit more than a dozen or so scientists to be capable of reviewing any given proposal. The COSPUP experiment enabled us to test this hypothesis. If the number of eligible reviewers were, in fact, small, we would expect that a fairly high proportion of the original NSF reviewers would also have been selected by the experimental selectors. In each of the three programs, about 80 percent of the NSF reviewers were not selected by either of the two COSPUP selectors, about 15 percent were selected by one of them, and about 5 percent were selected by both. These data suggest that the pool of eligible reviewers for most proposals is at least ten, and given the low overlap rates we found, we would predict that if other equally qualified selectors were employed we would find it to be substantially larger than twenty-five.[7] Of course, in actual practice there is not a clear distinction between eligibles and noneligibles, and the numbers certainly vary according to subspecialties.

Given that the pool of scientists available to evaluate a given proposal is generally considerably larger than the group selected, we want to see how the program director's choice from the group of eligibles affects the funding decision reached. To do this we will compare the evaluations made by the NSF selected reviewers with those made by the COSPUP selected reviewers.

In general, the COSPUP reviewers tended to give slightly lower scores than did the NSF reviewers (see table 22.1). For example, for the fifty chemical dynamics proposals, the grand mean of the NSF reviewers' ratings was about 38 on the 10-to-50 scale, and that of the COSPUP reviewers was about 35. The experimental reviewers may have been slightly harsher in their evaluations because they knew their ratings would have no effect on the careers of the applicants.

TABLE 22.1 Correlation of Mean Ratings of NSF Reviewers and COSPUP Reviewers on Grant Applications (N = 50)

Field	Mean Ratings NSF	COSPUP	Correlation Coefficient
Chemical dynamics	37.7 (±5.85)	35.0 (±6.45)	0.595
Economics	33.6 (±9.76)	31.5 (±9.28)	0.659
Solid-state physics	38.2 (±6.15)	35.5 (±6.33)	0.623

Note: Numerical values were assigned to the original ratings as follows: excellent = 50, very good = 40, good = 30, fair = 20, and poor = 10. Figures in parenthesis are standard deviations.

The correlations between the mean NSF rating and the mean COSPUP rating for each proposal are moderately high (0.60, 0.66, and 0.62). Proposals that are rated high by NSF reviewers tend also to be rated high by the independent sample of reviewers used by COSPUP. The match is, however, less than perfect.

The findings presented thus far do not address one of the fundamental questions for evaluating the peer review system at NSF: How many funding decisions would be reversed if they were determined by the COSPUP ratings rather than by the procedures followed by NSF?

Figures 22.1, 22.2, and 22.3 show the rank order of the proposals in each program according to the NSF mean ratings and the mean ratings of the COSPUP

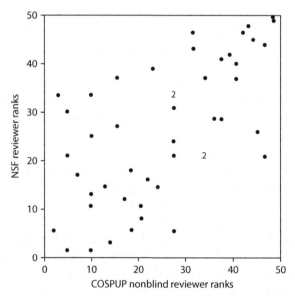

22.1 Rank order of chemical dynamics proposals as reviewed by NSF and COSPUP nonblind reviewers (N = 50).

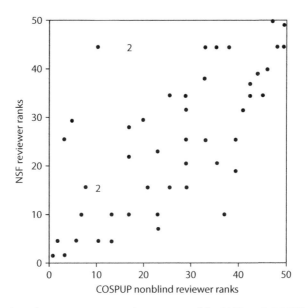

22.2 Rank order of economics proposals as reviewed by NSF and COSPUP nonblind Reviewers (N = 50)

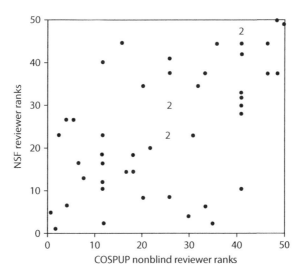

22.3 Rank order of solid-state physics proposals as reviewed by NSF and COSPUP nonblind reviewers (N = 50).

478   CONSENSUS IN SCIENCE

reviewers. (Half-integer ranks are the result of ties.) Because the mean ratings generally determine which proposals are funded, and half were funded and half declined, decisions on proposals that were ranked in one half of the range of scores by NSF reviewers and in the other half by COSPUP reviewers would have been reversed by the COSPUP ratings. There were differences of that degree in the ratings of approximately one-quarter of the proposals.

Reversals

The NSF is faced, of course, with a zero-one decision rule: to fund or not to fund a proposal.[8] It follows that proposals with mean rankings that are fairly close, or virtually identical, may fall on opposite sides of the dividing line. Therefore, it was almost inevitable that we would find some reversals.

In determining what should be classified as a reversal, we rank-ordered the proposals according to their mean COSPUP ratings and assumed that those with the top 25 scores would be funded and the bottom 25 would be declined. We then compared the COSPUP ratings with both the actual NSF decision and the decision NSF would have made if it had relied solely on mean ratings. The NSF funding decision was highly, though not perfectly, correlated with the mean ratings that NSF reviewers had given the proposals; hence, the two comparisons yield approximately the same results (see table 22.2).

If decisions on the fifty proposals were made by flipping a coin, we would expect to obtain a 50 percent reversal rate, on average. Correlatively, if the COSPUP reviewers were to rate the same proposals in such a way that there was complete agreement with the NSF reviewers on which proposals fell into the top half and which into the bottom, the reversal rate would be zero. Thus, we would expect to find a reversal rate somewhere between zero and 50 percent. In fact, the reversal rate turns out to be between 24 percent and 30 percent for each of the three programs computed in each of the two different ways. That is, for twelve to fifteen of the fifty proposals in a program, COSPUP reviews led to a "decision" different from that of NSF reviews.

We would expect to find some reversals around the cutting point, for example, to find that a proposal ranked twenty-fourth by NSF was ranked twenty-sixth or twenty-seventh by COSPUP. We want to examine the extent to which reversals were common not only at the midpoint but at a distance from it. This is shown in table 22.2 by the reversal rates within quintiles.[9] In chemical dynamics and solid-state physics we find, as expected, the highest reversal rate in the middle quintile. A 50 percent reversal rate for this quintile would not be surprising. In chemical dynamics it is 60 percent in both comparisons; in solid-state physics it is 49 percent and 43 percent. In economics, on the other hand, we find higher reversal rates in the second and fourth quintiles than in

TABLE 22.2 NSF Outcomes Reversed in COSPUP Rank-Order Quintiles and Overall (in Percentages)

NSF Outcome	Quintile Based on COSPUP Rating					Overall ($N = 50$)
	1	2	3	4	5	
Chemical dynamics						
Mean rating decision	26	24	60	20	20	30
	26	24	60	20	20	30
Economics						
Mean rating decision	20	45	30	45	0	28
	5	45	28	42	0	24
Solid-state physics						
Mean rating decision	23	22	49	34	6	27
	16	24	43	29	11	25

Note: Reversals are shifts from the top 25 positions in the COSPUP rank order to the bottom 25, or vice versa.

the third. In all three programs there are more than a few reversals in the first quintile. There are, in fact, proposals that were rated in the top quintile by NSF reviewers that would not have been funded had the decision depended on the appraisals of the COSPUP reviewers.

There are several possible explanations for the reversals. Differences between NSF procedures and COSPUP procedures will be considered first. If the two sets of reviewers used different criteria in appraising proposals, the outcome could have differed significantly, creating reversals, for example, if one group of reviewers based their ratings strictly on evaluations of the proposal and the other primarily on the track record of the applicant. Because the two groups of reviewers were given identical instructions about the criteria, it is unlikely that there were systematic differences of that kind.

Another procedural cause of reversals might be that NSF and COSPUP selected different types of reviewers. Reviewer differences rather than proposal differences could then result in reversals. Because a comparison of the characteristics of the two groups of reviewers showed few differences, it is likely that they were drawn from the same population.

Assuming that reversals did not result from the procedures employed in the experiment, we are left with two possible substantive explanations. Reversals could result from bias in the way in which the reviewers were selected by either the NSF program director or by the COSPUP experiment. If, for example, the NSF program director purposely selected reviewers who would give unrepresentative negative or positive ratings to a proposal, a reversal could result.

Second, reversals could have resulted from disagreements among fairly selected reviewers using the same criteria. If there is substantial dissensus in the population of eligible reviewers of a given proposal, it would be possible for equally qualified and unbiased groups of reviewers using the same criteria to differ in the mean rating.

Consider a hypothetical proposal for which there is a population of approximately one hundred eligible reviewers. If all of them totally agreed about its merits, any four or five selected at random from among them would agree among themselves, and two independently selected samples would not reach different conclusions. However, if the population of eligible reviewers had substantial disagreement about the proposal, two randomly selected samples could yield different mean ratings, possibly leading to different outcomes for the proposal. Our data indicate that the reversals in this experiment were a result of such disagreement.

Consensus

In order to determine the extent to which the reversals could be explained by bias or disagreement, we used analysis-of-variance techniques. Because we did not want to make the usual statistical assumptions (such as normality) that must be made in a standard two-way analysis of variance, we used a components-of-variance model that did not require some of these assumptions but that would be useful in answering the same substantive question.

In order to assess the relative magnitude of contributions to the proposal evaluation method and of the reviewer to the variation in ratings, we represent the rating y_{ijk}, given by the kth reviewer under method i to proposal j by $y_{ijk} = a_i + b_j + c_{ij} + e_{ijk}$. Where a_i is the overall average rating by evaluation method i ($i = 1$ for NSF, and $i = 2$ for COSPUP), b_j is the differential effect of proposal j, c_{ij} measures the extent to which the rating on proposal j depends on the evaluation method, and e_{ijk} is the effect caused by the kth reviewer of proposal j by evaluation method i.

We consider a_i to be a fixed quantity and the remaining terms to be random with means equal to zero. Then we can decompose the variance associated with proposals under evaluation method i into three terms: $\text{Var}(Y_{ijk}) = \sigma_p^2 + \sigma_I^2 + \sigma_{R,i}^2$, Where $\sigma_p^2 = \text{Var}(b_j)$ reflects the intrinsic variability of the proposals; $\sigma_I^2 = \text{Var}(c_{ij})$ is the variability associated with the interaction of proposals and evaluation method; and $\sigma_{R,i}^2 = \text{Var}(e_{ijk})$ is the reviewer variance for method i.

If σ_p^2 is large relative to σ_I^2, or $\sigma_{R,1}^2$, and $\sigma_{R,2}^2$, we interpret this to mean that it is relatively easy to distinguish the proposals independent of the evaluation method. However, if σ_I^2 is of the same order of magnitude as σ_p^2, this would suggest that dependence between proposal and evaluation method is masking

Experts' Consensus and Decision-Making at the National Science Foundation   481

some of the intrinsic proposal variability. As a consequence, the proposals would be ranked differently under the two evaluation methods. If, as actually occurs in these data, $\sigma_{R,1}^2$ and $\sigma_{R,2}^2$ dominate σ_I^2 and are of the same magnitude as σ_p^2, reviewer variability will be so pronounced that two different evaluations will give dissimilar rank orders.

The estimates of σ_p^2, σ_I^2, $\sigma_{R,1}^2$ and $\sigma_{R,2}^2$ are presented in table 22.3. The dependent variable for the analysis is the rating given the proposal by a reviewer. If we consider all the variance in an entire set of reviews (for example, all reviews done by both NSF and COSPUP reviewers for the fifty proposals), we want to know the sources of variance. There are four possible sources of variance, two of which turned out to be trivial in this study. Consider these four sources and the estimated effects for solid-state physics. The results for economics and chemical dynamics have parallel interpretations.

First, reviewers' responses to proposals differ because proposals differ in quality. That is easily dealt with statistically by taking as a rough indicator of the quality of a proposal the mean of all its ratings by both NSF and COSPUP reviewers. This leads to a measure of the variation in quality of proposals (σ_p^2) that can be compared with other sources of variation. The estimated proposal variance for the solid-state physics proposals was 24.43.

Second, the NSF review procedures and the COSPUP procedures were not identical. On average, there may be systematic differences between NSF reviewer responses to all proposals and COSPUP reviewer responses. In fact, this "method effect" can be observed in the differences in the mean ratings of proposals by NSF and COSPUP reviewers. As noted, COSPUP reviewers generally were slightly harsher than NSF reviewers. In the NSF-COSPUP comparison the estimated overall difference is 2.72 points, with NSF higher. Because

TABLE 22.3 Components of Variance of NSF and COSPUP Ratings

		Reviewer Variance			
Field	**Proposal Variance** $\hat{\sigma}_p^2$	**NSF** $\hat{\sigma}_{P,I}^2$	**COSPUP** $\hat{\sigma}_{R,2}^2$	**Interaction Variance** $\hat{\sigma}_I^2$	**Method Difference** $\hat{a}_1 - \hat{a}_2$
Chemical dynamics	23.67	55.91	56.67	1.18	2.73[a]
Economics	58.33	89.22	96.25	0.00[b]	2.14[a]
Solid-state physics	24.43	48.93	50.24	0.17	2.72[a]

[a] NSF is higher.
[b] Computed as −1.36.

482 CONSENSUS IN SCIENCE

funding decisions are based on rankings, this method effect is not important (but we did not ignore it in the mathematical analysis).

Even after compensating for the average methods effect, reviewers may disagree in their ratings of a proposal because they are members of two groups selected differently: NSF reviewers as opposed to COSPUP reviewers. This "interaction" effect (σ_I^2) between proposals and evaluation method is important. It is the key component in estimating whether there appears to be any systematic bias among NSF program directors in the selection of reviewers. If there were bias in the selection of NSF reviewers, or if the two groups of reviewers had significant differences in the way in which they evaluated the proposals *due to any reason*, we would expect the interaction effect to be large. If it were large, the NSF reviewer group and the COSPUP reviewer group would have evaluated proposals differently. If it were small, they would not have and we would not be able to detect any bias in the selection of the NSF reviewers. It turns out that the estimated interaction σ_I^2 is trifling for each of the three fields, so there is no evidence of disagreement between the two selection methods, aside from apparent disagreement resulting from the reviewer variability.

Finally, variation that remains is denoted by $\sigma_{R,i}^2$, which measures reviewer variation within a given evaluation method *i*. Reviewer variances were estimated to be 48.93 and 50.24 for solid-state physics. These numbers are higher than the estimated proposal variance of 24.43. Thus, the reviewer brings to this process a higher variance than does the proposal. Of course, the average of several reviewers will have a lower variance; indeed, the average of four reviewers will have a variance of 48.93/4 = 12.23 (NSF) or 50.24/4 = 12.56 (COSPUP), but these are still significant compared with the proposal variance. This hard fact explains why the data exhibit so many reversals; they reflect substantial reviewer variance, not any fundamental disagreement between NSF and COSPUP reviewing methods or substantive evaluations. We may therefore conclude that there was no systematic bias in the way in which NSF reviewers were selected or in the way the two groups of reviewers made their evaluations.

To explain the reversals, then, we must look at two sources of variance: differences among the proposals and differences among the reviewers of a given proposal (see table 22.3). In the two physical sciences the variance among reviewers of the same proposal is approximately twice as large as the variance among the proposal means; in economics the reviewer variances are about 50 percent higher than the proposal variance. If the pooled proposal mean (the mean of both sets of ratings in each comparison) is taken as a rough indicator of the quality of the application, we can see that the variation in quality among the fifty proposals is small compared with the variation in ratings among reviewers of the same proposal. We have treated the reviewer variances as rough indicators of disagreement among reviewers. In all three fields there is a substantial amount of such disagreement. It is the combination of relatively

Experts' Consensus and Decision-Making at the National Science Foundation   483

TABLE 22.4 Total Variance in Reviewers' Ratings Accounted for by Differences Among Reviewers of Individual Proposals (in Percentages)

	Total Variance	
Field	NSF	COSPUP
Chemical dynamics	60 (242)	53 (213)
Economics	51 (192)	49 (190)
Solid-state physics	43 (163)	47 (182)

Note: The number in parentheses is the total number of reviewers. For each field there were 50 proposals.

small differences in proposal means and relatively large reviewer variation that creates the conditions for reversals.

The substantial disagreement among reviewers of the same proposals can be shown by a simple one-way analysis of variance for each group of reviewers (see table 22.4). About half of all the variance in ratings is seen to result from disagreement among reviewers of the same proposals. We replicated this one-way analysis of variance for the ten research programs studied in the first phase. In each of these programs we found that reviewer disagreement accounted for the largest share of the total variance in reviewer ratings. The within-proposal variance accounted for 35 to 63 percent of the total variance in the ten programs. Contrary to expectation, there was no less consensus in the social science fields of anthropology and economics than there was in the natural sciences (see table 22.5).[10]

TABLE 22.5 Consensus Among Reviewers of Ten NSF Programs

	Percentage of Total Variance Due to Intraproposal Variance $(I - R^2)$
Algebra	45
Anthropology	43
Biochemistry	49
Chemical dynamics	45
Ecology	54
Economics	35
Fluid mechanics	43
Geophysics	57
Meteorology	63
Solid-state physics	55

484 CONSENSUS IN SCIENCE

Another way of conceptualizing the relatively low consensus among reviewers of the same proposal is to think of placing all the NSF reviews of the fifty chemical dynamics proposal in a hat and drawing two out at random. If we did this a large number of times we would find, on average, an expected absolute difference in the ratings of 10.08 points. Now, if we placed all the reviews of a single proposal in a hat and drew out two, we would find, on average after multiple trials, an expected absolute difference of 8.45 points.[11]

Number of Reviewers and Reversals

Statistical theory indicates that as the number of sampled reviewers approaches the size of the population, the probability of approximating the "true" rating of a proposal increases, and of course, the probability of reversals of decisions is reduced.

In order to examine the relationship between the number of reviewers of proposals and the probability of reversals, we pooled the reviews obtained on each proposal by the NSF and by the experiment. The pooling is supported by the absence of any significant "method effect" in table 22.3. The probability of reversals depends, of course, on two factors: the "true mean quality" of the proposal and the population variance in expert opinion about the proposal. Plainly, if a proposal has a very low "true quality" rating and no variance in opinion, it will not be funded. Correlatively, highly regarded proposals that also have virtually no variance in the opinion of reviewers in the population will be funded. Unfortunately, these two groups account for a relatively low proportion of the total number of proposals submitted to the NSF. Cases in which the mean from a small sample may give the impression of consensus, or of low variance in opinion, among the qualified experts in the population but where in fact the actual population variance is rather high are potential sources for reversals. This is especially so when the mean ratings from the small set of sampled reviewers lay close to the "cutting point." In short, there are some proposals that are "high risks" in terms of becoming reversals, while others are "low risks."

Figure 22.4 presents data on the expected number of decision reversals for samples of size n reviewers, given the actual means and variances in the 150 proposals examined in the COSPUP experiment. The total number of pooled ratings received for each proposal (that is, for both forms of the COSPUP reviews, plus the NSF reviews) is taken as indicative of the "true" distribution of expert opinion about the particular proposal in the population of eligible reviewers. We calculate the expected number of reversals out of the fifty proposals in each of the three programs studied, for sets of reviewers of size n.[12]

Let us examine results generated from the pooled data for chemical dynamics. The first set of results presents, in graphic form, the number of expected reversals as a function of the number of reviews obtained.[13]

Experts' Consensus and Decision-Making at the National Science Foundation

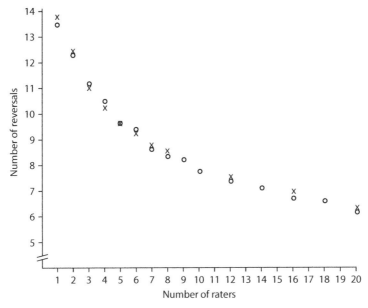

Note: X = jackknifed estimates; 0 = unjackknifed estimates.

22.4 Expected number of decision reversals based on pooled data and number of peer reviewers (475 random trials, chemical dynamics). X = jackknifed estimates; o = unjackknifed estimates.

These estimates are based upon 475 random trials for each category of number of raters. Thus, we find the average expected number of reversals for all fifty proposals that would result using, say, one, two, six, or twenty reviewers as the basis for creating a mean score and on the basis of 475 random draws based on the existing distribution of ratings.

Even a cursory glance at the shape of the curve produced by these numbers in figure 22.4 suggests that as the number of raters increases, there is a significant decline in the estimated number of reversals. Of course, if we had an infinite number of reviewers and a similar number of random trials, the curve presented here would approach the original set of mean ratings, and the number of expected reversals would move to zero. Therefore, the more interesting question is how rapidly does the reversal rate fall off, given a number of reviewers that might be conceivable within the framework of the NSF peer review system. Although these results are based upon many trials whereas the NSF cannot operate with multiple trials, they do suggest what the NSF might "buy" into the ways of reducing probabilities of "misclassifications," or reversals, by increasing the number of its raters or, correlatively, what it gives up by not increasing the number of raters for each proposal.

If we look at the resulting mean number of estimated reversals when four or five reviewers are used, which is the number obtained for 76 percent of the NSF proposals in chemical dynamics, we see that the number closely approximates those we obtained in our simple comparisons between mean COSPUP ratings and the NSF decision. The level of reversals using both this simulation and the actual data is approximately 20 percent. However, if we look at the ratio of estimated reversals obtained if five reviewers are used rather than one, we find that we reduce substantially the expected number of reversals—from 13.74 to 9.64, or to 70 percent of the number of reversals that result when only one reviewer is used.

If eight reviewers are used, we find a further reduction in the estimated number of decision reversals. As can be seen from the points in figure 22.4, the shape of the curve is not linear; the slope decreases as we add reviewers. In short, the NSF buys more for each additional reviewer at the low end than at the high end of sample sizes. For example, by drawing a sample of eight reviewers rather than one, the number of estimated reversals of the same proposals is reduced by 40 percent, but by adding twelve additional reviewers to bring the total to twenty reviewers for each proposal, the estimated number of reversals is reduced only by another 15 percent.

These curves suggest that there is an identifiable point of diminishing returns on increases in sample sizes. They also indicate, however, that proposals whose mean ratings are based on only one or two reviews are relatively unstable estimates, which might lead to a significantly higher number of reversals than if the larger samples of reviewers were employed.

Consensus About the Past Contributions of Applicants

As we pointed out earlier, Phase I of our research indicated that peer review ratings were not strongly correlated with indicators of past scientific performance or "track record." Would the peer review system work better if reviewers placed more emphasis on track record? What weights should be given to track record and to the content of the proposal in the decision-making process? This becomes a particularly important issue in light of the substantial disagreement among reviewers about the quality of proposals. If there is more agreement about the prior track record of the applicants than about proposals, perhaps prior track record should be emphasized more in the peer review process.

Before concluding that track record should be weighted more heavily, we must address a basic question: Is there evidence of more consensus about the quality of the past research contributions of the applicants than about the merits of their current research proposals, at least among referees of proposals?

To begin to answer this question we collected data on the assessments of the past research contributions made by the applicants whose proposals were under review during the second phase of this study.

Consider how these data were generated, the pattern of response, and the results obtained. A single-page, two-item questionnaire was sent to the reviewers, who had approximately a year before appraising the 150 research proposals in Phase II. A total of 1,598 questionnaires were sent to reviewers in chemical dynamics, solid-state physics, and economics. An additional seventy never reached reviewers because of address changes. The overall response rate after two mailings was 81 percent (1,290 of 1,598 sent). There was some variability among the three fields: 82 percent of the reviewers in chemical dynamics answered the questionnaire on reputations, 86 percent in solid-state physics, and 74 percent in economics.

Two questions appeared on the questionnaire. We asked scientists to assess the principal investigator's contribution through his published work over the past ten years both to his entire field (question 1) and to his specific specialty (question 2). Five substantive categories made up the rating scale:

> Is among the few most important contributors to the field [specialty].
> Has made very important contributions to the field [specialty].
> Has made above average contributions to the field [specialty].
> Has made average contributions to the field [specialty].
> Has made below average contributions to the field [specialty].

It is not an interval scale, but it allows for an ordering akin to the adjectival appraisals used by NSF for proposal evaluations: from "excellent" to "poor." Two additional categories allowed the reviewers to indicate that they were not familiar with the principal investigator's work but had heard of him or that they had never heard of the principal investigator.[14]

The results obtained from these questionnaires were analyzed similarly to those obtained for peer review proposal ratings. An analysis of variance of the evaluations excluded those cases in which the evaluator did not make an evaluation of the principal investigator's contribution. In this one-way analysis of variance we were interested in the amount of the total variation in "reputational standing" that could be accounted for by disagreements among evaluators of the same individual and, correlatively, what proportion could be accounted for by differences between individuals.

The data presented in table 22.6 show the level of consensus about the reputational standing of scientists whose proposals were reviewed in the Phase II experiment. For appropriate comparative purposes, we also present data on the amount of total variance explained by intraproposal disagreements when the unit of analysis is the research proposal.

488 CONSENSUS IN SCIENCE

TABLE 22.6 Consensus Among Peer Reviewers on Peer Review Ratings and Scientific Reputations

| | Reputations Experiment | | Peer Review Ratings: Proportion of Total Variance Due to Intraproposal Disagreements | | | |
| | | | | | COSPUP* | |
	Contribution to Field	Contribution to Specialty	Phase I	NSF	Nonblind	Blind
Chemical dynamics	0.50	0.45	0.45	0.60	0.53	0.63
	(285)	(263)		(242)	(213)	(212)
Economics	0.44	0.47	0.35	0.43	0.49	0.48
	(245)	(219)		(163)	(190)	(199)
Solid-state physics	0.50	0.45	0.55	0.51	0.49	0.51
	(234)	(202)		(192)	(190)	(203)

Note: The number in parentheses is the total number of reviewers.

* The COSPUP experiment involved two conditions: one attempted to replicate the NSF procedures exactly (here called "Nonblind"); the other attempted to eliminate all identification of the authors of the proposal (here called the "Blind" condition). See note 6 for a fuller explanation.

The results are instructive. Consider the level of consensus in chemical dynamics. When we examine the amount of the total variance in the ratings of a scientist's contribution to their field over the past decade (question 1), we find that 50 percent of the variance is a result of disagreements among evaluators about the individual scientist's reputation. For the more highly specified question about contribution to the scientist's particular research specialty, 45 percent of the total variance is accounted for by these disagreements about an individual.* These figures are very similar to those obtained when we examined the level of consensus among raters of NSF and COSPUP research proposals.

The story is much the same in both economics and solid-state physics. The level of consensus among these peer reviewers about the importance of the past contributions of the research scientists is roughly equal to their level of consensus about the merits of scientific proposals. Approximately one-half of the total variance in reputational ratings can be accounted for by disagreements among reviewers about the importance of the past research contributions made by the particular principal investigator. It is worth noting again that there do not seem to be any patterned differences in the level of

---

* Recall that the higher the figure in table 22.6, the greater the amount of dissensus about the reputational standing of the scientist.

consensus among the three research programs. There is as much consensus among economists about the merits of contributions by other economists as there is among chemists or physicists about the importance of contributions by scientists in their fields.

These findings on consensus in reputations can be usefully coupled with the fact that there is a substantial correlation between reviewer evaluations of a proposal and their evaluation of the reputational standing of the applicant. (The average correlation in the three fields was in the range of 0.5.) In short, there is consistency within individual reviewers in their evaluations of various aspects of an applicant's work, but there is substantial inconsistency among several reviewers both about the merits of proposals and about the scientific contributions made in the past by the principal investigator.

These findings have implications for science policy. Although additional study is called for, the results here suggest that we probably cannot expect greater consensus among qualified experts about the importance of a principal investigator's track record than we can about the quality of his proposal. It is therefore not at all clear that weighing subjective evaluations of track record more heavily will substantially reduce the probability of obtaining reversals with independently selected samples of reviewers. It has become increasingly clear from an examination of the data produced during Phase II that a central problem that confronts administrators of the peer review system is how to develop greater consensus among estimates of the merits of research proposals and of the prior track record of applicants. The problem is to find indicators of performance and of potential that produce relatively high levels of consensus without being artifactual, that is, indicators that accurately measure the actual opinions of reviewers but also allow for fewer potential reversals in decisions.

Peer Review and Self-Selection

Most studies of peer review concentrate on how decisions are made by experts to distinguish which among a group of applicants for limited resources should receive them. Little attention is paid to the equally important process of *self-selection*. The process of self-selection consists of all those factors which influence the decision of a scientist as to whether or not he or she should apply for a research grant. Our research on how peer review works at the NSF led us to hypothesize that the applicants for NSF funds are not anything like a representative sample of scientists. They are a highly self-selected group greatly over-representing the more active and creative members of the research community. The evidence to support this hypothesis is still limited. However, we have some data to support the idea that applicants to the NSF tend to be significantly more productive scientists, on average, than the rank-and-file within their own fields.

490 CONSENSUS IN SCIENCE

The proportion of "eligible" scientists applying for NSF funds, of course, varies from field to field depending upon several factors.[15] One of the most important is the availability of other sources of research support. In mathematics, for instance, there is apparently very little support available from other governmental agencies. Therefore, a relatively high proportion of university-based mathematicians apply for funds from the NSF. Nonetheless, a relatively small proportion of all scientists employed in PhD-granting institutions in the United States actually apply for research funds.

Interviews with fifty program directors at the NSF suggest that self-selection operates to screen out weaker candidates for NSF funds and to encourage the most productive and creative scientists to apply.

> Section head (S): The thing that surprises me quite a bit is the fact that there are a lot of people that don't try. In fact, it's amazing, if you talk to people about the success ratio at the Foundation that the batting average of the guys out there is really very good. But the reason is that there are a lot of people who never try us.
>
> Interviewer (I): We were wondering about this. It's our impression that the people who apply and get turned down are probably on the average better scientists than the people who don't apply, but we have no way of judging.
>
> S: I agree with your opinion, but I can't prove it. I think it's got to be the case, though, because I think in *many, many cases guys do not apply because they are afraid of being turned down*, and they would rather not try and not have that declination or two or three or whatever showing on their record because it would make their situation more obvious. (Emphasis added)

One of the most important functions of the peer review system may turn out to be its role in making self-selection work.

> I: What would be the consequence of eliminating the requirement that a scientist submit a proposal? Why couldn't all scientists who wanted to receive research funds just send in their name to the NSF and have the decision made on the basis of their past contributions to science?
>
> S: Well, the only trouble is I think if you did this, you'd find that a lot of people who don't now submit proposals to the Foundation would begin clamoring for money. I suspect if you went through the graduate directory of_____, I've never done this exercise but I'm sure that we support—oh, only about—20 to 25 percent of the people in the directory; hopefully the 20 to 25 percent of the best people in the graduate directory. If you did as you suggest [omit the requirement of a proposal], I'm sure you would have 100 percent of the people in the graduate directory clamoring into the Foundation for their cut.
>
> I: In other words, the requirement of submitting a proposal serves the function of preventing the less qualified people from asking for money.

S: There is the very important point that many people overlook: the fact that there is a self-screening process, and it's very real. It puzzles me that you would think people who have gone through all the trouble to earn a PhD and become researchers would not be screened out so easily, but just as a matter of fact there are a lot of people who just never come into the competition. They screen themselves, they take themselves out of the action, and I would say that, you know, that's fine. If they don't want to compete in this game, then chances are they are not top notch researchers, and so leave the money to the people who want to do the research.

Consider several pieces of quantitative data related to the self-selection of applicants for NSF research funds. We compared the distribution of citations to NSF applicants with the distribution of citations to random samples of scientists employed in PhD-granting institutions. We have data relating the distribution of citations in 1971 to work published between 1965 and 1969 for physicists, chemists, and geologists employed in the 1971 American Council on Education (ACE)–rated graduate departments.

Let us look first at the physicists. In 1971 physicists in the rated departments received, on average, twelve citations to work they had published between 1965 and 1969. Twenty-two percent of them had twelve or more citations of their recent work. The data on applicants to the NSF Solid State Physics Program show that in 1975 they had a mean of fourteen citations to work published between 1970 and 1974. Thirty-three percent had more than eleven citations. In fact, the difference between the NSF applicants and the random sample may be attenuated for several reasons.

(1) The random sample contains many scientists who are undoubtedly NSF applicants. We are really interested in the difference between applicants and nonapplicants.

(2) The sample of physicists located at ACE-ranked, PhD-granting institutions is itself a relatively elite group. We are really interested in a comparison between NSF applicants and *all* eligible nonapplicants.

(3) Solid-state physicists receive somewhat fewer citations, on average, than do physicists in such other specialties as elementary particles.

Turning to the chemists, the random sample of university chemists had a mean of fourteen citations, and 24 percent had that or more. The applicants to the Chemical Dynamics Program at NSF had a mean of twenty citations, and 46 percent had fourteen or more. Finally, the random sample of university geologists had a mean of five citations, and 27 percent had that or more whereas the applicants to the geophysics program had a mean of ten citations, and 48 percent had five or more. These comparisons provide rough and tentative evidence

492 CONSENSUS IN SCIENCE

that NSF applicants are a self-selected group of relatively productive and creative scientists.

These data tend to corroborate the impressions of the program directors at NSF. We plainly need more data to test the accuracy of the self-selection idea. If, however, additional data support the claim, the process clearly has problematic implications for expert decision-making at agencies such as the NSF. The job of evaluation is made both more and less difficult. It is less difficult in the sense that "mistakes" may be less costly. If the great bulk of scientists who apply for funds are doing good research, a mistake in awarding a grant to one person rather than another will be less costly than if a high proportion of the applicants are not doing good research. It also makes the process more difficult because it is harder to distinguish among the relative merits of the different applicants.

Understanding that applicants for NSF funds are a self-selected group helps us put the meaning of some of our findings in perspective. First, consider the finding that in Phase I we found a relatively low correlation between indicators of past track record and peer review ratings. If a substantial majority of NSF applicants have strong track records, this means that we have a truncation of range on our independent variables. When the range is truncated it is difficult to find high correlation coefficients. If scientists with very differing past track records were to apply for funds, we might find a considerably higher correlation between rating received and indicators of past performance.

Second, in the light of self-selection, consider the significance of our findings on reversals. The finding that reversals can occur because of disagreement among reviewers is disturbing if we believe that there are large differences in the potential contributions to be made by the different applicants. In such a case, failing to fund a potentially very important contribution because funding was given to a potentially less important contribution could mean that scientific advance would be impeded. But if in fact the great bulk of the applications represent proposals to do potentially important work, then the thought of funding one rather than the other as a result of chance is less disturbing.

The Significance of Disagreement

We have shown that the reversals observed in the COSPUP experiment can be explained by the substantial disagreement among reviewers of the same proposal. If getting an NSF grant were an entirely random process, we would have found a reversal rate of approximately 50 percent. If instead of conducting an independent evaluation of the proposals, we had simply flipped a coin to determine which of the fifty proposals evaluated by the NSF would be funded, we would have obtained a 50 percent reversal rate. The difference between what we would expect with the coin flip and what we observe with the data can be viewed as a measure of what we "buy" from the peer review process. Since the reversal rate is about 25 percent, we may conclude that the fate of a particular grant application is roughly half determined by the characteristics of the proposal and the principal investigator and about half by apparently random elements of chance.

Although we conclude that the funding of a specific proposal submitted to the NSF is to a significant extent dependent on chance elements in the program director's choice of reviewers, this should not be interpreted as meaning either that the entire process is random or that each individual reviewer is evaluating the proposal in a random way. In order to clarify the way in which chance works, we must look at the sources of reviewer disagreement.

Sources of Disagreement

Some of the observed differences among scores given to the same proposal by different reviewers are undoubtedly artifacts of what anthropologists refer to as intersubjectivity. That is, two reviewers may translate their substantively identical opinions differently; reviewer A's opinion is expressed as "excellent" and reviewer B's as "very good."

The great bulk of reviewer disagreement observed is probably a result of real and legitimate differences of opinion among experts about what good science is or should be. This became evident from the qualitative comments reviewers made both on the proposals studied for the COSPUP experiment and on those studied in the first phase of the peer review study. Contrary to a widely held belief that science is characterized by wide agreement about what is good work, who is doing good work, and what are promising lines of inquiry, our research both in this and other studies in the sociology of science indicates that there is substantial disagreement in all scientific fields concerning work in progress.[16]

The analysis of consensus and, correlatively, of dissensus presented thus far has been based entirely on numbers assigned to "adjectival ratings" made by peer reviewers. In order to convey the nature of the reviews and to suggest that we are not dealing here with simple statistical artifacts, we present excerpts from two case files: one indicating a great deal of consensus, the other sharply divergent views. We believe that any detailed examination of the substantive comments associated with proposals received by the NSF will turn up a significant number of "replications" of these two illustrations.[17]

Consensus: Positive Evaluations
The first example is a case that was not problematic for the program director. There was complete intellectual agreement among the reviewers about the scientific merit of the proposal. The panel gave the proposal a mean rating of 1.2,

494 CONSENSUS IN SCIENCE

and mail reviewers gave it an average rating of 1.7. (Recall that ratings range from 1 for "excellent" to 5 for "poor." In most fields, ratings tend to cluster more toward the "excellent" to "good" ratings.) After briefly describing the subject matter of the proposal, the first reviewer begins a terse evaluation:

> The quality of work, the insight, and the general productivity of programs headed by _____ are, in my judgment, unsurpassed in the nation. The methodology which they have developed and the kinds of questions which their research addresses are models for the rest of the work to follow. . . . In summary, I give my highest recommendation for the continued support on this pioneering research group. (rating: 1.0)

A second reviewer of this proposal sums up his reaction with the following comments, quoted here in their entirety, except for identifying information:

> I strongly recommend that this proposal be funded. The objectives of the proposal are important to both basic and applied _____. The experimental procedures have been designed with care and appear to be sound. The previous research of _____ indicates that he could conduct this research program in an excellent manner. Based upon my personal observations, I rate _____ as being possibly the most competent experimental scientist in this research area. The equipment and facilities that he has developed are excellent for the proposed research. I am a highly critical person, but I found nothing of substance to criticize in this excellent proposal. (rating: 1.0)

A third reviewer, somewhat more critical, offered this evaluation:

> Although I do not find the proposal to be particularly well written—particularly in terms of a sufficient explanation of experimental design and techniques to be employed, I would still strongly endorse funding of this proposal. This endorsement is based primarily upon the excellent work which has emanated from this group in the field of _____. I have carefully followed the progress of these people over the past several years, and I am rather well convinced that some of the most incisive and yet technically advanced work in the area of _____ in the nation has been performed by this group. . . . Had this proposal come from most any other group in the country I would not provide the same enthusiastic endorsement of what is specifically contained in the proposal. (rating: 2.0)

This last reviewer is clearly emphasizing one of the NSF's stated criteria: the track record of the principal investigator.

The comments of the review panel are not available, but from its rating, the comments must have been strongly favorable. Even a cursory examination of the comments made by reviewers suggests that this proposal should pose few problems for a program director, unless the program has no funds to distribute.

Dissensus Among Reviewers of the Same Proposal

The second example, which is drawn from a social science discipline but is not significantly different from many similar sets of reviews of proposals in the natural sciences, illustrates the kinds of disagreements often found among reviewers of the same proposal. There were ten reviews of the initial proposal: one "excellent-plus," one "excellent," three "very good," one between "very good" and "good," three "good," and one "poor." We have selected comments that represent the range and thinking of these ten reviewers:

> *Reviewer 1*: I give this proposal an "Excellent-*plus*" rating. The measurements to be developed will, if successful, provide extremely useful tools for the assessment of scientific progress in several dimensions, and I am fully confident that the effort will succeed. I do not think it an exaggeration to suggest that such new measures will afford advances in understanding science comparable to the advances in biology made possible by the substitution of the electron microscope for the optical microscope. . . . This reviewer is acquainted [with] both of the [principal investigators] and has admired their work over the past ten years. They have been at the forefront of _____ and this proposal testifies to their clear sense of what the field needs now in order to advance. . . . The proposal has my complete and enthusiastic endorsement. (rating: Excellent plus)

> *Reviewer 2*: This is a well-thought-through proposal which addresses itself to issues of great significance. . . . On the whole the proposal promises to look at very important problems—problems of great significance to the academic community and to policy makers. (rating: Very Good)

> *Reviewer 3*: This is an extremely competent and carefully designed project which should yield substantial scientific payoff. It is in the tradition of much good work in recent years in the area of _____. I basically expect that this project will produce the best measures yet available of the _____. Clearly this type of measurement is not and should not be the end of the story and additional work is needed to interpret the evidence of these measures. . . . But this is a very sound first step. The project passes with flying colors on criteria IV and V. (rating: Very Good).

> *Reviewer 4*: I am very impressed with the design and objectives of this proposal. It seems to me of the highest quality and importance and is a major

496 CONSENSUS IN SCIENCE

desideratum for further advances in _____. As to the applicants, their recent work on _____ has been, I suppose, the most impressive of anybody in the world. . . . I give the entire proposal full marks with absolutely no reservations. (rating: Excellent)

The enthusiasm displayed here is not matched, however, by several other of the mail reviewers. Consider the judgment of two reviewers who did not see merit in the proposal:

Reviewer 5: This proposal has to be evaluated in terms of the very highest scientific standards. . . . The proposed project would build an excessively elaborate, expensive, and questionable statistical analysis on a foundation of derivative empirical data that has not yet been shown to be extensive, and consistent enough to support significant findings, or have the resilience to provide reliable indicators in the future. . . . As a scientific inquiry, the proposed project has fundamental weaknesses that are not clearly acknowledged. . . . As it stands, this proposal is not methodologically sound, despite its preoccupation with ways to anticipate and correct for potential errors resulting from _____. The budget is very large in proportion to the new knowledge this project would produce. [This summarizes two full, single-spaced pages of comments.] (rating: Poor)

Reviewer 6: This proposal is long and detailed, indeed even prolix. It offers strong evidence of the industry, determination and tenacity of the principal investigators. [They] are well-respected students of _____. However, I am far from persuaded that it would be wise to commit _____ of NSF funds to the project they outline. . . . This proposal veers uncertainly between grandiose aims . . . and claims . . . and some rather modest specifics. . . . Perhaps a modest grant for a pilot project would be in order. (rating: Good)

After more than a year of deliberation and another round of reviews, the proposal was rejected for funding. This last proposal was very familiar to the authors of this paper. It was our own. We use it not to suggest that this was an incorrect decision but only to illustrate that very considered substantive differences often exist about the same proposed work and that these differences are based upon legitimate bases of disagreement.

As long as substantial reviewer disagreement, whatever its source, exists, the fate of a particular proposal will depend heavily upon which reviewers happen to be selected. The element of chance would, of course, be reduced by increasing the number of reviewers; the larger the sample of reviewers the less likely it is that the sample mean will differ significantly from the population mean.

Conclusions

It must be stressed that all of the data presented in this chapter apply only to the system of peer review as it is practiced at the National Science Foundation. It remains unclear what types of disagreements would be found if we examined other forms of peer review, such as the study section method used at the National Institutes of Health. If we found less reviewer disagreement in that context, would that indicate that study sections are a method for achieving intellectual consensus on the relative merits of research proposals, or would it simply reflect "artificial" consensus resulting from the influence of nonintellectual forces that are part of group dynamics?

The results of this study begin to raise serious questions about the structure of the current peer review system and indeed, more broadly, about the structure of scientific knowledge and its sociology. Putting to one side whether the peer review system should be replaced by some alternative mode of decision-making (which we would not advocate), does the current system at the NSF, where decisions do depend to a significant degree on chance, represent "the best of all possible worlds?" And, furthermore, what are some of the larger problematics about scientific activity that emerge from this study of consensus?

How much consensus should we expect in any system of peer or expert evaluation, whether it be of scientific research proposals or of the products of scientific research? There are differences of opinion among "experts" on this subject. Some audiences of experts have found the level of dissensus we found at NSF unexpectedly high and counterintuitive—at least for the natural science disciplines. Others were less surprised. Indeed, some claimed surprise at the "high" level of consensus. One "expert" recently observed,

> It is a common observation that experts disagree, and that careful objective studies of expert judgment typically find them disagreeing more than had been expected and more than the experts themselves find comfortable. Consider wine-tasting, economic prediction, demographic projection, medical diagnosis, judgment in beauty contests, diving contests, and cattleshows, and even Supreme Court decisions. Variability is usually considerable, the more so in close cases.[18]

We find several curiosities about these observations. The problem with dissensus in peer review and in other scientific appraisal systems is not, of course, with the fact that scientists disagree. The problem lies in the mythology that they do not or should not disagree. Some observers of science attempt to perpetuate the idea that science, at least the natural sciences, is marked by great consensus. We suspect that few scientists would like to think of peer review of NSF proposals, or refereeing of scientific papers at leading journals, as akin to the subjectivity in beauty contests or wine tasting. Furthermore, we suspect that the majority of scientists do not think of decision-making at the NSF as influenced significantly by the personal values of the judges—at least not as much as might be seen in a capital punishment case before the Supreme Court. Perhaps we are approaching an age of scientific realism akin to the "sociological jurisprudence" of Benjamin Cardozo or to the "legal realism" of Roscoe Pound and his colleagues in the 1930s. Are we ready to concede that personal values and ideology play a central role in decision-making about the distribution of the scientific research "pie?" Is it not true that "science is different?" Are the standards of evaluation in science undistinguishable from those employed in other institutional spheres? Should the level of chance in activities such as peer review be seen as immutable?

To frame the issue neutrally, what are, in fact, the similarities and differences in levels of consensus not only in different scientific disciplines but also between scientific and nonscientific activities? How does the structure of consensus, and its opposite, in science compare with levels of consensus in the types of activities previously noted? Perhaps more importantly, what are the various types of consequences, both functional and dysfunctional, of varying levels of expert consensus in science? Is it not healthy for science to have substantial disagreement among scientists who evaluate proposals, rather than a single, agreed-upon dogma? But, at what point does disagreement become dysfunctional for the development of science? A distinction must be made between the effects of randomness in systems such as peer review on individual applicants and the effect on science itself. Plainly, the random element can be frustrating and debilitating for individual scientists trying to obtain financial support for their work, and indeed, it may be dysfunctional for science by the discouragement and withdrawal it may cause, as Pardee so aptly points out. But these random elements just may not affect the rate of development of science as a whole.

The study of dissensus in evaluations in science must also consider the relationship between dissensus, decision-making, and the expected duration of the consequences of "wrong" decisions. We can speculate that "expected duration" plays a key role in understanding the consequences of dissensus in different institutional spheres.[20] Consider only in passing the domains alluded to here, beauty contests and Supreme Court decisions, as well as science. While there may be substantial expert disagreement in beauty contests, the expected duration of the consequences of a "wrong" decision is considered by most observers of the contest as of little import. And so differences in opinion are treated with equanimity. This is surely not the case with Supreme Court decisions or with science decision-making. And because the expected duration of the consequences of decisions in these two spheres is considered great, there is frustration and anxiety associated with "wrong" decisions and with dissensus that might place decisions in an ambiguous state. This was clearly perceived by the justices in 1954, when there was great "politicking" to obtain consensus in *Brown v. Board of Education*. Correlatively, one need only look at the first major death penalty case, *Furman v. Georgia*, to find the chaotic situation that follows dissensus in the Court (in this case, where nine separate opinions were filed). Similarly in science, decision-making on what research is to be done is perceived to have a lasting effect on the development of knowledge. Consequently, dissensus that can be related to high levels of chance over what research actually does get funded, that can lead to instability in the existence of scientific laboratories, or that can lead to instability in the careers of individual scientists becomes a phenomenon requiring inquiry.

In closing, there are additional aspects of consensus in science that need increased attention. Consider only several additional questions for which there are no good data. If there is substantial dissensus among experts in science about both the quality of new work to be done and of "old" work that has been done, how do scientific fields move from these levels of dissensus to consensus on "core theory" or methods? Why do some fields move from dissensus to consensus while others do not? What are the historical trends in levels of consensus within a particular field? How do social and intellectual conditions outside of science affect consensus within science? What social and intellectual processes attend the filtering of wheat from chaff, the signal from the noise? Indeed, how has the proliferation of scientific literature, the explosion of publications and journals (some of which is a result of pressures for continued scientific funding), affected levels of consensus and processes of identifying significant scientific contributions?

The data presented in this chapter, as well as in the larger study of peer review at the NSF, lead to the larger issues about consensus and its consequences in science. To better understand the development of scientific knowledge and the nature of scientific decision-making, it is time for us to address these issues of consensus.

Notes

1. Arthur B. Pardee, "On Adequately Funding Medical Research in 1982 and Beyond," private communication, 1982.
2. The renowned physicist Leo Szilard, in his book *The Voice of the Dolphins and Other Stories* (New York: Simon and Schuster, 1961), discussed the best way to retard scientific progress. He didn't believe that it would be difficult at all. One character in his mythical dialogue says,

    > You could set up a foundation, with an annual endowment of thirty million dollars. Research workers in need of funds could apply for grants, if they could make out a convincing case. Have ten committees, each composed of twelve scientists,

appointed to pass on these applications. Take the most active scientists out of the laboratory and make them members of these committees. And the very best men in the field should be appointed as chairmen at salaries of fifty thousand dollars each. Also have about twenty prizes of one hundred thousand dollars each for the best scientific papers of the year. This is just about all you would have to do. Your lawyers could easily prepare a charter for the foundation. As a matter of fact, any of the National Science Foundation bills which were introduced in the Seventy-ninth and Eightieth Congresses could perfectly well serve as a model.

But why would this foundation retard scientific progress? Szilard's character explains tersely:

It should be obvious . . . . First of all, the best scientists would be removed from their laboratories and kept busy on committees passing on applications for funds. Secondly, the scientific workers in need of funds would concentrate on problems which were considered promising and were pretty certain to lead to publishable results. For a few years there might be a great increase in scientific output; but by going after the obvious, pretty soon science would dry out. Science would become something like a parlor game. Some things would be considered interesting, others not. There would be fashions. Those who followed the fashion would get grants. Those who wouldn't would not, and pretty soon they would learn to follow the fashion, too (pp. 100–01).

3. There is an increasing literature on problems of consensus in science in general and in the refereeing of scientific papers and proposals. Here is a partial list of recent articles on this subject:

Barnett, A., D. J. Kleitman, D. Rosenbaum, and B. Singer. "A Statistical Procedure for Testing Nuclear Powered Cardiac Pacemakers." *Technometrics* 20 (August 1978): 221–26.Blaivas, A., R. Brumbaugh, R. Crickman, and M. Kochen. "Consensuality of Peer Nominations Among Scientists." *Knowledge* 4, no. 2 (1982): 252–70.

Blumfeld, Warren S., and Sidney Q. Janus. "Interrater Reliability of a Performance Criterion for a Very Homogeneous Group of Managers." *Psychological Reports* 35 (1974): 1076.Bowen, Donald D., Robert Perloff, and Jacob Jacoby. "Improving Manuscript Evaluation Procedures." *American Psychologist* 27: (March 1972): 221–25.

Carter, Grace. *Peer Review, Citations and Biomedical Research Policy: NIH Grants to Medical School Faculty.* Santa Monica, CA: Rand Corp. Report 1583- HEW, 1974.

Chan, James L. "Organizational Consensus Regarding the Relative Importance of Research Output Indicators." *The Accounting Review* 53 (April 1978): 309–23.

Cicchetti, Domenic V., and Harold O. Conn. "A Statistical Analysis of Reviewer Agreement and Bias in Evaluating Medical Abstracts." *The Yale Journal of Biology and Medicine* 49 (1976): 373–83.

Cole, Jonathan R., and Stephen Cole. *Social Stratification in Science.* Chicago: University of Chicago Press, 1973.

Cole, Jonathan R., Stephen Cole, and Committee on Science and Public Policy. *Peer Review in the National Science Foundation: Phase Two of a Study.* Washington, DC: National Academy Press, 1981.

Cole, Stephen. "Scientific Reward Systems: A Comparative Analysis." In *Research in Sociology of Knowledge, Sciences, and Art,* vol. 1, ed. Robert Alum Jones. Greenwich, CT: Jai, 1978, pp. 167–90.

Cole, Stephen, and Jonathan R. Cole. *Reviewing Peer Review at the National Science Foundation.* New York: Center for the Social Sciences preprint series, 1978.

Cole, Stephen, Jonathan R. Cole, and Lorraine Dietrich. "Measuring the Cognitive State of Scientific Disciplines." In *Toward a Metric of Science: The Advent of Science Indicators*, ed. Yehuda Elkana, Joshua Lederberg, Robert K. Merton, Arnold Thackray, and Harriet Zuckerman, 209–51. New York: Wiley, 1978.

Cole, Stephen, Leonard Rubin, and Jonathan R. Cole. *Peer Review in the National Science Foundation: Phase One of a Study.* Washington, DC: National Academy of Sciences, 1978.

Cole, Stephen, Jonathan R. Cole, and Gary A. Simon. "Chance and Consensus in Peer Review." *Science* 214 (November 20, 1981): 881–86.

Crandall, Rick. "Interrater Agreement on Manuscripts Is Not So Bad!" *American Psychologist* 33 (June 1978): 623–24.

Crane, Diana. "An Exploratory Study of Kuhnian Paradigms in Theoretical High Energy Physics." *Social Studies of Science* 32, no. 10 (February 1980): 23–54.

——. "The Gatekeepers of Science: Some Factors Affecting the Selection of Articles for Scientific Journals." *American Sociologist* 2, no. 4 (November 1967): 195–201.

Cutler, Robert S., Vincent A. Martino, and Alfred M. Webb. "Biomedical Research Revelance Assessment." In *Health Care Delivery Planning*, ed. Arnold Reisman and Marylou Kiley, 89–117. New York: Gordon and Breach, 1973.

Eaves, George N. "Who Reads Your Project-Grant Application to the National Institutes of Health?" *Federation Proceedings* 31 (January–February 1972): 2–9.

Everitt, B. S. "Some Properties of Statistics Used for Measuring Observer Agreement in the Recording of Signs." *British Journal of Mathematical and Statistical Psychology* 30 (1977): 227–33.

Frame, J. Davidson, and Francis Narin. "NIH Funding and Biomedical Publication Output." *Federation Proceedings* 35 (December 1976): 2529–32.

Garfield, Eugene. "The 100 Most-Cited SSCI Authors. 2. A Catalog of Their Awards and Academy Memberships." *Current Contents* 45 (November 1976): 5–9.

Gottfredson, Stephen D. "Evaluating Psychological Research Reports: Dimensions, Reliability, and Correlates of Quality Judgments." *American Psychologist* 33, no. 10 (October 1978): 920–34.

Greenberg, Daniel S. "Washington Report." *The New England Journal of Medicine* 300 (June 1979): 1495–96.

Harnad, Stevan, ed. "Peer Commentary on Peer Review." *The Behavioral and Brain Sciences* 5 (June 1982): 185–86.

Knorr, Karin D. "The Nature of Scientific Consensus and the Case of the Social Sciences." In *Determinants and Controls of Scientific Development*, ed. Karin D. Knorr, Hermann Strosser, and Hans Georg Zilian, 227–56. Dordrecht, Holland: D. Reidel, 1975.

Lakatos, Imre, and A. Musgrave, eds. *Criticism and the Growth of Knowledge.* Cambridge, MA: Cambridge University Press, 1970.

Lasagna, Louis, MD. "Consensus Among Experts: The Unholy Grail." *Perspectives in Biology and Medicine* 19 (Summer 1976): 537–48.

Laveck, G. D., L. R. Freedman, H. H. Walter, and F. S. Steinberg. "Recipients of Research Grants for NICHD: Do Age, Sex, Type of Degree Affect Funding Chances?" *Pediatrics* 53 (1974): 706–11.

Lewis, Gwendolyn Lorita. *Paradigms, Consensus, and Group Structure: A Comparison of Three Scientific Subfields.* PhD dissertation, Princeton University, 1975.

Lindsey, Duncan. "Comments." *American Sociologist* 14 (February 1979): 45–46.

—— "The Corrected Quality Ratio: A Composite Index of Scientific Contribution

502 CONSENSUS IN SCIENCE

to Knowledge." *Social Studies of Science* 8 (1978): 349–54.

Lodahl, Janice Beyer, and Gerald Gordon. "The Structure of Scientific Fields and the Functioning of University Graduate Departments." *American Sociological Review* 37 (February 1972): 57–72.

Maxwell, A. E. "Agreement Among Raters." *British Journal of Psychiatry* 118 (1971): 659–62.

McReynolds, Paul. "Reliability of Ratings of Research Papers." *Psychology in Action* 26, no. 4 (1971): 400–01.

Morgan, Peter P. "Peer Review and Scientific Method in Clinical Research." *Canadian Medical Association Journal* 124 (February 1981): 251–53.

Mulkay, Michael. "Consensus in Science." *Social Sciences Information* 17 (1978): 107–22.

Nystedt, Lars. "Consensus Among Judges as a Function of Amount of Information." *Educational and Psychological Measurement* 34 (1974): 91–101.

Patterson, C. H. "Evaluation of Manuscripts Submitted for Publication." *American Psychologist* 24: (1969) 73.

"Peer Review and Quality Assurance." *Professional Psychology* (Special issue, February 1982).

Peters, D. P., and S. J. Ceci. "Peer-Review Practices of Psychological Journals: The Fate of Published Articles, Submitted Again." *Behavioral and Brain Sciences* 5 (June 1982): 187–255.

Pfeffer, Jeffrey, Gerald R. Salancek, and Huseyin Leblebici. "The Effect of Uncertainty on the Use of Social Influence in Organization Decision Making." *Administrative Science Quarterly* 21 (1976): 227–45.

Rosenthal, Robert. "Estimating Effective Reliabilities in Studies That Employ Judge's Ratings." *Journal of Clinical Psychology* 29 (July 1973): 342–45.

Ruderfer, Martin. "The Fallacy of Peer Review—Judgement Without Science and A Case History." *Speculations in Science and Technology* 3 (1980): 533–62.

Scarr, Sandra, with Barbara L. R. Weber. "The Reliability of Reviews for *The American Psychologist*." *American Psychologist* 33, no. 10 (October 1978): 935.

Schneier, Craig Eric. "Multiple Rater Groups and Performance Appraisal." *Public Personnel Management* 6 (January–February 1977): 13–20.

Scott, William A. "Inter-referee Agreement on Some Characteristics of Manuscripts Submitted to the *Journal of Personality and Social Psychology*." *American Psychologist* 29, no. 9 (September 1974): 698–702.

Singer, Burton. "Exploratory Strategies and Graphical Displays." *Journal of Interdisciplinary History* 7 (Summer 1976): 57–70.

Small, Henry G. *Characteristics of Frequently Cited Papers in Chemistry*. Philadelphia: Institute for Scientific Information, 1974.

——. "Cited Documents as Concept Symbols." *Social Studies of Science* 8 (1978): 327–40.

—— . *Report on Citation Counts for National Science Foundation Grant Recipients and Non-Recipients*. Philadelphia: Institute for Scientific Information, 1974.

Smith, James M. "A New Rater Selection Technique for Use with Behavioral Rating Scales." *Journal of Clinical Psychology* 30 (1974): 40–43.

Symington, James W., with Thomas R. Kramer. "Does Peer Review Work?" *American Scientist* 65 (January–February 1977): 17–20.

Teasdale, Graham, Robin Knill-Jones, and Jaap Van Der Sande. "Observer Variability in Assessing Impaired Consciousness and Coma." *Journal of Neurology,*

Neurosurgery, and Psychiatry 41 (1978): 603–10.

Useem, Michael. "State Production of Social Knowledge: Patterns in Government Financing of Academic Social Research." *American Sociological Review* 41 (1976): 613–29.

Werts, C. E., K. G. Joreskog, and R. L. Linn. "Analyzing Ratings with Correlated Intrajudge Measurement Errors." *Educational and Psychological Measurement* 36 (1976): 319–28.

Winch, Robert F., and R. Bruce W. Anderson. "Two Problems Involved in the Use of Peer-Rating Scales and Some Observations on Kendall's Coefficient of Concordance." *Sociometry* 30, no. 3 (1967): 316–22.

Wolff, Wirt M. "A Study of Criteria for Journal Manuscripts." *American Psychologist* 25, no. 7 (1970) 636–39.

Yankauer, Alfred. "Editors Report: Peer Review." *American Journal of Public Health* 69 (March 1979): 222–23.

Zuckerman, Harriet. "Theory Choice and Problem Choice in Science." *Sociological Inquiry* 48 (1978): 65–95. Reprinted in *Sociology of Science*, edited by Jerry Gaston, San Francisco: Jossey-Bass, 1978.

Zuckerman, Harriet, and Robert K. Merton. "Age, Aging and Age Structure in Science." In *Aging and Society: A Sociology of Age Stratification*, vol. 3, ed. Matilda White Riley, Marilyn Johnson, and Anne Foner, 292–356. New York: Russell Sage Foundation, 1972.

——. "Patterns of Evaluation in Science: Institutionalisation, Structure and Functions of the Referee System." *Minerva* 9 (January 1971): 66–100.

4. Cole, Rubin, and Cole, *Peer Review in the National Science Foundation*. Peer review in the form used at the National Institutes of Health (NIH) was not studied. The NIH study section form of peer review differs significantly from the form of peer review at NSF. The results reported in this chapter apply only to the basic science programs at NSF and may not be generalized to any other form of peer review.

5. The report (Cole, Rubin, and Cole, *Peer Review in the National Science Foundation*) of the first phase of that study includes detailed description of the central role of the NSF program directors in the decision-making process, the role of peer review panels, the influence of budget size on decisions, the effects of "self-selection" processes on peer review ratings, an analysis of problematic and unproblematic cases, and the correlates of peer review ratings and funding decisions. See also S. Cole, L. Rubin, and J. R. Cole, "Peer Review and Support of Science," *Scientific American* 237 (October 1977): 34–41.

6. The 150 proposals were reviewed under two different conditions in the COSPUP experiment, only one of which is reported in this chapter. Under one condition (the one dealt with here), the COSPUP reviewers were sent the proposal exactly as it was received by NSF, with the name and institution of the principal investigator listed on the title page, and were given exactly the same evaluation criteria to follow as were NSF reviewers. The participants in the experiment were told that their opinions would not influence the funding decision, which had already been made.

Under the second condition, the proposal was altered so as to try to hide the identity of the principal investigator. A report on this part of the experiment can be found in Cole and Cole, *Peer Review in the National Science Foundation*. The principal result of this part of the experiment was that it proved almost impossible to remove all evidences of the applicants' identity without destroying the integrity of proposals.

The reviewer selectors received a copy of the proposal that did not include the name of the principal investigator and from which all references to the principal investigator's

504 CONSENSUS IN SCIENCE

prior work had been deleted. It was therefore possible for the reviewer selector to name the principal investigator as a possible reviewer. In these cases, of course, the principal investigator's name was removed from the list of reviewers. We also removed from the list anyone who was at the institution of the principal investigator and anyone who had reviewed the proposals for the NSF.

7. The size of the pool (N) can be estimated as follows: This approximation assumes that each COSPUP selector makes six equiprobable choices from the pool of N. Suppose that individual A had been selected by NSF. Then P (A will be selected by COSPUP selector) = $6/N$. Let $P = 6/N$. Then P (no overlap) = $(1 - p)^2$; P (one overlap) = $2p (1 - p)$; P (double overlap) = p^2. The proportions $(1 - p)^2$: $2p (1 - p)$: p^2 correspond closely to the proportions observed when $p = 0.1$. This suggests that $N = 60$. The approximation is not perfect, of course. The field of economics produces a surprising number of double overlaps.

8. NSF may decide to fund a piece of the proposed scientific work, reducing the amount of the grant accordingly. Here we do not differentiate such a grant from a full grant.

9. We classified reversals within quintiles as follows: The proposals were grouped into quintiles based on COSPUP rank, the first (best) quintile containing proposals with ranks 1, 2 . . . 10. A proposal was counted as reversed if it was in the upper half by one set of ratings and in the lower half by the other set. Where there were ties in the mean ratings crossing quintile boundaries, proposals were apportioned among the categories involved. This rule results in noninteger numbers of reversals.

10. J. R. Cole and S. Cole, "Which Researcher Will Get the Grant?" *Nature* 279, nos. 575–76 (1979).

11. If one takes two independent observations from a normal population with standard deviation sigma, the expected absolute difference is $2\sigma / \sqrt{\pi}$. This statement is a reasonable approximation even when the population values do not follow a normal distribution. The numbers given in the text reflect a reviewer standard deviation of 7.49 and a proposal standard deviation of 4.36. Note that $(2 / \sqrt{\pi})7.49 = 8.45$ and $(2 / \sqrt{\pi}) (4.36^2 + 7.49^2)^{1/2} = 10.08$. The value of $(4.36^2 + 7.49^2)^{1/2}$ reflects the fact that a randomly selected review incorporates both the proposal standard deviation and the reviewer standard deviation.

12. Detailed description of the methods used to generate the 475 random trials, the sampling from the existing peer review ratings, and the jackknifing procedures appear in draft form of the Phase II peer review study and will be published shortly. Professor Thomas DiPrete wrote the computer programs to perform these analyses. For discussion of the jackknife technique, see F. Mosteller and J. W. Tukey, "Data Analysis, Including Statistics," in *The Handbook of Social Psychology*, vol. 2, ed. G. Lindzey and E. Aronson (New York: Addison-Wesley, 1954), 134ff.

13. Similar analyses were performed for both solid-state physics and economics. To make the graph numbers clear, we found the following expected number of reversals for proposals with raters: 1 = 13.74, 2 = 12.37, 3 = 10.95, 4 = 10.20, 5 = 9.64, 6 = 9.23, 7 = 8.76, 8 = 8.53, 12 = 7.47, and 16 = 7.03. For twenty raters the expected number of reversals was 6.39. All of these figures are jackknifed estimates. Figure 3.4 plainly shows that there were only minimal differences between the jackknifed estimates and estimates that did not use the jackknife procedure.

14. There is a potential problem with our measurement of reputation. The scientists who produced assessments of the reputations were the same scientists (at least drawn from the same sample of scientists) as those who produced evaluations of proposals. More significantly, many of the reviewers had made a prior review of the proposal a year

before assessing the scientist's reputation. The evaluator's prior experience with the proposals could have influenced their judgment of reputation. Assuming that the strain toward consistency is great, reviewers might have either recalled their earlier proposal ratings or looked up the ratings before judging reputations.

Although this remains possible and could only be definitively handled by obtaining appraisals from an entirely independent sample of reviewers, the data suggest that confounding is unlikely to have occurred. The evidence for this derives from the availability of the NSF ratings of proposals, which after all, were all by raters different from those used in our experiment. In that sense, they represent an independent sample. It turns out that the NSF ratings are just as highly correlated with the reputational measures as are those ratings derived from COSPUP reviewers. Because the reputational ratings were not made by NSF raters, the ratings could not be a result of an attempt by the reviewers to be consistent in their proposal and reputational evaluations. Thus, the only possible confounding factor would be if both the NSF ratings and the reputational ratings were equally influenced by a reading of the proposal.

Further evidence suggests that the reputational measures are not confounded by the prior evaluation of the proposal. If a respondent's judgment of the reputation of a scientist had been influenced by his previous reading of the proposal, then the differentials between the average peer review ratings produced by the NSF reviewers and by the COSPUP reviewers should be correlated with the reputational ratings that they produce. In order to test this idea, we examined the relationship between the difference in the mean NSF proposal ratings and the mean ratings for the nonblind COSPUP proposals, and the mean reputational score received by the proposal.

We examined the zero-order correlation coefficients between the differences in mean ratings and the mean reputational score for the three programs studied in Phase II. There were no substantive or statistically significant correlations between these differences in mean ratings and the reputational scores received by the authors of the proposals. The strongest correlation can be found in the chemical dynamics program, where the differences between the NSF and the COSPUP nonblind mean ratings were correlated with the mean reputational score of a scientist's contribution to his field, $r = -0.14$ (n.s.). The respective correlations were 0.07 for economics and 0.01 for solid-state physics. The results are similar whether we relate the difference scores to appraisals of contributions to the discipline as a whole (question 1), to the particular specialty of the principal investigator (question 2), or to the visibility of the principal investigator (that is, simply whether the evaluator had heard of the principal investigator). They are all weak associations.

15. Cole, Cole, and Dietrich, "Measuring the Cognitive State of Scientific Disciplines," 162.
16. For detailed description of different types of cases involving both agreement and disagreement, see Cole, Rubin, and Cole, *Peer Review in the National Science Foundation: Phase One of a Study*, 86–109.
17. Professor William Kruskal, private letter to Dr. John Slaughter, director of the National Science Foundation, February 4, 1982.
18. The sociological concept of "socially expected duration" is developed fully in recent work by Robert K. Merton.

CHAPTER 23

TESTING THE ORTEGA HYPOTHESIS

Milestone or Millstone? (1987)

STEPHEN COLE AND JONATHAN R. COLE

A comment on "Testing the Ortega Hypothesis: Facts and Artifacts" by M. H. MacRoberts and B. R. MacRoberts, Scientometrics *12 (1987): 293.*

(Received March 6, 1987)

Introduction

It is gratifying to know that fifteen years after its publication, sociologists of science find the ideas presented in our paper "The Ortega Hypothesis" still of interest.[1] It is mildly distressing, however, to see that the field has not progressed very much in these years, as is evidenced by the MacRoberts and MacRoberts (M&M) paper.[2] The MacRobertses argue that citations are not a valid or true measure of scientific influence.[1] This argument has been made ad infinitum and although it is true that citations remain far from an ideal indicator of quality of work or influence, the measure has proven to be of great use to sociologists of science and bibliometricians. No better indicator has been discovered.

We will first review what the Ortega hypothesis paper did and did not claim, go on to discuss the criticism made by M&M, and end with a discussion of the policy implications set forth in the Ortega hypothesis paper.

The Basic Question in the Ortega Hypothesis

First, in the 1972 paper, we were interested in this question: "How many scientists are contributing *through their published research* to the movement of science, and how many are not?" (369; emphasis added). If this is the central question, the first approximation of the answer is self-evident by even a cursory examination of some very simple and well-known statistical patterns of scientific productivity. As Derek Price pointed out in *Little Science, Big Science* (and as noted in the 1972 paper) about half of all scientists with PhDs publish nothing.[3] If a scientist publishes nothing, then by definition that scientist cannot contribute through his or her published work to scientific advance. Price goes on to show of the remaining 50 percent who do publish, half of them publish only one paper (Lotka's law). In short, the overwhelming majority of scientific papers are published by only a small majority of scientists. Put in another frame, approximately 10 percent of the scientist population produce 50 percent of all research publications. This skewed distribution of scientific production has been validated now in scores of papers and books.[4]

Since most scientists do not publish very much, if papers that influence a scientist were to be picked at random from a list of all published papers, a high proportion of these randomly selected influences would be authored by the few scientists who do most of the publishing—those scientists who by definition constitute an elite or upper tier of publishers. But there is substantial research to show that there is a significant correlation between the quality of a scientist's publications and the quantity of that scientist's publications: the work of the heavy producers has greater impact and is *on average* deemed better by peers than the work of those scientists who publish one or two articles. This further increases the probability of any scientist being influenced by the published work of the elite. Therefore, if one accepts the assumption that we are only looking at influence through published work and one examines the distribution of publications, little further evidence is needed to disprove the Ortega hypothesis.

The MacRobertses' Millstone

Let us turn to some of the specific claims made by M&M. They are incorrect in saying that we compare the citations made by a group of "average" scientists with those made by highly cited authors. We argue that all scientists, rank-in-file or elite, disproportionately cite the work of elites. The important and relevant point here is that the validity of this claim is not dependent on citations being a complete and unbiased indicator of influence. Of course, the process of citing prior influences is not completely rationally structured. We have pointed out in our own content analyses of citations that a high proportion of citations are "ceremonial" in nature and that citations serve many functions, including the identification of the author with prestigious figures in the field.[5] Elites are, in fact, more apt to receive ceremonial citations than others. But one must ask how a particular person became a member of the elite tier. Generally, contributions judged significant by authorities and peers in the field lead to the social emergence of "stars" and they then gather increased citations through the process of accumulative advantage. Furthermore, given the overall distribution of citations, which like publications is of course highly skewed, the citations to elites that represent "ceremonial" citations are not "marginally important" in the definition of their work as influential in the growth of knowledge.

The M&M paper published in this issue is largely influenced by a prior paper they published in *Social Studies of Science*.[6] In that paper, the authors argue that scientists frequently do not make explicit reference to much of the work that has influenced them. This is a strong claim since their rejection of our results in the 1972 Ortega paper is based upon the rejection of citations as a measure of influence. No direct tests are made of our results, but the measures used are questioned. This important question is raised by their analysis: What is an "influence"?

The problematics with this question are made clear by the self-exemplifying misuse of the concept found in the M&M paper. The authors discuss "influences" on our 1972 Ortega paper that, they claim, we failed to cite or reference. M&M state, "Thus the key proposition of the paper—the functional theory of stratification—and the myriad individuals who were involved in its development, are not referenced."[7]

To us this is a very curious statement indeed. We are being told that the "key proposition" of our paper is the functional theory of stratification. When we wrote this paper we had no conscious interest in the functional theory of stratification nor were we interested in proving or disproving the functional theory. In developing the Ortega hypothesis paper we were not "influenced" by the authors and commentators on the functional theory. But the way M&M read the paper, they see the functional theory not only as an "influence" but as "the key proposition" in the paper. To the authors of the Ortega hypothesis this is quite simply a gross misreading of the influences on us in the construction of that paper.

What then are the "influences" on an author writing a paper: those which the author is conscious of or those which some analyst says she or he should have been conscious of? The approach of M&M is strangely antiquated—an approach which can be found among positivist intellectual historians. These historians look for connections between ideas even if such connections were not evident to the authors. Often, this approach leads to the "sin of adumbrationism," that is, finding in intellectual history all manner of antecedents for ideas and downplaying the originality of the more recent contribution. Sociologists of science have for some time been more concerned with how science is actually done, and what the actual and perceived influences are on research activities, than on the normative question of what the scientist ought to have been influenced by or should have been aware of.[8]

Setting themselves up as definers of "true" influence, and neglecting wholly any effort at validating their own procedures and results, the M&M paper takes to scolding scientists for not citing those people that they "should" have. This normative position is given substance when they take Einstein to task for not citing in his 1905 paper the "many others who obviously influenced" him.[9] They scold Gibson for not citing Kantor.[10] Gibson apparently did this as a result of "ignorance" of the literature; but nonetheless he is taken to task for not knowing what he should have known. They point out that sociologists of science influenced by Lotka's law sometimes fail to make an explicit reference to Lotka's paper but base their reading on Price's discussion of this paper, referring to this latter practice as "lifting" a citation. Using this normative approach we would chide M&M for their failure to cite (out of ignorance we are sure) the 1981 paper by Adatto and Cole on the role of "intermediaries" as transmitters of ideas.[11]

Throughout their analysis M&M point out how all of the failures on the part of scientists to be accurate in their citing practices make it difficult or impossible to use the citations in a paper as an indicator of the influences on that paper. By their unsystematic and unvalidated criteria (never made explicit) only roughly 30 percent of all references that should enter a text actually do. If one defines influences as that work which *should* have influenced an author, perhaps M&M are correct. But if one were to accept as a designation of influence only those works and people that the author is conscious of, how serious are the practices pointed out by M&M in testing the Ortega hypothesis?

For any *particular* paper the problems might be serious, but as we shall see below, even here the differences between ideal and practice might not lead to different results. As sociologists of science we wanted to know simply what proportion of active scientists influenced the development of scientific knowledge in a field, not a particular paper. We would argue that despite the "flaws" of citing scientists, in the *aggregate* citations represent a fairly accurate index of the extent to which ideas in a field have been influenced by groups of scientists. In short, the M&M paper has misspecified the proper unit of analysis. We are interested in the proportion of scientists whose work influences advances in knowledge and whether in this set of all scientists how many would not be captured by using citations as an indicator of influence. For the MacRobertses' claim to be supported by evidence, the composition of the set of "influencers" and "non-influencers" would have to change significantly were we to use indicators of influence on research other than citations.

510 CONSENSUS IN SCIENCE

The significant question for the validity of the Ortega hypothesis is not whether scientists cite people who have not had a significant influence on them but whether they fail to cite people who have had an important (conscious) direct or indirect influence on them. All of our research has suggested that scientists tend to over rather than under cite other scientific works. The limited data presented by M&M in their 1986 paper simply illustrate who they believe should have influenced the authors of the papers they analyzed. Of course, it will be possible to find here and there *an example* of a scientist failing to cite a particular clear-cut influence, but it is unlikely that if that scientist and his or her work is truly influential in the growth of knowledge that most of the scientists who should have recognized the influence will totally ignore it—thus rendering the impact of the work invisible. Furthermore, it is unlikely that a significant portion of influences will go uncited, and it is even more unlikely that those uncited influences will be scientists who have published one or two papers rather than a great number. Indeed, the possibility that there exists "intellectual blacklisting" of significant contributions to the advance of knowledge is an interesting empirical question that is worth serious inquiry. Could we identify conditions under which we see a conscious effort on the part of a community of scientists to systematically exclude from visibility the work of others? We might find such "intellectual blacklisting" practices in fields which have strong ideological and value components.

But for the sake of argument, let us assume that the claim of the MacRobertses is correct and that there is only a 0.3 probability that an influential work will actually be cited in a paper. With the proper specification of unit of analysis as the index of whether or not a scientist has or has not contributed to scientific advance through research, we can calculate the probability that scientists will or will not be among the influentials. Suppose that a paper that is influential has only a 0.3 probability of being cited in any paper that *should* cite it. Consider that there are a hundred papers in a field that should cite this paper because it was a "true" influence. What is the probability that the paper will receive n citations? This probability can be obtained by considering a simple binomial probability distribution:

$$P(d/n, \pi) = \binom{n}{d} \pi^d (1-\pi)^{n-d},$$

where d = the number of citations;
    n = the number of papers that should have been influenced by an author; and
    π = the probability that the paper that should be cited is actually cited.[2]

In 1970, the average number of citations to any cited scientist (this of course representing a distinct minority in itself) was less than ten per year—and this was to the entire corpus of that scientist's work. If one hundred papers published in a field in a given year should have cited this paper, the probability is extremely high that this individual paper will receive at least ten citations from those hundred papers, given the 0.3 probability of any single paper including the citation. This exercise with the binomial probability distribution can be carried out for different probabilities of "errors"; for different sized literatures; and for different definitions of "high impact" or "influential" papers, or more importantly, for the aggregate number of papers produced by a scientist.[3] We did this for literatures of one hundred and one thousand papers and for definitions of influential that ranged from above the average number of citations to more than sixty (which was used as one group in the 1972 Ortega paper). Unless the number of papers that should have cited the paper in question is very low, the number of citations required for inclusion among "elites" is very high, and the probability of "errors" much higher than 0.3, there will rarely be cases where the "error" structure of citations will exclude papers that should be classified as "influential."[4] In short, while errors of underinclusion are made in individual cases, at the aggregate level truly influential scientists have, in fact, a very high probability of being identified and classified as such even though the absolute number of citations to their work will be lower than it "should" be.

Beyond the unlikely probabilities of misclassification using citations as a method for identifying influence, there are additional conceptual reasons why the argument by M&M is implausible. They tend to assume that the group of scientists who "suffer" most from the error structure in citations are those rank-in-file scientists who are invisible but who ought to be recognized for their actual influence. If we were to "clean up" the variety of acknowledged problems with citations as an indicator of influence, we would hypothesize that the citation distribution would be even more skewed than it is. This would result from the process that Merton has called "obliteration through incorporation." (Here we shall be self-exemplifying by not citing the source for Merton's idea. Nonetheless, he will surely remain within the "class" of influentials.) In short, many of the scientists whose work is omitted from formal recognition through citation are identified in the text by name but not by formal citation. The MacRobertses' paper acknowledges such "errors" but fails to provide any estimate of what proportion of all errors are of this type. If we added to the set of all citations those omitted because of obliteration through incorporation into the body of knowledge, would the distribution of influentials become even more skewed than it already appears to be?[5]

The MacRobertses' paper argues that citations are used primarily as a form of persuasion and therefore will be biased toward the inclusion of elites. This has simply not been adequately demonstrated with empirical data. While persuasion is undoubtedly one reason for including a reference to another's work, it surely does not exhaust the reasons for citations.

512 CONSENSUS IN SCIENCE

References to the MacRobertses' papers here are surely not intended to persuade any reader of the force of our argument. More work needs to be done on the processes by which scientists select references and, more interestingly, on the ways inclusion of citations can distort the content of the cited work. But this has little to do with the validity of the arguments presented in the Ortega hypothesis paper.

Policy Implications in the Ortega Paper

Even if the Ortega hypothesis is incorrect and it turns out that the majority of scientists do not contribute to the growth of knowledge through their research, does it follow that the policy implications suggested in the 1972 paper were correct—that we might be able to reduce the size of the scientific community without influencing the rate of advance of knowledge? First, contrary to the claim of M&M, we never argued "that the majority of scientists could be dispensed with" but simply that the number of PhDs in science *might* be reduced. In fact, we stated a possible policy implication of the results—but one which was fraught with problems. That policy conjecture has been translated by some in the field as a strong assertion of fact. Here is what we actually said in the 1972 paper:

> Although the conclusions of this paper may be reasonably clear, the implications of these data for the structure of scientific activity, at least in physics, need careful consideration. Consider only one problem emerging out of the findings that *needs a great deal of further research: the size of the research establishment of modern science.* If future research on other fields of science corroborates our results, we may inquire what it implies about the relationship between the number of scientists and the rate of advance in science, and whether it is possible that the number of scientists could be reduced without affecting the rate of advance.[12] [Emphasis added].

Roughly 25 percent of the original paper was devoted to a statement of this policy implication and the elaboration of several basic criticisms of it, including the fact that we dealt with only one generation of influence and that there would be great difficulty in identifying correctly the scientists who would go on to produce important scientific discoveries. In short, we were aware of the multiple difficulties of drawing strong policy inferences from those limited data.

In fact, the policy implication of the Ortega paper was examined in limited detail in a paper by S. Cole and G. Meyer.[13] The MacRobertses' paper cites this work but completely ignores its substance. This paper presents empirical evidence suggesting that the policy implication of the 1972 paper was incorrect.

The 1972 paper implied that the brightest young people would continue to pursue science careers even if the size of science was reduced. The data presented by Cole and Meyer suggest that this turned out not to be true. As opportunities in science declined, bright students chose other careers such as medicine and later business over science.[6]

The Cole and Meyer paper presented an agenda for future research. First, that paper contained data for only one field and for only one time period. They were able to show that as the demand for academic physicists declined in the 1970s the number of new young physicists publishing significant papers declined sharply. The data suggest that there is a direct correlation between the number of people going into science and the number of scientists who turn out to produce significant contributions. This hypothesis should be researched for other fields and for other time periods. If this conclusion is generalizable, it would be extremely important both theoretically and from a policy point of view. As Cole and Meyer point out, there is some evidence in the work of Merton and Ben-David to suggest that both cultural and structural variables influence the flow of talent into science, and this in turn affects the rate of scientific advance in a particular society.[14] Understanding the social determinants influencing the rate at which knowledge accumulates should be a major goal of the sociology of science.

Second, Cole and Meyer suggest, but have no direct evidence, that bright young people respond rapidly to changes in economic opportunity. If this is true, then as the demand for scientists in a particular society declines, the number of bright people seeking careers in science will decline. Direct evidence is needed to test this hypothesis.

It is unfortunate that more time is spent in rehashing old arguments about the nature of citations and what they do or do not measure than in taking up the multitude of substantive problems that would push forward our knowledge about how science advances. If we put aside the policy implications of the Ortega paper and attend solely to the question of what social and intellectual conditions influence the growth of knowledge, there are scores of old and new questions that need answering. No one, for example, has investigated the "filtering" of influence over time or has investigated the various forms of intellectual influence and their relative effects on the growth of knowledge. Few systematic empirical studies are being developed to study these problems. This is particularly unfortunate for a young specialty in need of both the application of talent to significant questions and a general sociological theory of the conditions under which science prospers.

Notes

1. Surely these references to the MacRobertses' paper are good examples of "overinclusion," references to work that has had no conscious influence on the development of our ideas.
2. When n is greater than 10, the binomial closely approximates the normal distribution. Consequently, computation of the probability of a paper receiving a specific number of citations or more or less than a specific number becomes a simple matter.
3. All assumptions of the binomial distribution are met in this example, with the possible exception of the assumption of independence. We simply do not know much about the extent to which citation practices are independent or dependent events.
4. It is important to emphasize here that in the Ortega hypothesis paper we were considering scientists' contributions as the unit of analysis, not individual papers. And we were considering influence within the community of scientists, not for any particular scientist. Nonetheless, carrying out the thought experiment for individual papers leaves little doubt that few papers and individuals who are actually influential by the MacRobertses' definition would be excluded from inclusion among influential scientists.
5. In the Ortega hypothesis paper we examined only physics journals. Others have since examined the Ortega hypothesis in other fields, as the MacRobertses' paper indicates. Inclusion of citations to scientists whose credit has been obliterated through incorporation might be offset by finding some portion of excluded citations to authors of few papers for whom it was difficult to obtain biographical data.
6. Of course, there are multiple reasons for the shift in interest to fields other than science. A reduced opportunity for academic careers was only one reason for the shift away from science; other cultural and social structural variables also played significant roles in this shift.

References

[1] Cole, J. R., and S. Cole. "The Ortega Hypothesis." *Science* 178 (1972): 368.
[2] MacRoberts, M. H., and B. R. MacRoberts. "Testing the Ortega Hypothesis: Facts and Artifacts." *Scientometrics* 12 (1987): 293.
[3] Price, D. de Solla. *Little Science, Big Science.* New York: Columbia University Press, 1963.
[4] For a review of that literature, see Cole, J. R., and H. Zuckerman. "The Productivity Puzzle: Persistence and Change in Patterns of Publication of Men and Women Scientists." In *Advances in Achievement Motivation*, ed. P. Maehr and M. W. Steinkamp, 217–58. Greenwich, CT: JAI, 1984.
[5] Cole, S. "The Growth of Scientific Knowledge: Theories of Deviance as a Case Study." In *The Idea of Social Structure: Papers in Honor of Robert K. Merton*, ed. L. Coser. New York: Harcourt, 1975; Adatto, K., and S. Cole. "The Functions of Classical Theory in Contemporary Sociological Research." In *Research in the Sociology of Knowledge Science, and Art III*, ed. F. Kuklick. Greenwich, CT: Johnson Associates, 1981.
[6] MacRoberts, M. H., and B. R. MacRoberts. "Quantitative Measures of Communication in Science: A Study of the Formal Level." *Social Studies of Science* 16 (1986): 151.
[7] MacRoberts and Macroberts, "Quantitative Measures of Communication in Science," 153.
[8] See for example, Latour, B., and S. Woolgar. *Laboratory Life*, Beverly Hills, CA: Sage, 1979; Knorr-Cetinam, K. *The Manufacture of Knowledge.* New York: Pergamon, 1981.
[9] MacRoberts and MacRoberts, "Quantitative Measures of Communication in Science," 154.

[10] MacRoberts and MacRoberts, "Quantitative Measures of Communication in Science," 157.

[11] Adatto and Cole, "The Functions of Classical Theory in Contemporary Sociological Research."

[12] Cole and Cole, "The Ortega Hypothesis," 372.

[13] Cole, S., and G. Meyer. "Little Science, Big Science Revisited." *Scientometrics* 7 (1985): 443.

[14] Merton, R. K. *Science, Technology, and Society in Seventeenth Century England.* New York: Harper & Row, 1970 (originally published in 1938); Ben-David, J., and A. Zloczower. "Universities and Academic Systems in Modern Society." *European Journal of Sociology* 3 (1962): 45–84.

CHAPTER 24

DIETARY CHOLESTEROL AND HEART DISEASE

The Construction of a Medical "Fact" (1988)

JONATHAN R. COLE

T he mass media play a critical role in transmitting health-risk information from knowledge producers to consumers. But do those reporting on health risks present an accurate picture of the state of scientific knowledge on these risks? More specifically, if there are biases and distortions of scientific information, what is the character of these problems in reporting? And what properties of the institutions of science and the mass media help us understand the types and sources of bias and distortion? To answer such questions we would need an extensive database of health-risk studies reported in the various media.[1] Such a set of data does not exist. Here I will present material on one case—the reporting of health risks purported to be associated with dietary cholesterol. The basic analytic problem of this paper can be stated quite simply: How is a questionable claim to truth, or medical "fact," transformed into an unquestionable one? What role do health scientists and the mass media play in this process?[2]

The focus is on reporting by scientists and reporters of recent studies on the relationship between cholesterol and coronary heart disease. The first section describes the recent results from a major, long-term study of cholesterol

---

A brief version of this paper appeared in *Columbia* magazine; a second was presented at the 1985 AAPOR Conference by N. J. McAfee on May 17, 1985. The work has been supported by a grant from the Josiah Macy Jr. Foundation. I thank Dr. John Bruer for his support throughout the project. For extensive comments on an earlier draft of this paper, I thank E. H. Ahrens Jr., Bernard Barber, Stephen Cole, Robert K. Merton, Hubert J. O'Gorman, Eleanor Singer, Stanley Schachter, and Harriet Zuckerman. This work was also supported by a grant from the National Science Foundation, NSF (SES-80-08609).

and notes what types of information were not transferred from the medical research community to the public via the media. The second section examines the recent history of the controversy over the health risks associated with dietary cholesterol and describes the development of a "scientific fact" that cholesterol intake is a major cause of heart disease and of the normative conclusion that it should be uniformly minimized in American diets. A final section discusses some possible sources of explanation for the distortion of health risks in the media.

The Cholesterol Case

In January 1984 the *Journal of the American Medical Association (JAMA* 251, no. 3) published the results of a recently completed ten-year, $150 million study of the link between lowering cholesterol levels and a reduction in heart attack deaths. Prior to this publication, several scientists from the Lipid Research Clinics (LRC) team who conducted the study on 3,806 subjects gave the media an extensive briefing on their research methods and the study's results (Brensike et al. 1982, 1984; Levy et al. 1984; Lipid Research Clinics Program, 1984*a*, 1984*b*).

Almost all the major newspapers and magazines in the United States carried reports on the scientists' findings, and the principal investigators were interviewed extensively on radio and television. The *Los Angeles Times* ran a typical headline: "Cholesterol Decisively Linked to Heart Attacks." The *Times* reported, "The average participant who received the [cholesterol-reducing] drug for between seven and 10 years had 19 percent lower risk of having a heart attack and was 24 percent less apt to die of a heart attack than those who received only the placebo."

It was plainly a breakthrough for medical science. Now doctors could recommend reducing cholesterol intake and feel confident that this would reduce risks to their patients. And those of us who rarely see a physician could make our own decisions about cutting down on eggs and beef. But we should consider first what major newspapers and network television news programs failed to discuss about this major medical study.

To this end, I took the 1984 *JAMA* publications of the lipid research group's major findings as a point of departure. I examined all news stories, columns, editorials, and opinion pieces appearing in the *New York Times*, the *Wall Street Journal*, the *Washington Post, Newsweek, Time*, and a less systematically collected set of additional reports by the wire services and television news programs.[3] I also traced back to 1980 the news treatment of the relationship between cholesterol and coronary heart disease and death and then traced it forward since January 1984. Portions of the scientific and scholarly literature (particularly the clinical trial and epidemiological research) and assessments of

518   CONSENSUS IN SCIENCE

the risks of dietary cholesterol by the National Research Council, the research arm of the National Academy of Sciences, were examined.[4]

Here, then, are seven features of the research that were given little or no space in the press reports following the 1984 Lipid Research Clinics' publications:

1.   Because the media reported the data as "% reduction in risk," the reader was not made aware of the difference in the number of coronary heart deaths (CHD) in the two groups studied. The control group's 1,900 members were matched with the 1,906 in the experimental group for a variety of factors but were not given cholestyramine, a cholesterol-reducing drug given to the experimental group. Over the ten-year period of the study, there were thirty-eight CHDs in the control group and thirty in the experimental group. In short, there was a difference of eight deaths in an experiment involving 3,806 people, or a death rate of 2 percent in the first group and 1.6 percent in the second. Of course, if we take the difference between the two—that is, 0.4 of 1 percent—and state it as a ratio of .4 to 1.6, we obtain, as was reported in the scientists' abstract, a 24 percent reduction in risk. Thus, a major conclusion of the study on the harmful nature of cholesterol rests on this 0.4 of 1 percent.[5]

2.   Although the scientists reported this difference to be significant at the .05 level for a one-sided test of their hypothesis, they neglected to report in the *JAMA* article that chi-square and other tests of significance do not show any significant difference between the experimental and control groups.

3.   Most reportage of the study failed to state that the overall mortality rate in the two groups was not significantly different.[6] In short, the control group members were no less likely than those in the experimental group to live through the ten-year period (3.7 versus 3.6 percent), a fact reported in the *JAMA* article but rarely picked up by the press.

4.   Virtually no news stories examined the gastrointestinal side effects of the treatment, which, though ostensibly not severe, did occur more often in the experimental group. Indeed, the data on gallbladder disease (one potential consequence from using lipid-reducing drugs) show insignificant effects, but if we apply the method of ratio of increased risk adopted by the scientists in presenting the data on coronary disease, it appears that the cholestyramine group had an increased risk of 46.1 percent over the placebo group of having "operations involving the gallbladder" (36 versus 25) (Lipid Research Clinics Program 1984*a*, p. 357).

5.   Although the *JAMA* article gave plenty of information about the criteria used to select subjects for the study, the press gave little or none. For instance, few reports cited the fact that the study concentrated on those in the top 5 percent of the population in terms of their cholesterol level—hardly a representative sample of the United States population or of those apt to read about the results.

Dietary Cholesterol and Heart Disease   519

6.   There was no discussion in more than one hundred news stories and analysis columns or in the *JAMA* papers of varying results among the score of lipid research clinics. In fact, there was substantial variability, making the pooling of the data for all of the clinics a somewhat questionable method of analysis.

7.   The news and health column stories did not deal with a set of other questions that, although not considered in the *JAMA* article, are relevant to readers attempting to make decisions about their cholesterol intake. For example, what is the relationship between dietary intake of cholesterol, on the one hand, and cholesterol plasma levels on the other? What scientific problems are there in extrapolating from dietary intake of cholesterol and cholesterol plasma levels? How do the rates of death within specific age groups from all causes in this study differ from those in the population of American men during the same period of time? How do women, children, Blacks, and whites react to the treatment?

Not having any training in medicine, I am not suggesting, of course, that any given level of cholesterol consumption is appropriate for all or any of us. I am saying that the conclusions about the risks of cholesterol intake were raised by some to a level of scientific certainty ("definitive," according to the principal investigators and quoted in the press) prematurely. And I am suggesting that this type of distortion might lead in turn to significant changes in the public's dietary habits—especially after it has been bombarded with "news" of the risks and newly spawned advertising campaigns about the dangers of cholesterol.

From my limited evidence, it appears that reporters of health risks working for leading newspapers, newsmagazines, and television network news programs have difficulty in incorporating scientific disagreement and uncertainty into their stories. In fact, among leading experts in the field of cholesterol research, there is substantial scientific disagreement about the value of changing dietary habits and about cholesterol-reducing drugs as means of influencing rates of coronary heart disease. None of the alternative views from the one calling for dramatic dietary change was given prominent airing in the news articles following the 1984 LRC publications.

Take the views of Edward H. Ahrens Jr., the Frederick Henry Leonhardt Professor of Rockefeller University, who has been working in the "murky business of lipids" since the 1940s, as an example. He has studied how cholesterol metabolism is regulated in different types of people. Cholesterol is necessary of course in normal body functioning: It is a good thing. But do we have "too much of a good thing"? As Ahrens puts it,

> Blood levels of cholesterol are useful indicators of people at risk . . . but the important question to ask about such a person, once identified, is *why* the level is elevated; is the body making or absorbing too much cholesterol or not excreting enough? Most treatment is aimed simply at lowering blood lipid

levels, which is not necessarily beneficial. (Rockefeller University Research Profiles 1984, p. 4)

Considering the ten-year cholesterol study from this angle of vision, Ahrens makes the point that it is "not scientifically sound to extrapolate drug results to advice on diet for people in whom synthesis-absorption feedback regulation works efficiently" (p. 6). In a recent paper in the *Lancet*, Ahrens questions some overly broad extrapolations from the results of clinical trial and epidemiological studies of cholesterol. The past evidence appears, in fact, weak:

> Over the past 25 years this hypothesis [that CHD rates will drop if plasma cholesterol levels are reduced] has been put to the test in more than 20 trials which attempted to lower plasma cholesterol levels by dietary manipulations or by the administration of plasma-cholesterol-lowering drugs. Only the Lipid Research Clinics coronary prevention trial (LRC-CPPT) . . . produced evidence for benefit that was any more than suggestive. (Ahrens 1985, p. 1085)

Extrapolations by scientists from limitedly supported hypotheses to proof is disquieting to Ahrens, not only because these extrapolations can mislead reporters who include them in their stories but also because major agencies in the scientific community, for example, the National Institutes for Health (NIH), are reinforcing these leaps. Ahrens makes several additional points.

He questions whether the results from research on cholesterol can be applied to all subgroups in the population.

> I cannot accept the recommendation that the prudent diet be adopted by everyone over the age of 2 years. That viewpoint was rationalized on two grounds: (1) the generalization by the CPPT authors that any 1 percent reduction in plasma cholesterol level will lead to a 2 percent reduction in CHD incidence in all segments of the population; and (2) the public health view that dietary interventions are more feasible if adopted by a whole family than singly by a high-risk member of a family. . . . I know of no evidence that the prudent diet will prevent the development of arterial atheroma at any age: the hypothesis is reasonable but not proven. (Ahrens 1985, p. 1086)

Perhaps most tellingly, Ahrens is also worried

> that drastic changes in fat consumption, both in quality and quantity, which have been observed to change cell-membrane structure, may have undesirable effects on an individual's immunologic responsiveness and susceptibility to other diseases. (Rockefeller University Research Profiles 1984, p. 6)

He would prefer the health community to focus its attention on

> the 20 percent or so of the population at the top end of the blood cholesterol level scale who need vigorous, *individualized* testing and treatment. It is not a popular theme song these days . . . but more and more experts are learning the tune. (p. 6)

This cautionary position, which would limit extensive treatment to the people at "real" risk, is shared by Michael S. Brown and Joseph L. Goldstein, who recently received the Nobel Prize for their work on LDL (low-density lipoprotein) receptors. These receptors bind particles that carry cholesterol and remove them from the body's circulatory system. Brown and Goldstein take a far more cautious stance toward sharp dietary changes than do the Lipid Research Clinics researchers:

> If the LDL-receptor hypothesis is correct, the human receptor system is designed to function in the presence of an exceedingly low LDL level. The kind of diet necessary to maintain such a level would be markedly different from the customary diet in Western industrial countries (and much more stringent than moderate low-cholesterol diets of the kind recommended by the American Heart Association). It would call for the total elimination of dairy products as well as eggs, and for a severely limited intake of meat and other sources of saturated fats. (Brown and Goldstein 1984, p. 65)

Then, the authors make the critical point about the consequences of drastic changes in our eating habits:

> We believe such an extreme dietary change is not warranted for the entire population. There are several reasons. First, such a radical change in diet would have severe economic and social consequences. Second, it might well expose the population to other diseases now prevented by a moderate intake of fats. Third, experience shows most Americans will not adhere voluntarily to an extreme low-fat diet. Fourth, and most compelling, people vary genetically. Among those who consume the current high-fat diet of Western industrial societies, only 50 percent will die of atherosclerosis; the other 50 percent are resistant to the disease. Some individuals resist atherosclerosis because their LDL level does not rise dangerously even though they consume a high-fat diet. (pp. 65–66)

Elsewhere, Goldstein and Brown specifically note that the positive benefits of lowering blood cholesterol remain unproven.

522 CONSENSUS IN SCIENCE

> In some subjects with high serum cholesterol levels . . . atherosclerosis to some degree always develops with time. In other people, cholesterol accumulates even when their blood cholesterol level is within the normal range; their arteries seem to be sensitized by unknown factors. It can be argued that such individuals would benefit from a lowering of their blood cholesterol levels, *but that has not yet been demonstrated scientifically.* (Goldstein and Brown 1985, p. 46; emphasis added)

Putting aside the questions of economic and social cost-benefit analysis associated with dietary change, substantial scientific disagreement exists about its consequences among highly reputable scientists. Ahrens, as well as Brown and Goldstein, view the scientific problems and related policy recommendation for cholesterol intake in different terms from the authors of the ten-year cholesterol study. The former group allows that dietary recommendations may be in order, but they emphasize that these should be highly individualized, dependent upon detailed knowledge of individual histories and other risk factors. The lipids research group was far more willing to generalize its results to the entire U.S. population, suggesting in several interviews that significant dietary changes should begin for all children beyond the age of two. The central point here is not who is right or wrong but that the reportage neglected these alternative perspectives and failed to give the reader any idea that there might be disagreement within the scientific community over the meaning and implications of the results of the ten-year study.

Today we find almost no discussion in the news media of the scientific disagreement over the hypothesized relationships between cholesterol intake, plasma lowering of cholesterol, and atherosclerosis. The harmful effects of cholesterol in the diet are not only taken as a given but, based largely on the Lipid Research Clinics findings, are given the following felicitous functional form: For every 1 percent reduction in cholesterol there is a 2 percent reduction in the risk of coronary heart deaths. This has become a scientific "fact"—at least in news stories and probably in the minds of many concerned Americans who attend to this news. Yet, as I have suggested, such consensus simply does not yet exist in the scientific community. In the following section, I take up the question of how scientific disagreement about the effects of dietary intake of cholesterol was transformed in the press into apparent consensus, how it became a "fact."

From Possibility to Probability to Fact: The Cholesterol Case in Detail

Elevating important but inconclusive scientific results to the level of "definitive fact" is often facilitated by certification through institutional authority.

For example, the National Institutes of Health recently have formed "consensus committees," on which a panel of "experts" sit and "weigh" evidence, attempt to reach a "consensus," and publish a set of conclusions and recommendations. Such a committee was formed to review cholesterol data.

Thorny problems can arise in selecting members of consensus committees, and these choices can affect what the media receive as "best evidence." The cholesterol consensus conference is a good example of this problem. Among the members of the planning committee who set up the NIH consensus panel were Dr. Basil M. Rifkind, the associate deputy director for atherogenesis lipid metabolism, Atherogenesis Branch of the NIH's National Heart, Lung, and Blood Institute. And it included Dr. Kenneth Lippel from the same NIH unit, as well as Dr. Charles Gluck of the Lipid Research Clinic in Cincinnati. Each of these three is undoubtedly an expert in this field. But they also happen to have been deeply involved in the ten-year cholesterol study. So were we really obtaining "peer review" from the most knowledgeable experts on cholesterol research?

If this consensus conference was composed of unbiased experts, it surely did not appear so to M. F. Oliver of the Cardiovascular Research Unit of the University of Edinburgh, who concluded,

> Clearly, the aims of both the consensus development conferences were to develop a consensus view and, not surprisingly, the final statements prepared at the end of each 2½ day meeting were biased. How could they have been otherwise? Those who initiated the idea were either naive or determined to use the forum for special pleading, or both. The panel of jurists . . . was selected to include experts who would, predictably, say that . . . all levels of blood cholesterol in the United States are too high and should be lowered. (Oliver 1985, p. 1088)

Regardless of their makeup, when NIH consensus panels talk, the media listen. And the *New York Times* on December 13, 1984, ran a front-page story on the conclusions reached. The National Institutes of Health, as a leading scientific authority, are now supporting the "fact" that there is a direct causal link between cholesterol and CHD. And few caveats enter into the news story. A paragraph in the *Times* story states, "The most recent study showed that reducing cholesterol levels in the blood could prevent deaths from heart disease, with every 1 percent reduction in cholesterol lowering the coronary risk by 2 percent." No mention any longer of the subpopulation on which this questionable result is based; no discussion any longer of the statistical or other problems we have already noted. This ratio of 1:2 is gaining "cognitive independence" from the original study.

The progression of scientific evidence from the status of possibility to probability to fact occurs, I suggest, with some frequency in the media. This would hardly be significant if the progression were coupled with notable improvement

524 CONSENSUS IN SCIENCE

in the quality of scientific evidence, but, in the case we are dealing with, it is not. By tracing briefly the historical journey of the standing of cholesterol vis-à-vis coronary heart disease, we can develop a picture of how the values and cultural ideology of scientists, scientific organizations, and members of the media influence the public's perception of scientific "facts."

An abbreviated sketch of the 1980 to 1985 media presentation of the link between cholesterol and coronary heart deaths, triggered by new epidemiological and clinical trial studies, will demonstrate how the construction of this fact is accomplished.

Phase I: Possibility

Toward Healthful Diets—In 1980 the National Research Council (NRC) published a report of the Food and Nutrition Board, *Toward Healthful Diets* (National Academy of Sciences 1980). The board evaluated existing scientific evidence on aspects of diet and obesity, cardiovascular disease, hypertension, cancer, and diabetes mellitus. The board took a cautious position on the link between cholesterol and atherosclerosis.

> The causes of atherosclerosis are unknown. . . . A number of risk factors for cardiovascular disease have been identified from epidemiological studies. . . . Risk factors are those factors found to be statistically associated with an increased incidence of disease. They cannot, without independent evidence, be considered to be causative agents of the disease. . . . Diet modification as recommended for the prevention of atherosclerosis is based upon the assumption, not yet adequately tested, that reduction of high serum cholesterol levels . . . will reduce the probability of cardiovascular disease. (pp. 8–9)

Now consider a truncated presentation of the critical reception of this work by the media.

1. May 28, 1980: Jane E. Brody of the *New York Times* writes a front-page story with the following lead: "In a sharp departure from recent dietary recommendations, the Food and Nutrition Board of the National Research Council said yesterday that it found no reason for the average healthy American to restrict consumption of cholesterol. . . . For human beings . . . the link between cholesterol and fat in the diet and heart disease is largely circumstantial."

2. On the same day, however, Lawrence K. Altman of the *Times* writes an article on an inside page entitled "Report About Cholesterol Draws Agreement and Dissent." He presents a scenario of disagreement within the scientific community. There is the spokesman for the American Heart Association: "We

stand firmly behind our dietary evidence to the American public—eating a maximum of 300 milligrams of cholesterol and 30 to 35 percent of calories from fat . . . These recommendations represent the work of hundreds of experts who have sifted carefully through the available scientific evidence." With a contrasting view is Dr. Norman Spitz, professor at the NYU Medical School and "a recognized expert on nutrition": "it turns out that our bodies manufacture much of our cholesterol and that the effect of diet is 'relatively small.' . . . Clinical studies that tried to lower cholesterol directly by dietary and drug means were not successful in lowering the death rates."

3.   On May 29, we have Susan Okie, writing in the *Washington Post*: "A scientific panel's finding that healthy Americans need not lower the amount of fat and cholesterol in their diet was welcomed by milk, meat and egg producers yesterday but caused some chagrin at the American Heart Association." Later she quotes the rising chorus of dissenters: "Dr. William Kannel . . . head of the Framingham study that first produced evidence on the relation of lifestyle to heart disease, said the latest report was inconsistent in its recommendations. . . . 'It makes me wonder . . . at their objectivity.' " Michael Jacobson, director of the Center for Science in the Public Interest is quoted: "It stinks. It reads as if it was written by the meat, dairy, and egg industries."

4.   May 29, 1980: Jane Brody of the *New York Times* writes a story headlined "Dispute on Americans' Diets." The story details the scientific disagreement on the effects of cholesterol.

5.   June 1, 1980: the *New York Times* has this front-page headline, "Experts Assail Report Declaring Curb on Cholesterol Isn't Needed." The counterattack on the National Academy of Sciences (NAS) begins in earnest. The following points are made. (*a*) *The NAS is attacked for its lack of representatives on the board of different perspectives.* Robert Levy, director of the National Heart, Lung, and Blood Institute of the NIH is quoted: "It's true that not all the facts are in. . . . But to recommend doing nothing in the meantime is inappropriate . . . Americans should hedge their bets and seek a diet lower in saturated fats and cholesterol, at least until more evidence is available." (*b*) *The Nutrition Board members had ties to the food industry.* (*c*) *The board's recommendations were inconsistent.*

6.   June 1, 1980: the *New York Times*'s "Week in Review" runs a story, "Sunny Side Up for Cholesterol," in which it briefly summarizes the NAS report.

7.   On the same day, Jane Brody, under the headline, "When Scientists Disagree, Cholesterol Is in Fat City," notes,

> Lacking ironclad proof that changing one's diet can prevent heart disease in otherwise healthy persons, the board recommended no restrictions in cholesterol intake. . . . The board's advice is contrary to that offered by at least 18 organizations concerned with nutrition and health.

526 CONSENSUS IN SCIENCE

Thus far, the presentation of information on cholesterol reflects the existing disagreement within the scientific and medical community. There is a possibility that large cholesterol intake is harmful, but it is neither proven nor beyond scientific dispute. But it is also plain that there is a growing interest in the press, at least in some quarters, in debunking the NAS findings altogether. For example, on Monday, June 2, 1980, a *Washington Post* editorial entitled "Cholesterol Does Count" states, "'Toward Healthful Diets' . . . not only has increased public confusion over proper diet. It has also soiled the reputation both of the board and of the academy for rendering careful scientific advice" (p. A-18). The *Post* seems to be asking that a subject that, *in fact*, generates real scientific disagreement and some confusion be simplified and made easily digestible for the American public, even if the reports are not true to the controversy.

The *New York Times* follows with an editorial on June 3 entitled "A Confusing Diet of Fact." It admonishes the academy:

> The National Academy of Sciences is supposed to be an authoritative, impartial source of scientific advice to both the public and government—a Supreme Court of Science. But its latest report on health diets is so one-sided that it makes a dubious guide to nutritional policies. . . . The Academy should be espousing more than a single view.

And in fact subsequent articles do reflect existing disagreement. Consider just a few additional headlines: "A Few Kind Words for Cholesterol" (*Time*, June 9, 1980, p. 51); "How Bad Is Cholesterol?" (*Newsweek*, June 9, 1980, p. 111), which presents the conflicting opinions of various authorities; "Lower Fat Diet Affirmed Despite Recent Findings" (*Washington Post*, June 10), which is based upon testimony at the National Academy by Dr. William Castelli, "medical director of the famed Framingham, Mass., study of lifestyle and disease." On June 11, 1980, Mimi Sheraton, a food and dining specialist for the *Times*, writes under the headline "Conflicting Nutrition Advice Bewilders U.S. Consumers." A variety of points of view are expressed, including Dr. Michael De Bakey's: "A good 60 percent of the people who have arteriosclerosis do not have elevated serum cholesterol levels. . . . Why then are we basing our assumptions on the 40 percent minority?"

The *Washington Post*, June 12, 1980, headline for a story by Victor Cohn: "Panel Quits to Protest Advice on Cholesterol." The *New York Times*, June 18, 1980, headline for a column by Jane E. Brody: "Hidden Fat: The Hazards." The *New York Times*, June 20, 1980, "Scientists Clash on Academy's Cholesterol Advice," reports that "scientists . . . agreed that there was no scientific correlation between lowering of cholesterol and a reduction in coronary disease, they disagreed on whether Americans should have dietary advice on cholesterol." Suffice it to say that in June 1980 the public was presented with a spate of news stories emphasizing the scientific disagreement over the effects of cholesterol.

Several points emerge from the plethora of stories. The NRC-NAS report's findings questioned the "received wisdom" about cholesterol and heart disease. Consequently, the media present a sense of a lively, and often confusing, debate over what constitutes scientific fact on this issue. They explicitly call for the presentation of diverse points of view and take the academy to task for its supposedly unbalanced presentation of this scientific conflict. But, as we shall see, when subsequent scientific studies present findings that appear to support the prevailing received wisdom, the press fails to present any similar discussion of conflicting points of view. And the fundamental point is that no critical experiments or results of scientific work were done in the intervening years to warrant a shift in position from a possible to a probable linkage between cholesterol and coronary heart deaths.

The Framingham Study—The controversial academy report was not the only cholesterol and atherosclerosis news in 1980. In that year, the results of the Framingham study, a massive twenty-four-year epidemiological effort "to determine the risk factors for coronary heart disease and other atherosclerotic disorders" (the results of which had been published in more than 150 scientific papers between 1950 and 1978), were pulled together and summarized in a book by Thomas Royle Dawber (1980), one of the study's principal investigators. The Framingham study followed 5,127 men and women between ages thirty and fifty-nine. Initial examinations of subjects began in the spring of 1950. Individuals who were free of coronary heart disease at the initial physical examination constituted the population to be followed (pp. 20–21).[7]

On the effects of blood cholesterol levels, Dawber states, "Observations from the Framingham study over 24 years clearly indicate that serum cholesterol plays a role in the incidence of coronary heart disease. . . . The average annual rate during the entire 24 years for men with cholesterol levels 260 mg or more was twice that of those with levels below 200" (1980, pp. 129–30).

Now consider these results in greater detail. For women, the findings are wholly inconclusive. "Analysis of the 24-year risk in women . . . shows [that], in both the youngest and oldest decades examined, no significant differences in incidence of coronary heart disease on the basis of cholesterol levels were observed" (Dawber 1980, p. 130).[8] Even in the middle-aged group there is no difference in incidence of coronary heart disease for every cholesterol category except for the highest cholesterol group, representing a relatively small proportion of the population (see Dawber 1980, fig. 8.4, p. 133). For most of the population of women there is no relation between cholesterol and heart disease.

Even for men the relationship is tenuous. Of course, men in all age groups are far more apt than women to develop heart disease. But for men with cholesterol levels in the lowest group, the average annual rate of coronary heart

528 **CONSENSUS IN SCIENCE**

disease was 7.4 per 1,000; 8.4 in the next group; 12.7 in the midgroup; 10.1 in the next-to-highest; and 14.6 in the highest. The incidence of heart disease is about double between the extremes, but the relationship is less marked when comparing the other groups—for example 1.7 persons per thousand between the second-lowest and next-to-highest cholesterol groups. These rates are simple bivariate distributions. Multivariate analysis describing the independent effect of cholesterol on coronary heart disease and mortality by cholesterol levels is not presented. Dawber never indicates the percentage of variance explained by dietary cholesterol. He does suggest that diet had little to do with coronary heart disease in the Framingham population:

> The expectation that the intrapopulation differences in cholesterol level could be attributed to individual dietary habits has not been fulfilled. The variability of blood cholesterol values within the Framingham population appears to reflect inherent or constitutional traits rather than differences in life habits. . . Differences in dietary intake that could affect the blood cholesterol level did not account for the intrapopulation differences nor for the effect of cholesterol level on the *relative* risk within this population. (Dawber 1980, pp. 138–39; emphasis in original)

The relatively small differences in risk associated with varying cholesterol levels in men; the virtual absence of any relationship among women; and that differences in cholesterol were unrelated to diet have been almost totally ignored in the media's presentation of the Framingham study. The message of Framingham has been that the intake of cholesterol in diet produces significantly increased risk of heart disease and death.

Phase II: From Possibility to Probability, 1981–1984

Between 1981 and the Lipid Research Clinics trial results in 1984, the media transform the possible linkage into a highly probable one. This progression is again associated with the publication of results of scientific research.

The Western Electric Study—In January 1981, the *New England Journal of Medicine* publishes the paper "Diet, Serum Cholesterol, and Death from Coronary Heart Disease: The Western Electric Study." Based on a twenty-year follow-up physical examination of 1,900 middle-aged men, this study investigated "the associations of dietary saturated fatty acids, polyunsaturated fatty acids, and cholesterol with serum cholesterol level and risk of death from coronary heart disease (CHD)" (Shekelle et al. 1981, pp. 65–70). There are design problems with this study. For example, there was only one follow-up over a twenty-year period and no monitoring of changing patterns of health, diet, and lifestyle variables in the intervening years. Women, children, members of varying occupations and social classes are not studied at all or are underrepresented. But more importantly, the percentage difference in coronary heart deaths between the lowest and highest third of the 1,900 men in terms of saturated fats was 10.9 versus 11.8 percent and in terms of dietary cholesterol, 10.9 versus 13.6 percent. In short, these small substantive differences mean that cholesterol explained little variance in death rates, although the authors choose to interpret the effects very generously indeed. "The correlations between dietary variables and serum cholesterol concentration in our study were small" (p. 69). But the media that I examined treat the results otherwise.

Time, January 19, 1981: "Cholesterol: The Stigma Is Back—New Report Reaffirms the Link to Heart Disease." The story states, "The main finding: those who had consumed large amounts of cholesterol and saturated fat suffered upwards of a third more deaths from heart disease than those who consumed relatively small amounts." Although this is a misleading summary, *Time* acknowledges that the debate over cholesterol will probably continue. A *Newsweek* article contends that the Western Electric study "demonstrated a specific link between individual eating habits and fatal heart attacks" (*Newsweek*, January 19, 1981, p. 74).

The "MRFIT" Study—In September 1982, the *Journal of the American Medical Association (JAMA)* published the results of a randomized primary prevention trial to examine the effects of changes in hypertension, cigarette smoking, and dietary advice for lowering blood cholesterol on mortality from coronary heart disease. Almost 13,000 high-risk men aged thirty-five to fifty-seven were chosen from 361,000 initially screened and randomly assigned to "special intervention" groups, who received counseling to alter existing habits, and a "usual care" group, which represented a control. To become a participant, a person had to be at "increased risk," that is, in the upper 15 percent of a risk distribution based upon cigarette smoking, serum cholesterol, and blood pressure (based upon data from the Framingham study). This Multiple Risk Factor Intervention Trial (MRFIT) encountered methodological problems, as had the previous major epidemiological and clinical trial studies. For one, the experimental and control groups differed little on a set of risk factors, even after attempted interventions. Members of both groups had lowered their smoking levels and their intake of cholesterol during the ten years of study.

The results were dramatic—but at variance with the wisdom being placed in print and on the air by the media. For example, the mortality rate from coronaries in the special intervention (SI) and usual care (UC) groups were almost identical, and the difference in overall mortality obtained went in the opposite direction from what had been predicted. The death rate was actually 2.1 percent higher in the special intervention group. The effects of serum cholesterol on CHD deaths were nonexistent or minimal. In short, cholesterol levels had no

530 CONSENSUS IN SCIENCE

significant effect on the death rates, even considering statistical interactions among the three risk factors.

This was a major, $115 million, ten-year study reporting a "negative finding" on cholesterol. MRFIT did not go unnoticed by major news organizations. As far back as 1980, there was keen interest in the potential of the MRFIT study.

Several months before the MRFIT results are published, Jane Brody discusses a Norwegian study of about 1,200 healthy men between forty and forty-nine years old with high cholesterol levels. Under the headline "Life-Saving Benefits of Low-Cholesterol Diet Affirmed in Rigorous Study," Brody begins her story:

> A major, well-designed study has shown more persuasively than any previous experiment that eating less fats and cholesterol can reduce the chances of suffering a heart attack or of dying suddenly from heart disease. . . . Though their blood pressure was normal, their cholesterol levels were considered high—from 290 to 380 milligrams of cholesterol per 100 milliliters of blood—and 80 percent of them smoked cigarettes. . . . The team [of scientists conducting the research] . . . calculated that dietary changes accounted for 60 percent of the difference in the number of heart attacks and heart deaths suffered by the two groups of men. (*New York Times*, January 5, 1982, p. C-1)

No mention whatsoever is made that men of this age with levels of cholesterol averaging well over 300 mg/100 ml would have been in the upper 1 to 5 percent of the Framingham distribution—hardly typical people. And, in the penultimate paragraph, nestled away in an inside page, Brody adds, "The researchers conceded that 'if this had been a diet trial only, the difference in MI [myocardial infarction, or heart attack] incidence in the two groups would probably not have reached statistical significance.'"

As it turns out, cholesterol intake (even in this extreme group) is not a strong predictor of subsequent coronary heart disease and deaths. That certainly is not the tone or direction of the article. Not one scientific expert, apart from the study director, is asked to evaluate the quality of the work or its limitations. The results support the probable relationship and the prevailing presuppositions.

In reporting the MRFIT study (*New York Times*, September 17, 1982, p. A-10), Brody's story details the methods used and several of the negative findings. She does a good job of this. But the article focuses on the various limitations of the study—a focus totally absent from the treatment of the Western Electric and other studies showing "positive" results. Brody notes the study director's own skepticism:

> Americans should not interpret the inconclusive results to mean it was all right to smoke, to be on a high fat diet and to have high blood pressure. "It is our judgment that the public should continue to reduce these risk factors associated

with heart disease," said Dr. Oglesby Paul, professor emeritus of medicine at Harvard Medical School, chairman of the study's steering committee. . . .

The researchers were also surprised to find that one treatment subgroup suffered a higher death rate than those getting regular treatment. The criticism of MRFIT continues on October 10, 1982. The *New York Times* reports methodological problems enumerated by study officials.

Although caution and skepticism are no less in order for MRFIT than for the other studies, why is a "negative" or "inconclusive" result presented as a failing? Indeed, it is puzzling why the reporters covering the cholesterol question seemed committed to the point of view that cholesterol intake was a substantial health risk. To be sure, they were following the lead of health scientists, but why was the opposing point of view given less space and weight in their stories— especially since reporters find reports of controversy appealing?

In 1983 the probable link is stressed once again. The *New York Times*, January 11, 1983: In a brief article, "Reversing Heart Disease," this probability is suggested. "A newly published study indicates that significant improvement in cardiovascular health can occur after just 24 days of a radically changed diet and lifestyle." This conclusion was reported on the basis of "a pilot study" of forty-six patients. Jane E. Brody, in her "Personal Health" column of March 16, 1983, writes,

> For a while it seemed as if the advice to lower cholesterol levels had fallen on hard times. . . . But before anyone could even digest a cholesterol rich meal, more meaningful data poured in. . . . The main recommendation is to eat less fat, in total, and especially less saturated fat, since this kind of fat raises cholesterol levels in the blood and increases the risk of heart disease. . . . So the heart-saving advice to reduce consumption of fats and cholesterol still holds. As a bonus, it may help protect you against cancer as well.

Brody fails to mention a single one of these new studies that "poured in"; no longer mentions skepticism about the results; omits all qualifiers about people at risk or the types of people who were enlisted in the research studies. The recommendation is apparently now applicable to all readers of the *Times*. Similar stories appear in other newspapers and on television news programs.

A week before *JAMA* publishes the Lipid Research Clinics' result, Peter Jennings on *ABC World News Tonight* anticipates the cholesterol publications when he and reporter George Strait interviewed Dr. R. Basil Rifkind of the NIH. Probability is now being transformed into fact:

> *Jennings*: Scientists now have conclusive proof that lowering the amount of cholesterol or fat that we eat can significantly reduce the chance of heart attacks.

532 CONSENSUS IN SCIENCE

George Strait: For more than a decade Americans have been warned to cut back on red meat, eggs and other foods high in cholesterol because scientists suspected that cholesterol was a prime cause of heart disease. Well, now they are sure. The new federal study confirms the definitive link between cholesterol and heart disease.

R. Basil Rifkind/NIH: The greater the reduction in cholesterol, the greater the reduction in coronary heart disease. And that a 25 percent reduction in blood cholesterol cuts risk by half. (*ABC World News Tonight*, January 12, 1984)

The Lipid Research Clinics papers are published at this point. The media assumption that these studies provide a conclusive causal link between cholesterol and mortality from coronaries has been documented. What has happened since the scientific papers were published? As far as the media are concerned, we now are dealing with fact. So what has happened to the "fact" once "established"?

Phase III: From Probability to "Fact"

Since the publication of the LRC results, the media have moved still farther away from discussion of the study's limitations, have generalized from the limited findings, and have rarely seen fit to present opposing views. The same sources that condemned the National Academy for not reporting the full spectrum of opinion in *Toward Healthful Diets* have now totally abandoned the presentation of scientific disagreement:

Time, January 23, 1984: "Sorry, It's True, Cholesterol Really Is a Killer." And on March 26, 1984, *Time*'s cover story reads "Cholesterol, and Now the Bad News." The *Reader's Digest* story is entitled "Cholesterol Is the Culprit." The statement that "every 1 percent cholesterol drop in the bloodstream means a 2 percent decrease in the likelihood of a heart attack in people with high cholesterol levels" is becoming a medical aphorism.

Victor Cohn in the January 29, 1984, *Washington Post*: "So now cholesterol is real. Government scientists have said so." The *Wall Street Journal*, March 15, 1984: "Cholesterol Levels in American Children Should Be Lowered, Heart Specialists Say." Jane Brody in the *New York Times*, May 16, 1984: "Diet to Prevent Heart Attacks Aims to Cut Blood Fat Levels." On March 27, 1984, Ted Koppel devotes ABC News's *Nightline* to cholesterol and heart disease. Dr. Robert Levy, a principal investigator in the Lipid Research Clinics study, states,

We've always known that cholesterol was a risk factor. We've always known that we've had diets and drugs that could lower cholesterol, but now for the first time in man, we have that conclusive evidence that if you lower your cholesterol your risk of heart attack and death . . . will decrease. . . . As you know, as with the tobacco story, there will be some who will never believe. But I think that the evidence that my family should act on, and I think all Americans should act on, is now in hand.

The views of other scientists who perceive the data differently are not presented or alluded to. The rule is now a general one. Finally, Jane Brody, in a 1985 story discussing "new" findings from the Framingham heart study, presents the health aphorism ("For every 1 percent change in cholesterol, the risk of heart disease changes by 2 to 3 percent") without any mention of the original source. It has now diffused into the culture.[9]

A Research Agenda for Understanding Distortions of Health Risks in the Media

The cholesterol case and others like it raise fundamental issues that need to be addressed before we can understand better what determines the distortion of health risks in the media. In this section questions are raised that might enter a research agenda on the presentation of health risks in the media. Specifically, processes at work in the development of scientific facts are examined, and a discussion follows on how the social organization of science and the media can influence the presentation of health-risk information.

Cholesterol and Coronary Heart Deaths: The Development of a Scientific Fact

The cholesterol case history is an exemplar for how values, intellectual con-flicts, and socially structured interests influence the development of a scientific fact. This is the hypothesis we are considering for its fact status: Dietary intake of cholesterol causes coronary heart disease and death. In its most graphic reportage form, the "fact" is that for every 1 percent reduction in cholesterol intake there is a 2 percent reduction in the risk of coronary heart deaths. Let us examine several elements in the struggle for fact status.[10]

The principal characters in this drama include a set of medical research scientists (the LRC group) who are proposing, in effect, that the results of their ten-year, $150 million study be accepted as a medical fact. Other actors include scientists who dispute this claim to fact, regarding the assertion as too strong an extrapolation from the available evidence and too sweeping in its claims to gen-erality. Still others include powerful organizations and interest groups, among others, the American Heart Association, the National Institutes of Health, the

National Academy of Sciences, and lobbies for the food industry, which had definite interests in the outcome of the fact dispute. The cast of characters is fleshed out by members of the press, science writers, editors, and reporters working the science beat, who transfer the content of this debate to the general public, and others with varying degrees of knowledge of the scholarly literature on which the fact claims are based. The contest is over the extent to which the fact claim is (a) accepted among influential members of the medical research community; (b) accepted for some duration of time by the research community to a degree that effectively eliminates disagreement from the scientific discourse; (c) accepted by science reporters as "definitive"; and (d) accepted by the public as a medical "fact."

The outcome of the contest can have significant effects on careers of scientists involved in the dispute, on the allocation of scientific resources for future research (and on processes of accumulating advantages and disadvantages), on the public's perception of health risks and what can be done about them, and on the economy of businesses whose products are affected by the public's perception of the effects of cholesterol.

How is the drama surrounding cholesterol related to current views about the social processes at work in the development of scientific facts? At least since Ludwik Fleck's work in 1935 on the development of the Wasserman reaction,[11] there is a recognition and growing acceptance that, at least in some measure, the choice of scientific problems, modes of presenting experimental evidence, acceptance or rejection of theories, and selective evaluation of experimental results as a function of selective perception are socially conditioned.

More recently, Bruno Latour and Steve Woolgar (1979), as well as Karen Knorr-Cetina (1981), have suggested that the processes of "persuasion" and "negotiation" are central elements in the construction of scientific facts. Each of these elements plays a role in the fact dispute over the effects of cholesterol in the diet. In analyzing sixteen drafts of a single scientific paper on potato protein concentrates, Knorr-Cetina suggests that the actual difference in content had little to do with changes in evidence but was largely a consequence of extended negotiations between author and critics, including, among others, collaborators, laboratory directors, reference individuals, and authorities in the scientific specialty (Knorr-Cetina 1981, pp. 94–135, esp. p. 104).

Knorr-Cetina suggests that the typical outcome from negotiation in science is the changing of modalities of certain assertions, toning down of claims, movement from over- to understatement, from more to less dramatic styles of presentation, from assertions to caveats. In short, the early or original draft is stripped typically of some of its dramatic content—without any changes in evidence. The normative structure of science enjoins scientists toward such outcomes of negotiations.[12]

The published news story or broadcast is the analog to scientific publication. The final content of this story is, surely to some extent, also a process of negotiation. The critics with whom the reporter must negotiate include possible collaborators, science editors, and general editors. However, the norms governing negotiations in the news business tend to run counter to those operating in science (Goodell 1975; Nelkin 1985; Winsten 1985). Unlike most scientists, competent science reporters face pressures in negotiation to tone their stories up rather than down. Consequently, health-risk stories hitting the newsstand or reaching the air tend toward fewer modalities, qualifications, caveats, and statements of limitations. To get stories in print or on the air, science reporters are apt to have to expunge rather than add qualifying statements.

The differences in normative structures between science and news become particularly problematic for the final story when the earlier negotiation over the scientific paper has resulted in a highly "dramatic" and possibly exaggerated claim about experimental results. This dramatic and possibly distorted presentation can be exacerbated still further by the negotiation process in the newsroom.

Akin to negotiation is what Latour and Woolgar (1979) describe as the process of persuasion, that is, an ongoing contest between scientists, one set trying to change the modalities of statements so that the statement holds an increasing "factlike status" and another trying to be more restrained about claims to fact status. Some scientists, generally the authors of papers, are pushing statements that have a greater factlike quality; critics are frequently pushing for greater caution and qualification. Latour and Woolgar discuss five types of statements found in scientific papers, ranging from statements of "facts" that everyone takes for granted (Type 5) to far less definite, limited, and more qualified statements, such as "A has a certain relationship to B," and finally to the fully qualified modality of conjectures and speculations (Type 1 statements). Latour and Woolgar characterize laboratory activity in terms of pushes and pulls toward fact status that result from interaction between fact claimants and their critics.

They are fully aware, of course, that sociological factors (such as power, authority, theoretical and methodological orientations, and personal values) play a critical role in the persuasion or negotiation process. But they do not consider the influence of structural features of science beyond the laboratory or more macrolevel factors that may influence the outcomes of contests and struggles about fact status.[13]

This work bears directly on the recent cholesterol papers and publicity because we see in the cholesterol controversy an ongoing contest about what should be granted fact status. We witness an effort by one set of researchers, most recently those associated with the Lipid Research Clinics' long-term clinical trial, who are attempting to persuade others—scientists, journalists, and the wider public—of the unadorned *fact* that dietary intake of cholesterol causes heart deaths. They are asserting with increasing frequency Type 5 statements that label cholesterol as a major cause of heart deaths for all members of the U.S. population. If they are the protagonists in this drama, the antagonists are those scientists, such as Ahrens as well as Brown and Goldstein, who are advocating scientific statements that continue to include strong qualifying modalities. It is important to emphasize that contests over fact status often take place with new empirical or experimental evidence continually influencing the dialogue, but in the recent cholesterol case the contest seems to be proceeding quite independently of additional evidence.

Throughout this paper examples have been presented of efforts by investigators associated with the LRC trial to reduce the number and types of qualifying statements, which have the effect of enhancing the fact claims by the authors. These actions suggest efforts by the principal investigators to persuade potential critics and the public of the fact status of their findings.

Here is where science reporters and the media in general enter the negotiation and persuasion process. They are recipients of the statements that emerge from scientific papers published in prestigious journals, such as *JAMA* or the *New England Journal of Medicine*, from personal interviews with scientists, and from formal news conferences and briefings. They are an important audience in the contest over fact status because indirectly they influence opinion formation, not only in the larger public but in the attentive publics of physicians who offer advice to patients (but who do not necessarily read the scholarly literature on cholesterol), other scientists, health administrators, staff members of congressmen and administrative agencies, and decision makers at federal funding agencies who oversee the funding of large-scale laboratory and clinical trial programs.

When dealing with the press, scientists interested in changing the modality of statements about research results are often not subject to the same mechanisms of social control that operate within their discipline.[14] A simplification may involve only the omission of a strategically placed "could" or "maybe" or "applies only to" from earlier, more cautionary statements. Reporters can hardly be expected to be familiar with much, or all, of the detailed work that comes out of a laboratory or with the details of the published data.[15] When acceptance of less tentative statements is coupled with the normative tendency within journalism to strengthen rather than weaken stories through further reduction of qualifying modalities, there is an increased probability of "premature facts" being elevated to "fact" status and being widely accepted as such.[16]

Institutional Properties Influencing the Distortion of
Health Risks in the Media

At least three institutional properties of both science and the media are apt to increase the probability of distorted representations of scientific "discoveries." The first of these is the structure of the reward and opportunity systems in *both* institutions; the second, inadequate training of reporters as critical observers of scientific information; and the third, ineffective mechanisms in the media and science to forestall the distribution of premature results or distorted news.

The Reward Systems—There are, of course, many forms of scientific recognition: prestigious positions; honorific awards, such as Nobel Prizes or election to leading academies of science; and visibility and peer esteem (see Cole and Cole 1973; Merton 1942, 1973; Zuckerman 1977). There is a substantial overlap among the recipients of these various forms: the few receive the lion's share. But with increasing frequency, scientific and public visibility, if not notoriety, can be attained through extensive media exposure—and sometimes even by conducting scientific controversies through the media (see Cole 1979; Goodell 1975).

Scientists, like others, seek recognition from important reference groups. Although precise figures do not exist, we may ask what proportion of scientists completely eschew media attention of their work; what proportion accept it without seeking it; and what proportion actively seek representation in the media. For the group, however small, who seek visibility through the press, what determines their interest? Some scientists may have a personal and professional interest in being "represented" in the press.

The consequences of representation in the press can extend, of course, beyond personal psychic gratification. Press exposure may aid scientists in obtaining continued resources for their research—especially if the research program is extremely expensive and is directly reviewed by congressional committees and directors of funding agencies.

The problem goes deeper than passivity among scientists because some (again, the exact proportion is surely unknown) contribute directly to media distortions by tailoring their own results through incomplete and misleading paper abstracts, news releases, and press conferences that give the media what the scientist has defined as "strong" news stories—even at the expense of complete summaries and inclusion of essential limitations. In the cholesterol case, we see the principal investigators in television interviews and newspaper stories making extreme claims for the results without acknowledging the limits to the data. But for the medical research and larger scientific community, we know little about the sources of distortion in reporting.

Jay A. Winsten (1985) recently completed focused interviews with twenty-seven science reporters, editors, and television producers at leading news organizations. Many of the reporters interviewed were authors of stories addressing health risks associated with cholesterol. The reporters repeatedly told Winsten that the competitive pressures of journalism and interest in recognition often led to altered stories, creating the appearance of either greater conflict and tension or greater consensus than in fact existed, and to other significant alterations in stories to meet the requirements of editors who were

538 CONSENSUS IN SCIENCE

gatekeepers in opening or shutting off opportunities for advancement and recognition. For example, one reporter told Winsten of tension between building a "strong" story while maintaining credibility:

> I'm in competition with literally hundreds of stories each day, political and economic stories of compelling interest. In science, especially, we sometimes have to argue [with editors], pound the table, and say, "This is an important story. It turns a key of understanding, it affects a lot of people," or "it's just interesting, it's part of the unfolding romance of science." But we have to make that clear in our copy. We have to almost overstate, we have to come as close as we can within the boundaries of truth to a dramatic, compelling statement. A weak statement will go no place. (Winsten 1985, p. 9)

Science editors, Winsten reports, are aware of the tension and the competing values associated with tension and accuracy.

The pressures for recognition among reporters and science editors are summarized in Winsten, but we surely cannot tell from the twenty-seven interviews, however instructive, how generalizable the results are, that is, with what frequency reporters and editors push beyond the boundaries of truth as they come to know it.

Structural Problems That Can Lead to Distortion

Problems of the Training Necessary to Report on Scientific and Medical Research—A potential source for the publication of biased or distorted health risk information lies in the quality and quantity of training of those on the "science beat." In fact, for most newspapers and broadcast journalism, general assignment reporters far outnumber science reporters in covering science topics. This was not true for the sources we used in tracing the cholesterol controversy, but it is the general case. Even if we focus exclusively on science reporters, what proportion of them and their editors can be sufficiently knowledgeable about the sciences covered in their stories to assume a critical posture toward the scientists who are making claims of fact? Given that science has become so highly specialized and technical, we can hardly expect that the majority of reporters would have specific knowledge of most of the sciences about which they write. Of course, it remains unclear how much detailed knowledge a reporter needs about the basic canons of scientific inquiry, of proof and evidence, of scientific skepticism, and of how facts and theories develop in the scientific community to write stories that correctly depict the fact claims and the extent to which these claims are matters of contention within the research community. Short of extraordinary knowledge, reporters must, of course, rely heavily on scientific authority, on the "word" and reputations of the principal investigators and scientific institutions, and on the peer review system. This can be problematic for several reasons, especially in dealing with health risks and medical research.

First, many flawed papers pass through the peer review system and are published without adequate attention to methodological or substantive inaccuracy—and this is so for the journals of the highest rank as well as those of lesser stature. Recently, DerSimonian et al. (1982) studied the reporting of clinical trials in the *New England Journal of Medicine, British Medical Journal, Journal of the American Medical Association*, and *Lancet.* They chose eleven basic topics and examined whether each one had been discussed in the paper reporting the clinical trial experiment.[17] Of the sixty-seven papers reviewed, only 1 percent gave information on all eleven items; more than 50 percent provided information on six or fewer. In short, many papers that have passed through the peer review process fail to provide basic information that allows readers to assess the quality of the study.

Second, the peer review system is marked by honest, intellectual disagreement among scientists about what constitutes "high-quality work" and what is a valuable paper. Published papers often had one or more referees who identified substantial problems with the papers (among many others, see Ingelfinger 1974; Stinchcombe and Ofshe 1969; Zuckerman and Merton 1971). Such disagreements in appraisals may actually be healthy for a science and, in any event, are common when people evaluate new scientific work at the frontier of knowledge, but as a consequence, acceptance or rejection will depend significantly on the "luck of the reviewer draw" (see Cole and Cole 1985; Cole, Cole, and NAS Committee on Science and Public Policy 1981; Cole 1983; Cole, Cole, and Simon 1981). Thus, peer review may represent the best of all possible but highly imperfect worlds. In sum, the fact that a paper appears in a prestigious medical journal offers no strong guarantee against the appearance of poorly analyzed or distorted results of medical and scientific research.

What is the media to do about such problems? They surely cannot develop an independent refereeing system, but it may be possible to extend the training of experienced reporters in the methods used in research and experiments as well as new analytic and statistical techniques.

Problems of Timing—Television and newspaper reporters are often under considerable pressure to get stories out in a timely way. Constraints of time and space in storytelling can work against adequate presentation of the limitations of scientific findings, which often require a longer period of time to investigate the adequacy of the evidence. However, data do not exist that address the relationship among the length of news stories, the speed with

540 CONSENSUS IN SCIENCE

which they must be developed, and the level of distortion in the final presentation of health risks.

Value Commitments of Scientists and Reporters in the Reporting of Science News

Personal values can also affect the handling of scientific data and preparation of science news stories. Work in the history and sociology of science over the past two decades supplies us with evidence that ideological presuppositions and values influence the perceptions of scientists; their choice of problems; their choices in examining one source of possible errors as opposed to another; and their reading and interpretation of the data and results (Barber 1961; Gould 1981; Holton 1978; Merton 1972; Polanyi 1963). These cognitive tendencies, which do not necessarily involve motivated deception, may be exacerbated when personal rewards and resources are more apt to accrue to those who conform to the received wisdom in the scientific community.

Pressure for the presentation of eye-catching and somewhat overblown results may also come today from the sources for funding science. With increased interest in fiscal "accountability," with populist pressures for more egalitarian distribution of scarce federal dollars for researchers, it is probably difficult for science programs that spend upward of $100 million on research to conclude, "The results are inconclusive." This is a hard position to argue and sustain, especially when entire laboratories and scores of scientists depend upon the government's support for their continuing research. Such pressures may produce a strain toward early or premature publication and an attempt to characterize results for the press and scientist colleagues more dramatically than they warrant.

Finally, if scientists and medical researchers have their set of presuppositions that influence the conduct of their research and its outcome, surely reporters and health columnists have their own presuppositions and ideological beliefs. Logically, this should influence their choice of stories and their selective perception and use of evidence, although little empirical work exists to specify this claim. Thus, the values, biases, and attitudes of both reporters and scientists can interact to create conditions under which the probability for significant distortion of health risks and other science news is increased.

The Absence of Effective Social Control

The absence of effective mechanisms of social control can also influence distortions about health risks. High-quality monitoring and feedback systems do not yet exist. Public "taste" for stories is plainly not a sufficient barometer of how well the media are doing.

One mechanism of social control is through complaints. However, the system of reporting health risks tends to reduce the effectiveness of this mechanism because scientists whose stories are covered tend to be a highly self-selected lot who are apt to be pleased by the coverage. Furthermore, those scientists who are skeptical about the likelihood of distorted reporting probably shy away from coverage.

Except for self-criticism within the media, there are few mechanisms to my knowledge that operate to minimize distortion. Winsten's interviews suggest that editors' insistence on producing "strong" stories has the frequent consequence of distorting them:

> The desk editors are the ones who always want you to push it a little harder. . . . I go to the desk, and they say, "Well, can't you make it a little stronger?" And then they give you an alternative. And I say, "No, you've just eliminated a qualifier, or you've added that word there that's just not true." And you go back and forth like that. So there is movement in the direction of a stronger statement—running out to the boundaries at which you've overgeneralized, at which you've just overdone it. (Winsten 1985)

Thus, the forces operating to produce an engaging and readable newspaper story or television news program appear to work against accurate and balanced reporting of health risks. More effective feedback mechanisms about balance and accuracy need to be constructed, if we are to reduce distortions.

Notes

1. For a discussion of the distinction between the frontiers of research and the "core" of knowledge, see Cole and Cole 1973; Cole 1983.
2. The paper represents one effort in a larger program of research focusing on the sociology of risk assessment. I have examined material for other cases, and although all of the cases examined to date lead to similar conclusions, we must be cautious about generalizations from a few cases before more extensive work on larger samples is completed.
3. In all, over one hundred cholesterol news stories were reviewed between 1980 and 1984. Twenty-two were located in the *Washington Post*, forty-six in the *New York Times*, ten in the *Wall Street Journal*; nineteen were from CBS News, five from ABC News, and more than twenty-five from various news magazines and news services. These stories varied, of course, in length, depth, and type—from straight news reports to opinion columns. References to "the media" and "the press" throughout this paper are restricted to these data sources.
4. David Gerwin, an undergraduate student at Columbia College, helped collect these materials.

542   CONSENSUS IN SCIENCE

5. Epidemiologists are quick to point out that very small fractions can represent large numbers of cases when extrapolated to the entire population. But this is so, of course, only if the small differences are to be treated as a real "signal" rather than "noise."

6. This conclusion, as well as others drawn throughout this paper, is limited to the journals and media that I examined, which were noted above (note 3). Thus, when I speak of the media with regard to the cholesterol case. I am limiting myself to these sources in the media.

7. The various methodological problems of the study, such as dropout rates and representativeness of the sample, require more elaborate discussion that can be found in Cole 1986.

8. The book omits discussion of mortality rates and discusses morbidity only.

9. Even the National Research Council's Committee on Nutrition in Medical Education had changed its tune by 1985. See the National Academy of Sciences report, *Nutrition Education in U.S. Medical Schools* (1985).

10. Space limits this section to the briefest elaboration of the relevance of ideas of Ludwik Fleck, Thomas Kuhn, Imre Lakatos, Bruno Latour and Steve Woolgar, Karen Knorr-Cetina, Robert K. Merton, and earlier work by Stephen Cole and myself as they bear on the development of "medical facts." See Cole 1986 for a broader discussion.

11. Fleck's (1935) work has become well known to American sociologists of science since its recent translation by Fred Bradley and Thaddeus J. Trenn, edited by Trenn and Robert K. Merton (1979).

12. It is not self-evident that toned-down outcomes of negotiation facilitate the development of knowledge. We know little about the incidence of truly important papers that go largely unnoticed because negotiations led to overly restrained claims and, correlatively, the incidence of exorbitant claims that led to significant false starts and unproductive research programs.

13. Knorr-Cetina (1981) and Latour and Woolgar (1979) concentrate on changes in texts that are attributed to different interests of authors and critics. They underplay changes that result from an awareness gained by authors through criticism that they have overgeneralized or have made statements which, upon reflection, are viewed by the authors as errors. Furthermore, they fail to see the act of writing as a creative process that represents an action quite distinct from the activities in the laboratory. This process of creating a text can itself lead to new ideas, some of which appear upon critical reflection to be too strong, too weak, imprecise, or in error. Thus, changing modalities of statements in scientific papers are not purely a result of a clash of interests, values, and so on.

14. We do not have any extensive data that can demonstrate this; there is plainly a need to develop systematic data to go beyond the impressions based upon more limited inquiries by Winsten and others.

15. There are probably a few specific research areas that each science reporter is familiar with. Based upon the reporters' own science background and subsequent interests, this may allow them to be highly informed critics in these specialty areas. To cite only one notable example, Walter Sullivan of the *New York Times* has extensive knowledge and has written important books on the geology of plate tectonics.

16. Ludwik Fleck has an extraordinary discussion of the simplification process and its consequences for fact development as we move from scientific papers to textbooks to popular accounts. See Fleck (1979), pp. 111–24.

17. The eleven items were eligibility criteria, admission before allocation, random allocation, method of randomization, patient's blindness to treatment, blind assessment of outcome, treatment complications, loss to follow-up, statistical analysis, statistical methods, and power (quoted in Mosteller 1985).

Bibliography

Ahrens, E. H., Jr. 1985. "The Diet-Heart Question in 1985: Has It Really Been Settled?" *Lancet*, May 11, pp. 1085–87.

Barber, B. 1961. "Resistance by Scientists to Scientific Discovery." *Science* 134: 596–602.

——. 1971. "Function, Variability, and Change in Ideological Systems." In *Stability and Social Change*, ed. Bernard Barber and Alex Inkeles, 244–62. Boston: Little, Brown.

Brensike, J. F., et al. 1982. "National Heart, Lung, and Blood Institute Type II Coronary Intervention Study: Design, Methods, and Baseline Characteristics." *Controlled Clinical Trials* 3: 91–111.

——. 1984. "Effects of Therapy with Cholestyramine on Progression of Coronary Arteriosclerosis: Results of the NHLBI Type II Coronary Intervention Study." *Circulation* 69, no. 2: 313–24.

Brown, M. S., and J. L. Goldstein. 1984. "How LDL Receptors Influence Cholesterol and Atherosclerosis." *Scientific American* 251, no. 5: 58–66.

Cole, J. R. 1979. *Fair Science: Women in the Scientific Community*. New York: Free Press.

——. 1986. *Health Risks in the Media: Some Food for Thought*. New York: Center for the Social Sciences, Columbia University, Preprint no. 108.

Cole, J. R., and S. Cole. 1973. *Social Stratification in Science*. Chicago: University of Chicago Press.

——. 1985. Experts' "Consensus" and Decision-Making at the National Science Foundation." In *Selectivity in Information Systems: Survival of the Fittest*, ed. Kenneth S. Warren, 27–63. New York: Praeger.

Cole, J. R., S. Cole, and the Committee on Science and Public Policy, National Academy of Sciences. 1981. *Peer Review in the National Science Foundation: Phase Two of a Study*. Washington, DC: National Academy of Sciences.

Cole, S. 1983. "The Hierarchy of the Sciences." *American Journal of Sociology* 89: 111–39.

Cole, S., J. R. Cole, and G. A. Simon. 1981. "Chance and Consensus in Peer Review." *Science* 214: 881–86.

Dawber, T. R. 1980. *The Framingham Study: The Epidemiology of Atherosclerotic Disease*. Cambridge, MA: Harvard University Press.

DerSimonian, R., L. J. Charette, B. McPeek, and F. Mosteller. 1982. "Reporting on Methods in Clinical Trials." *New England Journal of Medicine* 306: 1332–37.Fleck, L. 1979 [1935]. *Genesis and Development of a Scientific Fact*. Trans. F. Bradley and T. J. Trenn; ed. T. J. Trenn and R. K. Merton. Chicago: University of Chicago Press.

Goldstein, J. L., and M. S. Brown. 1985. "Familia Hypercholesterolemia: A Genetic Receptor Disease." *Hospital Practice*, November 15, pp. 35–46.

Goodell, R. 1975. *The Visible Scientists*. Boston: Little, Brown.

Gould, S. J. 1981. *The Mismeasure of Man*. New York: Norton.

Holton, G. 1978. *The Scientific Imagination*. Cambridge: Cambridge University Press.

Ingelfinger, F. J. 1974. "Peer Review in Biomedical Publication." *American Journal of Medicine* 56: 686 ff.

Knorr-Cetina, K. D. 1981. *The Manufacture of Knowledge. An Essay on the Constructivist and Contextual Nature of Science*. Oxford: Pergamon Press.

Latour, B., and S. Woolgar. 1979. *Laboratory Life: The Social Construction of Scientific Facts*. Beverly Hills, CA: Sage.

Levy, R. I., et al. 1984. "The Influence of Changes in Lipid Values Induced by Cholestyramine and Diet on Progression of Coronary Artery Disease: Results of the NHLBI Type II Coronary Intervention Study." *Circulation* 69, no. 2: 325–37.

544 CONSENSUS IN SCIENCE

Lipid Research Clinics Program. 1984*a*. "The Lipid Research Clinics Coronary Primary Prevention Trial Results. 1. Reduction in Incidence of Coronary Heart Disease." *JAMA* 251, no. 3: 351–64.

——. 1984*b*. The Lipid Research Clinics Coronary Primary Prevention Trial Results. II. The Relationship of Incidence of Coronary Heart Disease to Cholesterol Lowering. *JAMA* 251, no. 3: 365–74.

Merton, R. K. 1942. "Science and Technology in a Democratic Order." *Journal of Legal and Political Sociology* 1: 115–26. Reprinted in R. K. Merton, *The Sociology of Science: Theoretical and Empirical Investigations*, 266–78. Chicago: University of Chicago Press, 1973.

——. 1968. "Social Structure and Anomie." In *Social Theory and Social Structure*, enlarged ed., 185–214. New York: Free Press.

——. 1972. "The Perspectives of Insiders and Outsiders." *American Journal of Sociology* 77: 9–47.

——. 1973 [1957]. "Priorities in Scientific Discovery." In *The Sociology of Science*, 286–324. Chicago: University of Chicago Press.

Mosteller, F. 1985. "Selection of Papers by Quality of Design, Analysis, and Reporting." In *Selectivity in Information Systems*, ed. K. E. Warren, 98–116. New York: Praeger.

Multiple Risk Factor Intervention Trial Research Group. 1982. "Multiple Risk Factor Intervention Trial: Risk Factor Changes and Mortality Rates." *JAMA* 248, no. 12: 1465–77.

National Academy of Sciences. 1980. *Toward Healthful Diets*. Washington, DC: National Academy Press.

——. 1985. *Nutrition Education in U.S. Medical Schools*. Washington, DC: National Academy Press.

National Center for Health Statistics. 1984. *Monthly Vital Statistics Report* 33, no. 9 (December 20).

Nelkin, D. 1985. "Managing Biomedical News." *Social Research* 52, no. 3: 625–46.

Oliver, M. F. 1985. "Consensus or Consensus Conferences on Coronary Heart Disease." *Lancet*, May 11, pp. 1087–89.

Polanyi, M. 1963. "The Potential Theory of Adsorption." *Science* 141: 1010–13.

Rockefeller University Research Profiles. 1984. Edward H. Ahrens Jr. (Fall).

Shekelle, R. B., et al. 1981. "Diet, Serum Cholesterol, and Death from Coronary Heart Disease: The Western Electric Study." *New England Journal of Medicine* 304, no. 2: 65–70.

Stinchcombe, A. L., and R. Ofshe. 1969. "On Journal Editing as a Probabilistic Process." *American Sociologist* 4, no. 2: 116–17.

Winsten, J. A. 1985. "Science and the Media: The Boundaries of Truth." *Health Affairs* 3: 5–23.

Zuckerman, H. 1977. *Scientific Elite: Nobel Laureates in the United States*. New York: Free Press.

Zuckerman, H., and R. K. Merton. 1971. "Patterns of Evaluation in Science: Institutionalization, Structure, and Function of the Refereeing System." *Minerva* 9: 66ff.

CHAPTER 25

TWO CULTURES REVISITED

(1996)

JONATHAN R. COLE

The gulf in understanding between scientists and nonscientists may be traceable to an educational system that neglects the historical importance of scientific and technological developments.

I want to speak this morning about the recent past and the immediate future of science, technology, and society in twenty-first century America. Science and technology policy would appear to be in a state of crisis. There are many indicators that a crisis does exist in the partnership between the federal government and the American research universities—that the terms of the partnership are increasingly being questioned and reexamined. Even if it should turn out that this is not, in fact, the case, so many knowledgeable and informed members of the academic, scientific, and engineering communities believe that this is so that the perception of crisis is real in its consequences. It is probably not a matter of hyperbole to suggest that we are witnessing a number of fundamental changes in the relationship between the federal government and the scientific and technology research community. There are apt to be material changes in the national system of innovation in the years ahead.

To me this moment of crisis appears ironical. After all, each of us is aware of the extraordinary half-century of scientific and technological growth and achievement that we have witnessed in the United States. These years of extraordinary growth in knowledge, in terms of its diversity, sheer volume and

546 CONSENSUS IN SCIENCE

unquestionable quality, as well as its impact on our culture and everyday life, is perhaps unmatched since that glorious Newtonian period of scientific and technological development in seventeenth-century England. Yet, we confront in a wide array of places the emergence of strong antiscience and antitechnology movements—and not only among the limited number of fringe players but also among some very serious philosophers, poets, social scientists, and politicians.

Part of the explanation for these sentiments, I believe, lies in the American public's own ambivalence toward science and technology. Why is it that we in this country admire scientific genius and technological inventiveness, and can become truly obsessed by the fruits of science and technology, yet are so abysmally ignorant of facts and processes of scientific and technological discovery? Americans benefit in virtually every area of life from the fruits of these discoveries. The list of significant advances, in such areas as telecommunications, lasers, photonic systems, computer technology, synthetic materials, and biotechnology is long and constantly growing.[1]

Not all of these scientific and technological advances have been entirely for industrial or health gain; many supported changes in leisure time activities, from video games to new materials for tennis rackets and golf clubs. Indeed, science and technology have changed dramatically the nature of American culture and, in many ways, redefined the locus of what Schumpeter called "conspicuous consumption." Many reasonably well-to-do Americans have come to think of health as a middle-class consumer good that can be bought and from which we can expect straightforward returns from our investments.

For all of its appreciation of science and technology—to say nothing of its massive purchasing of the products of scientific and technological advances—the American public remains strikingly scientifically illiterate. Gerald Holton, one of our most distinguished physicists and historian of science, notes in his exceptionally powerful recent essays on science and antiscience that "public literacy [in the United States] is now at a level where 'half the adults questioned did not know that it took one year for the Earth to orbit the Sun.' Less than 7 percent of U.S. adults can be called scientifically literate by the most generous definition, only 13 percent have at least a minimum level of understanding of the process of science, and 40 percent disagree with the statement 'astrology is not at all scientific'" (Holton, 1993, p. 147).

Teacher Attrition, Qualifications

Furthermore, Holton says, our schools no longer are attracting young people who might correct this problem. "'We are losing thirteen mathematics and science teachers for each one entering the profession.' Only the following percentages of teachers meet the minimum established standards for coursework preparation at the high school level: 29 percent in biology, 31 percent in chemistry, 12 percent in physics. . . . 'In the most recent international science assessments, in comparison with students in 12 other countries, our high school students finished 9th in physics, 11th in chemistry, and last in biology. . . . In mathematics, our top 13 percent generally fell into the bottom 25 percent in comparison with other countries'" (Holton, 1993, pp. 147–48).[2]

The limited knowledge of the public—or even what William James called "acquaintance with knowledge"—is neither, as one might expect, monopolized by the poorly educated or found only among certain social classes, nor is it confined to individuals in a limited number of occupations. Most members of Congress with whom I have spoken in the past few years about national investments in science and technology, who by the way tend to hold generally positive views of the social and economic benefits of such investments, are almost wholly uninformed about the process of growth of knowledge and discovery, about the diffusion of knowledge, of the process of technology transfer, and of the historical links between fundamental discoveries and practical applications.

Widespread ignorance and illiteracy is, of course, not without its consequences. Out of the ocean of illiteracy flow ideas that challenge fundamental aspects of modern culture. One of the best examples of this is the recent challenges to the world of science produced by the extraordinary Czech poet, playwright, and political leader Vaclav Havel. Havel identifies the mental apparatus associated with scientific thinking (its "rational and cognitive thinking," its "depersonalized objectivity," and its belief that "there is a there out there") as a root cause of many of the true evils and atrocities that have marked the twentieth century's political history, from World War I to Nazism and Communist authoritarianism. Havel's attack on the epistemology of science is direct and unsparing:

Traditional science, with its usual coolness, can describe the different ways we might destroy ourselves, but it cannot offer us truly effective and practicable instructions on how to avert them. . . . The world today is a world in which generality, objectivity and universality are in crisis. . . . Many of the traditional mechanisms of democracy created and developed and conserved in the modern era are so linked to the cult of objectivity and statistical averages that they can annul human individuality. . . . [Ours is] an epoch which denies the binding importance of personal experience—including the experience of mystery and of the absolute—and displaces the personally experienced absolute as the measure of the world with a new, man-made absolute, devoid of mystery, free of the "whims" of subjectivity and, as such, impersonal and inhuman. It is the absolute of so-called objectivity: the objective, rational cognition of the scientific model of the world. Modern science, constructing its universally valid image of the world, thus crashes through the bounds of the natural world which it can understand only as a prison of prejudices from which we must

548 CONSENSUS IN SCIENCE

break out into the light of objectively verified truth. . . . With that, of course, it abolishes as mere fiction even the innermost foundation of our natural world; it kills God and takes his place on the vacant throne, so that henceforth it would be science which would hold the order of being in its hand as its sole legitimate guardian and be the legitimate arbiter of all relevant truth. (Quoted in Holton, 1993, pp. 175–77)

This powerful rhetoric, even if terribly incomplete and distorted, resonates in many quarters these days, including many of the best humanities and social science departments at some of our most prominent universities.

The level of American illiteracy about science and technology is disturbing not only because of the fertile ground it prepares for those antagonistic toward modern science and technology but also because we seem to be doing very little in our educational system to reverse the trend. Educators, including scientists and engineers, have not stepped up to the challenge of constructing a strong program for increasing scientific literacy. They have not assumed that educating the American public about the processes and fruits of discovery is one of their fundamental roles.

C. P. Snow, in his now famous 1959 Rede Lecture, "The Two Cultures," identified the wide gap that existed between the two worlds in which he lived—the world of writers, artists, and other literary intellectuals, and the world of science. Snow's analysis of the gulf between the two cultures suggested that literary intellectuals were in many ways more ignorant about science and its foundations than were scientists about literature. Notwithstanding Jerome Wiesner's reflection, "Better two cultures than none," Snow's analysis, which, of course, focused more on British than American society, remains cogent. He wrote,

It is obvious that between the two [cultures], as one moves through intellectual society from the physicists to the literary intellectuals, there are all kinds of tones of feeling on the way. But I believe the pole of total incomprehension of science radiates its influence on all the rest. That total incomprehension gives, much more pervasively than we realize, living in it, an unscientific flavour to the whole "traditional" culture, and that unscientific flavour is often, much more than we admit, on the point of turning anti-scientific. The feelings of one pole become the anti-feelings of the other. If the scientists have the future in their bones, then the traditional culture responds by wishing the future did not exist. It is the traditional culture, to an extent remarkably little diminished by the emergence of the scientific one, which manages the western world. . . . This polarization is sheer loss to us all. To us as people, and to our society. It is at the same time practical and intellectual and creative loss. . . . In our society (that is advanced western society) we have lost even the pretense of a common culture. Persons educated with the greatest intensity we know can no longer communicate with each other on the plane of their major intellectual concern. This is serious for our creative, intellectual, and above all, our normal life. It is leading us to interpret the past wrongly, to misjudge the present, and to deny our hopes of the future. It is making it difficult or impossible for us to take good action. . . . There is only one way out of all this: it is, of course, by rethinking our education. (Snow, 1959, pp. 11, 16, 60)

It seems to me that the divide between the two cultures is as pronounced today as it was in 1959, despite our awareness of the problem and despite the achievements of American science and technology, from the first manned space flight to the identification of genes that cause cancer. Despite the glorious history of science and technology under the model originally proposed by Vannevar Bush fifty years ago, we have failed as a culture to teach the most elementary aspects of the process of scientific growth and the mechanisms that enable technology and science to interact to the practical benefit of the society. Why is this still true today? Why have we strayed so far from the time in our Republic's history when the likes of Franklin and Jefferson revealed through their writing and discourse a profound knowledge about, and interest in, science and technology, as well as an ability to articulate reasons why knowledge of science and technology was a necessary aspect of developing a firm foundation for a precarious democracy? (See Cohen, 1995.) When we try to understand why we need to construct a new rationale for science and technology in the national interest, part of the reason lies in the absence of substantial knowledge about the achievements and failures of science and technology among the leaders of the nation.

Curious about what might partially sustain the divide between "the two cultures" and, derivatively, the level of scientific illiteracy in this country, I have recently begun a review of the way science and technology is represented in leading American history books, texts, and curricula. Some years ago, John Heilbron and Daniel Kevles (1988) carried out a similar study. Here are some rather unsystematic, preliminary results.[3]

One consequence of the current structure of education in America is that youngsters are labeled as scientifically "able" or "talented" at an early age. Most often this labeling is based upon how rapidly students can obtain correct answers to questions when the answers are already known. Most of our very talented students come to believe that they have little aptitude for science or engineering and so turn their attention to other subjects—history, the social sciences, the arts and literature—eventually moving into professions far removed from the sciences and engineering. Many never take a science course after they have completed the final requirement in high school. What they know about science and technology must, then, come from the various news media and from whatever teaching they may receive in American history during high school, college, or graduate school.

550 CONSENSUS IN SCIENCE

The Teaching of American History

Given the extraordinary contributions of science and technology to contemporary culture and the economic well-being of Americans—as perhaps the major cause of social and institutional change in contemporary society—one might expect some significant representation of the development of science and technology, and the processes of scientific discovery, in the works that describe recent American history. To explore what actually exists in those works, I asked three distinguished historians at Columbia to give me the titles of the leading books in American history covering the period from World War II to the present. I also obtained the titles of some leading American history textbooks that were used in colleges and high school courses on the subject. Finally, I examined the course offerings at the top ten departments of history, according to rankings in the recent National Research Council (NRC) (1995) evaluation of graduate PhD programs for courses focusing on science and technology. In examining references and discussions of science and technology in these historical works, I wanted some comparative reference points. So, I chose also to examine the works in light of the attention they give to several other themes, including religion, popular culture, and the counterculture.

In recounting my findings, let me emphasize that the historians whose work I have reviewed are leaders in their field; they have exceptional ability and a demonstrated record of publishing important books. And yet, to say that there is a paucity of references to and discussions of science and technology in the works could only be considered on inspection to be a gross understatement.

For example, in Michael Barone's (1990) *Our Country: The Shaping of America from Roosevelt to Reagan*, there is not a single reference to science and technology topics in the index. There is space devoted to the "upheavals" of the 1960s and some references to popular culture and counterculture, including Elvis Presley, premarital sex, and recreational drug use. There is some discussion of changing family patterns, such as lower fertility, more contraception, more divorce, and the declining authority of men in the family. But you won't find discussions of the discovery of the double helix by Watson and Crick or, for that matter, any aspect of the biological revolution, or of the plate tectonics revolution, or of the development of the laser, or of the transformational effects of computers and telecommunications on American society. Indeed, for the uninitiated young student examining the half-century of the American scientific revolution, science and technology don't exist.

William Chafe's (1986) exceptional work, well-known and widely used by colleges and high schools, *The Unfinished Journey: America since World War II*, offers another example of the striking neglect of the subject. In one chapter, Chafe has a two-page discussion of technological breakthroughs that were important for industrial development of aerospace, chemicals, and electronics, and he does say "the government served as a primary sponsor" for these breakthroughs—but that is all he says. He mentions the importance of the new computer technology and quotes Daniel Bell: "The knowledge revolution transformed the occupational structure of the country" (p. 320). That is it for science and technology during the post-war period. Chafe provides the reader with more material on television and popular culture (almost nine pages) than on science and technology.

William Leuchtenburg (1983), one of my former colleagues at Columbia, has but one indexed reference to research and development in his major work, *A Troubled Feast: American Society since 1945*. I'm afraid that reference must suffice for both technology and science. There is a brief section (pp. 40–41) devoted to a discussion of medical achievements, including five lines about antibiotics and a picture of a child receiving a polio shot, and a page and a half on computers, automation, and R&D generally (pp. 44–45). At one point, Herbert Simon is quoted saying that the computer represents "an advance in man's thinking processes as radical as the invention of writing" (p. 44). Leuchtenburg's discussion of technology advances attributes many of them to research going on at institutions like MIT and Caltech. He does devote an entire chapter to "The Consumer Culture" and another to the counterculture, called "The Greening of America," in which he discusses such topics as drugs, generational conflict, anti-consumerism, the reduction of sex differences, and pacifism.

No more than one more example is needed to bring the central message home: the young members of our society who become lawyers, businessmen, or members of Congress are surely not learning about the scientific and technology revolution in their reading of the best texts produced by American historians. James T. Patterson (1996), in his recently published 829-page megabook *Great Expectations: The United States, 1945–1974*, does in fact have thirteen references to technology in his index, grouped by decade. There is one reference to the National Institutes of Health (NIH) as well. But almost all of these references are simply cursory, without any effort to develop in the young reader an understanding of the processes of discovery, of technology transfer, or of the ways in which science and technology have helped bring about the economic and social transformation of modern American society in the post-war period. There is a twelve-line discussion of the effects of research and development on productivity and the expansion of various kinds of industry; there is a two-page discussion of advances in medical science, including the role of NIH, tranquilizers, vaccines (especially the polio vaccine); and another half-page mention of technological breakthroughs, including Xeroxing, heart transplants, and the moon walk. In contrast, there is a whole chapter on "Mass Consumer Culture," much of which is devoted to popular culture.

I won't belabor the point further but only note that other important works, including Frederick F. Siegel's (1984) *Troubled Journey: From Pearl Harbor to*

552 CONSENSUS IN SCIENCE

Ronald Reagan and Lawrence S. Wittner's (1978) *Cold War America: From Hiroshima to Watergate* include little, if anything, on our subject. Even the sixth edition of John A. Garraty and Robert A. McCaughey's (1991) long-standing, best-selling text for high school courses in American history, *The American Nation: A History of the United States since 1865*, has only two references to science, and those are to the study of science in the late 1800s. As for technology, there are references to the Manhattan Project, DDT, and television, and one page is devoted to the development of plastics, nuclear power, and computers—a good part of which stresses the adverse impacts of new technology. A brief reference to the space program under Kennedy, then under Reagan, can be found, but there is nothing on the developments of biomedicine.

Why is there so little to be found? Part of the answer may lie in the training that American historians receive at major graduate schools in the nation. In fact, here the answer is not quite as clear as in the examination of the output of leading American historians. The top ten history departments, according to the NRC assessment (National Research Council, 1995), are Yale, Berkeley, Princeton, Harvard, Columbia, UCLA, Stanford, Chicago, Johns Hopkins, and Wisconsin at Madison.[4] Most of these departments offer courses of some kind in the history of science and medicine; few devote much attention to technology. A number of these universities have separate departments or programs that focus on the history, philosophy, and, less often, the sociology of science. Harvard has a long-standing commitment to the history of science since the days of George Sarton, and Chicago, Stanford, Hopkins, and Wisconsin have what appear to be substantial offerings. Robert K. Merton, the putative father of the sociology of science, began his work at Harvard and later developed the specialty at Columbia. In short, there have been and continue to be many high-quality books and articles published on the history, sociology, and philosophy of science.

What remains unclear, however, is how much of the work at these major departments and programs is directed only toward highly self-selected students who are interested in science and technology—many of whom come with backgrounds in science—but who are not part of the mainstream programs of American history at the university. Are these course offerings part of the standard curricula of these graduate programs? Are all students of contemporary American history who will be teaching the subject and writing books about it required to study the influence of science and technology on American society? Even more particularly, are they asked to understand in an elementary way the nature of discovery and the forces that shape technological growth and development? The answers are negative. Unless the history and sociology of science and technology are, indeed, truly integrated into the core curriculum of American history graduate students, the teachers of the next generation of Americans are apt to be as scientifically illiterate as those in the classroom today.[5] Yet, from what I can gather, there is little interaction between graduate students in departments of history, that is, those graduate students who will eventually teach American history and write the books and articles read by the next generation of students, and the graduate students in small science-related departments operating on the periphery of the larger history departments.

A Self-Fulfilling Prophecy

I wonder how often, in discussions within academic departments about the need for new young faculty, the possibility of an appointment of a person who specializes in the recent history of science and technology even reaches the table. Almost never, I would conjecture. Moreover, American historians feel understandably very reluctant to discuss science and technology in their classes, texts, or scholarly books because they are acutely aware of how their training is inadequate to the task. Since these faculty and the students that come after them receive such dismal training in the history of American science and technology at the elementary and secondary levels, as well as at more advanced stages of training, publishers and authors feel (probably correctly) that there is no "market" for discussions of science and technology in new textbooks or books about recent American history. Conditions for a self-fulfilling prophecy fall into place—conditions that reinforce the existing definition of the situation.

It would not surprise me if a hundred years from now, historians of America in the post–World War II period devote far more attention to the role of science and technology in making the American nation a global power, the wealthiest nation in the world during the second half of the twentieth century, and the undisputed leader in the production of pathbreaking scientific and technological discoveries. Today, you surely would not get that impression from reading the best books narrating the critical forces in our nation's recent history.

Notes

1. I have benefited here from discussions and correspondence with Gerald Holton, who has written with extraordinary force and grace about problems with scientific literacy, the recent rise in "antiscience," and the absence of effective teaching of science and technology in American schools.
2. One might argue that it is a waste of national resources to train PhDs to teach in secondary school settings, but it seems to me that a strong counterargument to this position can be developed. One of our problems in graduate education is that we have led all PhD students at the major universities to believe that they can expect positions at similarly situated places. This is, of course, a practical impossibility. We have got to change the nature of expectations or reduce the size of these programs. There were times in the past when PhDs did populate secondary schools, in somewhat greater numbers than

today, and they do have places in some of the better private secondary schools. The need is substantial and means ought to be found to provide incentives for well-trained young scientists to migrate to the school system rather than to unrelated occupations.

3. Dr. Elinor Barber, affiliated with the Provost's Office at Columbia, helped collect the basic data reported here.

4. Although not included in the top ten rankings, MIT, Caltech, Georgia Tech, UC/San Diego, Rensselaer Polytechnic Institute, and the Illinois Institute of Technology have significant academic programs focusing on science and technology.

5. An important project at MIT, sponsored by the Sloan Foundation, is underway. It is an effort by a set of highly regarded American historians to create a textbook of American history that integrates fully and adequately the role of science and technology in the development of the nation. It is not a specialized text but a standard work that will attempt to correct some of the problems discussed here.

References

Barone, M. 1990. *Our Country: The Shaping of America from Roosevelt to Reagan*. New York: Free Press.

Chafe, W. 1986. *The Unfinished Journey: America since World War II*. New York: Oxford University Press.

Cohen, I. B. 1995. *Science and the Founding Fathers*. New York: Norton.

Garraty, J. A., and R. A. McCaughey. 1991. *The American Nation: A History of the United States since 1865*. New York: Harper & Row.

Heilbron, J., and D. Kevles. 1988. "Science and Technology in U.S. Textbooks." *Reviews in American History* (June): 173–85.

Holton, G. 1993. *Science and Anti-Science*. Cambridge, MA: Harvard University Press.

Leuchtenburg, W. 1983. *A Troubled Feast: American Society since 1945*. Boston: Little, Brown.

National Research Council. 1995. *Research-Doctorate Programs in the United States: Continuity and Change*. Washington, DC: National Academy Press.

Patterson, J. T. 1996. *Great Expectations: The United States, 1945–1974*. New York: Oxford University Press.

Siegel, F. F. 1984. *Troubled Journey: From Pearl Harbor to Ronald Reagan*. New York: Hill and Wang.

Snow, C. P. 1959. *The Two Cultures and the Scientific Revolution. The Rede Lecture*. New York: Cambridge University Press.

Wittner, L. S. 1978. *Cold War America: From Hiroshima to Watergate*. New York: Holt, Rinehart, and Winston.

CHAPTER 26

INTELLECTUAL DIVERSITY IN THE UNITED STATES

To What End? (2006)

JONATHAN R. COLE

Diversity of all kinds seems to be the new universal good. So, it must be a good thing when the conservative activist David Horowitz calls for "intellectual diversity" on American campuses to replace radical or liberal orthodoxy that is warping the minds of the nation's educated youth. Like much of what Horowitz has brought us lately—such as his Academic Bill of Rights he would have every state legislature adopt as law, or his recent book, which identifies the "101 most dangerous professors" on American campuses—there is in his work and proposals much factual error, doublespeak and conceptual muddle that poses as thoughtful, reasonable, and empirically validated statements of fact. But adoption of a Horowitz-like agenda for presenting the world to our students would be disastrous for American universities.

Here are just a few reasons why. Critics have a right to criticize, but they also have some responsibility to produce conceptual clarity. Horowitz and other like-minded critics, such as the American Council of Trustees and Alumni, fail to offer any clear idea, much less a definition, of what they mean by intellectual diversity and what would represent "balance" in an individual scholar's lectures or seminars, in a department's offerings, or in a university's curriculum and research agenda.[1] We are not offered any convincing evidence, beyond a few illustrative anecdotes and highly edited videos, that campus intellectuals are espousing orthodoxy. What is the size of the orthodoxy problem anyway? What remedy, if any, is necessary for this disease that Horowitz would have us cure?

Despite protestations to the contrary, political accountability to outside authorities is what Horowitz and his supporters really want. That is why they need state legislatures to pass the Academic Bill of Rights. They fear a system that resolves sharply divergent truth claims through a process of peer review rather than political review. In fact, they are most interested in substituting their orthodoxy for what they see as a misguided and dangerous one. How far would they go in producing "intellectual diversity"? Maybe we should teach "intelligent design" in science courses as a balance against the consensus scientific views about evolutionary biology. Or mandate a proper balance between those who criticize the Bush administration's forays into Iraq, which was based upon lies and false information, with an appropriate number of scholars who will defend the administration's actions regardless of what are considered now as facts.

Any ideas should be fair game for debate at universities, but those that fail over time to persuade appropriate experts in the field should lose out in the marketplace of ideas rather than be retained because political pressure has been put on universities to offer them as "representatives" of alternative points of view. I want to postulate that external political interference with academic life, free inquiry, and the open discourse that is essential to it has almost always had disastrous consequences for systems of higher learning.

Horowitz is a man who simply does not trust the basic system of knowledge creation and transmission that has produced in the United States the greatest system of higher education in the world. He neither trusts its professors nor its students. He believes that the professors are presenting a world not as proposed or tested theories but as ideology. He doesn't trust students because he sees them as naive and incapable of critical reasoning and of distinguishing between sound and shoddy evidence. He worries about universities that give safe harbour to radically different ideas that question existing institutional arrangements in our society. He does not believe the professoriate capable of self-policing or of distinguishing between claims to truth that do, or do not, measure up against rigorous methodological standards. At Horowitz's university, professors would hold their intellectual punches for fear of facing star-chamber tribunals for what they say or fail to say in class. It would be a university rich in balanced, but flaccid, curricula. Horowitz would produce a dangerous, as well as boring, university. In short, he would nullify the long-standing and highly successful compact between American society and its universities.

In fact, the problem with today's university lies less in the absence of intellectual diversity and more in the seeming unwillingness of most faculty members to engage in a civil but public clash of ideas. The absence of sustained intellectual contest and criticism of received wisdom and public policy—on the right and the left—is a far greater threat to the university than the problem of ideological imbalance. Where, in fact, is the sustained criticism among faculty and students of public policies that many at the universities privately consider to be wrongheaded? The deafening silence, and the absence of debate, at American universities about current domestic and foreign policies, compared with, say, the 1960s or 1970s, is cause for deep concern. Where are the critical voices and defenders of basic ideas—like the rule of law—when we need them most and that, historically, were found in significant numbers at our great universities? Too many faculty members, even those with strong opinions, now say, "I'd rather not get involved." What is tenure for, anyway? And, if it's the liberal-left professoriate that Horowitz fears, one can only conclude by looking at outcomes that it has done a terrible job of convincing college youth of the merits of its supposedly subversive ideas.

Horowitz is right about one thing: More academics identify themselves as politically liberal than as conservative. But Horowitz attributes this to a power elite of leftist liberal professors wielding their power to discriminate against those with more conservative ideas. He is wrong about this on two counts. First, the liberal orientation of faculty members at American universities is nothing new; there has been no sharp turn to liberal orthodoxy. Seymour Martin Lipset and Everett Carl Ladd Jr., among others, have shown that the American professoriate has consistently over the past fifty years identified more with a liberal political agenda than with a conservative one. Second, the life of the intellectual, the scientist, the scholar, tends to attract those who are critical of existing social institutions, inequalities, public policies, and dogmas, those more interested in change than stability, those who identify themselves as liberals. That process predominantly involves self-selection by these people into the academic life, not discrimination against conservatives in the academic reward system.

During my fourteen years as provost and dean of faculties at Columbia, I oversaw more than seven hundred tenure cases involving thousands of faculty members, and I found no evidence that the so-called liberal-left or conservative professoriate allowed their personal politics to influence hiring and promotion decisions. The personal political views of candidates were never raised as an attribute that would either qualify or disqualify candidates for tenure. The quality of research and teaching, as assessed by a set of qualified external and internal peers, was the overwhelming criterion used to determine who received tenure. Perhaps this is a case of windmills being mistaken for enemy soldiers.

Horowitz's confusion of process with outcomes is compounded when he draws the false and misleading inference that professors' personal political beliefs are correlated with the way they conduct their classes. Edward Said, the extraordinary literary critic and defender of the Palestinian people, was the bête noir of types like Horowitz, but in all of the years that I knew Said at Columbia he taught literature, not politics (with a point of view to be sure), challenged his students to think more clearly, and was revered by them. They flocked to his classes, and I never heard a single complaint that he was biased or intimidating or refused to listen to alternative views of his students.

558 CONSENSUS IN SCIENCE

If Horowitz fears radical professors, the American public does not, according to the Harvard sociologist Neil Gross and his colleague Solon Simmons of George Mason University, whose 2006 survey showed that the public has far greater trust in higher education than most other institutions and that fully 80 percent were opposed to the government controlling "what gets taught in the college classroom." The public has real concerns about universities and colleges, but political orthodoxy of radical professors is not high on that list.[2] Number one is the high cost of college tuition (43 percent)—a concern of roughly equal importance to liberals and conservatives. More than twice as many people thought that "binge drinking by students" (17 percent) rather than "political bias in the classroom" (8 percent) was the biggest problem facing universities. Although 12 percent of the public felt that the term "radical" appropriately describes professors, 40 percent preferred the term "professional." A majority believed that professors respected their students regardless of their political views. Older, conservative Republicans, who have had relatively little education themselves, are the group most concerned with biased professors.

Because of the way they label and classify professors and academic discourse, Horowitz and his followers fail to acknowledge the many crosscurrents of intellectual diversity in universities that cannot be captured through caricatures. The liberal left is hardly a monolithic, orderly group of academics conforming to a single orthodoxy. For example, in their orientation to the core ideas of the liberal state and the place of multiculturalism in it, professors sharply diverge. Some believe that sub-cultural groups in a larger society should be able "to sustain and perpetuate their cultural or religious differences . . . and their distinct communal identities." Others believe that individuals in sub-cultural groups should be offered the opportunity "to attain 'mainstream' educational, socioeconomic, occupational, and political status" that conforms to the larger value and cultural system and strives to eliminate group-based differences.[3] On a host of issues, from the prerogative of parents to sustain religious beliefs that do not conform to the interest of state-run schools to the legitimacy of customs related to circumcision, two proponents of liberal theory might well differ passionately—while classifying themselves as politically "liberal." In short, the complexities and nuances in academic discourse limit the value of typecasting professors as either ideologically liberal or conservative, and Horowitz fails to mention them.

Finally, there is a problem with Horowitz's idea of a fact. His book *The Professors: The 101 Most Dangerous Academics in America* is replete with factual errors and grossly misleading statements in every Columbia University case he cites. He claims Lisa Anderson, a political science professor and dean of Columbia's School of International Studies, selected Middle East studies professor Rashid Khalidi, to occupy the Edward Said Chair at Columbia. She did not. A blue-ribbon committee of scholars reviewed the credentials of scores of candidates before deciding Khalidi was the best person in the nation for the job. Horowitz, noting that the Columbia University president Lee Bollinger appointed Victor Navasky, a journalism professor and former editor and publisher of the *Nation*, to a group examining the future of journalism, asserts that Bollinger made no attempt at ideological inclusiveness. That is factually wrong. Although I doubt that Bollinger considered political ideology when inviting people to serve, he did include people like Karen House, the senior vice-president of Dow Jones and publisher of the *Wall Street Journal*, which is hardly known for its liberal editorial page. Finally, in uncovering the 101 most "dangerous" academics in the United States, Horowitz apparently could not find a single conservative professor that met his definition of "dangerous." Is the American academy totally free of dangerous conservatives? It defies statistical probability.

Before jumping on an illusory bandwagon labeled "Eliminate liberal or radical orthodoxy/Legislate intellectual diversity," we would be well advised to be skeptical. What is their actual intent? Are they accurately portraying a disease at our colleges and universities, or are they asking for a remedy for a non-existent disease that will undermine academic freedom and free inquiry at our institutions of higher learning? Jumping on that bandwagon will almost certainly contribute to the weakening of the system of higher learning in the United States that remains the envy of the world—one that dominates the list of the world's greatest 20 or 50 universities. If we allow political outsiders to undermine those values and structures that enable the teaching and discoveries that come from the very same universities that Horowitz identifies as exemplars of intellectual orthodoxy, but which actually suffer from no disease and that contribute mightily to the artistic, humanistic, scientific, social, and economic welfare of the nation, then the international preeminence of American universities will be at real risk. It will require active resistance, vigilance, and courage from those who understand the real idea of a university to see that this does not happen.

Notes

1. The American Council of Trustees and Alumni has concluded, "Throughout American higher education, professors are using their classrooms to push political agendas in the name of teaching students to think critically. In course after course, department after department, and institution after institution, indoctrination is replacing education" (May 2006).
2. Neil Gross and Solon Simmons, "Americans' Views of Political Bias in the Academy and Academic Freedom," working paper, May 22, 2006.
3. See Richard Shweder, "Conflicting Varieties of Liberal Expectancy: With Special Attention to Schooling in America." Prepared for the SSRC/RSF volume *Multicultural Schoolyard Fights: Is There a Conflict Between Pluralism and Inclusion in American Higher Education?* Forthcoming. Preprint, 3–4. Quoted with permission of the author.

PART IV

ACADEMIC FREEDOM AND FREE INQUIRY

THE ENABLING VALUE

CHAPTER 27

THE PATRIOT ACT ON CAMPUS

Defending the University after 9/11 (2003)

JONATHAN R. COLE

I want to talk this evening about "Defending the Idea of the University in Troubled Times." My guiding premise—in this talk and in my fourteen years as provost at Columbia University—is that research universities play an essential role in the social and economic development of our country and are worth defending. They are unique and fragile institutions that are admired around the world—so much so that higher education represents one of the few sectors of the U.S. economy with a favorable balance of trade. The contributions of these universities to the welfare of the nation are understood within the academy but are poorly understood in the larger public.

Since the events of September 11, 2001, and the close-on-its-heels passage of the USA PATRIOT Act and subsequent presidential directives, I have been struck by (and dismayed at) the near-deafening silence of the expected voices of dissent on the great university campuses and by the absence of a sustained debate over the fundamental issues and tension—the balancing act—between the needs for national security and the protection of basic, individual, constitutional liberties. In looking at the consequences of government policies toward universities since 9/11, we should not forget our own larger

---

This article is based on a talk presented as part of "The Futures of Higher Education," which was held at the University of Chicago's Graham School of General Studies on May 9, 2003. Bart Schultz organized the panel discussion. Geoffrey Stone, who participated in the discussion, has helped me gain a historical perspective on the balancing of individual rights and national security.

history and look to it for guidance. Periodically, in times of actual or perceived national crisis Americans have been asked to consider the appropriate balance between the rights of individuals and the need for national security. The Alien and Sedition Acts of 1798, President Lincoln's suspension of habeas corpus during the Civil War, the Espionage Act of 1917, the internment of Japanese-Americans after Pearl Harbor, and the Smith and McCarren Acts during the McCarthy period all stripped Americans (or some Americans) of some of their most basic civil liberties in an attempt to ensure national security. In each instance the curtailment of freedoms, which may have seemed necessary at the time, became in short order almost universally judged to have been excessive and overreaching, unnecessary if not futile, and a subject of national shame and regret.

Universities themselves have certainly succumbed from time to time to these moods of the nation. During the Cold War years of the 1950s some universities dismissed faculty members for their political beliefs, for their past political affiliations, and for "offensive" speech and publications. Even where such actions were not taken, the possibility that universities would bend to external pressures and make political beliefs a litmus test for academic employment had a chilling effect on discussion and research. Today, at the great research universities we face similar pressures to silence or influence speech by those who are offended or frightened by its content. Why are the tenured faculty at these great universities (who are protected by their tenure) not debating the wisdom of government policies that threaten the fabric of these institutions in the name of national security? Against this backdrop, and without a well-articulated defense of the idea of the research university, conservative organizations such as Campus Watch feel free to attack core values at the most renowned universities. These attacks are followed by demands from political and religious ideologues, as well as uninformed journalists and alumni, that universities fire faculty whose ideas they find repugnant, restrict access to foreign students, and limit or change research agendas.

Universities themselves bear some responsibility for the current situation. As educators we have failed to provide the public with any understanding of our full mission, particularly our research mission. In fact, the overwhelming majority of educated people, perhaps even our own graduates, could not tell us why tolerating opprobrious speech is linked to the conditions that allow us to maintain the vitality and creative energy of university communities. The prevailing views about universities have principally to do with undergraduate education, annual rankings, and intercollegiate athletics—mixed perhaps with some anecdotal knowledge about professional schools. But most Americans have very little idea about the place of research universities in improving the economic, social, political, and cultural life of the country. If these institutions are under threat, then the danger is very great, with enormous consequences for the broader society. Those of us who understand the importance of universities have a great deal of work to do.

In 1967 a group of professors from the University of Chicago wrote what came to be called the Kalven Committee Report—named for Harry Kalven, head of the committee and a distinguished professor of law at Chicago for thirty years. The committee report described the core mission of the University of Chicago and, I daresay, most great research universities, as well as the proper way for universities to address major political issues, particularly in times of great political conflict (like 1967). I would like to quote an excerpt from that brilliantly composed statement and then turn to some specific examples of what is going on at universities today:

> A university has a great and unique role to play in fostering the development of social and political values in society. . . . It is a role for the long term. The mission of the university is the discovery, improvement, and dissemination of knowledge. . . . A university faithful to its mission will provide enduring challenges to social values, policies, practices and institutions. By design and by effect, it is the institution which creates discontent with the existing social arrangements and proposes new ones. In brief, a good university, like Socrates, will be upsetting. . . . The neutrality of the university as an institution arises. . . . not from a lack of courage, nor out of indifference and insensitivity. It arises out of respect for free inquiry, and the obligation to cherish a diversity of viewpoints. And this neutrality as an institution has its complement in the fullest freedom for its faculty and students as individuals to participate in political action and social protest. It finds this complement, too, in the obligation of the university to provide a forum for the most searching and candid discussion of public issues.

This wonderful statement enjoins us to understand that in political matters there can be no single "university position" or voice. A university position tends to have a chilling effect on productive discourse within the community. The essence of the university is to enable debate about unconventional or unpopular views, whether they are political or challenges to received wisdom in any field. Today's opprobrious views may turn out to be tomorrow's received wisdom.

The ethos of the university can be represented, at least in part, in the application of four basic principles that are consistent with the fundamental idea contained in the Kalven Report:

1. *Universalism*: people should be rewarded on the basis of merit, not on the basis of any particular ascribed characteristic that they may have.
2. *Organized skepticism*: you should hold skeptical views toward almost anything that is proposed as fact or dogma or whatever.

3. *Disinterestedness*: individuals at universities shouldn't profit from their ideas directly.
4. *Open communication*: the results of research at universities should enter immediately into the public domain for public debate and, as it were, into the free market of ideas.

Since the Kalven Report was written universities have become truly international. We are recruiting extraordinary young students and faculty from all over the world. This global reach is a source of vitality and different perspectives for the major research universities, as it was earlier in the twentieth century. American science, in the middle of the past century, was the great beneficiary of Germany's purge of "Jewish science" from their universities. An extraordinary intellectual migration was triggered by the German abridgment of fundamental norms of free inquiry and meritocracy. American universities welcomed these extraordinary minds and they transformed many fields of inquiry after taking residence at American research universities. At Columbia today more than 30 percent of Arts and Sciences tenured faculty were born in countries other than the United States. So foreign nationals, both students and faculty, remain a critical source of talent for these universities.

American universities are global magnets for both students and faculty in part because the American research university has been preeminent among higher educational institutions in the world for the past fifty to seventy-five years. In a work called *The University: An Owner's Manual* (1990), Henry Rosovsky—formerly dean of Arts and Sciences at Harvard—estimated that perhaps two-thirds or more of the top twenty or thirty universities in the world were in the United States. I think that the ratio might be higher today. The vast majority of Nobel Prizes go to scientists working at American universities.

We should not forget the accomplishments of these universities. Of the ten leading industrial sectors of the United States today, 90 percent of what they are achieving results directly from discoveries that are coming out of research universities. (A century ago, that might have been true of two of the most important sectors.) University-based scientific and engineering research has produced the discoveries that are linked directly to the major advances in health care, high technology, and rates of economic growth that we have benefited from over the past half-century. The successful social policy interventions have also grown out of social science and public health research at these universities.

If we succumb to the pressure now put upon us, I believe that the preeminence of American universities may be at risk. And the risks come from a set of policies that have been implemented by the Bush administration and the Republican-controlled Congress in the name of national security. They threaten to undermine some of the core values that universities cannot abandon without significant negative consequences.

I will draw on the Columbia experience, as I know it best, to illustrate threats in a series of dimensions to the core values of the university.

At Columbia we have had a series of incidents over the past year that resulted in alumni, journalists, and others attacking the university for standing behind its faculty and defending the value of free inquiry. One can be sure that any public statement in support of the Palestinian people by the preeminent literary critic Edward Said will illicit hundreds of e-mails, letters, and journalistic accounts that call on us to denounce Said and to either sanction or fire him.

The Said story is familiar. But now it is taking on a new importance as a harbinger. This year Columbia was hosting a poet—an Oxford don named Tom Paulin. While he was visiting Columbia he was invited to give a lecture at Harvard about his poetry. Before he arrived, it came out that he had previously written poems and made some speeches that took a strong position against Israeli settlements in the West Bank. The revelations provoked a public outcry, in Boston and elsewhere. Harvard rescinded its invitation to Paulin (which was subsequently reinstated) and Columbia received hundreds of letters and e-mails calling for his immediate dismissal. A few trustees raised questions about the role, if any, that political positions had in the hiring process—more as an inquiry than as a request for Paulin's dismissal. In fact, the Columbia trustees have stood fast in supporting the defense of academic freedom against these external attacks. Some alumni withheld donations to the university; others threatened to do so.

In another case we were competing with the University of Chicago for a very distinguished historian of the modern Middle East, Professor Rashid Khalidi. He has taken positions on the Palestinian-Israeli conflict but has also been very active in the peace movement, for example as a representative of the Palestinians at the Oslo peace effort. Most importantly, Khalidi was enormously admired for the quality of his published work on Palestinian identity within the community of historians. But when it became known that we were recruiting Khalidi to Columbia the complaints started flowing in from people who disagreed with the content of his political views. Needless to say, we defended Khalidi and welcomed his acceptance of our offer to join the Columbia faculty.

Recently, when some documentary filmmakers offered a Palestinian film festival at Columbia, sponsored by Columbia faculty members, we received countless protests: "How can you possibly have these supporters of murderers and terrorists on the campus, representing Palestinians?" What they left out (or didn't know) was that the following month the same sponsor, the Department of Middle East Languages and Cultures, was holding an event to celebrate the retirement of an extremely distinguished Columbia historian of Jewish history.

Of course, those who protest these activities at Columbia have a right to do so, and they should be responded to in an appropriate way. But I want to emphasize that the letters, e-mails, and other communications are not simply from individuals who are spontaneously reacting to news of these events. They are coming from websites that are now supplying the text to be used in the protest messages. They are the result of organized efforts to pressure universities to take action against professors whose ideas they find repugnant.

The sources of protest are not limited to members of the public who object to the political views of professors or to activities taking place on the Columbia campus. At a recent antiwar teach-in a Columbia assistant professor of anthropology, Professor Nicholas de Genova, spoke against the U.S. war against Iraq and made a comment that was carried all over the country. He said that he wanted to see a million Mogadishus. De Genova's remarks were immediately—*immediately*—criticized as totally inappropriate by other distinguished faculty members who took part in the teach-in. Those refutations were largely ignored in the press. If I received a thousand emails over the Palestinian film festival, I must have received five times that many this time. But this case is important because the type of protest took on a different character than the receipt of irate e-mails from members of the public and alumni. In this case we received a letter—that is to say, Lee Bollinger, Columbia's president, received a letter—from 104 Republican members of the House of Representatives asking that de Genova be fired. The letter said that he "has brought shame on the great institution that Columbia University is. As an assistant professor, de Genova has not yet earned the promise of lifetime academic employment. We hope that you will take steps immediately to ensure that he never gets it." It is deeply troubling that nearly a quarter of the members of the House of Representatives should have such a profound misunderstanding of the basic principles governing a university—in particular, the process of self-policing through application of organized skepticism that actually worked at Columbia in this case through the criticism of this speech by colleagues.

Troubles, of course, are hardly confined to Columbia. This past winter university officials at the University of California, Berkeley, refused to allow a fundraising appeal for the Emma Goldman Papers Project because the appeal quoted Goldman on the subjects of suppression of free speech and her opposition to war. The university deemed the topics "too political" as the country prepared for possible military action against Iraq. After considerable protest around the country, they rescinded that position.

People who would have us fire or censure professors because of their political opinions and remarks often fail to understand that they are the current beneficiaries of a predominant point of view. But if content and ideology become the basis for hiring and firing decisions at universities, the tables can turn quickly. The moment has rarely failed to arrive when the prosecutors become the prosecuted. People must be able to imagine that their thoughts, beliefs, and speech might make them the victims of the unbridled power of the government of a university or of a nation. And we, in defending the idea of the university, must educate the public about why we defend the faculty whose ideas offend many people.

The troubles that universities face today are not limited to free speech on campus. The Bush administration and Republicans in Congress are intensifying their scrutiny of research projects that focus on sensitive subjects. They are intruding in the long-established practice of peer review, which is used to determine the scientific merit of research proposals. For example, a recently funded NIH project on sex trafficking and the transmission of HIV (with an interest in intervention and prevention) produced an inquiry from a staff member of the House Subcommittee on Criminal Justice, Drug Policy, and Human Resources. Here is an excerpt from the inquiry to the NIH:

> The Subcommittee strongly supports President Bush's efforts and is gravely concerned about the efforts at the National Institutes of Health that contradict the President's mission and instead seek to legitimize the commercial sexual exploitation of women. . . . The behaviors being examined by the NIH are immoral and illegal, as they should be, in the United States. Knowledge of such illegal exploitation should be reported to the appropriate legal authorities for investigation and prosecution. The NIH and its collaborator on this project are instead providing legitimacy and financial support to the continuation of the sex trade. . . . Please provide the Subcommittee with the following information: (1) Ethical reviews, if any, that NIH conducted for the San Francisco and Miami studies. (2) The name(s) of the NIH employee(s) who approved funding for the San Francisco and Miami studies, including the names of the individuals on the panels that reviewed the studies' applications. (3) A list of all efforts, if any, by the NIH and collaborators on these studies to notify law enforcement of illegal activities being conducted that were observed or witnessed. . . . (6) A full listing (including funding amounts) of all NIH funded studies over the past decade involving commercially sexually exploited women, including prostitutes or "commercial sex workers."

What I am suggesting is that in Washington there are increasing efforts to compromise the peer review system and to introduce ideological or political criteria into the selection process determining recipients of federally funded research—a system designed to be meritocratic and free of political influence.

The intrusion of ideology into the federal support for science can also be seen in President Bush's decision that federal funds could be used to support university-based research on human embryonic stem cells *only* for a very limited existing set of cell lines. In effect, for ideological reasons the administration put the brakes on one of the most promising lines of biomedical research. As Gerald D. Fischbach, Columbia's executive vice president for Health Sciences, wrote in a recent *Newsweek* opinion piece,

The cost in dollars of delaying new stem-cell research is difficult to estimate. It might measure in the hundreds of billions of dollars. . . . A less obvious, but real, cost is the damage to the fabric of America's extraordinary culture of inquiry and technical development in biomedical sciences. . . . A crippled research enterprise might add an unbearable stress with long-lasting effects on the entire system. If revolutionary new therapies are delayed or outlawed, we could be set back for years, if not decades. To steer clear of controversy, some investigators will redirect their research. Others will emigrate to countries where such research is allowed and encouraged. Some will drop out entirely.

National concerns over potential bioterrorism, as well as apprehension that the government could legislate restrictions on research publication, has led scientific societies and editorial boards of high-impact scientific journals to consider policies of self-censorship. The American Association of Microbiology recently issued a statement indicating that it may self-censor articles with possible bearing on bioterrorism—before publication and despite recommendations for publication from the peer review system.

Perhaps some form of internal policing is worth considering. There was useful discussion of the balancing of scientific publication and national security concerns at a recent meeting at the National Academies of Sciences. But few participated in this debate, the results of which could have very serious consequences for the fundamental value of open communication in science. Who should decide if scientific papers pose problems to national security if they are published? How should such extreme action be organized, and for what expected duration of time? Where should the burden of proof that publication could harm national security lie, and what forms of evidence should be required before restraints on publication are considered? I have heard almost no discussion of these issues among the faculty at Columbia.

Now let me refer to relatively new pieces of federal regulations, presidential directives, and Congressional legislation, including the USA PATRIOT Act, that allow increased surveillance of faculty and students and increased government intrusion into the substance and conduct of research at universities. First, as some of you know, provisions of the Patriot Act hold that, on account of their national origin and without any demonstration that they pose risks to national security, foreign students from about twenty-five nations must now be denied access to scientific research laboratories that use "select agents" (biological agents and certain toxins)—agents that might be usable for purposes of bioterrorism. The law prohibits these students from even entering these laboratories. The prohibition is based solely on national origin. National security trumps equal protection. And if a faculty member permits a prohibited student to enter a laboratory that is using select agents, that faculty member, too, is open to criminal penalties.

Second, in addition to classified research (which most research universities refuse to accept outside of affiliated laboratories), the federal government has created a new category: "sensitive but unclassified research." There is significant concern that federally funded research so designated may be subject to government scrutiny and that publication of research findings obtained under such contracts could be impeded or prevented. A number of research universities, including MIT, have refused to accept government contracts designated as "sensitive but unclassified." Other universities have accepted these contracts despite the inclusion of provisions for funding-agency review of findings prior to publication. The universities are not arguing that the government should not conduct classified research for defense purposes, but they are objecting to the creation of ambiguous categories that allows the government to implicate itself into the system of open communication of scientific results.

The Patriot Act also modifies the Family and Educational Rights Privacy Act and requires that educational institutions disclose educational records to federal law enforcement agents without notifying students that they are doing so and without the students' consent. In fact, the government—particularly Attorney General John Ashcroft and his office—has been employing what are known as "national security letters" that authorize the attorney general or a delegate, with no judicial approval, to compel production of substantial amounts of relevant information. The government at this time refuses to give us information about how extensively these "letters" are being used on university campuses.

Under provisions of these laws the government can now investigate library records (to learn who is taking out books from various libraries) without informing individuals, as well as examine the content of e-mail records. Librarians who are required to release this information may not report on these activities to the people whose records are being identified.

The new legislation places most foreign students under a microscope when they request student visas. Continuing students in good standing at their universities are fearful of returning home for holidays and summer breaks for fear of not being permitted reentry into the United States. The Enhanced Board of Security and Visa Entry Reform Act of 2002 requires that entering students be scrutinized through a system called SEVIS, which is supposed to have been put in place several months ago and which would gather information that would enable the government to track these individuals. But this system is not fully operational and is preventing students from getting visas or delaying them to the point that they either miss their time in school or must cancel their visits. It also is limiting access to universities by scholars who are traveling to the United States to take part in university activities.

I am also hearing about intrusions on privacy, increased surveillance, and personal intimidation in the name of national security. For example, I have now received several calls from Columbia faculty members—some are American citizens of Iraqi or Middle Eastern descent—reporting visits to their homes by FBI agents who are conducting investigations. While there have been no reports that the agents have acted improperly, these faculty members express fear and apprehension—fear about their privacy, apprehension that they may be under surveillance, and concern that they may be subjected to continuing visits from the FBI.

If these new laws and regulations stand up to judicial review, university faculty and students will have lost some of the degrees of personal privacy and intellectual liberty we now enjoy. Are we prepared to relinquish personal privacy and academic freedom to secure some vaguely articulated increase in national security? Do these new laws and regulations accomplish that? What effect will they have on the growth of knowledge and on the intellectual environment at universities? What, in fact, is the threat to national security that is posed by students and faculty at our universities? What evidence is there that select agents and toxins used in American scientific laboratories for legitimate purposes pose a real threat to national security and require that we deny students access to that research opportunity because of where they were born? Are there students with links to terrorist organizations studying at our universities? What evidence is there to support such a claim and do the probabilities of a vague potential threat warrant the types of measures being taken to limit free inquiry, open communication, and individual privacy? These are questions that must be considered by us as individuals and as members of American research universities. At the least, the academic community should not allow these measures to be put into place in silence.

The mission of a great university in our society is to create and disseminate new knowledge through research and teaching—and to lead debates that have broader implications for peoples' values, ethics, and behavior. Over decades research universities have evolved a value system to optimize their effectiveness in fulfilling their mission. Without freedom of expression, without open communication, without free inquiry, without open access to people of talent (regardless of their personal characteristics), we are doomed to accept received wisdom and current dogma. In our society the high calling of intellectuals and scholars is to challenge received wisdom, political correctness, and intellectual complacency; to be skeptical about claims of "fact" and "truth"; to question presuppositions and biases of others as well as their own. The growth of knowledge, insight, and understanding is better served through the clash of ideas than through the blind acceptance of dominant ideologies and the silencing of criticism. In fact, without free exchange, open communication, and meritocracy we cannot distinguish between truth and falsity. Those who believe they can define what speech or experiment is "good" or "evil," what speech or "fact" is "true" or "false," and what speech or experiment is causally related to specific violent acts in other parts of the world are mistaken about their own enterprise. Truth rests less in product than in process.

I believe it is time for the members of our faculties who believe in the values embedded in the research universities to engage in a debate on the wisdom of the laws that the government is enacting in the name of national security. And if they see these new constraints as deeply problematic, they should question them and try to have them changed. It is now time for members of the tenured faculty in particular to address these issues and to speak out. Silence betrays an acquiescence or indifference to the policies represented in the Patriot Act and subsequent "patriot acts." What is the protection of tenure for, if not to be used to voice one's opinions in these troubled times—to participate in the debate and in the defense of the university?

CHAPTER 28

ACADEMIC FREEDOM UNDER FIRE

(2005)

JONATHAN R. COLE

Today, a half-century after the 1954 House Un-American Activities Committee held congressional hearings on communists in American universities, faculty members are witnessing once again a rising tide of anti-intellectualism and threats to academic freedom.[1] They are increasingly apprehensive about the influence of external politics on university decision-making. The attacks on professors like Joseph Massad, Thomas Butler, Rashid Khalidi, Ward Churchill, and Edward Said, coupled with other actions taken by the federal government in the name of national security, suggest that we may well be headed for another era of intolerance and repression.

The United States paid a heavy price when the leaders of its research universities failed in the 1950s to defend the leader of the Manhattan Project, J. Robert Oppenheimer; the double Nobel Prize chemist Linus Pauling; and the China expert Owen Lattimore. But a wave of repression in American universities today is apt to have even more dramatic consequences for the nation than the repression of the Cold War.

Compared to today, universities during the McCarthy period were relatively small institutions that were not much dependent on government contracts and grants. In the early 1950s, Columbia University's annual budget was substantially less than $50 million. Its annual budget is now roughly $2.4 billion, and more than a quarter of this comes from the federal government, leaving research universities like it ever more vulnerable to political manipulation and control.

Universities today are also more deeply embedded in the broader society than ever before. They are linked to industry, business, and government in multiple ways. Their links to the larger society inevitably lead to public criticism of the university when faculty members or students express ideas or behave in ways that some in the public find repugnant.

Can the leaders and the tenured faculty of our great research universities rise to the challenge of rebutting such criticism? Can we do better at defending academic freedom than our predecessors did in the 1950s?

To do so, we must convince the public that a failure to defend dissenting voices on the campus places at risk the greatest engine for the creation of new ideas and scientific innovation the world has ever known. We must explain that one can never know the true worth of an idea unless one is free to examine it. We must explain that such freedom of inquiry is a key to innovation and progress over the long term in the sciences as well as the humanities. Above all, we must show that a threat to academic freedom poses a threat as well to the welfare and prosperity of the nation.

The preeminence of American universities is an established fact. It was recently reaffirmed in a 2004 study conducted in China at Shanghai Jiao Tong University that evaluated five hundred of the world's universities. The United States has 80 percent of the world's twenty most distinguished research universities and about 70 percent of the top fifty. We lead the world in the production of new knowledge and its transmission to undergraduate, doctoral, and postdoctoral students. Since the 1930s, the United States has dominated the receipt of Nobel Prizes, capturing roughly 60 percent of these awards.

Our universities are the envy of the world, in part because the systems of higher education in many other countries—China is a good example—do not allow their faculty and students the extensive freedom of inquiry that is the hallmark of the American system. As a consequence, our universities attract students from all over the world who either remain in this country as highly skilled members of our society or return home to become leaders in their own countries and ambassadors for the United States. The advanced graduates of the American research university populate the world's great industrial laboratories, its high-tech incubator companies, and its leading professions. Many of the emerging industries on which the nation depends to create new jobs and maintain its leading role in the world economy grow out of discoveries made at the American research university.[2] The laser, the MRI, the algorithm for Google searches, the Global Positioning System, the fundamental discoveries leading to biotechnology, the emerging uses of nanoscience, the methods of surveying public opinion, even Viagra—all these discoveries and thousands of other inventions and medical miracles were created by scholars working in the American research university.

Unfortunately, most leaders of higher education have done a poor job of educating the public about the essential values of the American research university. They have also failed to make the case for the research university as the incubator of new ideas and discoveries. As a consequence, when a professor comes under attack for the content of his or her ideas, the public has little understanding of why the leader of a research university, if he or she is to uphold the core principle of academic freedom, *must* come to the professor's defense.

Attacks on academics follow a clear pattern: A professor is singled out for criticism. This is followed by media coverage that carries the allegations to larger audiences. The coverage is often cursory and sometimes distorted. Some citizens conclude that the university harbors extremists who subvert our national ideals. Pressed by irate constituents, political leaders and alumni demand that the university sanction or fire the professor. This is an all-too-familiar story in our nation's history.

The recent attack on Professor Joseph Massad of Columbia University offers a perfect example of how this process unfolds. The drama began with a group called the David Project, which was launched in 2002 "in response to the growing ideological assault on Israel." The Project subsequently produced a one-sided twenty-five-minute film, *Columbia Unbecoming*, in which former students accused Professor Massad of inappropriate behavior in his elective course, "Palestinian and Israeli Politics and Societies." One former student alleged on camera that Massad used "racial stereotypes" and "intimidation tactics . . . in order to push a distinct ideological line on the curriculum"; another asserted that he had crossed the line "between vigorous debate and discussion, and hate."[3]

The David Project distributed this film to the media, and one-sided stories soon began to appear in conservative publications such as the *New York Sun*. This triggered follow-up stories in the *New York Times*, the *Chronicle of Higher Education*, and other local, national, and international news outlets. One, appearing in the Sunday edition of the *New York Daily News* on November 21, 2004, bore the headline "HATE 101."

Not every story about Massad was this crude. A correspondent for the *Jewish Week*, for example, interviewed an Israeli student at Columbia who strongly defended Massad. "The class was an incredible experience," this student reported. "It wasn't fun to be the only Israeli in class, but I never felt intimidated. Passionate, emotional, but not intimidated."

Unfortunately, these nuanced accounts could not compete with strident headlines about hate. At one point, Congressman Anthony D. Weiner, a New York Democrat, asked Columbia president Lee Bollinger to fire the untenured Professor Massad as a way of demonstrating Columbia's commitment to tolerance. The irony was seemingly lost on Mr. Weiner, who had the audacity to write, "By publicly rebuking anti-Semitic events on campus and terminating Professor Massad, Columbia would make a brave statement in support of tolerance and academic freedom."

Weiner's Orwellian ploy—of calling intolerance "tolerance"—must be seen in a broader context. There is a growing effort to pressure universities to monitor classroom discussion, create speech codes, and, more generally, enable disgruntled students to savage professors who express ideas they find disagreeable. There is an effort to transmogrify speech that some people find offensive into a type of action that is punishable.

There is of course no place in the American research university classroom for physical intimidation, physical assault, or violations of the personal space of students. There is no place for faculty members to use their positions of authority to coerce and cow students into conforming to their own point of view. No university will protect a professor's use of a string of epithets directed toward a particular student in a gratuitous manner that is unrelated to the substance of the course. There are workplace rules in place at universities that govern and control such forms of behavior. And there must be, by law, mechanisms for students or others at the university to lodge complaints against professors who violate these rules. This basic commitment to civility and professional responsibility is part of the code of conduct at Columbia and at every other major American research university.[4]

But the codes that place limits on conduct must *never* be directed at the content of ideas—however offensive they may be to students, faculty, alumni, benefactors, or politicians.

Critics of the university, such as those affiliated with the David Project, tend to blur the distinction between speech and action. They accuse professors of inappropriate action and intimidation when they are actually trying to attack the content of their ideas. They also tend to expropriate key terms in the liberal lexicon, as if they were the only true champions of freedom and diversity on college campuses.

Consider Students for Academic Freedom (SAF), an organization launched by veteran conservative activists. The group's very name implies a commitment to a core liberal value, just as the group's tactics promise to empower aggrieved students. Currently, it encourages students nationwide to organize and lobby university leaders, alumni, and members of state legislatures to adopt a "student bill of rights."

But SAF's language and tactics are misleading. Under the banner of seeking balance and diversity in the classroom, these students are trying to limit discussion of ideas with which they disagree. They want students to become judges, if not final arbiters, of faculty competence. They have supported the campaign against Massad at Columbia and have urged students to report "unfair grading, one-sided lectures, and stacked reading lists" as an abuse of student rights.

While I was provost at Columbia there were many efforts by outside groups to influence university policy and to silence specific members of the faculty. Repeated efforts were made to defame and discredit the renowned literary critic and Palestinian advocate Edward Said. External groups tried, but failed, to have Columbia deny an appointment to the eminent Middle East historian Rashid Khalidi. Sixty-two members of Congress wrote to Columbia calling on us to fire Nicholas de Genova, a professor of anthropology, after he made inflammatory remarks at an antiwar teach-in prior to the most recent Iraq War—–even though his remarks were immediately criticized at the same teach-in by other Columbia faculty members.

Even when nobody loses his or her job, these assaults take a toll. As Professor Massad explains on his website, "With this campaign against me going into its fourth year, I chose under the duress of coercion and intimidation not to teach my course ['Palestinian and Israeli Politics and Societies'] this year."

Most of the recent attacks on university professors have been leveled against social scientists and humanists. Many critics of the university seem to believe that sanctioning one group of professors will have no effect on those in other disciplines. This is dangerously naive, both in principle and in practice.

The stakes are high. The destruction of university systems has historically been caused by the imposition of external political ideology on the conduct of scholarly and scientific research. Defense of faculty members in the humanities and social sciences from external political pressure protects *all* members of the university community.

History suggests that the natural sciences, too, can be infected by political pressures to conform to ideological beliefs. German universities still have not recovered from the catastrophe of 1933 when Hitler began to dismantle German science and technology by purging those researchers who did "Jewish science." Japanese universities were damaged immeasurably in the 1930s by the purging of dissident intellectuals. Soviet biology never fully recovered from the imposition of Lysenkoism into the biological sciences.

Today, political pressure to include "creationism" and theories of "intelligent design" as alternatives to Darwinian evolution in the secondary school science curriculum has already led to a purging of Darwin's theory from the science curriculum in at least thirteen states. The National Academies of Sciences and the Union of Concerned Scientists have cataloged many examples of Bush administration interference with research and education. Consider just a few examples: Foreign students and scholars from "suspect" nations are harassed and even denied entry into the United States without a scintilla of evidence that they are security risks. American professors are prevented from working with gifted foreign scientists and students. Open scholarly communication is impeded by policies designed to isolate nations supporting terrorism; library and computer records are searched; political litmus tests are used by the Bush administration to decide who will serve on scientific advisory committees; and scientific reports whose content is inconsistent with the Bush administration's ideology have been altered. Even though the National Institutes of Health supported the research, some members of Congress almost succeeded in rescinding funding for projects on HIV/AIDS. Another recent bill, House Resolution 3077, almost succeeded in mandating direct government oversight of university "area studies" programs (the bill passed the House but died in the Senate).

These attacks should be related to still other threats to scientific inquiry. The USA PATRIOT Act and the Bioterrorism Defense Act have, for example, led to the criminal prosecution of Dr. Thomas Butler, one of the nation's leading experts on plague bacteria. Butler faced a fifteen-count indictment for violating the Patriot Act's provision requiring reporting on the use and transport of specific biological agents and toxins that in principle could be used by bioterrorists. Butler was acquitted of all charges related to the Patriot Act, except for a minor one—his failure to obtain a transport permit for moving the bacteria from Tanzania to his Texas laboratory, as he had done for the past twenty years. However, while investigating Butler's work with plague bacteria, the FBI combed over everything in his lab at Texas Tech University, reviewed all of his accounts, and added on fifty-four counts of tax evasion, theft, and fraud unrelated to the Patriot Act. His conviction was based on the add-on counts. The upshot of all of this was that he lost his medical license, was fired from his job, and now, if he loses his appeal, faces up to nine years in jail.

In another case, Attorney General John Ashcroft publicly targeted Dr. Steven J. Hatfill of Louisiana State University as "a person of interest" in the anthrax scare that followed 9/11. Although Hatfill has never been charged with any crime, LSU fired him because of the accusation and intervention of the Justice Department. Other faculty members at other institutions have suffered through unannounced and intimidating visits from the FBI to their homes or campus offices.

These crude efforts to enforce the Patriot Act have already had some serious consequences. Robert C. Richardson, whose work on liquid helium earned him a Nobel Prize in Physics, has described the atrophy of bioterrorism research at Cornell:

> The Patriot Act, which was passed after 9/11, has a section in it to control who can work on "select agents," pathogens that might be developed as bioweapons. At Cornell [before 9/11], we had something like 76 faculty members who had projects on lethal pathogens and something like 38 working specifically on select agents. There were stringent regulations for control of the pathogens— certain categories of foreign nationals who were not allowed to handle them, be

in a room with them or even be aware of research results. So what is the situation now? We went from 38 people who could work on select agents to 2. We've got a lot less people working on interventions to vaccinate against smallpox, West Nile virus, anthrax and any of 30 other scourges.[5]

Is our national security enhanced when the government turns our best immunology and biodefense laboratories into ghost towns?

In an atmosphere of growing fear and intimidation, we would be wise not to dismiss these attacks on the American research university as mere aberrations. Indeed, universities are fragile institutions, and they have historically caved in to external political pressure at key moments—as they did during the Red Scares that followed the two world wars.[6]

Periodically, often during times of national fear, political leaders and ideologues on the Right and the Left have silenced dissent and pressured universities to abandon their most fundamental values of free and open inquiry. Most university leaders and faculty members fell easily into line during the first Red Scare of 1919–1921 and the reign of Joseph McCarthy. As the historians Ellen Schrecker and Sigmund Diamond have shown, presidents and trustees of research universities often publicly espoused civil liberties, academic freedom, and free inquiry while privately collaborating with the FBI to purge faculty members accused of holding seditious political views.[7]

Some university leaders underestimated the gravity of the threat and bowed to wealthy benefactors who threatened to withdraw their support. Others dismissed professors out of fear of bad publicity. Still others supported these purges because they believed in them. For example, Cornell's president, E. E. Day, maintained that "a man who belongs to the Communist Party and who follows the party line is thereby disqualified from participating in a free, honest inquiry after truth, and from belonging on a university faculty devoted to the search for truth." Yale's president, Charles Seymour, proclaimed, "There will be no witch hunts at Yale because there will be no witches. We do not intend to hire Communists."

Robert Maynard Hutchins, chancellor of the University of Chicago, was one of the few great heroes during those perilous times. In 1949, testifying before a state commission investigating communists on campus, he boldly argued for tolerance:

> The danger to our institutions is not from the tiny minority who do not believe in them. It is from those who would mistakenly repress the free spirit upon which those institutions are built. . . . The policy of repression of ideas cannot work and has never worked. The alternative is the long, difficult road of education.

On another occasion, Hutchins observed that the problem with witch hunts was "not how many professors would be fired for their beliefs, but how many think they *might* be. The entire teaching profession is intimidated."

Hutchins's boss, Laird Bell, chairman of the University of Chicago's Board of Trustees, was equally outspoken: "To be great," he declared,

> a university must adhere to principle. It cannot shift with the winds of passing public opinion. . . . It must rely for its support upon a relatively small number of people who understand the important contributions it makes to the welfare of the community and the improvement of mankind: upon those who understand that academic freedom is important not because of its benefits to professors but because of its benefits to all of us.

What, then, are the defining principles that guide the work of the university? As scholars and scientists, we place a premium on openness, rigor, fairness, originality, and skepticism. We are part of an international community of scholars and scientists whose ideas transcend international borders. We collaborate and exchange ideas with Iraqis, Russians, Iranians, Chinese, and Israelis without considering politics or nationality. We hold that members of our community must always be free to dissent—to pursue and express new and even radical ideas in an environment of unfettered freedom.

By the same token, proponents of new ideas and their critics must be free to disagree. And this is especially true in the classroom, in which faculty and students must be free to explore and develop their ideas in robust and uninhibited debate. By encouraging independent thinking, no matter how preposterous or outrageous, the university promotes trust, creativity, collaboration, and innovation.

The goal is to establish an environment in which it is possible for the inquisitive mind to flourish. In contrast to private enterprise, the university places the welfare of the community above individual gain. The coin of the academic realm is the recognition that professors and students receive based on the quality of their contributions to the creation, transmission, and understanding of knowledge. The university is a meritocracy. Ideally, quality of mind expressed through teaching, research, and learning is rewarded without regard to race, religion, nationality, or gender.

This does not mean, of course, that real merit is always rewarded: like any complex institution, the modern university does not always function as it is meant to. But it is simply ridiculous to perpetuate the myth that research universities are rogue institutions that operate in an uncontrolled environment. Most of them are probably more accountable for their products and for their financial transactions than most large American corporations.[8]

Universities are evaluating themselves from dawn to dusk. State and regional accrediting agencies are continually reviewing the academic quality of university programs and faculties. Federal funding agencies conduct extensive peer reviews of grant applications that evaluate the quality of applicants' prior work, the quality of the proposals submitted, and the potential value of the work when completed; they use site visits to review elaborate proposals before funding large centers or university institutes. Obsessed with knowledge about their reputation and quality, research universities use ad hoc or standing committees of experts to evaluate the quality of the curriculum, the quality of the faculty, and the quality of departments and schools. The scientific and scholarly papers and monographs of faculty members are peer reviewed before they are accepted for publication and are assessed in terms of the potential impact of this work on the field. The results of course evaluations by undergraduate and graduate students are part of the "teaching portfolios" that are used in deciding on the promotion and tenure of junior faculty members. Finally, there is accountability for personal conduct: students and colleagues can file grievances of discrimination with deans, department chairs, ombuds officers, the university senate, and the EEOC, among other outlets for claims of inappropriate behavior.

The governing role played by peers makes universities different from most other American institutions. The research university was founded on the idea that professors should regulate their own affairs. This aspiration has never been fully realized. But it is plainly evident in the tradition that those who oversee the core academic work of the university—the president, the provost, the deans, and the department chairs—are themselves distinguished scholars and teachers who are respected members of the faculty. Moreover, university leaders govern by persuasive and delegated authority, not by the exercise of power.

Another essential feature of the American research university is that no one speaks "for" the university—not even its official leaders. While the president and the provost and the board of trustees have the responsibility and the authority to formulate and carry out university policies, the essence of a university lies in its multiplicity of voices: those of its faculty, its students, its researchers, and its staff. Presidents and provosts are often asked questions of the following kind: "What is the university's position on the writings, or remarks, or actions of Professor X?"

In fact, there is no "university position" on such matters. The university does not decide which ideas are good and bad and which are right and wrong. That is up for constant debate, deliberation, and discourse among the faculty and students. For the university to take such positions would stifle academic freedom and alienate those whose views differ from those of the institution's leaders. The responsibility of these leaders is not to decide whose ideas are best but to create an environment in which all ideas may be explored and tested.

First and foremost, the American research university is designed to be unsettling. Was this not Socrates's purpose as well? Because it is committed to the creation of *new* knowledge and the intellectual growth of its students, the university must nurture the expression of novel and sometimes startling ideas and opinions. Lionel Trilling, the preeminent literary critic, wrote in *Beyond Culture* about the contentious nature of the literature sometimes taught at the university:

> Any historian of literature of the modern age will take virtually for granted the adversary intention, the actual subversive intention, that characterizes modern writing—he will perceive its clear purpose of detaching the reader from habits of thought and feeling that the larger culture imposes, of giving him a ground and a vantage point from which to judge and condemn, and perhaps revise, the culture that has produced him.

Whether in 1965, when this was published and Trilling taught at Columbia, or today, the mission of the American research university is to encourage faculty and students to challenge prevailing values, policies, beliefs, and institutions. That is why the university will always have—and must welcome—dissenting voices and radical critics.

Researchers at America's universities do not generally investigate questions for which there are "right" or ready answers—answers at the back of a book. The goal of academic discourse is not merely to convey information but to provoke, to stimulate ideas, and to teach students to *think* and provide them with the intellectual and analytical tools that will enable them to think *well*. Great teachers challenge their students' and colleagues' biases and presuppositions. They present unsettling ideas and dare others to rebut them and to defend their own beliefs in a coherent and principled manner. The American research university pushes and pulls at the walls of orthodoxy and rejects politically correct thinking. In this process, students and professors may sometimes feel intimidated, overwhelmed, and confused. But it is by working through this process that they learn to think better and more clearly *for themselves.*

Unsettling by nature, the university culture is also highly conservative. It demands evidence before accepting novel challenges to existing theories and methods. The university ought to be viewed in terms of a fundamental interdependence between the liberality of its intellectual life and the conservatism of its methodological demands. Because the university encourages discussion of even the most radical ideas, it must set its standards at a high level. We permit almost any idea to be put forward—but only because we demand arguments and evidence to back up the ideas we debate and because we set the bar of proof at such a high level.

These two components—tolerance for unsettling ideas and insistence on rigorous skepticism about all ideas—create an essential tension at the heart of the American research university. It will not thrive without *both* components operating effectively and simultaneously.

Here we must acknowledge an area where the university today faces a real and difficult problem with the mechanisms it uses to evaluate the work of its scholars. For the threats to free inquiry do not come only from government policies, from local or national politicians, from external lobbying groups, or from lazy journalism. Some of the most subtle threats come from within the academy itself.

For example, an unspoken but widespread aversion to airing topics that are politically sensitive in various fields sometimes limits debates that ought to take place. The growth of knowledge is greatly inhibited when methodological thresholds for evidence are relaxed, and claims to truth are advanced on the basis of shoddy evidence or on the basis of supposedly possessing privileged insight simply as a result of one's race, gender, religion, or ethnicity.

Most scholars and scientists at leading universities would more than likely exercise their right to remain silent before placing on the table for debate any number of controversial ideas: for example, the idea that differences in educational performance between different racial groups are not a result of discrimination; that occupational differentiation by gender may be a good thing; that dietary cholesterol above and beyond genetic predispositions has only a minimal effect on coronary heart deaths; that the children of crack cocaine mothers will nevertheless experience normal cognitive development; or, until recently, that prions, as well as bacteria or viruses, can cause disease.

I have suggested that we entertain radical and even offensive ideas at universities because we simultaneously embrace rigorous standards in determining the adequacy of truth claims. But if scholarly skepticism is sometimes compromised by a lack of courage or an intolerance for competing points of view, then the primary mechanisms by which universities ensure the quality of research will not always reliably function. To complicate matters, different disciplines have evolved somewhat differently in institutionalizing mechanisms to ensure that rigorous standards exist to evaluate ideas and the results of research.

Biologists may broadly agree that advocates of creationism are simply in error and that the theory they defend is unworthy of serious scientific debate, while social scientists are more likely, for example, to disagree about the scholarly merits of theories that stress the influence of socialization rather than innate abilities on individual achievement. As new areas of research and inquiry appear in the modern university and begin to dominate their disciplines, the definition of acceptable research questions may well change, as may definitions of what is acceptable methodology, acceptable evidence, acceptable standards of proof, and also acceptable peer reviewers (who in turn will judge whether a given scholar's methodology and use of evidence is acceptable). As a statistician might put it, whoever owns the "null hypothesis" often determines what is taken for fact.

When skepticism falters or fails, does the academic community, even in the longer run for which it is built, have the mechanisms to correct its errors?

This has to be an open question. Currently, there is broader agreement about the appropriate corrective mechanisms in the natural sciences than in the humanities and social sciences, although in periods of what Thomas Kuhn called revolutionary rather than normal science we often also find sharp disagreements within natural science over standards of proof and truth claims. It is the very possibility of ongoing disagreement, however, that is a primary justification for protecting and promoting freedom of thought. John Stuart Mill put it this way:

> Truth, in the great practical concerns of life, is so much a question of the reconciling and combining of opposites, that very few have minds sufficiently capacious and impartial to make adjustment with an approach to correctness, and it has to be made by the rough process of a struggle between combatants fighting under hostile banners.[9]

Moreover, as Mill well knew, it is more important to tolerate an occasional error in the current appraisal of conflicting ideas than to risk compromising free expression. For in the long run, it is unfettered freedom of inquiry that ensures innovation, intellectual progress, and the continued growth of knowledge.

I have defended the right of academic freedom within the community of scholars. But what, if any, right to freedom of expression does a student have against his or her professor? The rise of groups like Students for Academic Freedom raises this important question.

Students clearly have the right—indeed, the obligation—to enter the general debate within the university community. They have the right to express their ideas forcefully in the classroom and to argue against their professor's views. I've made the point that professors in the classroom must never discriminate against students on the basis of their ascribed characteristics—simply, on the basis of who they are in terms of their race, ethnicity, religion, or gender.

At the same time, there is a clear differentiation of roles between professors and students. We expect professors, not students, to offer their own best judgment on competing truth claims. A student may argue for creationism or intelligent design, but that does not oblige his or her biology professor to take his or her views seriously as a rival to the evolutionary accounts favored by virtually all contemporary biologists. Similarly, a professor of Jewish history is under no obligation to take seriously the arguments of a student who denies the Holocaust.

586 ACADEMIC FREEDOM AND FREE INQUIRY

What, then, about a student who says he or she is being discriminated against by Professor Massad of Columbia because Massad declares the student's position on the 1982 Shatila massacres in Lebanon to be factually erroneous. Is that student therefore entitled to level formal charges against Massad?

If we are going to allow the biology professor and the Jewish historian a right to offer their best judgment on competing truth claims, and because of those judgments to take some students more seriously than others, then don't we also have to grant this right to Joseph Massad?

In any case, we should remember that the proper goal of higher education is enlightenment—not some abstract ideal of "balance." Indeed, those who demand balance on some issues never demand it on others. The University of Chicago's school of economics is admired widely for its accomplishments. Must Chicago seek balance by forcing its economics department to hire scholars with contrasting points of view?

Occasionally, students have to do the hard work of seeking alternative points of view across institutional boundaries. They cannot always expect "balance" to be delivered in neat packages. It is the professor's pedagogical role that grants him or her the authority and the right to judge which scientific theories or historical facts are presented in the classroom. We cannot deny the asymmetry in these roles. If we do, we fail to understand a legitimate goal of higher education: to impart knowledge to those who lack it. Of course, one can question the competence of a professor—that happens routinely in a good university. But the evaluation of that competence must be, and is, left to the professor's peers—not to students, and surely not to trustees, regents, congressmen, advocacy groups, or members of the press.

Over the past seventy-five years, the Supreme Court has expanded greatly the protection of free speech. Today, prevailing First Amendment doctrine holds that the government cannot restrict speech because of its content and that only forms of "low value" speech, such as "fighting words," libel, commercial advertising, or obscenity, can be regulated. Universities cannot act outside the law, but they can—and should—try to expand still further the limits placed on free expression when those constraints hamper inquiry and debate.[10]

Expression in the classroom requires virtually absolute protection. Absent such protection, professors will hesitate to discuss sensitive topics out of fear of retribution, suspension, dismissal, or litigation.

The university cannot and should not attempt to decide what ideas or perspectives are appropriate for the classroom. For one student, a professor's ideas may represent repugnant stereotypes or efforts at intimidation; for another, the same ideas may represent profound challenges to ostensibly settled issues. For example, a professor's discussion of our culture's bias against female circumcision may seem to one student an affront to what is self-evidently a basic human right; but to another student, it may seem a provocative illustration of cultural imperialism, raising serious moral questions that ought to be put on the table for debate. Are we to take seriously those who would have us sanction the professor for raising this subject in a seminar? And if we did, who would be cast in the role of the "Grand Inquisitor"?

The broadest possible protection of freedom of expression is of a piece with another important aspect of the academy. We have understood for some time now that the university is not a place where we exclusively house or train the kind of scientist or scholar who advises the prince—those who currently control the government. There are members of the faculty who sometimes voluntarily give advice to the prince—and there may even be academic programs (such as Russian studies during the Cold War) that exist in part to inform government policy—but it is not the point or the rationale of universities to furnish such advice nor to have the thematic pursuits of inquiry in the university shaped by the interests of the prince. That is why universities will often find in their midst those who air the most radical critiques of the prince and his interests. Were we to silence or even to inhibit such people, we would not only be undermining free inquiry, but we would also gradually reinforce the countervailing power of conformism.

Despite the commitment of the American research university to freedom of thought, the natural tendency of professors and students, as we have seen, is to avoid expressing views that may offend others. But the responsibility of the university is to combat this tendency and to encourage, rather than squelch, freewheeling inquiry. The university must do everything it can to combat the coercive demand for political litmus tests from the Right and the Left and the pressure to conform with established academic paradigms.

By affording virtually absolute protection to classroom debate, the university encourages the sort of open inquiry for which universities exist. Those members of the university community who are willing to take on prevailing beliefs and ideologies—be they the pieties of the academic Left or the marching orders of the politicians currently in power—need to know that the university will defend them unconditionally if they are attacked for the content of their ideas.

The defense of academic freedom is never easy.

It is understandable that university leaders will react to outside attacks with caution. There is always a risk that taking a public position on a controversial matter may alienate potential donors or offend one of the modern university's many and varied constituencies. In response to negative publicity, it is entirely natural for presidents and provosts—and for trustees and regents—to work feverishly "to get this incident behind us" and to reach for an accommodation that calms the critics and makes the problem go away.

588 ACADEMIC FREEDOM AND FREE INQUIRY

However, to act on such understandable impulses would be a grievous mistake. There are few matters on which universities must stand on absolute principle. Academic freedom is one of them. If we fail to defend this core value, then we jeopardize the global preeminence of our universities in the production and transmission of new knowledge in the sciences, in the arts, and indeed in every field of inquiry. Whenever academic freedom is under fire, we must rise to its defense with courage—and without compromise.

For freedom of inquiry is our reason for being.

Notes

1. Many colleagues have provided useful comments on earlier drafts of this essay. I received particularly helpful comments from Akeel Bilgrami, David Cohen, Joanna L. Cole, Susanna Cole, Tom Goldstein, Eric Foner, James Miller, Richard Shweder, and Geoffrey Stone.
2. The role of industry is, of course, critical as well, but most of the great industrial laboratories are highly dependent on these same research universities for PhDs who join these companies. So, universities have both a direct and an indirect influence on the production of innovation.
3. The film has been shown in at least four or five different versions; its content is continually changing.
4. "Academic freedom implies that all officers of instruction are entitled to freedom in the classroom in discussing their subjects: that they are entitled to freedom in research and in the publication of its results; and that they may not be penalized by the University for expression of opinion or associations in their private civic capacity; *but they should bear in mind the special obligations arising from their position in the academic community* [italics added]." *The Faculty Handbook*, Columbia University, 2000, 184.
5. Quoted in Claudia Dreifus, "The Chilling of American Science: A Conversation with Robert C. Richardson," *New York Times*, July 6, 2004, D2.
6. For an exceptionally fine discussion of these failures, see Geoffrey Stone, *Perilous Times: Free Speech in Wartime from the Sedition Act of 1798 to the War on Terrorism* (New York: Norton, 2004).
7. Reviewing the now available archival material at Harvard University, Robert N. Bellah has confirmed the accounts of Diamond and Schrecker. Bellah reports his findings in "McCarthyism at Harvard," letter to the *New York Review of Books*, February 10, 2005, 42–43.
8. I'm not focusing here on financial accountability. In fact, universities have many ways of reviewing and accounting for their financial transactions. Fund accounting at universities allows auditors to review every research grant or contract in minute detail. Full-time federal auditors are fixtures at universities. Major accounting firms audit the books of the universities on an annual basis. Bonds floated by universities to finance construction projects are brought to market only after bond-rating agencies evaluate the credit worthiness of the university and rate the bonds. The ratings depend almost as much on qualitative factors of the quality of university schools and departments as on financial ratios and other indicators.
9. John Stuart Mill, *On Liberty* (Cambridge: Cambridge University Press, 1989), 49.
10. Here I'm putting aside the distinction between public and private universities.

CHAPTER 29

THE NEW McCARTHYISM

(2005)

JONATHAN R. COLE

A rising tide of anti-intellectualism and intolerance of university research and teaching that offends ideologues and today's ruling prince is putting academic freedom—one of the core values of the university—under more sustained and subtle attack than at any time since the dark days of McCarthyism in the 1950s.

As professors are publicly savaged for their ideas, often by outside groups, colleges are coming under pressure to fire them or control what they say in the classroom. Witness the furor last year over a purported "documentary" by the Boston-based David Project, *Columbia Unbecoming*, that charged professors with anti-Israel bias, or the Orwellian efforts by the national group Students for Academic Freedom that—in the name of ending the alleged politicization of the academy—attempt to limit legitimate scholarly discourse.

As political ideology trumps scholarly consensus, the government is undermining the peer-review system and the norms of scholarship. Conservative ideologues in Congress, for example, are trying to place political appointees on committees to monitor area-studies programs; the Bush administration and its followers on Capitol Hill and in statehouses are trying to intimidate professors whose work on topics like global warming or the transmission of HIV calls into question administration priorities. Such arbiters of truth are selectively bullying professors by investigating their work or threatening to withdraw federal grant support for projects whose content they find substantively offensive. In

---

The Chronicle Review 52, no. 3 (2005): B7.

resisting stem-cell research and supporting teaching intelligent design along with evolution, they have cast doubt on scientific expertise and legitimated the latest form of anti-intellectualism in America.

The USA Patriot Act allows the government to secretly monitor what students and faculty members read or transmit over the Internet; and the Public Health Security and Bioterrorism Preparedness and Response Act of 2002 places such extraordinary constraints on laboratory scientists that some of our most distinguished immunologists are abandoning important research—for example, on vaccines to prevent smallpox, anthrax, and West Nile virus—that could help deter terrorism. Foreign students and researchers from scores of nations are finding it increasingly difficult to obtain visas to study or work in the United States, disrupting the flow of the best talent to American universities.

The current attacks on academic freedom are not the only threats to free discussion in the university: Too many subjects, like those related to identity politics and challenges to reigning academic dogma, are also considered off limits. The result is that it has become increasingly difficult within the academy itself to have an open, civil debate about many topics. Scholars and scientists are often exercising their right to remain silent rather than face the potential scorn, ridicule, sanctions, and ostracism that challenging shoddy evidence and poor reasoning on politically sensitive topics can invite.

Why does that matter? Universities remain perhaps the last sanctuary for the relatively unbridled and unfettered search for truth and new important ideas. Without a climate of free inquiry, creativity and discovery will suffer. Today American research universities are the single most important source for major new discoveries that improve the health and social and economic welfare of people around the world. Tie a tourniquet around that free flow of intellectual energy, and we will halt the production of knowledge that is necessary for conquering disease and poverty and for improving the quality of everyday life.

The sad fact, however, is that few academic leaders and prominent members of their faculties are rising to the defense of academic freedom. Where is the Robert Hutchins of today, who protected the idea of the university against ideological foes during the 1940s and 1950s? As Hutchins said, it is "not how many professors would be fired for their beliefs, but how many think they *might* be." It is time to recognize the seriousness of the current attacks, analyze carefully the bases for them, scrutinize evidence on their incidence and consequences, and organize a defense of the university against those intent on undermining its values and quality.

CHAPTER 30

DEFENDING ACADEMIC FREEDOM
AND FREE INQUIRY

(2009)

JONATHAN R. COLE

A critical inflection point in the history of modern universities came in January 1933. What rapidly followed Hitler's rise to power was a disaster for Germany and its university system—one from which its research universities still have not fully recovered—and an enormously valuable, if unwanted, gift to the increasingly strong, but still not preeminent, American research universities searching for leadership. That great intellectual migration created a chemistry at these leading academic institutions where the horizontally mobile, highly distinguished professors from Europe were combined with an increasing number of young, exceptionally talented scholars and scientists (many of whom were Jewish) who were vertically mobile in the United States. It was the beginning of the rise of the American research university to preeminence. Alvin Johnson at the New School and people like Lord Rutherford in England did much to provide safe havens for many of these fleeing intellectuals, and the New School's role in that deliverance is worthy of our continued deep respect and celebration.

In this paper, however, I want to focus attention on a few principles that I believe guide great universities and to discuss and elaborate several propositions. My remarks will not attend to the particulars of the history of that great migration and its impact on the United States. A far more extended discussion of the consequences of that journey for American universities and the nation can be

Social Research 76, no. 3 (Fall 2009): 811–44.

592 ACADEMIC FREEDOM AND FREE INQUIRY

found elsewhere and in my book, *The Great American University* (Cole 2010). Here I want to suggest that my country has not distinguished itself particularly well in preventing episodes of repression and attempts to silence dissent at universities, nor has it produced an extraordinary number of courageous leaders over the past seventy-five years who have come forward to defend the principles of academic freedom. We have never reached the level of repression that Germany felt in the 1930s, nor that which was felt by Soviet geneticists at roughly the same time during the Lysenko years, but we have done significant damage to our system of higher learning because we have failed to understand fully the role that academic freedom and free inquiry play in creating the knowledge that societies depend on for their social and economic, as well as humanistic, progress.

The American research university today is the engine of discovery and innovation that is at the center of the nation's effort to create better lives for its citizens. And my emphasis will be on George W. Bush's presidential years, when we once again had to deal with significant threats to the core value of academic freedom and free inquiry—from both the government and organized interest groups as well as from within the belly of the academic community. Barack Obama's election in 2008 produced great hope on American college and universities campuses that a new enlightenment is at hand. Based on his action to open up stem cell research, to demand of his executive offices a respect for the integrity of science, and from his remarks to the National Academy of Sciences on the role that our great universities play in the process of discovery and economic innovation, there is a growing sense that President Obama understands the necessity of academic freedom and free inquiry, and that a new day has dawned. However, such a day cannot be created by the White House alone, I'm afraid, and we should not assume that even a heavy majority of Democrats in Congress, or in state governments, will see the world as the president apparently does.

At the outset, I want to assert what I believe to be true about great universities. They are designed to be unsettling. Great universities will, as the University of Chicago's famous 1967 Kalven Committee report said, "provide enduring challenges to social values, policies, practices, and institutions. . . . It is the institution which creates discontent with the existing social arrangements and proposes new ones" (Kalven Committee). So distinguished universities must entertain and not suppress the most radical thoughts—whether they are from scientists who challenge the long-standing belief that only bacteria and viruses cause disease or social scientists and humanists who attack the foreign policy of the United States. This is not an easy thing to do, as Justice Oliver Wendell Holmes Jr. reminded us in his memorable dissent in the 1919 *Abrams* case:

> Persecution for the expressions of opinions seems to me perfectly logical. If you have no doubt of your premises or your power and want a certain result with all your heart you naturally express your wishes in law and sweep away all

opposition. . . . But when men have realized that time has upset many fighting faiths, they may come to believe even more than they believe the very foundations of their own conduct that the ultimate good desired is better reached by free trade of ideas—that the best test of truth is the power of the thought to get itself accepted in the competition of the market, and that truth is the only ground upon which their wishes safely can be carried out. (259 U.S. 616 [1919])

The encouragement of radical thinking is accompanied by another strongly held value at our best institutions of higher learning—the value of skepticism about claims to truth or fact. Juxtaposed with the tolerance of radical thought, this conservative bias by design creates an essential tension within the university. The liberality of its intellectual life and the conservatism of its methodological demands allow great universities to challenge the prince, or other orthodoxies, while maintaining its commitment to the role of evidence as judged by experts in establishing facts.

Now I want to make another strong claim: it is impossible to create or sustain a truly great university system without a society's deep appreciation and commitment to the idea of academic freedom and free inquiry. Show me a counterexample. Remember as well that "academic freedom" is not another term for "free speech" and the concept is not even principally built on defense of free expression—although its abuses are most often discussed in terms of the right to free and unfettered speech. It is not simply a replacement at universities for individual rights to free speech. It also is *not* a bonus for employees of academic rather than financial institutions or a philosophical luxury without which universities would be no worse off. As Louis Menand has aptly said, "It is the key legitimizing concept of the entire enterprise. Virtually every practice of academic life that we take for granted . . . derives from it. The alternative is a political free-for-all" (Menand 1996: 4). In its most fundamental form, academic freedom leaves the structure of decision-making about what constitutes quality thought and work, quality research and teaching, and the quality of potential up to a set of academic peers. It wrests control for such decisions away from government, presidents and trustees, and boards of regents, and places them in the hands of those who have the background, training, and judgment to evaluate quality in specific areas of expertise.

Before turning to the actual threats to this core value in the United States from 2001 to 2008—threats that some of us very much hope that President Obama will get around to attenuating after he has fixed the world's economy and financial systems—let me place my observations in one particular context.

My remarks on the threats to academic freedom in the United States may appear as a joke, or as naive, for most academics working in the overwhelming majority of nations in the world. The people who teach, do research, or are enrolled as students at those places only wish that they could be guaranteed the freedom to pursue their own scholarly and scientific interests, the right to dissent and criticize their government's policies, as we have *even* during times of relative repression. When we think of levels of academic freedom throughout the world's universities, of course, even in troubled times American academics have relatively high levels of freedom. That said, I am concerned here with academic freedom in systems of higher education that either have been or aspire to be the most distinguished in the world, particularly in the United States.

Let me begin by simply noting, without expansion, the periods when the United States has experienced unusual and unfortunate interference with our universities' relatively high level of autonomy. Repression of speech, thought, and scholarship were abundantly evident in the United States during the period surrounding the American entrance into World War I and during the Cold War and the McCarthy period. Hardly a generation has gone by without efforts by the government to muzzle, and even prosecute, those at universities who were believed to be subversive or who associated with alleged subversives. In the historic tension between national security and civil liberties, more often than not national security has won out—much to our later embarrassment. And so it has been since the attacks on the United States on September 11, 2001.

I want to enumerate the forms that the most recent repression has taken in the United States. I note some of the external threats in part I, while I suggest in part II that the virus that causes attacks on academic freedom continues to be found within the academy itself. Taken together, they threaten the continued preeminence of our research universities. Finally, in part III, I offer a few thoughts about how we might begin to rethink and extend the principles of academic freedom, which have not changed much since the declaration of principles in 1915.

Part I: Threats to Preeminence Related to Academic Freedom

The preeminence of the American research university was not built overnight; it will not deteriorate overnight.[1] But when the government tries to censor scientists, and in the face of scientific consensus tries to create contrary scientific facts, tries to intimidate scholars, and unleashes other politically motivated advocacy groups to target individuals and specific universities, then we are going down a path that could lead over time to killing the goose that laid the golden eggs. The attacks that I focus on here are on the core values, on the structures and prerogatives, and on the faculty members at our centers of academic excellence.

In broad brushstrokes, the policies have had the following effects: social scientists and public health specialists conducting research on prevention of HIV supported by the National Institutes of Health (NIH) and its peer review system have been subjected to congressional inquiries. Members of Congress have threatened to rescind peer-reviewed and -approved research grants through congressional legislation and have required the director of the NIH to explain why these grants were funded. The White House and political appointees at NASA have tried to censor scientific reports and muzzle scientists from speaking about their scientific findings about global warming.

Federal employees at the Centers for Disease Control (CDC) have been ordered to alter information on the CDC website that focuses on the prevention of sexually transmitted diseases—alterations that eliminate the best knowledge we have about prevention in light of political sensitivities about its content. The changes, which produced erroneous information, were made. At variance with what they know are facts, CDC researchers are told, for example, to emphasize sexual abstinence and to eliminate or downplay information about the use of condoms in preventing spread of HIV. The scientists at CDC objected but complied.

Universities are being asked to discriminate against students on irrelevant criteria; to restrict their search for the best talent in the world; to alter laboratory work and potentially to restrict publication of ideas; to accept increased surveillance on campus by federal law enforcement personnel in the name of national security; to accept a government role in defining what is "good" or "bad" science; to allow the government to review the content of curricula; to limit research in areas with great potential for scientific and technological discovery; and to acquiesce to appeals from the government to fire faculty members who pose no threat to the safety of the nation. And, in its effort to justify its policy decisions, the Bush administration tried, to an extent unheard of in recent American history, to "reshape" scientific and technical facts on which there exists virtual scientific consensus. The federal government reconstructed scientific knowledge to fit its political decisions rather than base decisions on informed scientific knowledge. Censoring science and going after scientists for political purposes has triggered other individuals and nongovernmental organizations with their own axes to grind to attempt to influence the content of ideas expressed and pursued at our great universities. That, as we know from history, is a dangerous path to go down—for the universities and the larger society. It is cause for concern among those at universities and among those in the wider public who depend on what they produce.

Let me elaborate on these threats with more specific examples of tensions between government policies and the transcendent values that we are discussing here. I have written about many recent efforts in the United States to harass and censor, fail to promote, and expel those who offended the powers that be through their speech. I am well aware of the organized efforts to defrock professors who publicly offer challenges to dominant ideological opinions and beliefs. Here, I will not consider the affronts to academic freedom illustrated in the well-publicized attacks on Palestinian and other Arab scholars—many of exceptional quality—for criticizing Israeli government policies. Whether the challenges to Israeli policies are made by renowned scholars like Columbia's Edward Said, New York University's Tony Judt, or prominent scholars such as Chicago's John Mearsheimer, Harvard's Stephen Walt, and Columbia's Rashid Khalidi, or lesser known but highly respected scholars like Joseph Massad, you can be sure that any public (and increasingly private) statement of criticism, however well argued and reasoned, will bring forth the wrath and demands for their sanctioning by government representatives, university alumni, trustees, faculty, students, and private advocacy groups. And there is no reason to believe that this kind of assault will not continue during the Obama administration, regardless of his personal views about such attacks.

Almost all of these prior cases focus our attention on the boundaries of academic freedom and free expression. I want to concentrate principally on the "mutated" forms of the insidious virus that produce attacks on research, which you did not see even in the McCarthy period, because they are less visible, but in some ways they are at least as destructive to the body of great universities.

State Intrusion into University Affairs

Scientists who work to discover the causes and possible cures for highly infectious diseases often work with what are called "select agents," which are bacteria, toxins, and viruses that can produce the disease. Long before 9/11, biologists feared that these agents could cause pandemics and that there was a national need to increase our research that focused on these lethal bacteria and viruses. After 9/11 many people were concerned that these agents might get into the hands of terrorists who could use them as biological weapons. Hence, the use and movement of these agents, even for potentially beneficial research, became cause for legitimate heightened concern and greater control. Two post-9/11 pieces of American legislation were designed, in part, to restrict use of these agents: the USA Patriot Act and the Public Health and Bioterrorism Preparedness and Response Act (2003). A number of the provisions of these acts are important for university research and study. First, the Patriot Act allowed federal agents to enter university libraries and computer systems and explore the behavior of students and faculty members without warrants and without any evidence there was a credible threat from those under surveillance. Moreover, the act stipulated that university officials, like librarians, were prohibited under threat of indictment from informing the targets of surveillance that their records were being investigated by the FBI or other government officials. Other features of these acts included a prohibition on faculty members from having anyone who came from one of roughly twenty-five nations suspected of supporting terrorism, including graduate and postgraduate students, physically enter a laboratory that used select agents for biological research. So, for example, no Iranian student could work in a laboratory that did research on a vaccine for plague. Faculty members who violated this provision were subject to arrest and indictment. Federal agencies, including the FBI, had to be informed of any movement of certain research materials on the list of "select agents." All of these provisions are still on the books, even after the reauthorization of the Patriot Act in 2006. They are being acted upon, and there seems to be little inclination in Congress to change them.

Fear among scientists that caused many of them to abandon work on select agents was reinforced by the highly visible case within the scientific community of one of the nation's leading microbiologists, Professor Thomas C. Butler of Texas Tech University. Those restrictions on transporting pathogens, such as plague bacteria, which Butler studied, had grown even greater since the new war on terrorism and the passage of the Patriot Act, as he was about to find out.

> On 15 January [2003], 2 days after reporting that 30 vials of plague bacteria were missing from his lab, Butler was [arrested by the FBI,] shackled and thrown into a Lubbock jail, charged with lying to federal agents about the fate of the [30] vials [of plague bacteria that he reported missing] and illegally importing the Tanzanian samples into the country. . . . Seven months after his arrest, the government indicted Butler on 69 charges (Enserink and Malakoff 2003: 2054).

Butler was apparently following practices that he had used before 9/11 and the new anti-terrorism laws were passed. When Butler arrived at the Dallas airport he failed to declare plague bacteria samples as "commercial merchandise" for U.S. Customs.

To grossly truncate this story, while the FBI was investigating possible violations of anti-terrorism legislation, they were also investigating other aspects of Butler's scientific life at Texas Tech. They combed over materials related to his grants and other activities at the university and reviewed his tax returns. Ultimately, the indictment was expanded to include fifty-four unrelated charges of embezzlement, tax evasion, and mail fraud. He was accused of defrauding the university on clinical trial fees and cheating on his taxes. At the end of the day, at age sixty-two and at the peak of his career after thirty years of research and "on the verge of becoming the United States' hottest plague scientist," Thomas Butler was placed on trial facing a sixty-nine-count indictment (fifty-four of which had nothing to do with violations of the Patriot Act) "that carried a maximum of 469 years in jail and $17 million in fines" (Enserink and Malakoff 2003: 2062). To add to his reversal of fortune, Texas Tech placed Butler on a paid leave, denied him access to his laboratory, and began proceedings to fire him (Chang 2003: 32). On December 1, 2003, the jury convicted Dr. Thomas

Butler of all but twelve of the counts against him. He faced up to 240 years in jail and millions of dollars in fines. However, none of the forty-seven convictions were directly related to the original incident. These convictions, with the notable exception of his failure to obtain transport permits, had nothing to do with his plague research in Tanzania. After the verdict and facing dismissal from Texas Tech, Butler resigned from his position, repaid the university more than $250,000, and lost his medical license. Butler appealed the verdict, lost his appeal and his request for a certificate of certiorari by the Supreme Court, and faced nine years in jail. At the end of the day, Butler was sentenced to two years in jail.[2]

The actions of scientists working with select agents speak louder than their words. "Let me give you an interesting example from Cornell," Nobel Prize winner Robert C. Richardson recently said.

> At Cornell, we had something like 76 faculty members who had projects on lethal pathogens and something like 38 working specifically on select agents. . . . So what is the situation now? We went from 38 people who could work on select agents to 2. We've got a lot less people working on interventions to vaccinate against smallpox, West Nile virus, anthrax and any of 30 other scourges. (Dreifus 2004: D2)

These scientists simply abandoned this type of research given the conditions that they had to meet under the act.

Restricting the Flow of Talent to American Universities: Restrictive Visa Policies

Over the past fifty years, thousands of the most able students and scholars have yearned to come to the United States to be trained. Remarkably high proportions of those intellectual migrants have taken positions at our great universities and have remained among the most productive people in American society. Higher education, particularly at the graduate level, has been among the few American industries with a heavy favorable balance of trade. Many science, engineering, and professional school students have returned to their countries with exceptionally favorable impressions of the United States and have taken up leadership positions in their home society. In fact, we have become increasingly dependent on this exceptional talent for work in science and engineering. Disruptions in the flow of talent will have many unfortunate consequences for the United States. If a presidential administration could do more harm to cut into this pipeline of talent, it is hard to imagine one improving on the Bush administration in the post-9/11 years. At least until the end of the Bush administration, the government targeted foreign scholars and students as potential security risks, almost invariably without a scintilla of evidence that they in fact are threats.[3] It has become far more difficult to receive student visas and travel visas for scholars then it had been in the past, and many cases exist where visas are being denied without reasons provided. For scholars invited to participate in conferences or teach at American universities, the content of their ideas, not whether they are security risks, seems to influence the probability of their obtaining visas.[4]

The Value of Open Communication versus National Security

Open communication is essential for the growth of knowledge and remains one of the fundamental values at distinguished universities. The results of research at universities should promptly enter the public domain. Open communication allows results of experiments and assertions of fact to be critiqued. Only through publication, with the accompanying detailed description of techniques and methods, can work be replicated or falsified. An absence of detailed description of how experiments were conducted prevents the growth of knowledge since it limits scientists' understanding of the basis of claims of novel discoveries and it impedes the process of building on the work of others. Consequently, withholding discussions of methods or tools used in experiments undermines the value of the scientific contribution. During the Bush years the government tried to impose prior restraint rules on publication of scientific papers (mostly in the biological sciences that have used lethal viruses and bacteria, that is, those "select agents") that it argued could fall into the hands of terrorists who were trying to create biological weapons. Of course, the government and the scientific community have legitimate interests in knowing where certain dangerous bacterial agents and toxins are stored and who is working with them. While bioterrorism must be taken very seriously as a threat to national security, the real question is whether the censorship desired fit the size and nature of the problem identified. At least according to the conclusions of the National Academy of Sciences Fink Committee, the specific cases that led the government to threaten passing legislation that would censor scientific publications would not have prevented "enemies" from obtaining the information censored, nor was it likely to have done them any good. Responding to what might prove disastrous for scientific communication, and trying to forestall the legislation, editors of leading science journals set up their own review system. For biologists, more specifically, the problem is how they should handle "sensitive research" results and who should decide whether their results represent a security threat.[5] But the answer does not lie in government censorship.

The Integrity of Scientific Facts

I could provide scores of detailed illustrations of the ways that scientific inquiry was compromised during the Bush administration. Scientific integrity was under assault on many fronts. Some of the cases illustrative of these violations are better known than others.

The most visible one has been corrected by President Obama: the unnecessary restriction on the development of new lines of embryonic stem cells that could be used to study a plethora of ailments and diseases. While this is a moral issue for many Americans, roughly 60 to 70 percent of the public believes in the potential value of embryonic stem cell research and wants to see more of it. In fact, the embryos that were to be used for research would have come from those that were being stored and were going to be discarded with the consent of the donors. The Bush administration had limited federal funding support for work on the twenty lines already in existence when Bush took office. Early action by President Obama opened this field, but it was only after other nations had already taken the lead in this potentially critical area of biological research.

One of the more pernicious assaults on scientific integrity resulted from efforts by President Bush's political appointees to warp, distort, and censor scientific facts. As already noted, there are important but relatively obscure efforts by the administration to shape people's thinking about reproductive health, birth control, and the use of condoms rather than abstinence as a birth control method.

Another example of an egregious effort to politicize science came when NASA tried to control the content of James Hansen's speeches and scientific talks and publications about the human influence on global warming. If there are better climatologists in the world than Jim Hansen, who works for the Goddard Space Center, which is run by NASA and Columbia University's Earth Institute, you would be hard pressed to find them. Hansen has studied the processes of global climate change for decades and has been one of the most respected voices in assessing models of this change. He was a close adviser on the subject to Vice President Albert Gore. When he, like almost all respected scientists, concluded that the data overwhelmingly supported the conclusion that global warming was rapidly altering our environment and that strong international intervention was necessary, that conclusion conflicted with official Bush policy. Efforts were made to have Hansen's speeches and talks edited prior to presentation by political appointees and on occasions edits were made in his work without his consent. Hansen was not about to cave to such political pressure and went public. The administration backed away from this censorship, denying that it represented a flagrant attempt to have science comply with ideology. But the evidence overwhelmingly suggests that ideology was the basis for the efforts to muzzle him. When scientific facts become "negotiable," then the academic and scientific communities are in deep trouble.

The efforts to shape academic research reached into the social and behavioral sciences as well. When it came to light in the Republican-controlled Congress during the height of the Republican's assault on free inquiry that the NIH had funded through its peer review process a project in San Francisco that looked at the transmission of the HIV/AIDS virus through sex workers, staff members of powerful congressmen called NIH Director Elias Zerhouni on the carpet. They asked for explanations why the NIH had funded this and other projects that offended the political right of the party and suggested that the funding should be withdrawn and the researchers forced to provide federal authorities with the names of the sex workers who were participating in the study and, according to a committee staffer, committing "crimes."

The national peer review system and the long-standing belief, supported by Supreme Court decisions, to grant autonomy to universities to shape their own curricula, also were under attack. The Bush administration and Congress during the seven years following 2001 tried to shape the outcomes of government oversight panels and peer review efforts by adding political appointees to these committees. From the committee studying ethical questions associated with stem cell and other forms of biological research, the administration tried to add people to national committees whose ideological beliefs were consistent with its own, and, correlatively, tried to prevent highly qualified scientists from joining the committees after being recommended by members. The basis was simple: Were their stated positions consistent with the administration's policy positions? When Title VI National Resource Center Programs, which support area studies programs at major universities, came under attack from conservative interest groups and their leaders, like Daniel Pipes and David Horowitz, Congress tried to pass legislation embedded in the reauthorization of the Higher Education Act that would have external, nonacademic, monitors of the curricula of these programs to ensure that they did not have an anti-American bias. In short, the government began meddling in the content of the curriculum of area studies programs. After a tremendous amount of protest from the universities, the provision was watered down to eliminate monitors in the classroom, but it did not entirely remove some oversight of these programs by individuals who were hardly qualified to do so.[6]

One might conclude from this description that the Bush administration's policies were aberrational. They were the unfortunate outcome of a misguided political administration that was in power for eight years. All of that should change, one might think, now that there has been a change in administration. I have no doubt that President Obama and the science advisers that he has appointed will try to right many of the wrongs that took place over the past eight years. However, it will take more than Obama and his advisers to make meaningful change. And it is anything but clear that the huge Democratic majority in the 2009 Congress is inclined to reverse many of the most insidious and dangerous policies that were put into place after 9/11. Thus far, few changes in the Patriot Act or the Bioterrorism Defense Act have been entertained, much less embodied in new legislation. The jury remains out on whether the Obama administration can persuade members of Obama's own party that these policies are damaging academic freedom and free inquiry at our universities—and that this damage disrupts the most significant engine of scientific and technological discovery and innovation that our nation possesses.

Part II: The Herd of Independent Minds: The Tendency Toward Intellectual Orthodoxy as a Threat to Academic Freedom

Academic freedom and tenure combat the tendency of professors and students to sit on their hands and remain mum when they believe that their ideas and research may offend those who adhere to an ideologically "correct" way of thinking. But there is huge pressure toward ideological conformity within universities, as elsewhere, and universities, like other institutions, tend not to be tolerant of those in their midst who are courageous enough to challenge prevailing systems of thought. If the essence of a university is to be tolerant of all points of view that can be supported by evidence, then the most basic beliefs must be open to questioning within the academy. Yet they often are not. The limits placed on free inquiry within the academy threaten the realization of that ideal as much as threats from beyond the university campus.

In truth, there is both intellectual and personal risk involved in challenging the presumptions of the group. The weight of the community on the individual scholar is found in the way those who challenge "groupthink" are treated. More often than not, it's the faculty, not administrators, who define and enforce dominant orthodoxies. Sometimes a scholar harboring an unorthodox view may not be able to obtain a position at a major research university as easily as one with a more orthodox outlook, regardless of the quality of his or her mind and evidence.

The tendency toward orthodoxy is also felt in campus life and in the treatment of students. At universities it has become difficult even to discuss certain topics or to suggest ideas that offend some significant part of the academic community. What academic leaders not looking for a good fight would tell "Take Back the Night" students, who were marching to protest the failure of the administration to show greater concern with sexual assaults and to toughen up its sanctions in the university's sexual harassment policies, that they question the validity of the group's data on the percentage of young women who were victims of date rape? When any group of students asserts that the university is not sufficiently protective of its rights, to say nothing of its feelings, most administrators think first about how to redress the grievance rather than to investigate whether there is a basis for grievance. Bad things do happen at universities, which are no more immune from the malicious and unsavory behavior of some of its community members than is the larger society. Rather than viewing the unconventional thinking as an appropriate challenge to received wisdom and ideology, those being challenged often become defensive, and these questions, even if posed in the most neutral of forms, get people into trouble.

The remarkable thing about these retreats from the ideal of freedom of expression and inquiry is that liberals have been as responsible for them as conservatives. In the 1990s, there were cases involving opprobrious speech by students on campus that led to the adoption of speech codes, sensitivity training, and pro forma statements of moral outrage from deans, university presidents, and provosts, to say nothing of faculty members. They all tended to lose sight of the principles of academic freedom and protected speech. One case at Yale Law School followed the rape of a white female law student by two Black men in New Haven. Following the incident, according to one account, "ten Black law students found in their mailboxes a note about the incident which ended with the sentence: 'Now do you know why we call you NIGGERS?' "[7] The author of this letter was not identified. The law school faculty surely did the right thing by expressing its sympathy for the affected students and condemning the content of the letter. But the dean of the law school went further, linking the incident to the racism of the institutions in which we live and therefore suggesting that all of the Yale Law School community was implicated in this despicable act—transforming the situation of an unknown individual letter writer to collective guilt.

When asked what ought to be done to the letter writer, the dean replied, "For myself, I am convinced that there is no place in this school for such vicious cowards." Some three hundred students subsequently signed a petition to the same effect. At the time, the question of sanctions was put to Yale's president, Benno Schmidt, himself a former dean of the Columbia School of Law, who aptly responded: "Freedom of speech protects cowards, too." I use this illustration not only to convey the power of collective thinking in subverting the principles of free speech but also to point out its coercive effects on dissenting views. Schmidt may have been right, but I am sure he was not a popular man at Yale for his comment. And what about the rush to judge and expel three lacrosse students at Duke University in 2006 who attended a stripper party and were accused by one of the strippers of rape? The North Carolina attorney general later dropped the charges against them, finding that the allegations were false, and the media exposed a series of missteps by law enforcement authorities in the case.

The diversification of the university community that was brought about by opening its doors to talent and by special legislation, often opposed—ironically—by liberal professors in the 1950s and 1960s who wanted nothing of affirmative efforts to diversify the university student bodies and professoriate, produced dogma about privileged knowledge. This knowledge was not privileged because of the depth of one's research and expertise but on the basis of social status and group identities. The diversification of student bodies and faculties should have led to a more interesting collaboration in problem solving and collective thought, but for a good deal of time it has discouraged debate. As the Yale literary scholar David Bromwich, puts it,

> One response to the new demographics of universities is to pay constant attention to the different beginnings of the new students and use the university as a place for diversified social reinforcement. What it says the students are, on their race-class-gender chart, they will now learn themselves to be—but more proudly and resourcefully than before. A different response would be to treat the students as equally enlisting in an intellectual life and varying unpredictably in what they make of that life: they are taken to be equals in this above all. To raise the second view is I think to bring out the strangeness of the first: how socially oriented it is—always to a social result and a sociable feeling—and by the same token how anti-intellectual. (Bromwich 1994: 41)

It is, in fact, far easier for students and faculty to be part of group thinking than to question the prevailing wisdom. At universities, scholars, scientists, and their students must be free to break away from the "herd of independent minds" as the art historian Harold Rosenberg put it—to take risks, without fear of formal or informal sanctions. The aim should not be institutional compassion but truth seeking.

The academy's success at opening doors to students and faculty with different identities deserves praise of course—although it still far from achieving a true meritocracy. But in its fear of offending any of these groups, and in its resolution to reinforce distinct identities rather than to make a common effort to pursue truth that incorporates varying perspectives without privileging any identity, the university has often hindered open debate. The consequences of privileging groups on campus, according to Bromwich, is to restrict freedom of inquiry and thought:

> If academic life in America becomes less free in the near future, one way it may happen is by a series of concessions to the sensitivities of the advocacy groups. Divided by sex, race, class, or geography, these groups have little to say to each other: an educational address by Louis Farrakhan, solicited and admired by one group, will prove to be not what the others had in mind at all.

But communication is not what they seek in any case. Beneficiaries of institutional compassion, they want to control the scene of education to assure that nothing wrong, or strange, or possibly injurious to the group-esteem of their members, gets said in the public forum of the classroom or the quad. Success on their terms means that the liberal ideal of tolerance, which drew no comparable limits around permissible speech, will have been exposed as part of the imperial ethic of the West. The defeat of the latter entity will have been worth the sacrifice. But that is to look far ahead. In the meantime, sects like these in their present state can weaken the resources that make for uncoerced discussion at a university. For they naturally defend against one kind of knowledge— the kind that challenges the protective instinct of group identity. (Bromwich 1994: 45–50)

Bromwich was criticized for these comments by the powerful majority in the academy, despite the fact that he was simply enjoining us to risk giving up our primary identities and privileges for the possibility of gaining through knowledge generated by truly free discussion.

Tenure does provide limited protection from formal sanctions for scholars taking on generally ideologically prohibited subjects. But it does not protect those same scholars from contempt from their colleagues. Take the example of the topic of female circumcision, or female genital-cutting. There are strong ideological forces both inside and outside of the academy that conclude, without much evidence, that this widespread practice among African cultural groups is repugnant, morally despicable, and clearly an example of the oppression and coercion of women in those cultures. Nonetheless, Richard Shweder, a University of Chicago anthropologist, and others have had the intellectual courage to confront the prevailing ideology that attacks this custom without much evidence about local culture. Shweder has raised serious questions about this cultural practice, speculating about why millions of African women not only accept but also embrace it. Why have we passed laws against the female-circumcision practices engaged in by some subcultural groups, such as Somali immigrants, in America, despite the fact that we fully accept male circumcision? Regardless of whether Shweder and his colleagues are right, or whether you accept or reject his evidence, he is right to raise the questions and to expect that we will consider examining the evidence, trying to overcome whatever biases and presuppositions we brought with us to the discussion in order to understand his viewpoint and that of the African cultures engaging in the practice. But without tenure, I'm not sure that even a person with his intellectual courage would have made this project his first as a junior faculty member.

In fact, academic rank may have less to do with the willingness to take personal as well as intellectual risks than sheer intellectual courage. And intellectual courage, which is needed in abundance within the academy, is, unfortunately, in short supply these days. It takes a great deal of such courage for individuals within the academy to stand their ground and make their arguments, no matter how brilliant these arguments may be, in the face of overwhelming group pressure. And it takes personal and intellectual courage as well to come to the defense of those who raise such questions (even if you disagree with them), especially among academic leaders who may be able to use the opportunity presented by the situation to reinforce the value of free inquiry.

All of this suggests that despite the ideal of free inquiry at universities, there are numerous social pressures acting to limit or subvert it. Over the past decades we have witnessed a growing intolerance of tolerance itself. Part of this seems to be an impulse to construct a protective shield around our undergraduate students—*in loco parentis* carried to the extreme. The task of a committed and useful teacher is to force his or her students to recognize "inconvenient facts," as the brilliant sociologist Max Weber put it. The aim is not to offer a "balanced" view, or to present materials in such a way that no one is offended by the content, but to speak truth as the professor, as an expert in the field, knows it. It would be paternalistic, patronizing, and even insulting to treat very bright students to a benign presentation of difficult subjects—an insult to their ability to distinguish arguments that are nothing more than assertions of fact, poorly formulated hypotheses, or theories without evidence from ones that are grounded in logic and supported by evidence. Education is a hard thing to obtain; so is an independent point of view that relies on higher levels of critical reasoning and analytic skills. But it does not come more easily in an atmosphere that refuses to challenge students and their prior beliefs about what must be true or factual. As a former president of the University of California once said, "The university is not engaged in making ideas safe for students. It is engaged in making students safe for ideas." Harvard's former president, Derek Bok, asked the rhetorical question, "Whom will we trust to censor communications and decide which ones are 'too offensive' or 'too inflammatory' or too devoid of intellectual content?" (Bok 1985: 4, 6) The answer, of course, is that no one can be trusted to do this. Instead, there must be an open dialogue, with each person weighing the arguments against the evidence for him or herself.

If universities coddled their students and other community members and prohibited expressions or displays that could be taken as offenses, or as affronts to someone's self-esteem, much would be lost in the academy. Limits on expression have found their way into a host of codes designed to prohibit offensive speech on campus. None of these codes at public universities have stood up to judicial scrutiny, and for good reason: they prohibit speech that would be protected for any citizen of the country.[8] Moreover, the idea that people have a right to self-respect and self-esteem, as Ronald Dworkin has pointed out, is absurd (1996: 196–97).

Harvard's literary scholar and public intellectual Henry Louis Gates Jr., head of the W. E. B. Du Bois Institute at Harvard, has exposed the absurdity of some of the speech codes that have found their way into some of our greatest universities. Gates offered an example to demonstrate that prescribed limits on "hate speech" fail to address the real problems of stigmatization. He asked readers to contrast the following two statements addressed to a Black freshman at Stanford (which, like the University of Michigan, had a speech code):

A. LeVon, if you find yourself struggling in your classes here, you should realize it isn't your fault. It's simply that you're the beneficiary of a disruptive policy of affirmative action that places underqualified, underprepared, and often undertalented Black students in demanding educational environments like this one. The policy's egalitarian aims may be well intentioned, but given the fact that aptitude tests place African-Americans almost a full standard deviation below the mean, even controlling for socioeconomic disparities, they are also profoundly misguided. The truth is, you probably don't belong here, and your college experience will be a long downhill slide.
B. Out of my face, jungle bunny.

As Gates said, "Surely there is no doubt which is likely to be more 'wounding' and alienating to its intended audience. Under the Stanford speech regulations, however, the first is protected speech; the second may well not be, a result that makes a mockery of the words-that-wound rationale" (Gates 1996: 146).

All of this is to say that there is today at our great universities an insidious tone to a significant amount of discourse that avoids taking on orthodoxies and prevailing wisdom. In fact, there is false satisfaction in intellectual consensus and conformity that has not been earned. Conformity may sometimes occur because people are afraid to confront politically correct thinking; other times, it may be a calculated form of careerism, a way of pulling one's intellectual punches when one holds evidence to question beliefs that most in the academy take for granted. But either way, the result is a perversion of the ideal of a great university.

The growth of knowledge, insight, and understanding is better served through the contest between ideas than through the blind acceptance of dominant ideologies and the silencing of criticism. In fact, without those contests we cannot easily distinguish between truth and falsity. Truth rests less in product than in process. Great universities need to create a culture in which the brilliant intellectual maverick or iconoclast, who supports ideas with evidence, is not apt to be a social isolate, if not vilified, for questioning those "facts" and "truths" that are believed to be beyond doubt.

The trend toward ideological conformity and "group thinking" at the great universities has other disturbing consequences for the open discourse that we need and have come to expect. There is no room today for arguments and evidence that support, for example, that unequal representation of various groups (from their demographic proportions) in occupations is acceptable. To suggest that some groups are more likely to choose to be lawyers and doctors rather than scientists results from factors other than gender discrimination places the author of those ideas at risk. To suggest that affirmative action policies in university admissions is double-edged with a downside for minorities as well as an upside places the proponent of such an idea at risk. To suggest that Israeli government policies toward the Palestinians is morally repugnant and that those policies do not, in fact, further American interests in the region, will almost inevitably lead to the author of such a position, regardless of the evidence, being skewered within the university as well as in the media. He or she is apt to be intellectually ostracized and possible vilified by colleagues, students, and certainly by external advocacy groups, funding agencies, and local and national political leaders. If you are a renowned critic you are apt to survive such blunt attacks, but if you are a junior faculty member without tenure, you probably have put your career at risk.

All of this group pressure to conform to the ideological fashions of the day, what is typically referred to as "political correctness," has a chilling effect on free discourse and the development of truth based on analysis, evidence, and argument. These contemporary patterns of behavior at even the greatest universities threaten the norm of academic freedom and the testing of ideas in the marketplace.

Part III: Returning to Fundamentals

A remarkable aspect of the principles of academic freedom and free inquiry is how little these have changed since their original formulation in 1915 and then their updating in the 1940s. Almost seventy years have passed since we took a serious look at these principles and asked whether they stood the test of time or were in need of serious revision given the changes that have taken place in the role of universities in our society.

Most university faculty members closely link academic freedom with free expression. Yet, as Robert Post and others have repeatedly told us, academic freedom was formulated not so much as a free speech issue as an effort to change the employment arrangements between faculty members and their employers (Post 2006). At the time of their framing, during the so-called Lochner Age (Lochner v. New York 198 U.S. 45 [1905]), employers held virtually absolute power to hire and fire faculty members, and the causes for dismissal could be as simple as trustees finding a faculty member's ideas opprobrious. The aim was to fix this and to bring the faculty into the normative process of evaluation in hiring, promotion, and firing. While not initially very successful, over time the balance of university power shifted toward faculty governance and control of key decision-making. While these labor relations were linked to free expression, Post reminds us that we should not think of First Amendment jurisprudence and academic freedom doctrine as synonymous. Yet, as much as First Amendment doctrine has had a dynamic history since 1915, academic freedom doctrine has hardly evolved at all. And, if you follow Post, you would conclude that the normative structure for peer assessments of quality does not require any serious modification. In a very serious critique of Post's position, Judith Butler has raised questions about the very foundations of the traditional ideas about academic freedom and expert knowledge (Butler 2006). Are the norms that govern peer review and decision-making open to criticism and revisions themselves? How far can we go in questioning the very bases of competence and judgments about quality? I will return to these questions momentarily.

Academic freedom can be defended on intrinsic as well as pragmatic grounds. Many commentators on academic freedom see its value in the freedom to pursue ideas, theories, concepts, or whims that educators are interested in and that this is an essential part of the discovery and teaching process. Those who defend this point of view are less interested in the consequences of discovery and radical thinking than those who are particularly interested in the role that discovery and universities play in the national system of innovation. The more pragmatic defense rests on arguments about the good things that universities do for society. These defenders of academic freedom believe that the innovative process depends on this freedom and that the role that distinguished universities play in the production of social, economic, and scientific discoveries that transform our lives depends on defending that freedom. While I have argued elsewhere for the intrinsic value of academic freedom, I am particularly concerned how its abridgment can lead to a breakdown in the engine of change that universities represent.

In this sense, I think it may be time for us to move beyond the idea of academic freedom as attached principally to individuals, and from the relational concept of employer and employee rights, privileges, and obligations. We should consider academic freedom (and free inquiry) in broader sociological terms as an institutional and structural concept—one that is defined more precisely by what the profound philosopher and historian of ideas, Isaiah Berlin, called "negative liberty." Given the strides that we have made over the past seventy-five years in wresting decision-making control away from regents, presidents, and trustees and handing it over to competent faculty judges, we can use that as a point of departure. The role of the faculty in determining the contours of the curriculum, the standards of quality, and the evaluation of candidates for jobs and tenure is now firmly established in American universities. But what have not been well defined are the terms of the social compact between universities and the government and larger public, under which these external and often noncompetent judges stay out of the way of university decision-making. There is a good deal of acknowledgment by the courts, including the Supreme Court, and others of the importance of protecting the independence of universities from external intrusion, but a clear set of boundaries has never been articulated. On the contrary, as governments, both federal and state, have expanded their role in financing higher education they have taken it for granted that they may intrude on the university's freedom and business in a whole manner of policy and regulatory ways.

Not all of these intrusions are harmful for universities, such as legal efforts to increase equal employment opportunities, and of course, institutions of higher learning must be accountable not only to the sources of their food but to the larger public that expects this institution to deliver the goods in terms of the transmission and creation of new knowledge. But when the government or nongovernment interest groups that influence government are able to intrude excessively in the running of universities, the seats of higher learning will be damaged, as will the process of innovation. This is precisely what Vannevar Bush feared in his post–World War II design for government support of university-based research through the use of taxpayer money to foster the growth of fundamental knowledge. His design called for a National Research Foundation (which eventually morphed into the National Science Foundation after Congress got through with the proposals Bush put forth in *Science, the Endless Frontier*) that was freed from direct presidential control and congressional intrusion. Of course, at the end of the day, neither the executive nor legislative branches were about to permit such autonomy. But the question remains an open one: What should the rules of engagement be between universities who require this negative freedom and the powers that be? And who should decide on those rules?

This takes me back to Judith Butler's point. What happens to universities when their fundamental norms are themselves under assault and weakened— norms such as who is qualified to judge scholarship? Should we encourage such questioning of these norms from within the belly of the university, and is there any way of preventing it without ironically violating our own value of academic freedom? However, when the core values of the university are themselves challenged in ways that break down consensus, as they have been periodically in the visible humanities disciplines for the past twenty years or so, the robustness of the institution diminishes and the institutional commitments to certain values and norms, like the norms that guide the application of the rules of academic freedom and free inquiry, can become attenuated and broad support for the norms can waver. The core itself begins to be questioned. At such moments, the university can seem in disarray and it becomes more vulnerable to external intrusions and abuses. The prince is more apt to prey on the weak than the strong here too. If it is an appropriate thing not to take fundamental values and norms for granted, are we risking the welfare of the institution both by not engaging in appropriate self-criticism about our governing rules and by accepting the consequences of self-criticism?

This dilemma is the subject of an interesting essay on academic freedom by Richard Rorty, who spent a good deal of his philosophical career questioning the nature of objective truth and consequently the essential norms that gave rise to the modem university—the enlightenment ideals about truth and objectivity (Rorty, 1996). But suppose those in strategic positions of power construct the "truth" and they control as well the peer-review system that metes out the rewards and recognition in the academic community. Who then are the qualified experts into whose hands we place tenure decisions, the future of the disciplines, and indeed the future of the university? And if there is no "there out there," why shouldn't those external to the university decide what constitutes facts, truth, and qualifications for tenure as well as those who have been certified to do so by their powerful professions? But it is clear, even for Rorty, that

> one of the things that accumulated experience has taught us is that universities are unlikely to remain healthy and free once people outside the universities take a hand in redrawing this line [between academic politics and the disinterested pursuit of truth]. The one thing that has proved worse than letting the university order its own affairs—letting its members quarrel constantly and indecisively about what shall count as science or as scholarship—is letting somebody else order those affairs (1996: 28–29).

Rorty also accepted the idea that at crunch time, one had better get behind the idea of academic freedom. "Dewey, I think," says Rorty, "would say that if it should ever come down to a choice between the practices and traditions which make up academic freedom and anti-representionalist theories of truth and knowledge, we should go for academic freedom. We should put first things first." Rorty agrees with this, saying: "Nothing, including the nature of truth and knowledge, is worth worrying about if this worry will make no difference to practice." Then he goes on to say that one way of making a difference is to change what Wittgenstein called "the pictures that hold us captive" (1996: 35–36). I agree with this because in the history of our universities, this slow, often laborious process, of changing the "pictures" actually works. Fields are transformed; power is relinquished; new ideas, concepts, and theories take hold; new vitality is injected into the system. Yet turning away from the very tough challenges to the legitimacy of those who are anointed as the "competent" leaves one with a sense of copping out—of not being willing to be appropriately critical of our own enterprise and how it operates as a social system.

612 ACADEMIC FREEDOM AND FREE INQUIRY

This remains, then, an open question within the academy, but how to address such complex epistemological questions is not the central threat to academic freedom today. That threat lies elsewhere—in the power of the prince to subjugate those in our universities and undermine their search for original ideas that offend those in power.

If academic freedom is a necessary condition for true creativity, originality, and innovation, what limits should we place on government or other intrusions on our institutional freedom? The definition of those limits is, it seems to me, an important project that requires the engagement of those both inside and outside the academy. For if we are going to set such limits we will need those who are not insiders to recognize the critical nature of the decision: the very continued vitality of our great universities may depend on it.

Notes

1. Here I assume the reader is familiar with the various rankings of the standing of the world's universities. Studies suggest that as of 2009 the United States has roughly 80 percent of the world's top twenty, 75 percent of the top fifty, and about 55 percent of the world's top hundred research universities. Recognition for research accomplishments by the faculty at these universities, which weighs heavily discoveries and recognition for them, tends to place greater emphasis on the sciences and engineering than on the social and behavioral sciences and humanities. Germany, which dominated the set of the very best until that fateful day that Hitler came to power in January 1933, no longer can boast one in the top fifty. Other indicators of our world leadership and preeminence are receipt of honorific awards, such as Nobel Prizes, which we continue to dominate; or the proportion of the most highly cited research papers in most academic fields. In short, the United States today is the envy of the world in higher education and, in a world that increasingly recognizes that in the twenty-first century knowledge will determine economic welfare and growth through discovery and innovation, this leadership is of the greatest societal value.

2. Another case worthy of note is the saga of Dr. Steven J. Hatfill, whom Attorney General John Ashcroft labeled as a "person of interest" in the FBI investigation of the anthrax mail attacks in the fall of 2001, which caused the death of five people and illness in others. You may recall that several letters containing anthrax were sent to members of the media and to Senators Thomas A. Daschle and Patrick J. Leahy. The FBI interviewed hundreds of scientists and others who worked in fields related to biological weapons. Hatfill had never worked with anthrax and willingly cooperated with the FBI. He volunteered to take a lie detector test—which he passed, according to the examiner, suggesting that he had nothing to do with the anthrax attack. Nonetheless, in the months that followed, the attorney general focused the public's attention on Hatfill as if he were a suspect. Hatfill was fired from his job at Louisiana State University when its chancellor, with some prompting from the Justice Department, said,

> After careful thought and consideration, I have decided that it is in the best interest of Louisiana State University to terminate its relationship with Dr. Steven Hatfill. . . . In taking this action, the university is making no judgment as to

Dr. Hatfill's guilt or innocence regarding the FBI investigation. Our ultimate concerns are the ability of the university to fulfill its role and mission as a land-grant university, to fulfill its contractual obligations to funding agencies and to maintain academic integrity. In considering all of these objectives, I have concluded that it is clearly in the best interest of LSU to terminate this relationship. (Louisiana State 2002)

The chancellor's statement makes one wonder what the role and mission of LSU actually is, if it is not to defend principles of freedom of inquiry and attacks on a member of its faculty or research staff against whom no charges have been made. The university seems to have succumbed to the pressure placed on it by the Justice Department to terminate Hatfill's employment. His career in ruins, Hatfill sued the government and eventually settled the suit for over $5 million.

3. Here I refer to the Enhanced Border Security and Visa Entry Reform Act of 2002 and the USA Patriot Act. The Bush administration also used immigration policy without legislation in its attempts to limit access to American universities for people who came from a set of countries that it designated as supportive of terrorism, but it also included countries such as China and India.

4. The Bush administration restricted active participation of foreign scholars in the American academic community through its embargo on trade with countries allegedly linked to terrorist activities. One example will give you the idea of the absurdity of some measures taken.

   Thousands of Iranian scholars, scientists, and engineers have studied over the past four or five decades at American universities. They have become members of the international community—and this is particularly so among engineers. When Iranian engineers attempted to publish their scholarly papers in an American journal, the government refused to allow the editors to make any comments, suggestions for revisions, or changes in the manuscripts submitted for publication. This constituted an offense to the embargo on providing these suspect nations with technology and information that could aid them in their support of terrorism. The conditions were so restrictive that the editors of the journal refused to allow Iranian authors to submit papers for publication until this restriction was lifted. An Iranian engineering community, which held highly positive views about the United States, was being systematically alienated.

5. Not since the Cold War and the work by nuclear physicists during that era has the tension between national security and the rights of scientists to publish freely been so dramatic. During the early phases of developing nuclear fission and the atomic bomb, American physicists faced the dilemma of whether they should publish results of experiments that were important and clearly of use to German scientists. The great physicist Leo Szilard and his colleagues debated this issue. There were significant differences of opinion among the physicists whether their results could, in fact, be kept secret and whether scientists ought to be censoring the reports of their discoveries. The government clearly wanted to restrict the release of the important discoveries that were being made at the universities. At the end of the day, the small band of extraordinary physicists decided to restrict access to some of the knowledge that they had developed.

6. The assault by the government and private interest groups on the peer review system was not limited only to government-sponsored research. The pressure to conform to the political agenda of conservative organizations influenced the work of private foundations as well. One stark example was an attempt to shape the grant policies of the Ford and Rockefeller Foundations because of their alleged support of a Palestinian group that purportedly was linked to terrorist activities. Without any substantive

614 ACADEMIC FREEDOM AND FREE INQUIRY

evidence, a claim appeared in a Jewish weekly publication that the Ford Foundation had supported groups that had been highly critical of Israeli government policies. This activated some members of Congress, particularly those who had large Jewish constituencies in New York and elsewhere, to threaten legislative hearings regarding the policies of nonprofit foundations. As a result of blatant pressure, Ford publicly offered a mea culpa and ordered its grants officers to require all recipients of Ford grants to sign a statement that no funds (not limited to Ford funds) were passing from the university to any group that could be considered by the government as associated, either directly or indirectly, with terrorist activities. The policy was deeply offensive to the Ivy League institutions, whose provosts negotiated a limitation of this policy for grants awarded to their universities. But the policy remained in place and suggested that Ford had simply caved to the political pressures of the day—pressures from sources that covered a wide ground of political territory. Of course, no institution need accept Ford Foundation funding, but creating a policy that brazenly affronts the basic idea of free inquiry and academic freedom does as much harm to the foundation as it does to the universities who refuse to comply with Ford's wishes. For a much more detailed discussion of this incident and its effects on academic freedom, see Cole (2010).

7. Here I follow closely the description of the case in David Bromwich (1994: 33–34).
8. When we deal with the First Amendment in academic settings, we must remember that state and private universities are subject to very different constraints. Private universities are not regarded as "state actors," and consequently, constitutional constraints on free speech are not applicable to them in the legal sense. Public universities are subject to the same free-speech doctrine that any citizen or legislative body must uphold. Thus, we have few rulings on speech codes enacted at private universities. Of course, the moral or ethical dimensions of First Amendment doctrine for speech codes at private universities are reasonable subjects for discussion. I do not see the argument for distinguishing public and private universities as outweighing the value of applying First Amendment doctrine to private institutions. An example of the constraints placed on expression at a public institution can be seen in the policy on Discrimination and Discriminatory Harassment at the University of Michigan in 1989. It subjected people to discipline if, in educational and academic centers, they engaged in (1) any behavior, verbal or physical, that stigmatizes or victimizes an individual on the basis of race, ethnicity, religion, sex, sexual orientation, creed, national origin, ancestry, age, marital status, handicap or Vietnam-era veteran status, and that a) involves an express or implied threat to an individual's academic efforts, employment, participation in University sponsored extra-curricular activities or personal safety; or b) has the purpose or reasonably foreseeable effect of interfering with an individual's academic efforts, employment, participation in University sponsored extra-curricular activities or person safety; or c) creates an intimidating, hostile, or demeaning environment for educational pursuits, employment or participation in University sponsored extra-curricular activities (Greenawalt 1995: 72).

References

Bok, Derek. "Reflections on Free Speech: An Open Letter to the Harvard Community." *Educational Record* (Winter 1985).

Bromwich, David. *Politics by Other Means: Higher Education and Group Thinking* (New Haven, CT: Yale University Press, 1994).

Butler, Judith. "Academic Norms, Contemporary Challenges: A Reply to Robert Post on Academic Freedom." In *Academic Freedom after September 11*, ed. Behara Doumani. New York: Zone, 2006.

Chang, Kenneth. "30 Plague Vials Put Career on Line." *New York Times*, October 19, 2003, 32.

Cole, Jonathan R. *The Great American University: Its Rise to Preeminence; Its Indispensable National Role; Why It Must Be Protected* (New York: Public Affairs, 2010).

Dreifus, Claudia. "The Chilling of American Science: A Conversation with Robert C. Richardson." *New York Times*, July 6, 2004, D2.

Dworkin, Ronald. "A New Interpretation of Academic Freedom." *The Future of Academic Freedom*, ed. Louis Menand. Chicago: University of Chicago Press, 1996.

Enserink, Martin, and David Malakoff. "The Trials of Thomas Butler." *Science* 302: 5653 (December 2003).

Gates, Henry Louis, Jr. "Critical Race Theory and Free Speech." *The Future of Academic Freedom*, ed. Louis Menand. Chicago: University of Chicago Press, 1996.

Greenawalt, Kent. *Fighting Words: Individuals, Communities, and Liberties of Speech.* Princeton, NJ: Princeton University Press, 1995.

Kalven Committee. Report on the University's Role in Political and Social Action: Report of a Faculty Committee, under the Chairmanship of Harry Kalven Jr. Published in the *Record* 1, no. 1 (November 11, 1967).

Louisiana State University, Office of University Relations. "Chancellor Announces Decision to Terminate Steven J. Hatfill," September 3, 2002.

Menand, Louis, ed. "The Limits of Academic Freedom." In *The Future of Academic Freedom*. Chicago: University of Chicago Press, 1996.

Post, Robert. "The Structure of Academic Freedom." In *Academic Freedom after September 11*, ed. Behara Doumani. New York: Zone, 2006.

Rorty, Richard. "Does Academic Freedom Have Philosophical Presuppositions?" In *The Future of Academic Freedom*, ed. Louis Menand. Chicago: University of Chicago Press, 1996.

CHAPTER 31

THE CHILLING EFFECT OF FEAR AT AMERICA'S COLLEGES

(2016)

JONATHAN R. COLE

The coddling of students' minds has resulted in grave restrictions of free speech on campus—but academic leaders are also to blame.

No great universities exist in the world without a deep institutional commitment to academic freedom, free inquiry, and free expression. For the past sixty years, American research universities have been vigilant against external and internal attempts to limit or destroy these values. The First Amendment scholar Geoffrey Stone has noted that free expression, in one form or another, has been continually under attack on campuses for the past hundred years. Today, these core university values are being questioned again but from a new source: the students who are being educated at them.

What explains this recent outcry against free expression on campus? Multiple possible explanations exist, of course, including the hypothesis that parents have coddled a generation of youngsters to the point where students feel that they should not be exposed to anything harmful to their psyches or beliefs. Whether or not these psychological narratives are valid, there are, I believe, additional cultural, institutional, and societal explanations for what is going on. And the overarching theme is that today's youngsters, beginning in preschool, are responding to living in a contrived culture of fear and distrust.

There's hardly consensus among students on the forms or appropriateness of these restrictions on speech. Today, nearly half of a random sample of roughly three thousand college students surveyed by Gallup earlier this year are supportive of restrictions on certain forms of free speech on campus, and 69 percent support disciplinary action against either students or faculty members who use intentionally offensive language or commit "microagressions"—speech they deem racist, sexist, or homophobic. According to a free-speech survey conducted by Yale last year, of those who knew what trigger warnings are, 63 percent would favor their professors using them—by attaching advisories to the books on their reading lists that might offend or disrespect some students, for example—while only 23 percent would oppose. Counterintuitively, liberal students are more likely than conservative students to say the First Amendment is outdated.

Consider a few recent cases: Brown University, Johns Hopkins University, Williams College, and Haverford College, among others schools, withdrew speaking invitations, including those for commencement addresses, because students objected to the views or political ideology of the invited speaker. Brandeis University began to monitor the class of a professor who had explained that Mexican immigrants to the United States are sometime called "wetbacks," a comment about the history of a derogatory term that outraged some Mexican American students. Black students at Princeton University protested against the "racial climate on campus" and demanded that Woodrow Wilson's name be removed from its school of Public and International Affairs. The chilling effect of these kind of restrictions on speech were not lost in 1947 on Robert Hutchins, the president of the University of Chicago, who opined during the McCarthy period, "The question is not how many professors have been fired for their beliefs, but how many think they *might* be."

Born in the mid-1990s, seniors in my Columbia University undergraduate seminars today likely have not experienced major national threats, except for their vague memories of the 9/11 terrorist attacks. Yet these "millennials" might better be labeled "children of war and fear." During their politically conscious lifetime, they have known only a United States immersed in protracted wars against real and so-called terrorists, a place where fear itself influences their attitudes toward other civil liberties. Students are asked to pit freedom of expression or privacy against personal security. During times when elected officials have exploited the public's fear of terrorism for political gain, students seem more willing to trade civil liberties for a sense of security.

Since the 9/11 tragedy, the use of fear is still pervasive in the United States. Indeed, the distortion of fear pervades today's students' thinking—they tend to overestimate, for example, the probability of a terrorist attack affecting them. When this fear is combined with the rapid expansion of social media and the prevalence of government surveillance, students often dismiss concepts like "privacy" as old-fashioned values that are irrelevant to them. In fact, my experience at Columbia suggests that many students believe that the very idea of privacy is obsolete; most of my students don't seem to mind this loss when it's weighed against uncovering potential terrorists.

Add to this apprehension the fears that so many students of color experienced before college—a rational fear of the police, of racial stereotypes, of continual exposure to epithets and prejudice—and it is no wonder that they seek safe havens. They may have expected to find this safe haven in college, but instead they find prejudice, stereotyping, slurs, and phobic statements on the campuses as well. Additionally, many of these students employ the classification of "the insider." Believing that "outsiders" cannot possibly understand the situation that faces these groups of offended individuals, by virtual of race, gender, ethnicity, or some other category, the students often dismiss the views of their professors and administrators who can't "get it" because they are not part of the oppressed group.

Many of the young adults at highly selective colleges and universities have been forced to follow a straight and narrow path, never deviating from it because of a passion unrelated to school work, and have not been allowed, therefore, to live what many would consider a normal childhood—to play, to learn by doing, to challenge their teachers, to make mistakes. Their families and their network of friends and social peers have placed extreme pressure on them to achieve, or win in a zero-sum game with their own friends. While it's difficult to assess the cases, and while myriad factors likely contribute to the poor mental health among college students, in 2015 roughly 18 percent of undergraduates reported being diagnosed or treated for anxiety in the past year, according to the American College Health Association's 2015 annual survey; the rate was 15 percent for depression. Many are taking anti-depressants and anti-anxiety medication upon entry into college.

But there is a different, though equally important, reason many students today are willing to suppress free expression on campus. And the fault largely lies at the feet of many of the country's academic leaders. Students and their families have been increasingly treated as "customers." Presidents of colleges and universities have been too reluctant to "offend" their customers, which may help explain why they so often yield to wrong-headed demands by students. Courage at universities is, unfortunately, a rare commodity—and it's particularly rare among leaders of institutions pressured by students to act in a politically correct way.

It seems that the vast majority of presidents and provosts of the finest U.S. universities have not seized this moment of concern voiced by students as a teaching moment—a moment to instruct and discuss with students what college is about. Too many academic leaders are obsessed with the security of their own jobs and their desire to protect the reputation of their institution, and too few are sufficiently interested in making statements that may offend students but that show them why they are at these colleges—and why free expression is a core and enabling value of any higher-learning institution that considers itself of the first rank. Of course, there are strong academic leaders who do encourage open discussions of issues raised by students while also speaking out against restrictions on campus speech, against speech codes, safe-space psychology, and micro-aggressions. But they are too few and far between.

Students want to be protected against slurs, epithets, and different opinions from their own—protected from challenges to their prior beliefs and presuppositions. They fear not being respected because of a status that they occupy. But that is not what college is about. While some educators and policymakers see college primarily as a place where students develop skills for high-demand jobs, the goal of a college education is for students to learn to think independently and skeptically and to learn how to make and defend their point of view. It is not to suppress ideas that they find opprobrious. Yet students are willing to trade off free expression for greater inclusion and the suppression of books or speech that offend—even if this means that many topics of importance to their development never are openly discussed.

Of all of America's great universities, the University of Chicago seems to have come the closest historically to getting this right. The school's well-known 1967 Kalven Committee report was, I believe, correct when it stated, "The mission of the university is the discovery, improvement, and dissemination of knowledge. Its domain and scrutiny includes all aspects and all values of society. A university faithful to its mission will provide enduring challenges to social values, policies, practices, and institutions. *By design and by effect, it is the institution which creates discontent with the existing social arrangements and proposes new ones. In brief, a good university, like Socrates, will be upsetting.*" Almost fifty years later, at the request of its president, Robert Zimmer, the University of Chicago again articulated its position on "freedom of expression." The short document quotes the historian and former Chicago president Hanna Holborn Gray: "Education should not be intended to make people comfortable; it is made to make them think. Universities should be expected to provide the conditions within which hard thought, and therefore strong disagreement, independent judgment, and the questioning of stubborn assumptions, can flourish in an environment of the greatest freedom." "In a word," the report goes on, "the University's fundamental commitment is to the principle that debate or deliberation may not be suppressed because the ideas put forth are thought by some or even by most members of the University community to be offensive, unwise, immoral, or wrong-headed." Yet students may be signaling that their commitment to "community" values may take precedence over this core value that many administrators have seen as essential for truly great institutions of learning.

A physically safe environment is an absolutely necessary condition for heated debate over ideas. The university cannot tolerate violations of personal space, physical threats, sustained public interruptions of speakers, or verbal epithets directed toward specific students; that lies beyond the boundaries of academic freedom. That doesn't mean, however, that a college or university should introduce policies that will curtail or chill debate, that adhere to the politically correct beliefs of the moment, or that let their leaders off the hook through capitulation to "demands" that stifle discourse and conversations about what a university education aims to produce.

CHAPTER 32

THE TRIUMPH OF AMERICA'S RESEARCH UNIVERSITY

(2016)

JONATHAN R. COLE

Since the nineteenth century, the institutions' thinkers have discovered a dazzling array of new knowledge—yet attacks on academic freedom mean all their potential is now at risk.

Most members of the educated public probably think of America's greatest universities in terms of undergraduate and professional education—in terms of teaching and the transmission of knowledge rather than the creation of new knowledge. This point of view is completely understandable. They are concerned about the education of their children and grandchildren or relate to their own educational experience.

But what has made American research universities the greatest in the world has not been the quality of their undergraduate education or their ability to transmit knowledge, as important as that is. Instead, it's been their ability to fulfill one of the other central missions of great universities: the production of new knowledge through discoveries that change our lives and the world.

The teaching of undergraduate and graduate students is critically important and an integral part of the mission of great universities; some do this very well, others in a less distinguished manner. In many ways, such teaching is the faculty member's first calling. At least at the graduate level, many of the great research discoveries are produced through a collaboration between a faculty member and her students. At the undergraduate level, there is, in fact, contrary to popular belief, a modest positive association between the quality of a professor's research and the assessed quality of her teaching. But, the fulfillment of the undergraduate teaching mission is not what has made America's universities the best in the world.

When educated Americans think of their best universities, they probably don't think that lasers, FM radio, magnetic resonance imaging, global positioning systems, bar codes, the algorithm for Google, the fetal monitor, the nicotine patch, antibiotics, the Richter scale, Buckyballs and nanotechnology, the discovery of the insulin gene, the origin of computers, of bioengineering through the discovery of recombinant DNA, transistors, improved weather forecasting, cures for childhood leukemia, the pap smear, scientific agriculture, methods for surveying public opinion, the concept of congestion pricing, human capital, or "the self-fulfilling prophecy" all had their origins in the country's research universities. Even the electric toothbrush, Gatorade, the Heimlich maneuver, and Viagra had their origins at these great universities. These institutions have become the engines of innovation and discovery that now drive a large part of the economic growth and social change in the United States.

Of course, there are other great universities in Europe and Asia, but there is arguably no system of higher learning that matches that in the United States—as determined by the number of Nobel Prize winners it's produced (more than 350), the impact of its discoveries, the multiple international rankings that have American research universities at about 75 percent of the top fifty institutions and about 60 percent of the top one hundred. What, then, has made these universities the envy of the world? And how, in less than seventy-five years, have they become the preeminent backbone of higher education internationally—the place where many of the brightest and most able young people want to attend and work?

The American research university was born a century after the American Revolution, when Johns Hopkins University opened its doors in 1876. It was an amalgam of the British Oxbridge undergraduate system and the German emphasis on research; Hopkins's focus on inquiry and experimentation drew the attention of some of the late nineteenth century's great academic minds—people like Henry A. Rowland, who became the first president of the American Physical Society. America's research universities, even in their early years, were far more open and democratic than their European counterparts.

Perhaps more importantly, the system was, from the beginning, fiercely competitive. Even in its neonatal state it represented the beginning of academic free agency; it was always competing with other institutions to be the best. The renowned Columbia University physicist I. I. Rabi in the 1930s said that the United States had young people as talented as any in the world, but that it lacked the established leadership found at great German, English, and French universities. As it turned out, the nightmare that was created by Hitler's National

Socialism represented, ironically, a boom for American research universities. The migration of Jews and others at risk in Europe—including intellectuals in virtually all fields—propelled the American higher education system forward.

When this talent was combined with the internalization of the extraordinary value system that derived from the science revolutions of seventeenth-century England, the United States created the foundation on which great research universities could be built. Those core values included meritocracy; organized skepticism (the willingness to entertain the most radical of ideas, but subject the claims to truth and fact to the most rigorous scrutiny); the creation of new knowledge; the belief that discoveries should be available to everyone and that those that make discoveries should not profit from them; the peer-review system that relies on experts to judge the quality of proposed research that's seeking funding; and academic freedom and free inquiry, without which no great university can be established. These values became part of the culture of America's great research universities.

To be sure, these were ideals that have yet to be reached. Some of these values have, in fact, eroded over the past twenty-five years with the increased commodification of the university and the desire by universities and their faculty members to gain income from things like patented discoveries and from contracts with business that want to be associated with a university's football or basketball team. But essentially, these values have represented the foundation on which these universities have been built.

Add to these values the implementation after World War II of the most enlightened federal science policy that the world has ever produced—one that used taxpayer money to fund research; that outsourced the work to the great universities on a competitive basis; that linked research and teaching by concentrating the training of advanced students with laboratory work with a leading professor; that produced funding for veterans to return to school and to those who could not afford college without financial aid; and that granted great autonomy to universities in exchange for the production of new discoveries, increased human capital, and more enlightened citizens—then you have some of the conditions that led to the international preeminence of the American research university system.

Looking at the people who live in this community of scholars offers perhaps the best way to convey how research, combined with teaching, is the principal determinant of a university's greatness. Take Bonnie Bassler, a molecular biology professor working with students from around the world who's doing fundamental science at an extremely high level in her laboratory at Princeton University. She is charismatic and has won almost every prestigious award a scientist can hope for. Bassler works with bacteria that can cause lethal diseases such as anthrax, and her goal, of course, is to make fundamental discoveries that will lead to cures or treatments for these diseases.

She wants to develop molecules that will act as antimicrobial drugs aimed at bacteria that can cause lethal diseases. It turns out that these lethal bacteria are impotent against the human immune system when they attack it alone—but they have the ability to talk to each other (chemically), to strategize, and to attack the immune system at its weakest point. When they attack in great numbers, they can overwhelm a human's immune system; Bassler and her students and colleagues want to find a way to stop the bacterial from "talking."

Bassler is, of course, only one of thousands of extraordinarily talented scientists in America's research universities who train a multinational group of students—who effectively become their extended family members—in how to conduct the research necessary to make profound discoveries. She and other world-class scientists and engineers and humanists show just how deeply embedded teaching is in the research function of America's great universities. Postdoctoral fellows, graduate students, and occasionally undergraduates are learning through doing—they are rubbing minds with some of the most original thinkers of the era. Even if this is not the teaching that the public hears most about, it is going on in abundance at the country's major universities and accounts for the research discoveries and innovations that they're spearheading. Moreover, these research communities have become even more important amid the demise of so many prominent industrial research laboratories, such as the Bell Labs or Xerox laboratories.

Yet all of this research potential is now at risk. Academic freedom and free inquiry are under attack from many political leaders who endorse anti-science positions and who interfere with the relative autonomy of these institutions. The compact between the federal government and the universities is frayed, with weakening trust on both sides leading to thousands of federal government regulations, the stagnation of research budgets, unnecessary controls on research, false claims that the universities are bilking the taxpayer, and, ultimately, the loss of American supremacy in some important research areas.

The universities, looking for new sources of revenue while competing for students and faculty, are increasingly becoming commodities—businesses to be run. Visa policies are far too restrictive—preventing unusual and creative talent from staying in the United States and working at its distinguished postsecondary institutions. There are few great, outspoken, and courageous higher education leaders who have a vision for the future and the capacity to introduce significant change at their institutions. There are even fewer leaders who have "quitting issues"— who would be willing to resign when others undermine his or her core values. These threats represent real challenges for American universities' continued preeminence and to the character of these great institutions of higher learning.

CHAPTER 33

ACADEMIC FREEDOM AS AN INDICATOR OF A LIBERAL DEMOCRACY

(2017)

JONATHAN R. COLE

A cademic freedom and free inquiry are at the very core of the value systems of our great universities. These values enable other values, such as meritocracy or the open communication of ideas, to take shape at universities and, at least potentially, operate as an aspirational system on which greatness can be built. In fact, I have asserted elsewhere that you cannot have a truly great research university without a deep, abiding commitment to and defense of academic freedom (Cole, 2010). It must be woven tightly into the very fabric of a university. I have yet to be offered an existence proof of a truly world-class institution of higher learning that does not demonstrate a vigilant defense of academic freedom and the rights of its faculty and students to engage in free, unfettered, inquiry.[1] These values are foundational.

I want to take this argument a step further. The institutionalization and commitment to academic freedom and free inquiry by those both inside and outside a university is, I believe, a key indicator (of course not the sole indicator) of the existence and form of a liberal democracy.[2] Its existence will allow us to measure whether democratic ideals and adherence to principles of individual liberty and free expression really exist within a society. I shall also argue, in blueprint form here, that we are witnessing attacks throughout the world on these core values and that these attacks and their consequences

---

Globalizations 14, no. 6 (2017): 862–68.

suggest that democratic ideals, core values of a civil society, and innovation and discovery are at significant risk in major, advanced nations today, including the United States.

Let us begin with some historical and contemporary facts that suggest how the repression of these values is indicative of the state of the society in which these universities are embedded. Scholars at Risk (SAR) is an organization that monitors violations of academic freedom in nations around the world. Its mandate is to expose these violations, provide advocacy for scholars who have been victims of abuses of academic freedom, teach others about the existence of these violations of core academic norms, and where possible offer a safe haven for the oppressed. In reporting the results of its 2016 Monitoring Project in *Free to Think*, it analyzed 158 attacks on higher education in thirty-five countries and reported on killings, violence, and disappearances, as well as imprisonments, losses of position, improper travel restrictions, and cases of closing of universities or of military occupations of university campuses (SAR, 2016, p. 4).

In this year's report special attention is given to Turkey, where a series of criminal and administrative investigations were launched from January 2016 against some of the signatories of what has come to be known as the "peace petition";[3] many have since been suspended and/or dismissed from their positions, while other have been detained, arrested, and prosecuted. Pressures on the higher education space in Turkey were compounded in July and August, when thousands of higher education professionals were caught up in sweeping actions taken in response to the failed July 15 coup attempt (SAR, 2016, p. 5).

Similar accounts could be cataloged for Egypt, China, Pakistan, Ethiopia, India, South Africa, Syria, Thailand, and scores of other nations monitored by SAR. Attacks have also been launched against student protests in various countries. And the United States is hardly immune these days to threats against academic freedom and free inquiry (such as attacks by followers of President Donald Trump on scientists who work on global climate change) (Kolbert, 2016).[4] We have seen recently the creation, by followers of the president, of a "new enemies list" of so-called dangerous professors, who have been placed on a watch list because, supposedly, they "advance leftist propaganda in the classroom" (Yancy, 2016). What does this form of oppression tell us about the nature of the societies in which these forms of repression take place?

Twentieth-century U.S. history suggests an answer to this question. In periods where the society has repressed free speech and restricted other constitutional rights, we have witnessed simultaneous threats to academic freedom and free inquiry at our best research universities. Consider just a few examples. After the passage of the Espionage and Sedition Acts of 1918, supported by President Woodrow Wilson, we find suppression of the university's core values.

For example, Columbia's president, Nicholas Murray Butler, would tell a group of alumni in a commencement address in 1917:

> In your presence, I speak for the whole University . . . when I say . . . that there will be no place at Columbia University, either on the rolls of its faculty or on the rolls of its students, for any person who opposes or who counsels opposition to the effective enforcement of the laws of the United States, or who acts, speaks or writes treason. The separation of such person from Columbia University will be as speedy as the discovery of the offense. (cited in Cole, 2010)

Butler carried out his pledge, firing a number of professors who opposed the draft openly. Others, like the historian Charles Beard, left Columbia University because of these threats. Still others followed and opened the New School for Social Research, which was dedicated to supporting academic freedom and freedom of expression within the academic community.

During the years when the House Un-American Activities Committee was active in the 1940s and 1950s, and Joseph McCarthy was, following World War II, producing a "red scare," academics were targeted and legislation was passed that required teachers and professors to sign loyalty oaths, testify about their political commitments, and "name names" of others they knew had similar commitments. Many lost their jobs. There were champions of academic freedom and free inquiry as well. None was more articulate and vigilant and courageous than Robert Hutchins, president of the University of Chicago. As Hutchins put it, and this applies as much today as it did in his day, "The question is not how many professors have been fired for their beliefs, but how many think they *might* be. The entire teaching profession is intimidated" (cited in Cole, 2010).

The anti-intellectualism in the United States that combined ambivalent attitudes toward intelligence, which most people admired, and the intellect or the expert (which was the focus of hostility), continued to simmer and occasionally rise to a boil. Nixon had his enemies list. The Vietnam War had its repression of free speech—witness the shootings at Kent State University as a symbol of that set of national attitudes toward student protests. President Reagan did what he could to eliminate resources for the social and behavioral sciences—cutting the NSF budget for those fields by roughly 75 percent. And, more recently, President George W. Bush, pushed through the USA Patriot Act (October 2001) and Bioterrorism Preparedness and Response Act (June 2002) following the terrorist attacks on the World Trade Center and the Pentagon, which had many provisions that undermined the ability of scientists to work freely without review by agencies like the FBI. Under the threat of criminal prosecution, faculty members were not permitted to allow graduate students from certain nations that supposedly "sponsor terrorism" to even enter their laboratories. All of this produced apprehension and a chilling effect on discourse and debate—and some substantial

628 ACADEMIC FREEDOM AND FREE INQUIRY

limits on biological and environmental research (for a far more extensive discussion of the repressive aspects of these pieces of legislation and their effects on free inquiry and academic freedom, see Cole, 2010). In fact, as Robert C. Richardson, Cornell's 2004 Nobel Prize winner, opined, before the Patriot Act, there were 33 people working with "select agents" in an effort to find cures or antidotes for lethal diseases. After the enforcement of the Act, that number plummeted to 2 (here I paraphrase Richardson; for the full quote, see Cole, 2010, p. 399).

It has always seemed to me that in the 1960s, during times of great social progress and unrest in the United States, the University of Chicago got it right. In a brilliant, lucid, and short statement of principles about what a university stands for, a committee chaired by the renowned law professor Harry Kalven authored a three-page statement about the role of the university in difficult times. Without a university that is permitted, even under trying external circumstances, to reign free and despite harsh criticism of government policies by faculty and students, to be supported by that government, then society has lost its democratic soul. The 1967 Kalven Committee Report opined,

> A university faithful to its mission will provide enduring challenges to social values, policies, practices, and institutions. By design and by effect, it is the institution which creates discontent with the existing social arrangements and proposes new ones. In brief, a good university, like Socrates, will be upsetting. (The Kalven Committee, 1967)

But consider what a remarkable expectation this is coming from a great research university such as Chicago. The university is arguing that it has a social compact with the nation—a liberal democracy. It will produce independent, critical thinking among its students and faculty; it will prepare students for more highly skilled jobs than had previously existed; and perhaps most importantly, it will produce, because it is unfettered, a set of fundamental innovations and discoveries that will help propel the American economy forward, that will create new, higher-skilled jobs that graduates will be prepared for, and will produce programs for greater justice and equality than would otherwise be obtained. And in return, the federal and state governments will, despite all of the criticism they may receive, grant these universities substantial autonomy and freedom to criticize the very social arrangements constructed by government representatives. This is a huge, but noble, compact if it is entered into. As Justice Oliver Wendell Holmes Jr. said in his famous dissent in the 1919 Abrams case,

> Persecution for the expression of opinions seems to me perfectly logical. If you have no doubt of your premises or your power and want a certain result with all your heart you naturally express your wishes in law and sweep away all opposition. . . . But when men have realized that time has upset many fighting faiths,

they may come to believe even more than they believe the very foundations of their conduct that the ultimate good desired is better reached by free trade in ideas. (*Abrams v. United States*, 1919)

Increasingly, we are observing governments that seem not to believe that the rule of law and the ideas expressed freely at great universities may yield something more and different than what they believe with all their power to be right. When that happens, academic freedom and free inquiry are in peril. That is what seems to be happening in many societies today.

This exercise of power and repression is becoming increasingly symbolized in Hungary, Poland, Russia, China, and elsewhere, of their support for what Hungary's Viktor Orbán has called, in true Orwellian fashion, "illiberal democracy." But what is meant, in practice, is that governments will withdraw basic civil liberties from its citizens (often in the name of stability or economic gain), such as freedom of expression, freedom of the press, privacy rights, the right to petition their government, protection of minorities, and freedom to criticize their policies. It will close its borders to foreigners; it will use fear as a political instrument. And, if individuals in certain positions, including university positions, do not fall in line behind these ideas, they are purged from their jobs—or silenced out of fear of the prospect of losing their position or worse. It is, essentially, a movement toward authoritarian rule.

As in the past, one of the biggest targets, as seen by the governments that adopt "illiberal democracy" principles and values, is the intellectual, the scientist, the scholar, or the critical essayist who works at a major university. And research, which has the potential for so much benefit for these societies, is curtailed for ideological reasons. Anti-science becomes the brother of anti-intellectualism. Professors and students become threats and, as happened so heinously under National Socialism in the early 1930s, they are dispensed with in one way or another—often, as Holmes suggested, under the penumbra of the legal system. "Illiberal democracy" has rejected citizen participation in electing a representative government, governmental pluralism involving the division of power among multiple branches of government, basic civil liberties and the protection of the government of those who are disadvantaged, the concept of the rule of law and an *independent* judiciary that has the ultimate authority to interpret statutes and constitutional amendments, and an independent civil society, and individual rights, particularly freedom of expression by individuals and the press, freedom from government censorship, and freedom of artistic expression, as John Shattuck, the former rector and president of the Central European University has recently articulated (Shattuck, 2016).[5] The stakes are actually quite high in this move toward authoritarian rule. If we simply look at the existential threat of how global climate change could affect the future of our planet for those who live here—and we push back the clock under the mistaken idea that the threat of global climate change is merely the whining of a bunch of scientists and intellectuals who should be ignored, then the potential for a bleak future for our planet can be set before the anti-science forces come to their senses or lose power.[6] By then, it may be too late.

The political correlates of the repression in an authoritarian fashion of academic freedom and free expression were well expressed at a recent SAR sponsored conference by the President of Ireland, Michael D. Higgins, at Trinity College Dublin on November 29, 2016, echoing almost a hundred years later the words of Justice Holmes:

> Ideas, the free discussion of ideas, the critique and questioning of received ideas and the articulation of new ones are activities that are fundamental to the shaping of public discourse and to the vitality of democratic life.

Yet, around the globe, thousands of scholars are currently under attack because of their peacefully expressed thoughts and their words. Indeed, the attacks we are witnessing on educational communities do not simply reveal the vulnerability of the university space in societies that are plagued by violence and political instability. Scholars and universities are not simply the collateral casualties of conflict; they very often are the very focus of such conflict.

> Those who, today as yesterday, use violence to repress scholarly research, teaching and writing seek to subdue, or even eliminate, the spaces in which citizens are free to think, share ideas, and challenge the status quo. Such authoritarian abusers of freedom in any of its senses see the open, pluralist space of intellectual inquiry as a threat to their power, whether this power and its projects invoke a distorted and hateful version of religion and faith, or an authoritarian conception of the state—or both, for that matter. (SAR Ireland, 2016)

If the status of academic freedom and free inquiry in a nation's flagship universities provide us with an indicator of the extent to which that society abides by fundamental democratic principles, what is lost when those values in our universities are abridged or placed in peril? The critical voice is silenced and creativity, imagination, and the path to discovery are lost. Without academic freedom and free inquiry we would not have had cures for diseases, like childhood leukemia, or the GPS system, the FM radio, buckyballs, breakthroughs in biology like the discovery of prions and more recently gene editing techniques such as CRISPR/CAs9, which are leading us toward cures and treatments of multiple diseases. We might not have had the knowledge that prions determine diseases, we might not have had the imagination to pursue ideas such as gravitational waves, the Higgs boson, or the laser—often fundamental discoveries made at universities or by a league of creative minds that ultimately have had or may have huge practical implications despite originally being thought of as "discoveries in search of applications," Nor would we be working on new ways of controlling carbon emissions that threaten our planet. As L. Rafael Reif, president of MIT, put it in a recent opinion piece,

> From smartphones to supercomputers to LED lighting, today's innovations emerged from discoveries that were "new" decades ago. If we hope for technological solutions in the future to some of humanity's great challenges—Alzheimer's, cancer, infectious disease, cybersecurity, safe nuclear power, climate change, water and food for the world—we much renew our national commitment to supporting basic science. That reinvestment is also vital if we want more emerging opportunities in fields like advanced materials and manufacturing, renewable energy, photonics, quantum computing, synthetic biology and space exploration to benefit our society with jobs and quality of life. (Reif, 2015)

We would silence the critical voices of a Trilling, Chomsky, Said, Kahneman and Tversky, Merton, Elster, Piketty, or Schelling. We would not be able to understand behavior and thought processes that are involved in decision-making. Nor would we be able to look critically at our society in the belief that knowledge can lead to a more just and fair society for all—even if thus far we have failed to take the analysis of growing inequality of income and wealth and translate that into social policy. But we can begin to analyze what are the consequences of these inequalities or consider ways of addressing them. These ideas, and thousands of others, came from individuals located at great universities where they were given the freedom to pursue their ideas in a relatively unencumbered way. Investments in research universities and a commitment to the belief that only with great institutional autonomy and by leaving scientists, engineers, and scholars to pursue freely their wildest, and yes, sometimes very radical ideas, will we make remarkable discoveries that contribute greatly to the nation's welfare. But this kind of work cannot exist in an illiberal state, or a quasi-authoritarian state, that hides behind a veil of so-called democratic elections.

Disclosure Statement

The author is a member of the Board of Scholars at Risk (SAR).

Notes

1. This is not intended as a tautological statement. I would like to see evidence of greatness at an institution of higher learning that does not abide by these values.

632 ACADEMIC FREEDOM AND FREE INQUIRY

2. While I am suggesting that the state of academic freedom and free inquiry will provide us with strong clues as to whether the nation in which these universities exist are liberal states, respecting individual rights, I am not making a causal statement. If there is causal direction here it probably is that the degree of suppression of individual liberty leads to the repression of academic freedom and the closing down of creativity and discovery at that nation's research universities.

3. See Özkırımlı's essay in this issue for more details.

4. Thus, scientists belonging to the American Geophysical Union have recently created a new publication, *Handling Political Harassment and Legal Intimidation: A Pocket Guide for Scientists*.

5. To put this in a slightly different framework, "'illiberal democracy,'" as John Shattuck has discussed it, attempts to foster ethnic, racial, and linguistic homogeneity, portrays the nation as a victim of external forces (which leads to a nationalistic rhetoric, arguing for the virtues of a centralized authority that will, supposedly, foster efficiency, and perhaps most importantly, espouses government control—even ownership—of print and television media—of sources of a citizen's information. I want to thank John Shattuck for sharing these ideas with me in private conversation.

6. And if we enter the emotional realm of discussing the treatment of Palestinians by the Israeli government or the behavior of American groups and the U.S. Congress toward academics who question, not the existence of Israel, but its policies and American policies related to Israel, we face the open question of whether the treatment of Palestinians and their opportunities for full citizenship and equal rights by the Israeli government represents such a gross violation of freedom that it suggests the existence of an illiberal state and one that violates, as well, basic tenets of academic freedom and free inquiry? Are Arab citizens living in Israel, who have extraordinary talent, being denied equal access to the best Israeli universities by virtue of their Palestinian origins? Is that an indicator of the nature of Israeli society today? We might also consider just how "liberal" our democracy has been, given that the United States codified a form of the Orwellian term "illiberal democracy." The U.S. Constitution codified, of course, the absence of full rights for American women and its slave population of African-American people who were considered chattel; it did not produce equal opportunities for all, and despite the rhetoric of equal protection and due process, these ideal values were not extended to all Americans. On a relative scale we may represent a society that protects individual freedoms, but even today many of our citizens, including minorities, immigrants, women, and members of the LGBTQ (lesbian, gay, bisexual, transgender, queer) community are not treated equally despite our rhetoric.

References

Abrams v. United States, 250 U.S. 616. (1919).

Cole, J. R. (2010). *The Great American University: Its Rise to Preeminence, Its Indispensable National Role, Why It Must Be Protected*. New York: PublicAffairs.

Kolbert, E. (2016). "Why Scientists Are Scared of Trump: A Pocket Guide." *New Yorker—Daily Comment*, December 8. Retrieved from http://www.newyorker.com/news/daily-comment/why-scientists-are-scared-of-trump-a-pocketguide

Reif, R. L. (2015). "The Dividends of Funding Basic Research." *Wall Street Journal*, December 5, 1015.

Scholars at Risk. (2016). *Free to Think 2016*. Retrieved from https://www.scholarsatrisk.org/wpcontent/uploads/2016/11/Free_to_Think_2016.pdf

Scholars at Risk Ireland (Universities Ireland). (2016). Special event featuring address by President of Ireland, Michael D. Higgins, November 29. Retrieved from http://crossborder.ie/scholars-at-risk-ireland-universities-irelandspecial-event-featuring-address-by-president-of-ireland-michael-d-higgins/; http://www.president.ie/en/media-library/speeches/speech-at-a-conference-organised-by-scholars-at-risk-ireland

Shattuck, J. (2016). "Resisting Trumpism in Europe and the United States." *The American Prospect*, December 2. Retrieved from http://prospect.org/article/resisting-trumpism-europe-and-united-states

The Kalven Committee. (1967). Report on the university's role in political and social action, November 11. Chicago, IL: University of Chicago.

Yancy, G. "I Am a Dangerous Professor." *The New York Times*, November 30, 2016.

CHAPTER 34

ACADEMIC FREEDOM UNDER FIRE

(2021)

JONATHAN R. COLE

Academic freedom is under fire across the world—from Hong Kong, where the Chinese government's crackdown on open discussion has spurred an exodus of scholars from universities, to the United States, where there is a rising tide of anti-intellectualism and assaults on free inquiry. The effort to undermine this cornerstone of American democracy has been driven by former President Trump and his administration, members of Congress, state governors, and legislators. Alas, it also comes from faculty and students on the ideological right and left, and even from some presidents of the country's research universities. These are powerful forces that must be overcome.

The defense of academic freedom is never easy because the university is not a cloister and is no more a safe space from criticism or conflict than the society at large. Thus, there will be unpopular opinions expressed in scholarly research and in the classroom. But it is often unpopular opinions that shift paradigms, fell orthodoxies, and advance knowledge and inquiry to new levels of achievement. As the Kalven Committee stated in its 1967 report on the university's role in political and social action in the United States, "A good university, like Socrates, will be upsetting." De facto speech codes, the idea of "privileged knowledge," and weaponizing misunderstood concepts such as "critical race theory" all limit research on political grounds and threaten academic inquiry from within its own environment. To cancel speeches, to sanction students or faculty for their views, and to protect oneself from the upsetting ideas and language that are encountered in great texts and novels may "purify" universities but only by diminishing them. The American academician Robert Hutchins, perhaps the greatest defender of academic freedom, said that the problem with witch hunts was "not how many faculty would be fired for their beliefs, but how many who think they might be."

Academic freedom can slip away slowly, and if ignored, could leave society in a climate of censorship, mistrust, and even fear. Now is the time for bold and courageous leadership. Presidents and provosts of universities in the United States need to act collectively by, for example, issuing a public statement that defends academic freedom. Boards of trustees or regents should sign on as supporting this statement of principle. Moreover, the political and overly commercial influence on governing boards must end. Members of these boards too often consist of political appointments or elected representatives with a political point of view. Thus, the tendency is for special interests to cloud their notions of how the university should be governed—from developing the institution's mission and strategic goals to establishing policies. Rather than reinforcing university values—foundational academic, intellectual, and artistic freedoms—university governance is vulnerable to economic and political forces. People across this country, no matter what side of the political divide, want excellent education for their families. Universities must make the case that excellence is threatened by political interference and either restructure their boards or more closely scrutinize member influence.

In 1915, and again in 1940, the American Association of University Professors released a statement on the principles of academic freedom. It is time for these principles to be examined and fortified. Courts in the United States have defended academic freedom as falling in the penumbra of the Constitution's First Amendment, which enforces the freedom of speech, among other rights. An update to these principles that ensures the free search for truth and its free exposition in the name of the common good would fortify academic freedom and potentially be codified into federal law. This effort could involve a presidential blue-ribbon panel of experts that is open to new thinking about the principles. It should include discussion of free expression by students as well as faculty and protect remarks made outside their formal capacity as faculty members.

The university has a central role in the growth of knowledge, the exposure of students to diverse thoughts and differing views, and the preparation of a next generation for civic life. If that role is being undermined from outside or even within the university, its leaders should rise to its defense, even if it means jeopardizing their own jobs or removing those leaders who are a threat. It is a public principle worth defending at a high personal cost. When called for, such courageous action will educate the public and legislators on the importance of academic freedom.

INDEX

AAA (American Association for the Advancement of Science), 147n26, 471
AAP (Association of American Physicians), 159
ABC World News Tonight, 531–32
Academic Bill of Rights, 555, 556
academic freedom: absolute protection for, 586–87; censorship and, 595, 635; codes of conduct and, 577; components of, 582–83; concept of, 18–19; David Project and, 576–77, 589; defending, 587–88, 609, 634–35; diversity of student bodies and, 604; employee protections and, 608–9; encouragement of, 592–93; external groups attacking, 578–79; free speech compared to, 593; goals of, 581; ideological conformity and, 607–8; importance of, 625, 630–31; intellectual courage and, 605–6; intellectual orthodoxy as threat to, 602–8; intrinsic value of, 609; as liberal democracy indicator, 625–31; "negative liberty" and, 609–10; open communication *vs.* national security and, 599, 613n5; political correctness and, 608; political litmus tests and, 578–79; pragmatism of, 609; professor and student roles in, 585–86; public attacks on, 576; Red Scare and, 580–81, 627; restrictive visa policies and, 598–99; returning to fundamentals of, 608–12; rigorous skepticism of all ideas for, 582–84; SAR monitoring, 626; scientific integrity and, 600–602; speech codes and, 209, 577, 603, 607; state intrusion into university affairs and, 596–98; Students for Academic Freedom, 577, 585, 589; study of, 19–20; tenure and, 602, 605; threats to, 574–75, 589–90, 593–602, 630, 634–35; tolerance of unsettling ideas for, 582–84; USA PATRIOT Act's impact on, 579–80. *See also* universities after 9/11

access: affirmative action and, 253, 289, 318; in efficient communication system, 58; gate-keepers and rewards, 62; gender differences in output and, 290, 299; gender discrimination and resource, 288; Jewish scientists and, 392; student visa limitations denying, 571; tuition costs for research universities and, 242; women in science and, 275, 290, 298–99, 313, 391

638 Index

accumulative advantage: citation counts and, 508; gender differences in output and, 402; inequality and, 352; NSF peer-review system and, *427*, 429–30; women in science and, 282

ACE (American Council on Education), 154, 155, 181, 420

acquaintance with knowledge, 84, 272, 275, 281

ADD *(American Doctoral Dissertations)*, 291, 322n4

admission system: affirmative action and, 253, 289, 318; Common Core and, 238; criteria for, 237; curiosity and, 238–39; diversity and, 237; goals of, 234; importance of, 237; officers of, 234–35; standards of, 235–36; students left behind by, 236; test scores and, 235; transformation of, 236–37; USA PATRIOT Act and, 570; women in science and, 250–51

Adorno, Theodor, 408n30

advances. *See* scientific progress

affirmative action, 253, 289, 318

Age and Achievement (Lehman), 179

age: of authors in sociology of science, 127–28; awareness of work and, *59*, 59–60, *66*, 66–67; elitism and, 448n12; eminent female scientist productivity and, 336–37, *340*; gender differences in output and, 292; journal reading and, *66*, 66–67; marriage, motherhood and, 330–31; in NSF peer-review system, 427; patterns of influence in scientific progress and, 98n18; of U.S. medical schools, 169, 175n31; visibility of work and, 54–56, *55*

age and scientific progress: citation counts and, 181–85, *184, 185*; codification hypothesis of, 191–94, 196; current output and, *182*, 182–83; early recognition and, *189*, 189–90; field differences in, 191–95, *194, 195*; financial resources and, 190; history of science and, 179–81, *180*; Lehman on, 179–81, *180*; in mathematics, 183–88, *184–86, 188*, 193; mean age at first important discovery, 193–95, *194, 195*; old beliefs on, 178, *179*; productivity and, 181–88, *182, 184–86, 188*; reward system and, 188–91, *189*, 195–96; in single-authored

compared to multi-authored papers, 190–91, *191*; tenure and, 196n2

Age of Little Science, 44n18, 382, 405n3

Ahrens, Edward H., Jr., 519–21

Alien and Sedition Acts of 1798, 564

Allison, Paul, 171, 197n13

Altman, Lawrence K., 524–25

American Academy of Arts and Sciences, 240

American Association for the Advancement of Science (AAA), 147n26, 471

American Association of Microbiology, 570

American Behavioral Scientist, 147n26

American Council of Trustees and Alumni, 555, 559n1

American Council on Education (ACE), 154, 155, 181, 420

American Doctoral Dissertations (ADD), 291, 322n4

American Heart Association, 524–25

American history, teaching of, 550–53

American Journal of Sociology, 142

American medical schools. *See* U.S. medical schools

American Men of Science (AMS), 32, 42n3, 45n20; patterns of influence in scientific progress data from, 96n7; systematic differences in authors in, 448n9

American Nation, The (Garraty and McCaughey), 552

American research universities: challenges facing, 624; discoveries made at, 380–81, 622; German scientists fleeing for, 591–92, 622–23; preeminence of, 594; scientific progress at, 380–81, 575; talent at, 623–24; triumph of, 621–24; in world rankings, 612n1, 622. *See also* research universities; U.S. medical schools

American Sociological Association, 142; Merton's address to, 2, 230; sociology of science's growth within, 120–21

American Sociological Review, 142, 194

AMS. *See American Men of Science*

analysis-of-variance techniques, in COSPUP NSF peer-review study, 456, 480

Andersen, Charles J., 154, 171

Anderson, Lisa, 558

anticipation of future events, theory of limited differences and, 373 appointment to academic department. *See* professorial rank; rank of department

Ashcroft, John, 571, 579, 612n2

Assessment of Quality in Graduate Education, 42n3

assistant professorship, of women in science, 276

Association of American Physicians (AAP), 159

Astin, H. S., 277

astronomy, gender differences in output in, 294

atherosclerosis, 527–28

atomic and molecular physicists: awareness of work and, 64, *65*; size of field of, 107–8; visibility of work and, 54, *54*

authority, at research universities, 203, 207

Avery, Oswald, 383

awards: awareness of work and, *61*, 61–62, 72n42; exaggeration of knowledge of, 72n42; patterns of influence in scientific progress and, 85–86, *86*; in physics, 32–33, *33*; for quality of sociological research, 102–3; quantity of, 61, *61*; rank of department correlated to, 71n36, 81; in reward system of science, 32–35, *33*, *34*, 45n21, 46n25; theory of limited differences beyond first major, 362; visibility of work and, *51*, 52, *53*, 102–3; women in science and, 258–59, 277, 385. *See also* Nobel Prize; reward system of science

awareness of work: age and, *59*, 59–60, 66, 66–67; awards and, *61*, 61–62, 72n42; definition of, 58; geographical location and, 62, 62–63, *63*, 71n40; gross awareness, 62; journal reading and, 66, 66–67, 72n43; physicist types and, 64, *65*, 66; professorial rank and, *59*, 59–60; rank of department and, *60*, 60–61, 64, *64*; scores of, *58*, 58–59, 71n34; specialties and, 64, *65*, 66; visibility compared to, 49, 67

awareness scores, *58*, 58–59, 71n34

Bailyn, Bernard, 408n30

balance, enlightenment compared to, 586

Barber, Bernard, 3, 120, 137, 144

Barnes, Barry, 14

Barone, Michael, 550

Bartók, Béla, 408n30

Bassler, Bonnie, 623–24

Bayer, Alan E., 182–83, 277, 376n17

Beadle, George, 383, 388–89, 400, *400*

Beard, Charles, 627

beauty contests, subjectivity in, 498

Beauvoir, Simone de, 380

Becker, Gary, 12

Bell, Daniel, 551

Bell, Laird, 581

Bell Labs, 11

Ben-David, Joseph, 141, 150n60

Benedict, Ruth, 386

Bentley, Eric, 16

Berlin, Isaiah, 609

"best" papers, patterns of influence in scientific progress and, 88–90, *89*

Beyond Culture (Trilling), 582

Big Test, The (Lemann), 230

binomial probability distribution, 510–11

biochemistry: age and scientific progress in, 194, *194*; gender differences in output in, 294

Bloor, David, 14

Bok, Derek, 219, 606

Bollinger, Lee, 559, 568, 576–77

Brandeis University, 617

Brecht, Bertolt, 408n30

Brinton, Crane, 95n1

Brody, Jane E., 524–25, 530–33

Bromwich, David, 604–5

Brown, Michael S., 521–22

Brown v. Board of Education, 499

Brustein, Robert, 16

Bureau of Applied Social Research (Bureau), 6–7

Bush, George W., 592, 600, 627

Bush, Vannevar, 221, 389, 407n27, 549, 610

Butler, Judith, 609, 610

Butler, Nicholas Murray, 201, 400, 627

Butler, Thomas, 579, 597–98

Caltech, 389

Cam, Helen Maud, 386

Campus Watch, 564

Candolle, Alphonse de, 273

Cardozo, Benjamin, 498

Carnegie Commission, 252

640   Index

Cartter, Allen: limitations of school quality study by, 172; on rank of department in reward system of science, 33, 42n3, 96n5, 97n12, 171–72; on self-aggrandizement, 158
Castelli, William, 526
censorship, academic freedom and, 595, 635
Centers for Disease Control (CDC), 595
Centra, J., 277
ceremonial citations, 77, 508
Chafe, William, 550–51
chain reactions, theory of limited differences and, 375n15
chance, in NSF peer-review system, 451, 459–60, 465, 467–68, 498
chemistry: age and scientific progress in, 192, 193; biochemistry, 194, *194*, 294; differential reinforcement of men and women in, 313; gender differences in output in, 295
cholesterol. *See* dietary cholesterol and CHDs
Chomsky, Noam, 14
citation counts, 10–11; academic physicist practices and, 435–36; accumulative advantage and, 508; age and scientific progress and, 181–85, *184, 185*; "best" papers and, 88–90, *89*; ceremonial, 77, 508; clerical errors and, 111; of collaborative papers in SCI, 110–11; communication and, 103; contemporaneity and, 108–9; critical citations and, 104–5; on deviance, 125; elitism and, 81, 83; errors in, 104, 511; gender differences in output and, 287, 300–306, *302, 303, 305, 306*, 310; halo effect and, 435–36; of "highly-cited" papers, 90–91, 111; incentives of, 310; "intellectual blacklisting" and, 510; of less influential authors, 92, 92–93; of Merton, 133–36; M&M paper on, 506, 511–12; of Mössbauer effect, 109–10, 113n9; Nobel Prize and, 29, 43n11, 447n7; NSF peer-review system and, 424, 426–27; omitted, 436, 509, 510; Ortega hypothesis and author characteristics of, 436–40, *437*; output compared to impact and, 301; pairs of male and female scientists and, 306–9, *307*; pathfinding papers and, 438–40; patterns of influence in scientific progress and, 77–78; on perceived quality of U.S. medical schools, 172; *Physical Review* and, 44n16; of *Physical Review* "super" papers and Ortega hypothesis, 440–42, *441*; physicists and, 103; quality of sociological research and equal weight of, 105–6; quantity and quality of output and, 29, 71n37, 78, 102, 301; recent, 126; reward system of science and, 38–39, *39*; in SCI, 42n8, 77–78, 88–89; of self-citations, 91, *91*; self-selection and, 491–92; significance and interpretation of, 308–9, 323n7; size of field correlations to, 108; for sociology of science, 123–26, *124*, 148n27; straight and complete, 323n7; of "super" papers, 91–92, *92*; tenure and, 11; total number of papers and, 43n13; types of papers and, 314–15; visibility of work and, 435; weighing quality of, 29–39; weighted, 30, 39, *39*, 44n18, 69n23
civil rights movement, diversity and impact of, 201
Clark, Kenneth E.: on citation counts as quality indicator, 102; scientific output study of, 29, 102
classified research, 571
clerical errors, citation counts and, 111
climate change research, 600
clinical psychology, 194, *194*
clusters of papers, gender differences in output of, 299
codes of conduct, academic freedom and, 577
codification hypothesis, of age and scientific progress, 191–94, 196
cognitive consensus, in sociology of science, 123–26, *124*
cognitive development: selected aspects of, 123–40; of specialties, 118–19, 146n13
cognitively conforming specialties, 117–18
cognitively radical specialties, 117–18
cognitive structures, sociology of science and, 119–20
Cohen, I. Bernard, 135
Cohn, Victor, 526, 532
Cold War, 564
Cold War America (Wittner), 552
Cole, Stephen, 3, 4, 15, 98n21, 125, 467–70, 512–13
collaboration: citations to collaborative papers, 110–11; Columbia University and proposal for multi-university, 9–10;

name ordering and, 56, *57*; in sociology of science publications, 128; technology and, 9–10; visibility of work and, 56–57, *57*; of women in science, 267n55. *See also* pairs of male and female scientists

colleague relations, of women in science, 259–60

Columbia Unbecoming, 576–77, 589

Columbia University: academic priorities at, 227n12; budget concerns of, 201; budget growth of, 574; codes of conduct at, 577; David Project and, 576–77, 589; Department of Sociology, 3, 231; free inquiry incidents at, 567; Gluecksohn-Waelsch at, 393–94; great minds at, 16; medical school, 157; Morgan's "fly group" at, 9; multi-university collaboration proposed at, 9–10; Palestinian film festival incident at, 567; political views of professors at, 557; protests at, 567–68; return on investments at, 223; School of Library Service closure at, 205–6

Committee on Science and Public Policy (COSPUP). *See* COSPUP NSF peer-review study

Common Core, 238

communication: access in efficient system of, 58; citation counts and, 103; Kalven Report on open, 566; national security *vs.* open, 599, 613n5; patterns of influence in scientific progress and, 96n3; of scientific progress, 49; sociology of science in formal, 142–43; visibility of work and, 48–49. *See also* awareness of work; visibility of work

community: science as, 11–12, 94; scientific progress from, 75–76; universalism and, 5

compartmentalizing, marriage, motherhood and, 341

competence: knowledge for, 192; of professors and faculty, 577, 586; in sociology of science, 120–21; theory of limited differences and, 375n15

competition: for academic resources and funds, 204, 222, 420, 430; for priority, 134, 138; at research universities, 217–18; theory of limited differences and, 369–70

complete citation counts, 323n7

Comprehensive Dissertation Index, 121

computers, science and, 7–8

Comte, Auguste, 273

conflict of interest policies, at research universities, 224

Conlan, John B., 417, 420

Conley, Patricia, 235

consensus: articles on, 500n3; case study on COSPUP NSF peer-review study and, 493–95; in COSPUP NSF peer-review study, 456–59, *457, 459,* 480–84, *481, 483*; mythology of, 497; reward system in science and, 17; in science, 498–99

conservative professors, intellectual diversity and, 557

consolidation of research fronts, 126–27

contemporaneity, quality of sociological research and, 108–9

Cooper Union, 237

core of knowledge, 15

Cori, Gerty, 259

Cornell, 388–89

coronary heart deaths (CHDs). *See* dietary cholesterol and CHDs

correspondence theory of truth, 212, 216

COSPUP NSF peer-review study, 5; analysis-of-variance techniques in, 456, 480; case study on consensus in, 493–95; on chance, 451, 459–60, 465, 467–68, 498; concerns with, 497; conditions of, 461n3, 503n6; dissensus case study in, 495–96; on elitism, 460–61; first phase results of, 451, 461n2, 474, 503n5; "interaction effect" and, *457,* 457–58, *481,* 482; "method effect" and, 457, *457, 481,* 481–82, 484; number of reviewers and reversals in, 484–86, *485*; on past performance, 486–89, *488*; on personal values and ideology, 498; on precision, 461; on proposal review panels, 465–66, *469*; proposal variance in, 481, *481*; rank order of proposals in, 476, *476, 477,* 478; reputation and, 487–89, *488,* 504n14; reversals in, *454,* 454–56, 459–60, 462n6, 470, 478–80, *479,* 484–86, *485,* 492; reviewer comparisons in, 452–53, *453,* 475–76, *476*; on reviewer disagreement, 460, 468, 479–80, 492–93; reviewer selection bias and, 479; reviewer variation and, *481,* 482; second phase of, 451–53, 475; on self-selection, 489–92; variance and consensus in, 456–59, *457, 459,* 480–84, *481, 483*

642   Index

counter-sorters, 7
Crane, Diana, 10, 69n16
creationism, 578, 584
Crick, Francis, 443
critical citations, 104–5
Cronbach, L., 468
Crowther, J. G., 95n1
culture: endogenous and endogenous forces on, 402–3, 409n46; of excellence at research universities, 219; null hypothesis and, 403–4; of resisting women in science, 274, 402–3; socioanthropological approach to, 409n44; unfreedom in science and, 401–4
cumulative advantage. *See* accumulative advantage
cumulative productivity differentials, 348–49
Curie, Irene Joliot, 259
Curie, Marie, 249, 259; Nobel prize of, 470–71
Curie, Pierre, 470–71
curiosity, admission system and, 238–39
current output: age and scientific progress and, *182*, 182–83; definition of, 181

DA *(Dissertation Abstracts)*, 121, 322n4
Darwin, Charles, 179; theory of limited differences and, 375n15
Daschle, Thomas A., 612n2
David Project, 576–77, 589
Dawber, Thomas Royle, 527–28
Day, E. E., 580
De Bakey, Michael, 526
Delbrück, Max, 383, 408n30
Demerec, Milislav, 389
departmental prestige: ACE study on, 181; citation counts, Ortega hypothesis and, 438; NSF peer-review system and, 426. *See also* rank of department
Department of Health, Education and Welfare (HEW), 159, 167
depression, 618
DerSimonian, R., 539
Development of Sex Differences, The (Maccoby), 267n59
deviance: citation counts on, 125; Merton on, 138, 230–31; sociology of, 123–25, *124*
Dewey, John, 238
Diamond, Sigmund, 580
Dictionary of Scientific Biography, 10

dietary cholesterol and CHDs, 516; disagreements over, 519–20, 522; features missing in reporting on, 518–19; Framingham study on, 527–28; *Toward Healthful Diets* on, 524–27; LDL receptors and, 521–22; linking of, 517; mass media distortion on, 519, 523–24, 533–36; MRFIT Study on, 529–31; NAS report on, 525–26; NIH consensus committee on, 523; persuasion and negotiation in fact dispute over, 534–35; possibility phase of, 524–28; possibility to probability phase of, 528–32; probability to "fact" phase of, 532–33; unproven hypotheses on, 520–21, 533; Western Electric Study on, 528–29. *See also* health-risk information
differential reinforcement, pairs of male and female scientists and, 309–13
dilemmas of choice, facing research universities, 200; federal government partnership and, 221–25; framework for, 206–7; governance and, 202–7; teaching and research balance and, 217–21; "who owns the null," 207–17. *See also* research universities
DiPrete, Thomas, 504n12
discoveries. *See* scientific progress
discretionary time, marriage, motherhood and, 341–43
discrimination: gender differences in output and, 288, *346*, 347; inequality compared to, 275; pairs of male and female scientists and, 298; in promotion, 284; structural constraints and, 353; tenure and, 276; theory of limited differences and, 378n32; against women in science, 259–61, 274–76, 284
disinterestedness, Kalven Report on, 566
dissensus case study, on COSPUP NSF peer-review study, 495–96
Dissertation Abstracts (DA), 121, 322n4
diversity: admission system and, 237; civil rights movement impact on, 201; defining, 234; guidelines for, 235; political pressure for, 210–11; of student bodies, 604. *See also* intellectual diversity
doctoral dissertations, sociology of science and, 121

Double Helix, The (Watson), 94, 96n3, 443
Douglass, Carl D., 269n84
Dunn, L. C., 393, 394
Duran, Jane, 407n20
Durkheim, Emile, 19, 273
Dutton, Jeffrey E., 182–83
Dworkin, Ronald, 606

early events module: eminent female
    scientists and, 366; eminent male
    scientists and, 365–66; in theory of
    limited differences, 359–61
early recognition: age and scientific progress
    and, *189*, 189–90; gender differences
    in output and, 310–13, *311, 312*, 315, *316*,
    324n22; marriage, motherhood and, 337,
    338, 339
earth science, gender differences in output
    in, 294
economic development, research
    universities and, 223
economics: psychology and, 12; of research
    universities, 203, 204, 224; sociology
    and, 12; of tenure, 204; at University of
    Chicago, 586. *See also* financial resources
educational reform, women in science
    with, 383
Education and Employment (Sharp), 263n4
egalitarianism, universalism compared to, 94
Einstein, Albert, 179, 408n30
Eisenberg, David, 470–71
elementary particles physicists: awareness of
    work and, 64, *65*; size of field of, 107–8;
    visibility of work and, 54, *54*
Eliot, Charles William, 398–99
elitism: age and, 448n12; ceremonial
    citations and, 508; citation counts and,
    81, 83; COSPUP NSF peer-review study
    on, 460–61; M&M paper on, 507–8; NSF
    peer-review system and, 416–17; Ortega
    hypothesis and, 443–44; in patterns of
    influence in scientific progress, 76; rank
    of department and, 81, 83; in science,
    5–6, 17; scientific progress and, 443–44;
    taste and, 88
embryonic stem cell research, 600
eminent female scientist productivity: age
    and, 336–37, *340*; early events module
    and, 366; motherhood, marriage and,

333–37, *335, 340*; single women and, 336;
    theory of limited differences case study
    on, 366–68
eminent male scientists: early events
    module and, 365–66; productivity
    of married, 336; theory of limited
    differences case study on, 364–66
employee protections, academic freedom
    and, 608–9
employment patterns, of women in science,
    252–53, 266n45
endogenous forces, on culture, 402–3,
    409n46
engineering, PhDs of women in, *332*
Enhanced Board of Security and Visa Entry
    Reform Act of 2002, 571, 613n3
enlightenment, balance compared to, 586
equity, NSF peer-review system and,
    430–31, 474
Ernst, Max, 408n30
errors: citation counts and clerical, 111; in
    evaluation and quality of sociological
    research, 104; M&M paper on citations
    and, 511; NSF peer-review system, talent
    measurement and random, 464–65;
    perceived quality of U.S. medical schools
    and standard, 160, 174nn21–22
Espionage Act of 1917, 564
event list, productivity puzzle and primary,
    354, 354–55
evolving social processes, theories of, 351–52
exogenous forces, on culture, 402–3, 409n46
experimental psychology, 194, *194*

"factlike status," 535
faculty attitudes, toward women in science,
    251–52
faculty eminence, perceived quality of U.S.
    medical schools and, *165*, 166–67
faculty governance, at research universities,
    204, 207, 227n11
failure, in learning, 238
fairness in science: peer review and, 4;
    universalism and, 15–17
Fair Science (J. Cole), 4
Family and Educational Rights Privacy
    Act, 571
Faraday, Michael, 287, 319, 472
fear, universities after 9/11 and, 617–20

644   Index

federal government, research universities' partnership with, 221–25
Fermi, Enrico, 408n30
field differences: in age and scientific progress, 191–95, *194, 195*; gender differences in output and, 294–95
Fields Medal, 385
filtration of ideas, Ortega hypothesis and, 444
filtration theory, of utilization of scientific progress, 97n13
financial resources: age and scientific progress and, 190; competition for, 204, 222, 420, 430; of research universities, 202, 226n7; of U.S. medical schools, 159, *165*, 167, 170
Fink Committee, NAS, 599
First Amendment, 586, 613n8, 617
first authorship, gender differences in output and, 287, 304, 315
Fischbach, Gerald D., 569–70
Fleck, Ludwik, 534, 542n16
Fleming, Donald, 408n30
Florey, Lord: Royal Society address of, 95n1; on scientific progress, 433
Folger, John, 376n17
Foner, Eric, 382
Ford Foundation, 613n6
14th Amendment, 384
Fox, Mary Frank, 405n3
Framingham study, 527–28
Frankfurter, Felix, 290
Franklin, Benjamin, 220, 243
Franklin, Daria, 18
freedom, 17–18; Americans clinging to idea of, 382–83; defining, 382; McClintock on, 390–91; non-coercive threats to, 381. *See also* academic freedom; unfreedom in science
free inquiry: concept of, 18–20; defending, 567; importance of, 572, 590, 625; for innovation, 575; original formulation of, 608; political interference with, 556, 601; as structural concept, 609; threats to, 559, 584, 592, 601, 624, 626; for truth, 572, 580, 607; value of, 6, 606. *See also* academic freedom
free speech: academic freedom compared to, 593; illiberal democracy repressing,

629; political correctness threats to, 209; racism and, 209–11, 603; at research universities, 209–11, 603; speech codes and, 209, 577, 603, 607; student support for restrictions on, 617; Supreme Court expanding protections of, 586; Vietnam War and suppression of, 627. *See also* academic freedom
Free to Think, 626
Friedlaender, Walter, 408n30
full citizenship, for women in science, 279–81
Furman v. Georgia, 499
future events, theory of limited differences and anticipation of, 373

Galison, Peter, 9, 14
gallbladder disease, lipid-reducing drugs and, 518
Gambrill, Bessie Lee, 386
Gardner, Howard, 237
Garfield, Eugene, 10, 42n8, 43n12, 101, 112n2, 447n7
Garraty, John A., 552
gate-keepers, reward system of science and, 62
Gates, Henry Louis, Jr., 607
Gell-Mann, Murray: "highly-cited" papers of, 91, 111; as irreplaceable, 443; *Physical Review* citations of, 440
gender. *See* women in science
gender differences in output: access and, 290, 299; accumulative advantage and, 402; age and, 292; authorship distinctions for, 322n6; citation counts and, 287, 300–306, *302, 303, 305, 306*, 310; in clusters of papers, 299; constancy of, 286; data on, 291–93; discrimination and, 288, *346, 347*; early recognition and, 310–13, *311, 312*, 315, *316*, 324n22; empirical studies on, 320–21, 322n1; field differences and, 294–95; findings on, 293–97; first authorship and, 287, 304, 315; gradual emergence of, 347; historical contexts for, 289–91; impact compared to, 301; inequality and, 302; IQ scores and, 376n17; later productivity and, 315, *316*, 324n22; marriage, motherhood and, 288, 329–30; in pairs of male and female scientists, 297–313; period effects and, 318–19; PhDs and, 291–92; primary authorship and,

304; of prolific physicists, 295–96, 298; publication norms and, 317; published productivity and, 287–88, 293–97, *294–96*, 304–6, *305, 306*, 309–13, *311, 312*; sex role attitudes and, 291; of silent physicists, 295; in solo authorship, 299, 303–4, 315; in strings of papers, 299; structural determinants on, 317; during "tenure-relevant" years, 293–94; working hours and, 317. *See also* pairs of male and female scientists; productivity puzzle; theory of limited differences

Genova, Nicholas de, 568, 578

geographical location: awareness of work and, *62*, 62–63, *63*, 71n40; NSF peer-review system and, 421; perceived quality of U.S. medical schools and, 168–69

geographic mobility, women in science and, 282

geology, age and scientific progress in, 193

Germany: Gluecksohn-Waelsch in, 391–93; Hitler dismantling science in, 578; scientists leaving, 408n30, 591–92, 622–23

Giamatti, A. Bartlett, 218–19, 226n10

G.I. Bill, 243

Gieryn, Thomas, 6

Gilfillan, S. C., 121

Gillispie, Charles, 9–10, 135

Gini-concentration ratios, for sociology of science, 123, *124*, 125

Glaser, Donald A., 377n28

Gluck, Charles, 523

Gluecksohn-Waelsch, Salome, 386, *400*, 401; at Columbia, 393–94; family of, 393; in Germany, 391–93; migration to America, 392; NIH funding of, 394; scientific contributions of, 392; unfreedom in science stories of, 391–94

Goldstein, Joseph L., 521–22

Gore, Al, 600

governance dilemma, facing research universities, 202–7

graduate school, women in science and, 247, 250–52, 263n4, 276

Granovetter, Mark, 280

Gray, Hanna Holborn, 619

Great American University, The (J. Cole), 592

Great Depression, 384

Great Expectations (Patterson), 551

Griffith, Belver C., 117

Gropius, Walter, 408n30

Gross, Neil, 558

gross awareness, 62

Guerlac, Henry, 134–35

Guillemin, Roger, 379n41

Hagstrom, Warren, 111

Hahn, Otto, 273–74, 408n33

Hall, A. Rupert, 135

Hall, G. Stanley, 290

Hall, Stuart, 409n44

halo effect, citation counts and, 435–36

Hansen, James, 600

Harding, Sandra G., 405n3

Harvard: Eliot on women at, 398–99; free inquiry incidents and, 567; history of science at, 552; medical school, 168

Hatfill, Steven J., 579, 612n2

Havel, Vaclav, 547–48

Hazard, Caroline, 399

health-risk information: development of "fact" in, 533–36; institutional properties influencing distortion of, 536–38; in mass media, 516; peer review issues with, 539; personal values conflicts in reporting, 540; social control mechanisms for, 540–41; structural problems influencing distortion of, 538–40; timing issues reporting on, 539–40; training for reporting, 538–39. *See also* dietary cholesterol and CHDs; mass media

health science divisions, at research universities, 202

heart attacks, 530

heart disease. *See* dietary cholesterol and CHDs

Heckman, James, 12

Heilbron, John, 549

Henderson, L. J., 121, 137

HEW (Department of Health, Education and Welfare), 159, 167

Higgins, Michael D., 630

higher education, public trust and, 558

Higher Education Act, 601

"highly-cited" papers, patterns of influence in scientific progress and, 90–91, 111

Hindemith, Paul, 408n30

Histoire des sciences et des savants depuis deux siècles (Candolle), 273
history of America, teaching of, 550–53
history of science: age and scientific progress and, 179–81, *180*; at Harvard, 552; internalist, 135; Merton's influence on, 134–35; normal science in, 76, 434; Sarton and, 116, 133; teaching, 553
Hitler, Adolf, 391, 393, 408n30, 578, 591
HIV research, NIH-supported, 594–95, 601
Hodgkin, Dorothy, 259
Hofstadter, Richard, 16, 20
Hollander, Jack M., 470
Holmes, Oliver Wendell, Jr., 592–93, 628–30
Holton, Gerald, 11, 546, 553n1
honorific societies membership, 32–33, *33*, 45n20
Hopper, Grace, 394–96
Horowitz, David, 555–59, 601
House, Karen, 559
House Un-American Activities Committee, 574, 627
Hower, John, 465–66
Hughes, Raymond M., 154
humanities: defense of faculty in, 578; free inquiry for innovation in, 575; null hypothesis and, 211; science's relationship with, 13
Human Stain, The (Roth), 410n48
Humphreys, Lloyd, 464–65
Hungary, 629
Hutchins, Robert Maynard, 580–81, 590, 617, 627, 635
Hyde, William De Witt, 399

IBM computers, 7–8
ideological conformity, academic freedom and, 607–8
ideology, NSF peer-review system and, 498
illiteracy, on science and technology, 547–49
impact, output compared to, 301
individual particularism, 397–401, *400*
individual universalism, 398, *400*, *400*
industrial science, 11
inequality: accumulative advantage and, 352; discrimination compared to, 275; gender differences in output and, 302; theory of limited differences and, 344, 362–64; women in science and, 276, 280–81

informal social networks, women in science and, 283–84
initial conditions, theories of, 350–51
Inman, Cullen, 198n16
institutionalization of specialties, 117–19
institutional legitimacy, of research universities, 218
institutional particularism, 397–401, *400*
institutional universalism, 397–401, *400*
integration of basic ideas, quality of research and, 109–10
integrity, scientific, 600–602
"intellectual blacklisting," 510
intellectual courage, academic freedom and, 605–6
intellectual diversity: Academic Bill of Rights and, 555, 556; lack of sustained debate and, 556–57; limits of, 556; political views of professors and, 557; public trust in higher education and, 558
Intellectual Migration (Fleming and Bailyn), 408n30
intellectual orthodoxy, as academic freedom threat, 602–8
intelligent design, 578
"interaction effect," COSPUP NSF peer-review study and, *457*, 457–58, *481*, 482
interest-theory, 14
internalist history of science, 135
International Encyclopedia of the Social Sciences, 151n61
International Encyclopedia of Unified Science, 136
intersubjectivity, perceived quality of U.S. medical schools and, 172
"invisible college," 10, 117
IQ scores: gender differences in output and, 376n17; of women in science, 276
Iranian scholars, American embargo of, 613n4

Jacobson, Michael, 525
JAMA (Journal of the American Medical Association), 517–19, 529
James, John C., 269n84
James, William, 84, 272
Japanese-American internment camps, 564
Japanese university purges, 578
Jennings, Peter, 531–32
Jewish immigrants in science, 392

Jiao Tong University, 575
Jobs, Steve, 8
John Bates Clark Medal, 385
Johns Hopkins medical school, 156, 160
Johnson, Alvin, 591
Joliot-Curie, Irene, 470–71
Journal of Experimental and Theoretical Physics, 44n17
Journal of the American Medical Association (JAMA), 517–19, 529
journals: awareness of work and, 66, 66–67, 72n43; in-group control of, 40; peer review and, 4; sociology of science in, 142. *See also specific journals*
Judt, Tony, 596
Julliard School, 237

Kahneman, Daniel, 12, 404
Kalven, Harry, 565, 628
Kalven Committee Report, 565–66, 592, 619, 628, 634
Kandinsky, Wassily, 408n30
Kannel, William, 525
Kaplan, Norman, 132, 141
Keller, Evelyn Fox, 387, 405n3, 407n20
Keller, Suzanne, 386
Keniston, Hayward, 154
Kessler, M. M., 30, 44n17, 445
Kevles, Daniel, 549
Keywords (Williams), 409n44
Khalidi, Rashid, 558–59, 567, 578, 596
kick-reaction pairs, 347–50, 352, 357–58, 371–73, 374n13
Kinoy, Arthur, 16
Klein, Joel, 241
Knorr-Cetina, Karen, 534, 542n13
knowledge: "about," 84, 272; acquaintance with, 84, 272, 275, 281; codified fields of, 191–94, 196; for competence, 192; core of, 15; growth of, 1–2, 6; research universities and expansion of, 202–3; social construction of, 216; sociology of, 120–21, 150n44; technology advancing, 8–9
Kuhn, Thomas, 14, 191, 447n5, 585; impact of, 117; influential work of, 11–12; Merton and, 136–37, 151n61; on normal science, 76, 434; on scientific revolutions, 8–9, 117
Kush, Polykarp, 16
Kyburg, Henry E., Jr., 467

Laboratory Life (Latour and Woolgar), 2
Ladd, Everett Carl, Jr., 557
Lakatos, Imré, 116, 145n8
Lancet, 520
Larson, Lyle E., 121
later productivity, gender differences in output and, 315, 316, 324n22
Latour, Bruno, 2, 534–35, 542n13
Lattimore, Owen, 574
Lawrence, E. O., 8
Lazarsfeld, Paul F., 408n30; leadership of, 231; renown of, 6–7; in sociology of science most cited authors, 130
LDL (low-density lipoprotein) receptors, 521–22
Leahy, Patrick J., 612n2
learning, failure in, 238
Lederberg, Joshua, 10
legislation, for women in science, 290
Lehman, H. C., 179–81, *180*
Lemann, Nick, 230
Leonhardt, Frederick Henry, 519
less influential authors, patterns of influence in scientific progress and, 92, 92–93
Leuchtenburg, William, 551
Levin, Richard, 226n10
Levy, Robert, 532–33
liberal democracy indicator, academic freedom as, 625–31
liberal professors, intellectual diversity and, 557
life sciences, PhDs of women in, *332*
limited differences, theory of. *See* theory of limited differences
Lincoln, Abraham, 564
Lincoln Project, 240
lipid-reducing drugs, gallbladder disease and, 518
Lipid Research Clinics (LRC), 517–18, 523, 531–32
Lippel, Kenneth, 523
Lippman, Walter, 16
Lipset, Seymour Martin, 557
Little Science, Big Science (Price), 44n18, 405n4, 507
"Little Science, Big Science Revisited" (S. Cole and Meyer), 512–13
Lochner v. New York, 384, 608
Lonsdale, Kathleen, 268n82

Los Angeles Times, 517
Lotka's Law, 97n11, 509
low-density lipoprotein (LDL) receptors, 521–22
LRC (Lipid Research Clinics), 517–18, 523, 531–32
Lysenko, Trofim, 67n2

Maccoby, Eleanor, 267n59
Macintosh, 8
MacLeod, Colin, 383
MacRoberts, B. R., 506
MacRoberts, M. H., 506
MacRoberts and MacRoberts (M&M) paper: on ceremonial citations, 508; on citation counts, 506, 511–12; on elitism, 507–8; on errors in citations, 511; on "influences," 508–9; "intellectual blacklisting" and, 510; omitted citations and, 509, 510; Ortega hypothesis and, 509–13; on persuasion, 511–12; policy implications of, 512–13
Making Science (S. Cole), 15
maladaptive behavior, in sociology of science, 139
male scientists. *See* eminent male scientists
Mamet, David, 212–16
Manhattan Project, 574
manifest and latent outcomes, in sociology of science, 138–39
Mann, Thomas, 408n30
marriage: age and, 330–31; beliefs on careers and, 331, 333; compartmentalizing and, 341; discretionary time and, 341–43; early recognition and, 337, *338,* 339; eminent female scientist productivity and, 333–37, *335, 340;* eminent male productivity and, 336; gender differences in output and, 288, 329–30; historical dynamics of, 384; to other scientists, 341; rank-and-file female scientist productivity and, 337, 341, *342;* status sets and, 339–43, *342;* structural constraints and, 353; timing of, 339–40; women in science and, 249, 252–54, 256, 278
Massad, Joseph, 576–78, 586, 596
mass media: on CHDs linked to dietary cholesterol, 517; competitive pressures of, 537–38; dietary cholesterol and

CHDs and distortion of, 519, 523–24, 533–36; health-risk information in, 516; institutional properties influencing distortion by, 536–38; peer review and distortion in, 539; personal values conflicts in, 540; reward system of science and distortion of, 537–38; self-criticism within, 541; social control mechanisms and, 540–41; structural problems causing distortion in, 538–40; timing problems in, 539–40; on *Toward Healthful Diets,* 524–27; training issues in, 538–39
mass producers, 31, 41, 51
master-apprentice chains, in sociology of science, 132
mathematics: age and scientific progress in, 183–88, *184–86, 188,* 192, 193; codified field of, 192; gender differences in output in, 294
Math Reviews, 197n9
Matthew Effect: development of, 397; Merton and, 6, 13, 127, 138–39, 397; reward system of science and, 13; sociology of science and, 138–39; Zuckerman and, 397
Mayer, Maria, 259
McCarthy, Joseph, 574, 580, 627
McCarthyism, 589–90
McCarty, Maclyn, 383
McCaughey, Robert A., 552
McClintock, Barbara, 383, *400,* 401, 407n20, 407n27; background of, 388; Beadle's career compared to, 389; colleague respect for, 386–87; at Cornell, 388–89; on freedom, 390–91; genetic discoveries of, 387, 389; National Academy of Sciences membership of, 387; Nobel Prize won by, 389; unfreedom in science stories of, 386–91; on women in science, 387–88, 390
Mearsheimer, John, 596
media. *See* mass media
medical imaging technology, 9
medical schools. *See* U.S. medical schools
Meitner, Lisa, 273–74, 408n33
memory effects, in theory of limited differences, 358–62
Menand, Louis, 593
Mendel, Gregor, 104
mental health, 618

meritocracy: definition of, 406n5; egalitarianism and, 94; research universities and, 209–11, 216; unfreedom in science and, 382; universalism compared to, 5; women in science and, 275–76

Merton, Robert K., 6, 16, 67n4; American Sociological Association address of, 2, 230; Bureau and, 7; citation counts of, 133–36; codification hypothesis of, 191–94, 196; on competition for priority, 134, 138; on cumulative advantage, 352; death of, 233; on deviant behavior, 138, 230–31; at Harvard, 552; history of science influence of, 134–35; influence and legacy of, 2–3, 134, 229–30, 233; Kuhn and, 136–37, 151n61; on manifest and latent outcomes, 138–39; Matthew Effect and, 6, 13, 127, 138–39, 397; on obliteration through incorporation, 511; *On the Shoulders of Giants* by, 232; on overconformity and maladaptive behavior, 139; perfectionism of, 133; *Physical Review* study of, 377n22; "Priorities of Scientific Discovery" by, 127, 133, 230; publication standards of, 232; on reward system of science, 31; on scientific genius, 443; scientific norms of, 3–4; on self-reinforcing processes, 139–40; "Social Structure and Anomie" by, 230–31; sociology of science contributions of, 116, 120, 128, 132–40, 144, 149n42, 150n60, 230; sociology study and, 2–3; on stratification in science, 134; students of, 132–33, 229–30; teachers of, 132; teaching style of, 231–32; on universalism, 405n2

"method effect," COSPUP NSF peer-review study and, 457, 457, 481, 481–82, 484

Metzger, Walter, 16, 19–20

Meyer, G., 512–13

MI (myocardial infarction), 530

mid-career dynamics, in theory of limited differences, 361–62

Mies van der Rohe, Ludwig, 408n30

military, research universities' investment rationale of, 222

Mill, John Stuart, 585

Minerva, 142, 147n26, 377n22

Mirzakhani, Maryam, 385

MIT, 554n5

M&M paper. *See* MacRoberts and MacRoberts paper

Moebius, Paul, 273

Morgan, Thomas Hunt, 9, 383

Morrill Act, 243

Mössbauer effect, 109–10, 113n9

motherhood: age and, 330–31; beliefs on careers and, 331, 333; compartmentalizing and, 341; conflicts from, 339; discretionary time and, 341–43; early recognition and, 337, 338, 339; eminent female scientist productivity and, 333–37, 335, 340; gender differences in output and, 288, 329–30; number of children in, 341; rank-and-file female scientist productivity and, 337, 341, 342; status sets and, 339–43, 342; structural constraints and, 353; tenure and decisions of, 340–41; time commitment of, 334; timing of, 340–41; women in science and, 249, 252–54, 256, 278; work strategies with, 334

MRFIT (Multiple Risk Factor Intervention Trial) Study, 529–31

Muller v. Oregon, 384

Mullins, Nicholas C., 117, 131, 132, 137, 146n13

multi-authored papers, age and, 190–91, 191. *See also* pairs of male and female scientists

Multiple Risk Factor Intervention Trial (MRFIT) Study, 529–31

multiplier effects, 375n15

Murdoch, John, 135

myocardial infarction (MI), 530

name ordering, visibility of work and, 56, 57

National Academy of Sciences (NAS): on dietary cholesterol and CHDs, 525–26; election to, 17, 181; Fink Committee of, 599; membership, 32–33, 33, 45n20, 387; NSF peer-review system study of, 416; women elected to, 258, 387, 389. *See also* COSPUP NSF peer-review study

National Institutes of Health (NIH), 159, 167, 170, 269n84, 394, 461n1, 503n4; dietary cholesterol consensus committee of, 523; HIV research supported by, 594–95, 601; ideological and political intrusion into, 569–70; peer review at, 369

650   Index

national meetings, sociology of science in, 142–43

National Opinion Research Center, 249

National Research Council (NRC), 524–27

National Science Foundation (NSF), 368; percentage of federal funds from, 415, 472; proposals for, 610; sociology of science support from, 141–42; staff of, 416. *See also* COSPUP NSF peer-review study; NSF peer-review system

national security *vs.* open communication, academic freedom and, 599, 613n5

Navasky, Victor, 559

negative kick-reaction pairs, 357–58

"negative liberty," academic freedom and, 609–10

negotiation, in construction of facts, 534–35

nepotism, theory of limited differences and, 376n19

neutral kick-reaction pairs, 357–58

Newsweek, 569–70

Newton, Isaac, 1, 179

New York Times, 523–26, 530–31, 542n15

Nicolson, Marjorie, 385

NIH. *See* National Institutes of Health

9/11 attacks. *See* universities after 9/11

Nixon, Richard, 627

Nobel Prize: aspirations for, 33; citation counts and, 29, 43n11, 447n7; laureates in SCI, 102–3, 447n7; McClintock winning, 389; of Pierre Curie, Marie Curie, Irene Joliot-Curie, 470–71; prestige of, 32, 33; as quality indicator, 102; women in science and, 259, 385; Zuckerman's work on winners of, 6, 17, 97n13, 325n25

normal science, in history of science, 76, 434

Northwestern University, 235

NRC (National Research Council), 524–27

NSF. *See* National Science Foundation

NSF peer-review system: accumulative advantage and, *427*, 429–30; age in, 427; budget and, 465; chance in, 451, 459–60, 465, 467–68, 498; characteristics of successful applicants in, *422*, *423*; citation counts and, 424, 426–27; congressional hearings on, 415–16; departmental prestige and, 426; elitism and, 416–17; equity and, 430–31, 474; evaluation criteria of, 416; expert consensus in, 473;

geographical location and, 421; Hower on, 465–66; Humphreys on, 464–65; Kyburg on, 467; lack of agreement of reviewers in, 424–25; National Academy of Sciences study on, 416; number of reviewers in, 484–86, *485*; old-boy hypothesis and, *418*, 420–21; panel ratings and success in, 426, *426*, *427*; past performance and, *426*, 428, 486–89, *488*; peer-review types in, 466; perceptions of, 369; personal values and ideology in, 498; precision of, 461; process of, 473–74; program directors in, 405, 416, 417, *419*, 425–26; proposal review panels in, 465–66, 469; qualitative and quantitative analysis of, 419–20; random error in talent measurement in, 464–65; rank of department and, 420–21, 428; reviewer disagreement in, 455–56, 460, 468, 479–83, 492–93; rich-get-richer hypothesis and, *422*, 422–24, *423*; secrecy of, 417; self-selection and, 461n2, 464, 489–92; Simon, Stephen Cole, J. R. Cole, on, 467–70; statistical analysis of, *418*, *431*; Westley on, 466–67; zero-one decision rule in, 478. *See also* COSPUP NSF peer-review study

nuclear physicists: awareness of work and, 64, *65*; size of field of, 107–8; visibility of work and, 54, *54*

null hypothesis: culture and, 403–4; definition of, 403; formulating, 210; free speech and, 209–11; *Oleanna* and, 212–16; overturning, 217, 403, 404; political power and, 211; relinquishing control of, 212–13; social construction of knowledge and, 216; truth and, 208–9

Obama, Barack, 592, 600–602

objectivity, research universities and, 212, 216

occupational prestige estimates, self-aggrandizement and, 158–59

occupation choice, women in science and, 281–83

Ogburn, W. F., 121, 145n3

Ogilvie, Marilyn, 383

Okie, Susan, 525

old-boy hypothesis, NSF peer-review system and, *418*, 420–21

Oldenburg, Henry, 1
Oleanna (Mamet), 212–16
Oliver, M. F., 523
omitted citations, 436, 509, 510
On the Connexion of the Physical Sciences (Somerville), 406n5
On the Origin of Species (Darwin), 375n15
On the Shoulders of Giants (Merton), 232
open communication: Kalven Report on, 566; national security *vs.*, 599, 613n5
Oppenheimer, J. Robert, 19, 574
Orbán, Viktor, 629
organizational infrastructure, for sociology of science, 140–44
organized skepticism, Kalven Report on, 565
Ortega hypothesis: basic question in, 507; cited author characteristics and, 436–40, *437*; departmental prestige, citation counts and, 438; elitism and, 443–44; filtration of ideas and, 444; future research on, 442; "intellectual blacklisting" and, 510; M&M paper and, 509–13; omitted citations and, 510; pathfinding papers and, 438–40; PhD supply and demand and, 446–47; *Physical Review*-cited "super" papers and, 440–42, *441*; policy implications and, 512–13; recruitment challenges and, 446–47; replaceability of scientists and, 443–45; stratification in science and, 433–44; teaching function and, 445
Ortega y Gasset, José, 75, 91
Our Country (Barone), 550
output, impact compared to, 301
outstanding research, patterns of influence in scientific progress and, 88–90, *89*
overconformity, in sociology of science, 139

Page, Scott E., 237
pairs of male and female scientists: access and, 299; authorship distinctions for, 322n6; citation counts and, 306–9, *307*; comparisons of, 297, *297*; differential reinforcement and, 309–13; discrimination and, 298; distribution of differences for, *297*, 297–98; early recognition and, 310–13, *311*, *312*; primary authorship and, 304; productivity distribution changes among, 298–300;

prolific physicists and, 298; solo authorship compared to, 299
Palestinian film festival incident, at Columbia, 567
Panofsky, Erwin, 408n30
"Paradigm for Functional Analysis" (Merton), 138
Pardee, Arthur B., 378n40, 473
Parsons, Talcott, 3, 121, 132
particularism: individual and institutional, 397–401, *400*; universalism dominating, 94; women in science and, 13
part-time employment, of women in science, 256, 261, 269n87
past performance, NSF peer-review system and, *426*, 428, 486–89, *488*
pathfinding papers, Ortega hypothesis and, 438–40
patterns of influence in scientific progress: age and, 98n18; awards and, 85–86, *86*; "best" papers and, 88–90, *89*; citation counts and, 77–78; communication and, 96n3; community contributing to, 75–76; conclusions on, 93–95; conflicting theories on, 74–75; data set for, 77, 96nn5–8; elitism in, 76; "highly-cited" papers and, 90–91, 111; less influential authors and, *92*, 92–93; marginal distributions of characteristics in, 79, *80*; quantity and quality of output and, 84–87, *85*, *87*; rank of department in, 81–84, *82*, *87*, 87–88; self-citations and, 91, *91*; stratified system of, 76; "super" papers and, 91–92, *92*; taste and, 88; under-used research and, 94; usefulness in, 76–77
Patterson, James T., 551
Paulin, Tom, 567
Pauling, Linus, 19, 443, 574
Pearl Harbor attack, 564
Pearsonian correlation coefficients, 187, *188*
peer review: fairness in science and, 4; health-risk information issues with, 539; journals and, 4; mass media distortion and, 539; at NIH, 369; science and, 4–5; terminology of, 415. *See also* NSF peer-review system
Pell grants, 241

652 Index

perceived quality, of U.S. medical schools, 159; citation counts on, 172; correlation matrix of variables on, 169–70, *170*; faculty eminence and, *165*, 166–67; financial resources and, *165*, 167, 170; findings on, 160, *161–63*; geography patterns and, 168–69; impact of studies on, 171–72; intersubjectivity issues with, 172; productivity and, 164–66, *165*, *166*, 168, 175nn27–29; quality compared to, 172; standard errors on, 160, 174nn21–22; stratification of, 163–64; variables of, 169–70, 175n32
perfectionists, 31, 41
period effects, 318–19
personal values: mass media and conflicts of, 540; in negotiation of construction of facts, 535; NSF peer-review system and, 498; tenure and, 557
persuasion, in construction of facts, 534–35
PFL. *See* Lazarsfeld, Paul F.
PhDs: gender differences in output and, 291–92; ranking programs for, 550; supply and demand of, 446–47; teaching expectations of, 553n2; theory of limited differences and, 345, 374n12; women in science and, 247–48, 253, 263n5, 263n9, 266n45, 272–73, 314, *332*
Phillips, Melba, 394–95
Philosophical Transactions, 1
Physica, 44n17
Physical Review, 30, 38, 194, 377n22, 445; citation counts and, 44n16; less influential authors in, 92, 92–93; Ortega hypothesis and "super" papers cited in, 440–42, *441*; readership of, 66; "super" papers in, 91–92, *92*
physical sciences, PhDs of women in, *332*
physicists: awareness of work and, 64, *65*, 66; citation count practices of academic, 435–36; citation counts and, 103; quantity and quality of output and, 31, 41; size of field of, 107–8; visibility of work and, 51, 54, *54*
physics: age and scientific progress in, 192, 193; awards in, 32–33, *33*; collaborative research in, 56; eminence in, 32; gender differences in output in, 294; in-group control of journals on, 40; knowledge

growth in, 6; rank of department in, 33, 36; series of papers contributing to, 29. *See also* specialties
Physics Today, 45n21
Physiological Feeble-Mindedness of Women, The (Moebius), 273
Pickering, Andrew, 14
Pipes, Daniel, 601
political correctness, 212; academic freedom and, 608; free speech threatened by, 209
political litmus tests, academic freedom and, 578–79
political power, null hypothesis and, 211
political views of professors, intellectual diversity and, 557
Popper, Karl, 116
positive kick-reaction pairs, 357–58
Post, Robert, 608–9
Pound, Roscoe, 498
Price, Derek de Solla, 10, 201, 226n4, 292; on Age of Little Science, 44n18, 382, 405n3; on contemporaneity and citation count, 109; on growth parameters, 122; on "invisible college," 117; on recent citation counts, 126; on stratification in science, 76, 434, 507
primary authorship, gender differences in output and, 304
primary event list, productivity puzzle and, *354*, 354–55
Princeton University, 617
"Priorities in Scientific Discovery" (Merton), 127, 133, 230
prison costs, research university costs compared to, 240–41
privacy, surveillance and, 571–72, 618
probability specifications, in theory of limited differences, 358–62
productivity: age and scientific, 181–88, *182*, *184–86*, *188*; competition and, 369–70; cumulative productivity differentials, 348–49; data improvements on, 316–17; gender differences and later, 315, *316*, 324n22; gender differences and published, 287–88, 293–97, *294–96*, 304–6, *305*, *306*, 309–13, *311*, *312*; measuring, 373n4; pairs of male and female scientists and change in, 298–300; perceived quality of U.S. medical schools and, 164–66, *165*, *166*,

168, 175nn27–29; period effects on, 318–19; reward system of science and research, 38–41, *39, 40*, 47n32; skewed distribution of, 345–47; of women in science, 254–55, 267n55, 277–78, 282–83. *See also* eminent female scientist productivity; gender differences in output; rank-and-file female scientist productivity

productivity puzzle, 344; disparities prompting, 346–47; ongoing, 314; persistence of, 288–89; predicting solutions to, 317; primary event list as basis of, *354*, 354–55; prior explanations for, 350–54; structural constraints and, 353; studies demonstrating, 345–46, *346*; theoretical questions connected to, 319; theories of evolving social processes and, 351–52; theories of initial conditions and, 350–51. *See also* gender differences in output

professional identity, sociology of science and, *143*, 143–44

professorial rank, 220; awareness of work and, *59*, 59–60. *See also* rank of department

Professors, The (Horowitz), 558

program directors, in NSF peer-review system, 405, 416, 417, *419*, 425–26

Progressive Movement, 384

prolific physicists, 31, 41, 295–96, 298

promotion: discrimination in, 284; research universities and decisions of, 219–20; of women in science, 257–60, 284

proposal review panels, in NSF peer-review system, 465–66, 469

proposal variance, in COSPUP NSF peer-review study, 481, *481*

protests, universities after 9/11 and, 567–68

Prusiner, Stanley B., 6, 408n29

psychological traits, theories of initial conditions and, 350–51

psychology: age and scientific progress in, 193; Clark's scientific output study and, 29, 102; clinical, 194, *194*; economics and, 12; experimental, 194, *194*

publication norms, gender differences in output and, 317

publication processes, theory of limited differences and, 355–57, *356*

Public Health Security and Bioterrorism Preparedness and Response Act of 2002, 19, 590, 596–97, 627

public trust, in higher education, 558

published productivity, gender differences and, 287–88, 293–97, *294–96*, 304–6, *305, 306*, 309–13, *311, 312*. *See also* citation counts

"publish or perish" doctrine, 28, 35, 40

quality of sociological research: awards for, 102–3; clerical errors in citation counts and, 111; collaborative paper citations and, 110–11; contemporaneity and, 108–9; critical citations and, 104–5; equal weight of citations for, 105–6, *106*; errors in evaluation and, 104; integration of basic ideas and, 109–10; quantity and quality of output in, 106–7; SCI for measuring, 101–2; size of scientific fields in, 107–8

quality of U.S. medical schools, perceived quality compared to, 172

quantity and quality of output: citation counts and, 29, 71n37, 78, 102, 301; data measures on, 28–31, *30*; Nobel Prize and, 29; patterns of influence in scientific progress and, 84–87, *85, 87*; physicist types and, 31, 41; "publish or perish" doctrine and, 28, 35, 40; reward system of science and, 27–28, *30*, 30–34, *34*, 41–42, 46n27, 47n36; in SCI, 106–7; visibility of work and, *51*, 51–52, *53*, 69n21; of women in science, 254–55, 278. *See also* gender differences in output; quality of sociological research

Rabi, I. I., 622

race, unfreedom in science and, 382, 410n51. *See also* diversity

racism: fear and, 618; free speech at universities and, 209–11, 603

random error in talent measurement, in NSF peer-review system, 464–65

rank-and-file female scientist productivity: marriage, motherhood and, 337, 341, *342*; pre-childbirth, 337; theory of limited differences case study on, 368–69

654   Index

rank of department: awards correlated to, 71n36, 81; awareness of work and, *60*, 60–61, 64, *64*; elitism and, 81, 83; NSF peer-review system and, 420–21, 428; in patterns of influence in scientific progress, 81–84, *82*, *87*, 87–88; in physics, 33, *36*; in reward system of science, 33, 35, *36*, 42n3, 46n25, 96n5, 97n12, 171–72; U.S. medical schools and, 156–57; visibility of work and, *51*, 52–53, *53*, 63–64, *64*, 83–84; women in science and, 255–57, 276. *See also* departmental prestige; professorial rank
rationality, research universities and, 212
Ratner, Sarah, 394–96
reaction reports, theory of limited differences and, 372
Reader's Digest, 532
recent citation counts, 126
recognition, of women in science, 257–60, 284
recruitment challenges, Ortega hypothesis and, 446–47
Red Scare, 580–81, 627
Reif, L. Rafael, 631
reinforcement processes, 351–52
reinforcement theory, 351
Reitz, Jeffrey, 44n18
replaceability of scientists, Ortega hypothesis and, 443–45
reputation: COSPUP NSF peer-review study and, 487–89, *488*, 504n14; of research universities, 199–200; in reward system of science, 33, *36*, 36–38, *37*, 46n25; sociology of science and, 154; of U.S. medical schools, 154–55, 159–60, *161–63*; of women in science, 277. *See also* perceived quality, of U.S. medical schools
Research! America, 240
research productivity, reward system of science and, 38–41, *39*, *40*, 47n32
research support, sociology of science and NSF, 141–42
research universities: authority at, 203, 207; bidding wars for academic stars at, 219; budget concerns of, 201, 226n6; censorship and, 595; challenges to organizational principles of, 207–8; classified research and, 571; competitiveness at, 217–18; conflict of interest policies at, 224; crises facing, 199;

culture of excellence at, 219; defining, 225n3; dilemmas of choice facing, 200; discoveries made at American, 380–81, 622; diversity of student bodies at, 604; economic development and, 223; economics of, 203, 204, 224; employee protections at, 608–9; enlightenment compared to balance at, 586; evaluations and accountability at, 581; faculty governance at, 204, 207, 227n11; federal government partnership with, 221–25; financial resources of, 202, 226n7; free speech at, 209–11, 603; German scientists fleeing for American, 591–92, 622–23; governance dilemma of, 202–7; growth and change patterns of, 200–202; health science divisions at, 202; hierarchical structure of, 203–4; increased costs of, 222; institutional legitimacy of, 218; knowledge expansion at, 202–3; meritocracy and, 209–11, 216; military rationale for investment in, 222; objectivity and, 212, 216; open communication *vs.* national security at, 599; peer-regulation of, 581–82; pillaging of state, 240–43; preeminence of American, 594; present day, 592; priorities of, 202; prison costs compared to, 240–41; public support of, 240; rationality and, 212, 216; reputation of, 199–200; resistance to women in, 385–86, 397–401; restrictive visa policies and, 598–99; return on investments at, 223; reward system of science and, 218–19; rigorous skepticism of all ideas at, 582–84; scientific integrity and, 600–602; scientific progress at American, 575; social construction of knowledge and, 216; speech codes at, 209, 577, 603, 607; state intrusion into affairs of, 596–98; structural reorganization at, 224–25; tax increases to pay for, 242; teaching and research balance facing, 217–21; teaching loads at, 219; tenure and promotion decisions at, 219–20; tolerance of unsettling ideas at, 582–84; triumph of American, 621–24; tuition costs for, 241–42; uncertain investments in, 223–24; unfreedom and shortcomings of, 381; US dominance of, 575; "who owns the null" dilemma facing, 207–17; world rankings of, 612n1, 622. *See also* academic freedom; intellectual diversity; universities after 9/11

Reskin, Barbara F., 301, 313, 315

responsibility, with visibility of work, 70n29

restrictive visa policies, 571, 598–99

reversals, in COSPUP NSF peer-review study: bias and, 455; causes of, 455–56, 459–60, 479–80; classifications of, 454, 462n6; criteria differences and, 455, 479; at cutting point, 478; number of reviewers and, 484–86, *485*; proposal review panels and, 470; rates of, *454*, 454–55, 469, *479*; reviewer differences and, 455–56, 479; reviewer disagreement and, 455–56, 479–80; self-selection and, 492

reviewer disagreement, in NSF peer-review system, 455–56, 460, 468, 479–83, 492–93

reviewer selection bias, COSPUP NSF peer-review study and, 479

reviewer variation, COSPUP NSF peer-review study and, *481*, 482–83

revolutions, scientific, 117

reward system of science, 4; age and scientific progress and, 188–91, *189*, 195–96; awards in, 32–35, *33*, *34*, 45n21, 46n25; citation counts and, 38–39, *39*; consensus and, 17; elitism and, 17; gate-keepers and, 62; mass media distortion and, 537–38; Matthew Effect and, 13; Merton on, 31; Nobel Prize and, 29; "publish or perish" doctrine and, 28, 35, 40; purpose of, 27; quantity and quality of output and, 27–28, *30*, 30–34, *34*, 41–42, 46n27, 47n36; rank of department and, 33, 35, *36*, 42n3, 46n25, 96n5, 97n12, 171–72; research productivity and impact of, 38–41, *39*, *40*, 47n32; research universities and, 218–19; scientific reputation in, 33, *36*, 36–38, *37*, 46n25; women in science and, 255–57. *See also* awards

Rhoades, Marcus, 386, 388–89

Rice, Condoleezza, 241

Richardson, Robert C., 579–80, 598, 628

rich-get-richer hypothesis, NSF peer-review system and, *422*, 422–24, *423*

Rifkind, Basil M., 523, 531–32

RKM. *See* Merton, Robert K.

Roose, Kenneth D., 154, 171

Rorty, Richard, 227n17, 611

Rosenberg, Harold, 604

Rosovsky, Henry, 566

Rossiter, Margaret W., 405n3

Roth, Philip, 410n48

Rowland, Henry A., 622

Royal Society of London, 1; Florey's address to, 95n11; as "invisible college," 10, 117; women in, 268n82

Rutherford, Lord, 591

"sacred spark," theories of initial conditions and, 350

SAF (Students for Academic Freedom), 577, 585, 589

Said, Edward, 557, 567, 578, 596

salaries, of women in science, 256–57, 268n76, 277

SAR (Scholars at Risk), 626, 630

Sarton, George, 76, 121; history of science and, 116, 133; student demands of, 133

Schally, Andrew, 379n41

Schapiro, Meyer, 16

Scharrer, Berta, 395–96

Schiebinger, Londa, 405n3

Schmidt, Benno, 603

Schoenberg, Arnold, 408n30

Schoenheimer, Rudolph, 393

Scholars at Risk (SAR), 626, 630

School of Library Service, Columbia's closure of, 205–6

Schrecker, Ellen, 580

SCI. *See Science Citation Index*

science: as community, 11–12, 94; competition and, 369–70; computers and, 7–8; consensus in, 498–99; elitism in, 5–6, 17; free inquiry for innovation in, 575; growth of, 1–2, 6; humanities' relationship with, 13; idealization of, 381; industrial, 11; Jewish immigrants in, 392; Merton's norms of, 3–4; peer review and, 4–5; period effects on, 318–19; policy crisis facing, 545–46; public ambivalence toward, 546; public illiteracy on, 547–49; taste in, 88; truth in, 381; uncertain business of, 95; value system of, 94. *See also* fairness in science; reward system of science; sociology of science; unfreedom in science; women in science

656 Index

Science, 147n26
Science, Technology and Society in Seventeenth Century England (Merton), 2, 230
Science Abstracts, 42n3, 68n12, 110
Science and the Social Order (Barber), 144
Science Citation Index (SCI), 10, 42n3; citation counts in, 42n8, 77–78, 88–89; clerical errors in citation counts and, 111; collaborative paper citations in, 110–11; errors in evaluation in, 104; growth of, 185; measurement tools of, 28–29; Mössbauer effect in, 109–10, 113n9; Nobel laureates in, 102–3, 447n7; NSF peer-review system and citation counts in, 424, 426–27; origin of, 101; quality measure complications of, 43n15; for quality of sociological research, 101–2; quantity and quality of output in, 106–7; sociology journals added to, 101–2, 107; *Source Index* of, 292; U.S. medical school faculty data from, 159
Science magazine, 5
Science Studies, 142
scientific genius, Merton on, 443
scientific integrity, universities after 9/11 and, 600–602
scientific progress: American research universities and, 380–81, 575; communication of, 49; competition and, 369–70; defining moments in, 383; elitism and, 76; filtration theory of utilization of, 97n13; Florey on, 433; number of scientists needed for, 94; Ortega hypothesis and, 433–34; pathfinding papers and, 438–40; PhD supply and demand and, 446–47; theories of, 74–75, 95n1. *See also* age and scientific progress; patterns of influence in scientific progress
scientific reputation, in reward system of science, 33, *36*, 36–38, *37*, 46n25
scientific revolutions, 8–9, 117
scientist, terminology of, 406n5
Scientist's Role in Society, The (Ben-David), 141
Searle, John, 208
Second Sex, The (Beauvoir), 380
Sedofsky, Jan, 320
self-aggrandizement, 173; occupational prestige estimates and, 158–59; types of, 157; U.S. medical schools and, 157–59, *158*

self-citations, patterns of influence in scientific progress and, 91, *91*
self-fulfilling prophecies, women in science and, 249–50, 256, 282, 407n17
self-reinforcing processes, in sociology of science, 139–40
self-selection: citation counts and, 491–92; NSF peer-review system and, 461n2, 464, 489–92; reversals and, 492; of women in science, 12, 261, 281
September 11, 2001 attacks. *See* universities after 9/11
SEVIS system, 571
sexism, free speech at universities and, 209–11
sex role attitudes, gender differences in output and, 291
Seymour, Charles, 580
Sharp, Laure M., 263n4
Shattuck, John, 629, 632n5
Shenton, James, 16
Sher, Irving H., 43n12
Sheraton, Mimi, 526
Shils, Edward, 227n11
Shirley, David A., 470
Shweder, Richard, 605
Siegbahn, Kai, 470
Siegel, Frederick F., 551–52
silent physicists, 31, 41, 51, 295
Simmons, Solon, 558
Simon, G. A., 467–70
Sinclair, Upton, 399–400
Singer, Burton, 13, 467
single women, eminent female scientist productivity and, 336
size of scientific fields, in quality of research, 107–8
Sloan Foundation, 554n5
Smelser, Neil J., 210–11
Smith and McCarren acts, 564
Snow, C. P., 548–49
social construction of knowledge, 216
social control mechanisms, for health-risk information, 540–41
socialization patterns, theories of initial conditions and, 350–51
social learning, 351–52
social media, 617–18
social networks, women in science and informal, 283–84
Social Science Citation Index (SSCI), 172

social sciences, PhDs of women in, *332*
social selection, women in science and, 281, 318
"Social Structure and Anomie" (Merton), 230–31
Social Studies of Science, 508
sociology: age and scientific progress in, 193; of deviance, 123–25, *124*; economics and, 12; errors in evaluation and, 104; of knowledge, 120–21, 150n44; Merton and study of, 2–3; SCI adding journals on, 101–2, 107. *See also* quality of sociological research
sociology of science: age of authors in, 127–28; citation counts for, 123–26, *124*, 148n27; cognitive consensus in, 123–26, *124*; cognitive structures and, 119–20; collaborative publication on, 128; competence in, 120–21; consolidation of research fronts and, 126–27; continuity of influential contributors to, *130*, 130–32; deviant behavior and, 138; doctoral dissertations and, 121; emergence of, 115–16; in formal communication and formal organization, 142–43; Gini-concentration ratios for, 123, *124*, 125; growth parameters for, 120–23; institutionalization of specialties and, 117–19; in journals, 142; manifest and latent outcomes in, 138–39; master-apprentice chains in, 132; Matthew Effect and, 138–39; Merton's contributions to, 116, 120, 128, 132–40, 144, 149n42, 150n60, 230; most cited authors in, *129*, 129–30; in national meetings, 142–43; NSF research support for, 141–42; organizational infrastructure for, 140–44; overconformity and maladaptive behavior in, 139; professional identity and, *143*, 143–44; publication rates on, 122–23; reputation and, 154–55; scientific revolutions and, 117; self-reinforcing processes in, 139–40; sociology of deviance compared to, 123–25, *124*; specialties in, 116–20; structure of influence in, 128–32, *129*, *130*; students recruited into, 132–33
solid state physicists: awareness of work and, 54, *54*, 64, *65*; size of field of, 107–8; visibility of work and, 54, *54*
solo authorship: age and multi-authored papers compared to, 190–91, *191*; gender differences in output of, 299, 303–4, 315

Somerville, Mary, 406n5
sophisticated residualism, 275
Sorokin, P. A., 121
Source Index, SCI, 292
Spearman rank-order correlation coefficient, 71n37, 110
specialties: awareness of work and, 64, 65, 66; cognitive development of, 118–19, 146n13; cognitively radical and cognitively conforming, 117–18; cognitive structures and, 119–20; consolidation of research fronts and, 126–27; growth parameters for, 121–22; institutionalization of, 117–19; organizational infrastructure for, 140–44; in sociology of science, 116–20; visibility of work and, 53–54, *54*
speech codes, 209, 577, 603, 607
Spemann, Hans, 392
Spencer, Herbert, 273
Spitz, Norman, 525
SSCI *(Social Science Citation Index)*, 172
standard errors, on perceived quality of U.S. medical schools, 160, 174nn21–22
Stanford medical school, 160
state intrusion into university affairs, 596–98
state universities, pillaging of, 240–43
status quo bias, 404
status sets, motherhood, marriage and, 339–43, *342*
Stehr, Nico, 121
stem cell research, 600
Stephenson, Margery, 268n82
Stewart, Paul A., 197n13
Stigler, Stephen M., 471
Stigler, Virginia L., 471
Stone, Geoffrey, 616
Storer, Norman W., 47n32
straight citation counts, 323n7
Strait, George, 531–32
stratification in science: historical development of, 397; Merton on, 134; Ortega hypothesis on, 433–44; patterns of influence and, 76; PhD supply and demand and, 446–47; power and, 345, 432; Price on, 76, 434, 507; recruitment challenges and, 446–47; replaceability of scientists and, 443–45; theories on, 432; visibility of work and, 76
Stravinsky, Igor, 408n30

strings of papers, gender differences in output of, 299

strong publishers, *186*, 186–88

structural constraints, 353

structural determinants, on gender differences in output, 317

Structure of Scientific Revolutions, The (Kuhn), 136, 447n5

Students for Academic Freedom (SAF), 577, 585, 589

student visas, 571, 598–99

Suicide (Durkheim), 273

suicide rates, of women, 290

Sullivan, Daniel, 9

Sullivan, Walter, 542n15

"super" papers, patterns of influence in scientific progress and, 91–92, *92*

Supreme Court decisions, subjectivity in, 498–99

surveillance, privacy and, 571–72, 618

Szilard, Leo, 408n30, 499n2, 613n5

"Take Back the Night," 602

talent measurement, NSF peer-review system and random error in, 464–65

taste, in science, 88

Tatum, Edward, 383, 389

tax increases, for research universities, 242

teaching: of American history, 550–53; at graduate level, 621; history of science, 553; loads at research universities, 219; Ortega hypothesis and function of, 445; PhD expectations with, 553n2; portfolios, 219; qualifications, 546–47; research universities balancing research and, 217–21; science and technology challenge of, 547–49; at undergraduate level, 621–22; USA PATRIOT Act and, 570

technology: collaboration and, 9–10; knowledge advanced through, 8–9; medical imaging, 9; policy crisis facing, 545–46; public ambivalence toward, 546; public illiteracy on, 547–49

tenure: academic freedom and, 602, 605; age and scientific progress and, 196n2; citation counts and, 11; decisions of, 611; denial of, 368, 369; discrimination and, 276; economic obligations of, 204; gender differences in output during

years relevant to, 293–94; motherhood decisions and, 340–41; personal views and, 557; research universities and decisions of, 219–20; women in science and, 255, 276, 386

test scores: admission system and, 235; IQ, 276, 376n17; labels from, 237–38; skewing of, 238; women in science and, 250

Texas Tech University, 579, 597–98

textbooks, women in science in, 410n51

Thaler, Richard, 12

Theories and Theory Groups in Contemporary American Sociology (Mullins), 146n13

theories of evolving social processes, 351–52

theories of initial conditions, 350–51

theory of limited differences, 13; anticipation of future events and, 373; chain reactions and, 375n15; competence and, 375n15; competition and, 369–70; concept of, 344; cumulative productivity differentials and, 348–49; Darwin and, 375n15; discrimination and, 378n32; early events module in, 359–61; eminent female scientist case study on, 366–68; eminent male scientist case study on, 364–66; beyond first major award, 362; general outline of, 347–50; hypothetical examples of, 364–69, *365*; inequality and, 344, 362–64; kick-reaction pairs and, 347–50, 352, 357–58, 371–73, 374n13; measurement issues and, 372–73; at micro level, 370; mid-career dynamics in, 361–62; multiplier effects and, 375n15; nepotism and, 376n19; outcome processes and, 355–57, *356*; PhDs and, 345, 374n12; primary event list and, *354*, 354–55; probability specifications and memory effects in, 358–62; rank-and-file female scientist case study on, 368–69; reaction reports and, 372; sources of disparity between groups in, 362–64; structural constraints and, 353; theories of evolving social processes and, 351–52; theories of initial conditions and, 350–51

Thomas, Dorothy, 145n3

Thomas, M. Carey, 399

Time, 529, 532

timing: of marriage, 339–40; mass media problems of, 539–40; of motherhood, 340–41

Toward Healthful Diets (NRC), 524–27

track record. *See* past performance

Trilling, Lionel, 16, 582

triple penalty, women in science and, 248, 260–62, 274

Troubled Feast, A (Leuchtenburg), 551

Troubled Journey (Siegel), 551–52

Trump, Donald, 626, 634

truth: correspondence theory of, 212, 216; defining, 209, 403; free inquiry for, 572, 580, 607; in nature, 14; null hypothesis and, 208–9; in science, 381

tuition costs, for research universities, 241–42

Turkey, 626

Tversky, Amos, 12, 404

"Two Cultures, The" (Snow), 548–49

UCLA medical school, 156, 168

under-used research, patterns of influence in scientific progress and, 94

Unfinished Journey, The (Chafe), 550–51

unfreedom in science: common conditions and shared experiences of, 394–97; culture and, 401–4; definition of, 381; Gluecksohn-Waelsch and, 391–94; McClintock and, 386–91; meritocracy and, 382; overview of, 17–18; particularism, universalism and, 397–401, *400*; race and, 382, 410n51; research universities' shortcomings and, 381; short stories of, 386–97; universalism and, 382; women and, 382, 402, 404–5

universalism: community and, 5; definition of, 405n2; fairness in science and, 15–17; individual and institutional, 397–401, *400*; Kalven Report on, 565; meritocracy compared to, 5; particularism dominated by, 94; unfreedom in science and, 382; women in science and, 13

universities after 9/11: challenges facing, 572–73; classified research and, 571; fear and, 617–20; free inquiry incidents and, 567; hiring and firing issues at, 568–69; historical precedent and, 563–64; ideological and political intrusion into, 569–70; institutional responsibility and,

564–65; Kalven Report and, 565–66; open communication *vs.* national security at, 599; privacy, surveillance and, 571–72, 618; protests and, 567–68; scientific integrity and, 600–602; state intrusion into affairs of, 596–98; student visas and, 571, 598–99; threats facing, 566–67; USA PATRIOT Act and, 19, 563, 570–72, 590, 596–97. *See also* academic freedom

University, The (Rosovsky), 566

University of California Berkeley: protests at, 568; tuition costs of, 241

University of Chicago, 565, 567, 580–81, 586, 617, 619, 627

University of Michigan, 613n8; tuition costs of, 241

University of Missouri, 389

University of Pennsylvania, 243

University of Southern California medical school, 174n20

University of Wisconsin, Madison, 242

upward mobility, visibility of work and, 57

USA PATRIOT Act, 19, 563, 570–72, 579–80, 590, 596–97, 627

U.S. medical schools: age of, 169, 175n31; characteristics of quality, 153; citation counts on studies on, 172; correlation matrix of variables on, 169–70, *170*; faculty eminence and, *165*, 166–67; financial resources and, 159, *165*, 167, 170; findings on, 160–70, *161–63*; geography patterns and, 168–69; impact of studies on, 171–72; intersubjectivity issues with studies on, 172; method for studying, 155–60; perceived quality of, 159–60, *161–63*; productivity and, 164–66, *165, 166*, 168, 175nn27–29; quality compared to perceived quality of, 172; rankings of, 156; rank of department and, 156–57; reputation of, 154–55, 159–60, *161–63*; SCI data on faculty in, 159; self-aggrandizement and, 157–59, *158*; standard errors on perceived quality of, 160, 174nn21–22; stratification of perceived quality of, 163–64; variables of perceived quality of, 169–70, 175n32; visibility of, 159–60, *161–63*. *See also* American research universities

660 Index

value system of science, 94
variance, in COSPUP NSF peer-review study, 456–59, *457, 459*, 480–84, *481, 483*
Veblen, Thorstein, 399
Vetter, Betty M., 471
Viazovska, Maryna, 385
Vietnam War, 627
visas, student, 571, 598–99
visibility of work: age and, 54–56, *55*; awards and, *51, 52, 53*, 102–3; awareness compared to, 49, 67; citation counts and, 435; collaboration and, 56–57, *57*; communication and, 48–49; consequences of, 56–57; name ordering and, 56, *57*; physicist types and, *51, 54, 54*; quantity and quality of output and, *51*, 51–52, *53*, 69n21; questionnaires on, 49–50, 68n7, 68n9; rank of department and, *51*, 52–53, *53*, 63–64, *64*, 83–84; responsibility with, 70n29; scores on, 50, *50*, 63; specialties and, 53–54, *54*; stratification in, 76; upward mobility and, 57; U.S. medical schools and, 159–60, *161–63*
visibility scores, 50, *50*, 63
Voice of the Dolphins and Other Stories, The (Szilard), 499n2
voting rights, 384

Waddington, C. H., 375n15
Waelsch, Heinrich, 393
Walt, Stephen, 596
Warren, Elizabeth, 241
Washburn, Margaret Floy, 268n82
Washington Post, 525, 526
Wasserman reaction, 534
Watson, James D., 94, 96n3, 443
weak publishers, *186*, 186–88
Weber, Max, 272, 606
Weicker, Lowell, 242–43
weighted citation counts, 30, 39, *39*, 44n18, 69n23
Weiner, Anthony D., 576–77
Western Electric Study, 528–29
Western Rationalistic Tradition, 208
Westley, John, 466–67
Whewell, William, 406n5
"who owns the null" dilemma, facing research universities, 207–17

Wiesner, Jerome, 548
Wigner, Eugene, 70n29
Williams, Raymond, 409n44
Wilson, E. O., 375n15
Wilson, Woodrow, 617, 626
Wilson cloud chamber, 9
wine tasting, subjectivity in, 498
Winsten, Jay A., 537–38, 541
Wittner, Lawrence S., 552
women in science, 12–13, 18; access and, 275, 290, 298–99, 313, 391; accumulative advantage and, 282; admission system and, 250–51; affirmative action and, 253, 289, 318; assistant professorship of, 276; awards and, 258–59, 277, 385; collaboration of, 267n55; colleague relations of, 259–60; cultural resistance to, 274, 402–3; discrimination against, 259–61, 274–76, 284; domestic responsibilities of, 256; educational reforms and rising number of, 383; employment patterns of, 252–53, 266n45; faculty attitudes toward, 251–52; fears of increasing numbers of, 262; full citizenship for, 279–81; future research on, 281–84; geographic mobility and, 282; graduate school and, 247, 250–52, 263n4, 276; historic marginalization of, 272–74, 385–86; improvements for, 279–81; increasing presence of, 271–72; inequality and, 276, 280–81; informal social networks and, 283–84; IQ scores of, 276; legislation protecting, 290; marriage, motherhood and, 249, 252–54, 256, 278; McClintock on, 387–88, 390; meritocracy and, 275–76; NAS election of, 258, 387, 389; NIH funding for, 269n84; Nobel Prize and, 259, 385; occupation choice and, 281–83; particularism and, 13; part-time employment of, 256, 261, 269n87; PhDs and, 247–48, 253, 263n5, 263n9, 266n45, 272–73, 314, *332*; present day, 274–79; productivity of, 254–55, 267n55, 277–78, 282–83; promotion and recognition of, 257–60, 284; quantity and quality of output of, 254–55, 278; rank of department and, 255–57, 276; recruitment of, 248–50; reputation of,

277; research universities and resistance to, 385–86, 397–401; reward system of science and, 255–57; in Royal Society of London, 268n82; salaries of, 256–57, 268n76, 277; self-fulfilling prophecies and, 249–50, 256, 282, 407n17; self-selection of, 12, 261, 281; social selection and, 281, 318; as survivors, 330; tenure and, 255, 276, 386; test scores and, 250; in textbooks, 410n51; triple penalty and, 248, 260–62, 274; undergraduate education and, 247; unfreedom and, 382, 402, 404–5; universalism and, 13; Woodrow Wilson fellowships and, 251; Zuckerman interviewing, 4, 12–13, 18, 406n16, 409n37. *See also* eminent female scientist productivity; gender differences in output; marriage; motherhood; pairs of male and female scientists; rank-and-file female scientist productivity

Woodrow Wilson fellowships, 251

Woolgar, Steven, 2, 534–35, 542n13
working hours, gender differences in output and, 317
Wozniak, Steve, 8

Yacovone, Donald, 410n51
Yale: freedom of speech issues at, 602; Giamatti Proclamation at, 226n10; medical school, 160; Red Scare and, 580
Yang, C. N., 193

Zerhouni, Elias, 601
zero-one decision rule, in NSF peer-review system, 478
Ziman, John, 137
Zimmer, Robert, 619
Zuckerman, Harriet, 3; codification hypothesis of, 191–94, 196; Matthew Effect and, 397; on Nobel Prize winners, 6, 17, 97n13, 325n25; *Physical Review* study of, 377n22; women interviewed by, 4, 12–13, 18, 406n16, 409n37

Printed and bound by CPI Group (UK) Ltd, Croydon, CR0 4YY
02/04/2024
14478040-0005